SODIUM–CALCIUM EXCHANGE AND THE PLASMA MEMBRANE Ca^{2+}-ATPase IN CELL FUNCTION

Fifth International Conference

ANNALS OF THE NEW YORK ACADEMY OF SCIENCES
Volume 1099

SODIUM–CALCIUM EXCHANGE AND THE PLASMA MEMBRANE Ca^{2+}-ATPase IN CELL FUNCTION
Fifth International Conference

Edited by André Herchuelz, Mordecai P. Blaustein,
Jonathan Lytton, and Kenneth D. Philipson

Published by Blackwell Publishing on behalf of the New York Academy of Sciences
Boston, Massachusetts
2007

Library of Congress Cataloging-in-Publication Data

Sodium-calcium exchange : fifth international conference / edited by
Andre Herchuelz ... [et al.].
    p. ; cm. – (Annals of the New York Academy of Sciences, ISSN
0077-8923 ; v. 1099)
    Includes bibliographical references and index.
    ISBN-13: 978-1-57331-649-1 (paper : alk. paper)
    ISBN-10: 1-57331-649-0 (paper : alk. paper)
    1. Sodium channels–Congresses. 2. Calcium channels–Congresses.
3. Sodium cotransport systems–Congresses.
    I. Herchuelz, Andre. II. New York Academy of Sciences. III. Series.
    [DNLM: 1. Calcium Channels–Congresses. 2. Sodium
Channels–Congresses. 3. Biological Transport, Active–Congresses.
4. Sodium-Calcium Exchanger–physiology–Congresses. W1
AN626YL v. 1096 2007 / QU 55.7 S679 2007]

    QP535.N2S58 2007
    572'.5238224–dc22

                                                        2006101537

The *Annals of the New York Academy of Sciences* (ISSN: 0077-8923 [print]; ISSN: 1749-6632 [online]) is published 28 times a year on behalf of the New York Academy of Sciences by Blackwell Publishing with offices at 350 Main St., Malden, MA 02148 USA; 9600 Garsington Road, Oxford, OX4 2ZG UK; and 600 North Bridge Rd, #05-01 Parkview Square, 18878 Singapore.

Information for subscribers: For new orders, renewals, sample copy requests, claims, changes of address and all other subscription correspondence please contact the Journals Department at your nearest Blackwell office (address details listed above). UK office phone: +44 (0)1865 778315, fax +44 (0)1865 471775; US office phone: 1-800-835-6770 (toll free US) or 1-781-388-8599; fax: 1-781-388-8232; Asia office phone: +65 6511 8000, fax; +44 (0)1865 471775, Email: customerservices@blackwellpublishing.com

Subscription rates:
Institutional Premium   The Americas: $4043   Rest of World: £2246
The Premium institutional price also includes online access to full-text articles from 1997 to present, where available. For other pricing options or more information about online access to Blackwell Publishing journals, including access information and terms and conditions, please visit www.blackwellpublishing. com/nyas
*Customers in Canada should add 6% GST or provide evidence of entitlement to exemption.
**Customer in the UK or EU: add the appropriate rate for VAT EC for non-registered customers in countries where this is applicable. If you are registered for VAT please supply your registration number.

Mailing: The *Annals of the New York Academy of Sciences* is mailed Standard Rate. Mailing to rest of world by DHL Smart & Global Mail. Canadian mail is sent by Canadian publications mail agreement number 40573520. **Postmaster:** Send all address changes to *Annals of the New York Academy of Sciences*, Blackwell Publishing Inc., Journals Subscription Department, 350 Main St., Malden, MA 02148-5020.

Membership information: Members may order copies of *Annals* volumes directly from the Academy by visiting www.nyas.org/annals, emailing membership@nyas.org, faxing 212-298-3650, or calling 800-843-6927 (US only), or 212-298-8640 (International). For more information on becoming a member of the New York Academy of Sciences, please visit www.nyas.org/membership. Claims and inquiries on member orders should be directed to the Academy at email: membership@nyas.org or Tel: 212-298-8640 (International) or 800-843-6927 (US only).

Printed in the USA. Printed on acid-free paper.

Annals are available to subscribers online at the New York Academy of Sciences and also at Blackwell Synergy. Visit www.blackwell-synergy.com or www.annalsnyas.org to search the articles and register for table of contents e-mail alerts. Access to full text and PDF downloads of *Annals* articles are available to nonmembers and subscribers on a pay-per-view basis at www.blackwell-synergy.com and www.annalsnyas.org.

The paper used in this publication meets the minimum requirements of the National Standard for Information Sciences Permanence of Paper for Printed Library Materials, ANSI Z39.48-1984.

ISSN: 0077-8923 (print); 1749-6632 (online)
ISBN-10: 1-57331-649-0 (paper); ISBN-13: 978-1-57331-649-1 (paper)

A catalogue record for this title is available from the British Library.

ANNALS OF THE NEW YORK ACADEMY OF SCIENCES

Volume 1099
March 2007

SODIUM–CALCIUM EXCHANGE AND THE PLASMA MEMBRANE Ca^{2+}-ATPase IN CELL FUNCTION
Fifth International Conference

Editors
ANDRÉ HERCHUELZ, MORDECAI P. BLAUSTEIN, JONATHAN LYTTON,
AND KENNETH D. PHILIPSON

This volume is the result of a meeting entitled **5th International Conference on Na/Ca Exchange 2006**, held on August 23–27, 2006, at the Hotel Bedford, Brussels, Belgium.

CONTENTS

Part VI. NC(K)X and PMCA KO Mice

Part VII. Cardiac Function

Part VIII. NCX and PMCA in Neuronal and Smooth Muscle Function

Part X. Na/Ca Exchange in the Endoplasmic Reticulum, the Mitochondria, and the Nuclear Envelope

Part XI. Na/Ca Exchange Inhibitors: Therapeutic Opportunities

Position Paper

Financial assistance was received from:

- Fonds National de la Recherche Scientifique, Belgium
- Fonds voor Wetenschappelijk Onderzoek-Vlaanderen, Belgium
- Pfizer Pharmaceutical Group, Belgium
- Therabel Pharma, Belgium
- Taisho Pharmaceuticals, Japan
- Merck & Co. Inc - USA

Preface

Ca^{2+} plays a crucial second-messenger role in many physiological processes, including signal transduction, stimulus-contraction, and stimulus-secretion coupling, cell-to-cell communication, gene regulation, cell growth, and cell death. In order to exert such a role, the concentration of the ion in the cytoplasm and in the different intracellular compartments (e.g., endoplasmic reticulum, mitochondrion, and nucleus) must be tightly controlled. Alterations in such control may lead to various pathological conditions and diseases, such as excess cell death and heart failure.

The plasma membrane Na^+/Ca^{2+} exchanger (NCX) and the plasma membrane Ca^{2+}-ATPase (PMCA) are the two ubiquitous mechanisms responsible for Ca^{2+} extrusion from cells and are essential for the control of Ca^{2+} homeostasis. Classically, NCX is considered to have a low affinity but high capacity for Ca^{2+}, whereas the PMCA has a high affinity but low capacity for the divalent cation. Hence, NCX eliminates significant rises in intracellular Ca^{2+}, whereas the PMCA finely regulates the cytosolic free Ca^{2+} concentration around its basal value (100 nM).

This volume contains papers presented at the Fifth International Conference on Sodium–Calcium Exchange, held in Brussels, Belgium, on August 23–27, 2006. It was the fifth gathering of international scientists from various fields studying NCX function and regulation. Previous meetings (Stowe, England, in 1987; Baltimore, MD, in 1991; Woods Hole, MA, in 1995; and Banff, Canada, in 2001) were all important focal points for the exchange of new information and ideas. Although this meeting was the fifth in a series of meetings devoted to NCX, it was the first to bring together experts in both the NCX and the PMCA fields. Hence the title of the volume: *Sodium–Calcium Exchange and the Plasma Membrane Ca^{2+}-ATPase in Cell Function*. Indeed, an increasing number of scientists are studying the two Ca^{2+} extrusion mechanisms in their fields, and it was thus logical to gather together the experts of the two systems, especially because there is no equivalent meeting for the PMCA. The experience was very rewarding for researchers from both fields, and it was suggested that this format should be repeated in future meetings.

The meeting also celebrated the 40th anniversary of the discovery of the two Ca^{2+} extrusion mechanisms. Na/Ca exchange was discovered simultaneously and independently in Mainz, Germany by Harald Reuter, and in Plymouth, England by Peter F. Baker and Mordecai P. Blaustein, in the autumn of 1966. Already in these early days, the exchanger was hypothesized to be the missing link between Na^+ pump inhibition and the cardiotonic action of digitalis. The

Ann. N.Y. Acad. Sci. 1099: xv–xvi (2007). © 2007 New York Academy of Sciences.
doi: 10.1196/annals1387.070

same year, Hans Jurg Schatzmann published his famous paper in Experientia on the plasma membrane Ca^{2+}-ATPase. (see historical notes on the discovery of NCX and the PMCA).

The cloning of the genes encoding NCX and the PMCA (17 and 19 years ago, respectively) has been a strong impetus for rapid progress in our knowledge of the roles of these two Ca^{2+} extrusion mechanisms in physiology and pathophysiology. This was clearly evident during the conference. At the end of the meeting, the general feeling was that the fields had developed in a remarkable way over the last 5 years. In addition to a fundamental understanding of the genetics and molecular processes, the detailed structure–function relationship, the fine regulatory mechanisms, and the multitude of physiological roles for the many different NCX and PMCA gene products, including their splice variants, we have promising leads for new therapeutic approaches. The interaction of the PMCA with several signaling molecules suggests that it may also act as a signaling protein.

As chair of the organizing committee for the Fifth International Conference on Sodium–Calcium Exchange, I would like to thank my colleagues on the committee—Mordecai Blaustein, Ernesto Carafoli, Ken Philipson, Jonathan Lytton, and Karin Sipido—for their help in setting up the scientific program and their advice in general matters. I would also like to express our sincere gratitude to the other members of the scientific committee—Lucio Annunziato, Denis Noble, Cesare Terracciano, David Eisner, Takahiro Iwamoto, Reinaldo DiPolo, Junko Kimura, John Reeves, and Larry Hryshko—for their suggestions, remarks, and advice.

I also thank the members of my group who contributed to the practical organization of the meeting, especially Julie Brunko for her outstanding secretarial work.

And lastly, the meeting would not have been successful without the outstanding contributions of all the participants. I am especially grateful to the PMCA experts who all accepted our invitation to participate in our meeting devoted predominantly to Na/Ca exchange.

—ANDRÉ HERCHUELZ
Free University of Brussels, Brussels, Belgium

Historical Note Regarding the Discovery of the Na/Ca Exchanger and the PMCA

ANDRÉ HERCHUELZ

Faculty of Medicine, Free University of Brussels, 1070 Brussels, Belgium

As indicated in the Preface to this volume, this meeting celebrated the 40th anniversary of the discovery of the two Ca^{2+} extrusion mechanisms. Na/Ca exchange was discovered simultaneously and independently in Mainz, Germany by Harald Reuter, and in Plymouth, England by Peter F. Baker and Mordecai P. Blaustein, in the autumn of 1966.

In September 1966, Richard Steinhardt and Mordy Blaustein traveled to the marine laboratory at Plymouth (with Baker, who worked across the hall) to explore the activation of the squid axon Na^+ pump by external cations. Within a month, they had uncovered a large ouabain-insensitive (and thus non-Na^+-pump-mediated) Na^+ efflux in Na^+-free Li^+-containing medium. This efflux was dependent on external Ca^{2+}, suggesting that it involved an exchange of Na^+ for Ca^{2+}.[1] Baker pointed out the parallels with Rolf Niedergerke's observations on Na^+-dependent Ca^{2+} movements in the frog heart.[2] The reading of the article by Niedergerke convinced Blaustein that cardiac muscle also had an Na/Ca exchange system and that this was the missing link between Na^+ pump inhibition and the cardiotonic action of digitalis.[3,4]

At the very same time, Harald Reuter was studying Na/Ca exchange in mammalian cardiac muscle.[5] Blaustein met Reuter at the International Physiological Congress in Washington in 1968. They subsequently agreed to work together in 1971 in Bern, Switzerland, where they studied Na/Ca exchange in vascular smooth muscle. Their manuscript, in which they alluded to the possible role of Na/Ca exchange in hypertension, was rejected by *Nature* and *Science*.[4]

Although he was on the sidelines, compared to Blaustein and Reuter, Denis Noble met Reuter in 1966 at a course in Homburg/Saar, Reuter being a student and Noble a lecturer.[6] It was at that course that Reuter was stimulated to look for an Na/Ca exchanger.[4]

That same year, Hans Jurg Schatzmann published his paper in *Experiencia*;[7] it was a little masterpiece, both from scientific and writing points of view.[8]

Address for correspondence: André Herchuelz, Universite Libre de Bruxelles, Laboratoire de Pharmacodynamie et de Therapeutique, 808 route de Lennik Bat GE, CP 617 Brussels 1070. Voice: 32-2-555-62-01; fax: 32-2-555-63-70.

andre.herchuelz@belgacom.net

Ann. N.Y. Acad. Sci. 1099: xvii–xviii (2006). © 2007 New York Academy of Sciences.
doi: 10.1196/annals1387.072

Perhaps the most important aspect of his work was that he was working essentially alone and yet managed to produce two sensational pieces of work: the discovery that digitalis had the Na^+ pump as a target, and the discovery of the plasma membrane Ca^{2+}-ATPase (PMCA) pump.[8] Schatzmann has had much less recognition than he deserves. What was interesting in those early days of the PMCA pump was the persistent attitude coming from the Ca^{2+} aficionados that PMCA was an erythrocyte phenomenon.[8] Even after the article by Pico Caroni and Ernesto Carafoli, showing conclusively that the PMCA pump also existed in heart sarcolemma,[9] opinions did not change much. Several years had to pass before the PMCA pump was accepted as a general exporter of Ca^{2+} alongside its bigger brother, the Na/Ca exchanger. Making these sorts of observations was, of course, complicated by the fact that the secrets of the PMCA enzyme were difficult to determine. Even now, the PMCA pump is the exclusive territory of a few people who have the ability and inclination to work very hard.[8] Personally, I remember that in order to adequately discuss this topic in the first paper I wrote on the PMCA pump,[10] I had to get away from the bench for a 2- or 3-week period and do nothing but read the published literature again and again.

ACKNOWLEDGMENT

The author thanks Mordecai Blaustein, Denis Noble, and Ernesto Carafoli for their important contributions in writing this note.

REFERENCES

1. BAKER, P.F., M.P. BLAUSTEIN, A.L. HODGKIN & R.A. STEINHART. 1969. The influence of calcium on sodium efflux in squid axons. J. Physiol. Lond. **200:** 431–458.
2. NIEDERGERKE, R. 1963. Movements of Ca in frog heart ventricles at rest and during contractures. J. Physiol. (Lond.) **167:** 515–550.
3. BLAUSTEIN MP. 1987. This week's citation classic. Curr. Content **35:** 14.
4. BLAUSTEIN MP. Personal communication.
5. REUTER H & SEITZ N. 1968. The dependence of calcium efflux from cardiac muscle on temperature and external ion composition. J. Physiol. (Lond.) **195:** 451–470.
6. NOBLE D. Personal communication.
7. SCHATZMANN, H.J. 1966. ATP-dependent Ca++ extrusion from human red cells. Experiencia **22:** 364–365.
8. CARAFOLI, E. Personal communication.
9. CARONI P. & E. CARAFOLI. 1980. An ATP-dependent Ca2+-pumping system in dog heart sarcolemma. Nature **283:** 765–767.
10. KAMAGATE, A. Herchuelz, A. BOLLEN & F. VAN EYLEN. 2000. Expression of multiple plasma membrane Ca^{2+}-ATPase mRNAs in rat pancreatic islet cells. Cell Calcium **27:** 231–246.

What We Know about the Structure of NCX1 and How It Relates to Its Function

DEBORA A. NICOLL, XIAOYAN REN, MICHELA OTTOLIA, MARTIN PHILLIPS, ALFREDO R. PAREDES, JEFF ABRAMSON, AND KENNETH D. PHILIPSON

Departments of Physiology and Medicine and the Cardiovascular Research Laboratories, University of California, Los Angeles, California 90095, USA

ABSTRACT: NCX1 is modeled to contain nine transmembrane segments (TMS) with a large intracellular loop between TMS 5–6 and two reentrant loops connecting TMS 2–3 and TMS 7–8. NCX1 also contains two regions of internal repeats. The α repeats are composed of TMS 2 and 3 and TMS 7 and 8 and are involved in ion binding and transport. The β repeats are in the large intracellular loop and are involved in binding of regulatory Ca^{2+}. Our studies on the structure/function analysis of NCX1 have focused on the α- and β-repeat regions and on how the TMS pack in the membrane. We have examined the $\alpha 1$ repeat by mutagenesis of residues modeled to be in the reentrant loop and TMS 3 and by determination of ion affinities of the mutants. Our results show that TMS 3 and not the reentrant loop is involved in Na^+ binding. No mutants demonstrated altered affinity for transported Ca^{2+}. We have synthesized a fusion protein composed of the $\beta 1$ repeat. This fusion protein was expressed in *Escherichia coli* and purified. The fusion protein binds Ca^{2+} and shows conformational changes on binding. The crystal structure of the $\beta 1$ repeat shows that it is composed of a seven-stranded β-sandwich with Ca^{2+}-binding sites located at one end of the sandwich. Four Ca^{2+} ions bind to the $\beta 1$ repeat in a manner reminiscent of Ca^{2+} binding to C2 domains. Packing of TMS in the membrane has been studied by cross-linking induced mobility shifts on SDS-PAGE. Interactions between TMS 1, 2, 3, 6, 7, and 8 have been identified.

KEYWORDS: sodium–calcium exchange; membrane protein; reentrant loop; calcium binding; transmembrane packing

MUTATIONS IN THE $\alpha 1$ REPEAT

The $\alpha 1$ repeat composes TMS (transmembrane segment) 2, extracellular loop c, and the first half of TMS 3 in NCX1. Some mutations in TMS 2 and

Address for correspondence: Debora A. Nicoll, 3645 MRL, 675 Young Dr. S., Los Angeles, CA 90095-1760. Voice: 310-825-5137; fax: 310-206-5777.
dnicoll@mednet.ucla.edu

Ann. N.Y. Acad. Sci. 1099: 1–6 (2007). © 2007 New York Academy of Sciences.
doi: 10.1196/annals.1387.014

3 decrease the level of exchanger activity expressed in *Xenopus* oocytes,[1] and some mutations in the reentrant loop alter affinity for extracellular Ca^{2+}.[2] Experiments involving cysteine scanning mutagenesis and modification have demonstrated that residues in extracellular loop c are accessible from the intracellular surface,[2] leading to a model where loop c forms a reentrant loop. Helix packing experiments have demonstrated that TMS 2 and 3 are near TMS 7 and 8[3] and suggest that the α1 and α2 repeats may interact to form an ion conduction pathway similar to those seen in the aquaporin family of proteins. However, the amino acid residues in loop c are poorly conserved compared to the residues in TMS 2 and 3 suggesting a less important role for the reentrant loop than TMS.

Single-site mutations were generated in the reentrant loop and TMS 3 portions of the α1 repeat, and the mutant exchangers expressed in *Xenopus* oocytes where the effects of mutations on affinity for intracellular, transported Na^+ or Ca^{2+} ions, and regulatory properties were measured.[4] The apparent affinities and Hill coefficients of mutant NCX1s for transported Ca^{2+} and Na^+ are summarized in FIGURE 1. The most striking observation is that none of the mutants

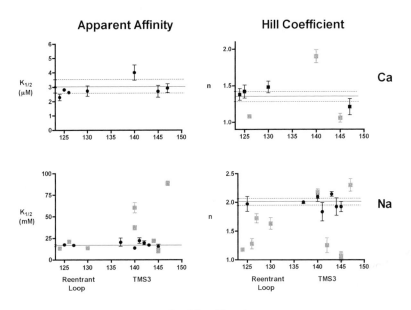

FIGURE 1. Apparent affinities and Hill coefficients for transported Ca^{2+} or Na^+. *Top*: apparent affinity (*left*) and Hill coefficient (*right*) for transported Ca^{2+}. *Bottom*: apparent affinity (*left*) and Hill coefficient (*right*) for Na^+. Numbers on the *x*-axis refer to the number of the residue mutated. *Solid* and *dotted lines* indicate the average and standard deviation, respectively, of the wild-type NCX1. *Gray symbols* refer to values that are statistically different from wild type.

show a change in Ca^{2+} affinity, while several show altered Na^+ affinity. The mutants with the most pronounced change in Na^+ affinity are at residues 140 and 147. Both are modeled to be in TMS 3. Only small changes in Na^+ affinity could be detected in mutations in the reentrant loop. These results suggest that TMS3 of the α1 repeat is involved in binding Na^+, not Ca^{2+}. This notion is supported by a canine–squid chimeric exchanger.[4] The chimera has identical TMS 2 and 3, but six amino acid changes in the reentrant loop. For the chimera, the $K_{1/2}$ for Na^+ is 13.4 ± 0.7 mM, and the Hill coefficient is 1.3 compared to 16.6 ± 1.2 mM and 2.00 ± 0.1 for the wild-type NCX1. For Ca^{2+}, the affinity of the chimera is 2.7 ± 0.3 mM with a Hill coefficient of 1.4 ± 0.1 compared to 3.1 ± 0.5 mM and 1.4 ± 0.1 for wild-type NCX1. Thus, the reentrant loop, unlike TMS 3, does not appear to play a major role in ion transport. Instead, the reentrant loop plays a role in entry into the I_1 inactivation state. The mutant with an asparagine at position 124, in the reentrant loop, displays a statistically significant increase in the time constant (8.8 ± 0.7 s) for entry into I_1 compared to the wild-type (3.3 ± 0.23 s). Likewise, the canine–squid chimera is also significantly slowed (4.1 ± 0.2 s).

CRYSTAL STRUCTURE OF THE β1 REPEAT

The β repeats are in the large intracellular loop between TMS 5 and 6. It has been shown that mutations in β1 reduce the affinity of NCX for regulatory Ca^{2+}.[5] A fusion protein encoding β1 can be expressed in *Escherichia coli* and can bind Ca^{2+}.[6] We have crystallized the fusion protein and determined its structure.[7] The crystal structure overlaps very well with the structure that has been determined by NMR[8] and is composed of a seven-stranded immunoglobulin-like domain with Ca^{2+} binding on one end and an unstructured loop at the other (FIG. 2 A).

Four Ca^{2+} bind the β1 fusion protein in a parallelogram-like configuration (FIG. 2 B). The carboxylate groups of eight different aspartate and glutamate residues provide most of the chelation sites for Ca^{2+}. There are also three water molecules and two backbone carbonyls assisting in Ca^{2+} chelation. In a previous study, we examined the effects of mutations at three of the residues now shown to be involved in Ca^{2+} chelation.[5] These mutants (at positions 447, 498, and 500) resulted in decreased affinity of regulatory Ca^{2+}. Interestingly, these residues are involved in binding Ca4 and Ca3. We also had examined the effect of a deletion in the Ca^{2+}-binding region. Mutant Δ450–456 has chelating residues 451 (Ca1 and Ca2) and 454 (Ca1) removed and this mutant also showed reduced affinity.[5] Preliminary data for new mutant D454K indicate that the mutant is like wild type with a $K_{1/2}$ for Ca^{2+} regulation of 0.4 μM Ca^{2+} (wild-type is 0.4 ± 0.2 μM Ca^{2+}).[5] Given the results to date, it appears that binding ions Ca2, Ca3, and Ca4 but not Ca1 affects Ca^{2+} regulation.

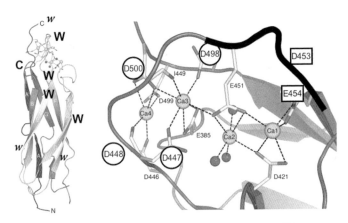

FIGURE 2. (A) Structure of the β1 repeat and location of Ca^{2+}-sensitive residues. β strands are indicated with arrows and strands are labeled A through G. W = sites where introduced tryptophans display Ca^{2+}-dependent changes in fluorescence. w = sites where introduced tryptophans are unaffected by Ca^{2+}. C = Cysteine 383, which is labeled with MIANS only in the absence of Ca^{2+}. (B) Structure of the Ca^{2+} chelation region of the β1 repeat. Ca^{2+} ions are shown as *labeled spheres* and water as smaller, *unlabeled spheres*. Mutations at *circled residues* result in a decrease in affinity for regulatory Ca^{2+} and mutation at *boxed residues* has no affect on Ca^{2+} affinity.

It is believed that Ca^{2+} binding to the regulatory site results in a conformational change that puts NCX1 into a more active state. We have previously published work showing that Ca^{2+} binding to the β1 fusion protein causes a conformational change that can be seen as altered mobility on SDS-PAGE.[6] Also, when two GFP analogs are fused to β1, Ca^{2+}-induced changes in FRET can be measured.[9] We find a decrease in the hydrodynamic radius after addition of Ca^{2+} indicating a more compact structure (FIG. 3 A). Hilge *et al.*[8] report that on removal of Ca^{2+}, the Ca^{2+}-binding half of the β1 protein becomes unstructured. In support of that observation, we find that cysteine 383, in the A′-B loop is accessible to labeling only in the absence of Ca^{2+} (FIG. 3 B). Also, introduced tryptophans in the Ca^{2+}-binding half of the protein show Ca^{2+}-dependent changes in fluorescence, while tryptophans introduced into the other half do not (FIG. 3 C). FIGURE 2 A summarizes these data by showing the location of Ca^{2+}-sensitive residues.

PACKING OF TMS

To determine which TMS are adjacent in the packed, native NCX1, we look for cross-linking induced shifts in SDS-PAGE between introduced cysteines. It has previously been shown that native cysteine 14 or 20 can cross-link to cysteine 792.[10] Under nonreducing conditions, the cross-linked residues

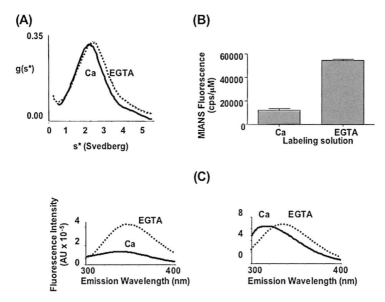

FIGURE 3. Conformational changes associated with Ca^{2+} binding to β1 fusion protein. (A) Sedimentation in the presence of 140 mM KCl, 25 mM MOPS, pH 7, 2 mM MgCl$_2$, and 1 mM Ca^{2+} or EGTA. In the presence of Ca^{2+}, the protein has a smaller Svedberg constant and hence a larger hydrodynamic radius. (B) MIANS labeling. Purified β1 fusion protein was treated with the sulfhydryl modfying reagent MIANS in the presence of Ca^{2+} or EGTA. Labeling is only observed in the presence of EGTA. (C) Ca^{2+}-induced changes in tryptophan emission. Single tryptophans were introduced into the β1 fusion protein and excited at 280 nm. Tryptophan emission was measured in the presence of Ca^{2+} or EGTA. An example of Ca^{2+}-induced quenching (left, Y422W) and Ca^{2+}-induced blue shift (right, F456W) are shown.

cause NCX1 to migrate at 160 kDa compared to the reduced form that migrates at 120 kDa. Thus, cross-linking between the first and second halves of NCX1 can show this change in mobility. Pairs of cysteines have been introduced into a cysteineless NCX1 background, and the effects of cysteine cross-linking reagents on SDS-PAGE mobility were examined. Using this approach,[3,11] we have found there to be proximity between TMS 1–6 and 2–6, 2–7, 2–8, and between TMS 3–7 (FIG. 4). No data are yet available for TMS 4, 5, or 9.

Two interesting aspects arise from these results. First, the α repeats are near one another in the folded protein suggesting that they may work in concert to form the ion conduction pathway. Second, TMS 6 appears to have considerable amount of flexibility since a residue in the middle of TMS 6 (768) can interact with residues at both the intra- and extracellular ends of TMS 1 and 2. Whether this is due to conformational changes associated with ion transport or NCX regulation remains to be determined.

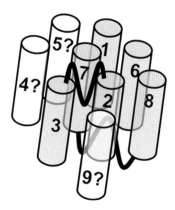

FIGURE 4. Helix packing of transmembrane segments in NCX1. Interactions between TMS have been measured for the darkened cylinders.

REFERENCES

1. NICOLL, D.A., M. OTTOLIA, L. LU, Y. LU & K.D. PHILIPSON. 1999. A new topological model of the cardiac sarcolemmal Na^+-Ca^{2+} exchanger. J. Biol. Chem. **274:** 910–917.
2. IWAMOTO, T., A. UEHARA, I. IMANAGA & M. SHIGEKAWA. 2000. The Na^+/Ca^{2+} exchanger NCX1 has oppositely oriented reentrant loop domains that contain conserved aspartic acids whose mutations alters its apparent Ca^{2+} affinity. J. Biol. Chem. **275:** 38571–38580.
3. QIU, Z., D.A. NICOLL & K.D. PHILIPSON. 2001. Helix packing of functionally important regions of the cardiac Na^+-Ca^{2+} exchanger. 2001. J. Biol. Chem. **276:** 194–199.
4. OTTOLIA, M., D.A. NICOLL & K.D. PHILIPSON. 2005. Mutational analysis of the α-1 repeat of the cardiac Na^+-Ca^{2+} exchanger. J. Biol. Chem. **280:** 1061–1069.
5. MATSUOKA, S., D.A. NICOLL, L.V. HRYSHKO, D.O. LEVITSKY, J.N. WEISS & K.D. PHILIPSON. 1995. Regulation of the cardiac Na^+-Ca^{2+} exchanger by Ca^{2+}. Mutational analysis of the Ca^{2+}-binding domain. J. Gen. Physiol. **105:** 403–420.
6. LEVITSKY, D.O., D.A. NICOLL & K.D. PHILIPSON. 1994. Identification of the high affinity Ca^{2+}-binding domain of the cardiac Na^+-Ca^{2+} exchanger. J. Biol. Chem. **269:** 22847–22852.
7. NICOLL, D.A., M.R. SAWAYA, S. KWON, D. CASCIO, K.D. PHILIPSON & J. ABRAMSON. 2006. The crystal structure of the primary Ca^{2+} sensor of the Na^+/Ca^{2+} exchanger reveals a novel Ca^{2+} binding motif. J. Biol. Chem. **281:** 21577–21581.
8. HILGE, M., J. AELEN & G.W. VUISTER. 2006. Ca^{2+} regulation in the Na^+/Ca^{2+} exchanger involves two markedly different Ca^{2+} sensors. Mol. Cell **22:** 15–25.
9. OTTOLIA, M., K.D. PHILIPSON & S. JOHN. 2004. Conformational changes of the Ca^{2+} regulatory site of the Na^+-Ca^{2+} exchanger detected by FRET. Biophys. J. **87:** 899–906.
10. SANTACRUZ-TOLOZA, L., M. OTTOLIA, D.A. NICOLL & K.D. PHILIPSON. 2000. Functional analysis of a disulfide bond in the cardiac Na^+-Ca^{2+} exchanger. J. Biol. Chem. **275:** 182–188.
11. REN, X., D.A. NICOLL & K.D. PHILIPSON. 2006. Helix packing of the cardiac Na^+-Ca^{2+} exchanger: proximity of transmembrane segments 1, 2 and 6. J. Biol. Chem. **281:** 22808–22814.

Structural Basis for Ca^{2+} Regulation in the Na^{+}/Ca^{2+} Exchanger

MARK HILGE,[a] JAN AELEN,[a] ANASTASSIS PERRAKIS,[b] AND GEERTEN W. VUISTER[a]

[a]*Department of Biophysical Chemistry, Institute for Molecules and Materials, Radboud University Nijmegen, 6525 ED Nijmegen, The Netherlands*

[b]*Netherlands Cancer Institute, Department of Molecular Carcinogenesis, 1066 CX Amsterdam, The Netherlands*

ABSTRACT: Binding of Na^{+} and Ca^{2+} ions to the large cytosolic loop of the Na^{+}/Ca^{2+} exchanger (NCX) regulates its ion transport across the plasma membrane. We determined the solution structures of two Ca^{2+}-binding domains (CBD1 and CBD2) that, together with an α-catenin-like domain (CLD) form the regulatory exchanger loop. CBD1 and CBD2 constitute a novel Ca^{2+}-binding motif and are very similar in the Ca^{2+}-bound state. Strikingly, in the absence of Ca^{2+} the upper half of CBD1 unfolds while CBD2 maintains its structural integrity. Together with a sevenfold higher affinity for Ca^{2+} this suggests that CBD1 is the primary Ca^{2+} sensor. Specific point mutations in either domain largely allow the interchange of their functionality and uncover the mechanism underlying Ca^{2+} sensing in NCX.

KEYWORDS: Na^{+}/Ca^{2+} exchanger; structure; calcium-binding protein; calcium sensor

INTRODUCTION

The Na^{+}/Ca^{2+} exchanger (NCX) in concert with the plasma membrane Ca^{2+} ATPase (PMCA) removes Ca^{2+} ions from the cell that entered the cytosol via L-type Ca^{2+} channels and/or internal Ca^{2+} stores during the action potential. Driven by the Na^{+} gradient generated through the Na^{+}/K^{+} ATPase the exchanger predominately expels one Ca^{2+} ion for the uptake of three Na^{+} ions,[1] may however, under certain conditions also operate in reverse mode.

A biochemically derived topology model of NCX predicts nine transmembrane α-helices[2,3] and a large cytosolic loop of approximately 500 residues (FIG. 1 A). Binding of Ca^{2+} ions to sites located in the cytosolic loop

Address for correspondence: Mark Hilge, Department of Biophysical Chemistry, Institute for Molecules and Materials, Radboud University Nijmegen, Toernooiveld 1, 6525 ED Nijmegen, The Netherlands. Voice: 0031-24-36-52-129; fax: 0031-24-36- 52-112.

hilge@nmr.ru.nl

Ann. N.Y. Acad. Sci. 1099: 7–15 (2007). © 2007 New York Academy of Sciences.
doi: 10.1196/annals.1387.030

(A)

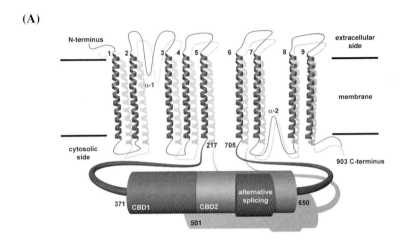

(B)

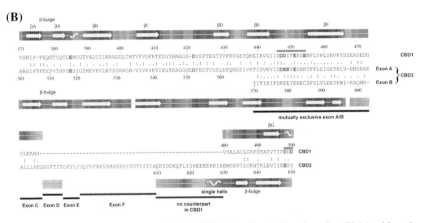

FIGURE 1. (**A**) Topology model for NCX displaying its four domains: TM (residues 1–216 and 706–903), CLD (residues 217–370 and 651–705), CBD1 (residues 371–500), and CBD2 (residues 501–650) with the numbering based on canine NCX1 AD-splice variant. (**B**) Sequence alignment of CBD1 and CBD2. Residues involved in Ca^{2+} binding are in bold and a bar above the letters indicates acidic segments. Italicized residues are not taken into account for the numbering. (Figure adapted from Reference 8.)

generally activates the exchanger,[4] whereas binding of Na^+ ions has been shown to deactivate NCX.[5] Residues 371–509 were reported to form a putative high-affinity Ca^{2+}-binding domain ($K_d \sim 140$–400 nM)[6] and appear to bind Ca^{2+} on a beat-to-beat basis during excitation–contraction coupling in cardiomyocytes.[7]

In order to explore Ca^{2+} regulation in NCX, we determined the structures of two Ca^{2+}-binding domains (CBD1 and CBD2)[8] that are located in the large exchanger loop. The existence of a second Ca^{2+}-binding domain (CBD2) was inferred by a sequence identity of 27% of residues 501–650 with CBD1 as well as the presence of two acidic segments, reported to be crucial for Ca^{2+} binding,[6] at homologous positions.

Structures of CBD1 and CBD2

We expressed three different constructs (residues 371–509, 501–657, and 371–657) of canine NCX1 in *E. coli* BL-21, which represent the stable forms of CBD1, CBD2, and CBD12, respectively. For CBD1 and CBD2 we determined the structures by high-resolution heteronuclear multidimensional NMR spectroscopy in the Ca^{2+}-bound form. Both domains display two antiparallel β-sheets that constitute a β-sandwich or Greek key motif with one β-sheet containing strands A, B, and E, and the other containing strands C, D, F, and G (FIG. 2). With the exception of the long FG loops (residues 468–482 in CBD1 and residues 599–627 in CBD2), the domains are well defined by the ensemble of NMR structures. Superposition of the central chains of CBD1 and CBD2 results in an average rmsd of 1.3 Å(on 101 C^{α} positions) with the biggest differences observed for the BC, CD, DE, and the FG loops as well as for parts of

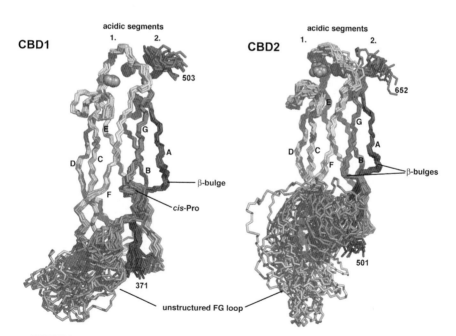

FIGURE 2. Overlays of the 20 NMR-derived structures of CBD1 (residues 371–509; accession code 2FWS) and CBD2 (residues 501–657; accession code 2FWU). C-terminal residues 504–509 (CBD1) and 653–657 (CBD2) are disordered and removed from the figure for clarity. (Figure adapted from Reference 8.)

the first acidic segment (residues 446–454 and 577–582, respectively). Both domains contain a β-bulge in strand A, but only CBD2 displays a second β-bulge in strand G. CBD1 instead, possesses a *cis*-proline at a structurally similar position.

Ca^{2+}-Binding Sites in CBD1 and CBD2

A threefold approach to define the Ca^{2+}-binding sites in CBD1 and CBD2 that included pseudo-contact shifts, structure calculations in the absence of Ca^{2+} ions as well as mutational data, suggested two Ca^{2+} ions bound within each of the two domains. The main contributors to the Ca^{2+}-binding sites are the two acidic segments, Asp446-Glu454 and Asp498-Asp500 in CBD1 (Fig. 1 B) as well as Asp577-Glu582 and Glu647-Glu648 in CBD2 (Fig. 1 B). Crucially, Glu451 in CBD1 and Asp578 in CBD2 coordinate Ca^{2+} ions on both sides of the EF loop. In CBD2, Lys585 occupies a position equivalent to Glu454 in CBD1 and forms a salt bridge with $O^{\delta 1}$ of Asp552 that constitutes an acidic cluster together with Asp578, Glu579, and Glu580. In the case of CBD1, however, Ca^{2+}-binding sites were better resolved by the recently published CBD1 X-ray structure[9] that showed four Ca^{2+} ions bound in an arrangement reminiscent of C2 domains. Notably, the two Ca^{2+} ions implemented in the NMR structure occupy the average positions of the two Ca^{2+} ion pairs found in the crystal structure.

Ca^{2+}-Free Forms of CBD1 and CBD2

In order to explore structural effects of Ca^{2+} binding in CBD1 and CBD2, we also recorded $[^1H,^{15}N]$-HSQC spectra for the two domains in the presence of 10 mM EDTA. Strikingly, the CBD1 spectrum displayed a strong decrease in peak dispersion with a large number of ^1H-resonances clustered in a spectral region, usually characteristic for unstructured residues. Mapping shifted resonances, likely to result from structural alterations or residues exposed to a changed chemical environment and disappearing resonances, probably reflecting unstructured residues onto the molecular structure of CBD1, reveals a loss of structural integrity in the upper half of the molecule (Fig. 3). In particular, residues in the second part of strand A, the initial parts of strand B and F, the end of strand G, as well as the AB, CD, and EF loops, which form the Ca^{2+}-binding sites, become unstructured. In sharp contrast, the $[^1H,^{15}N]$-HSQC spectrum of CBD2 does not show this degree of chemical shift changes, thus suggesting only limited structural alterations and preservation of the fold in the absence of Ca^{2+} (Fig. 3). This distinct behavior in the $[^1H,^{15}N]$-HSQC spectra of the individual domains in the absence of Ca^{2+} is also observed for CBD12, the construct containing both domains.

FIGURE 3. Antiparallel orientation of the two Ca^{2+}-binding domains, CBD1 and CBD2, in the regulatory exchanger loop. Residues in encircled areas show shifted resonances in the $[^1H,^{15}N]$-HSQC spectra and undergo conformational changes during Ca^{2+}-binding and release events. A movie based on the NMR structures as well as our biochemical data, illustrating the distinct behavior of CBD1 and CBD2, is available at http://www.markhilge.com.

Two Distinctly Different Ca^{2+} Sensors

The striking differences observed for CBD1 and CBD2 thus raise the question of their cause and functional consequences. Examination of the CBD1 and CBD2 Ca^{2+}-binding sites reveals the presence of three basic residues, Arg547, Lys583, and most importantly Lys585 in CBD2, which are well positioned to form salt bridges with Glu579, Glu580, and Asp552, respectively. Upon Ca^{2+} release these salt bridges may partly stabilize some of the negative charges and therefore prevent unfolding of CBD2. In sharp contrast, Ca^{2+}-binding sites in CBD1 do not contain any basic residues that could reduce repulsion between the numerous glutamate and aspartate residues.

To verify this hypothesis we aimed to invert the behavior of the two domains by mutating Lys585 in CBD2 to a glutamate and the orthologous residue in CBD1, Glu454 to a lysine residue. Indeed, contrary to their wild-type forms, $[^1H,^{15}N]$-HSQC spectra of the Glu454Lys and Lys585Glu mutants in the presence of EDTA are characteristic for a fully structured and disintegrated domain, respectively. Similarly, gel mobility shifts on native polyacrylamide gels in the absence and presence of Ca^{2+} are similarly indicative of the dramatic conformational differences. We therefore conclude that electrostatic repulsion is the major driving force for the unfolding of CBD1 in the absence of Ca^{2+}.

To further characterize Ca^{2+} binding we determined dissociation constants of wild-type CBD1 and CBD2 as well as of the Glu454Lys and Lys585Glu mutants by isothermal titration calorimetry (ITC). The ITC measurements demonstrate that CBD1 binds Ca^{2+} with K_d values of 120 and 240 nM, while the respective values of CBD2 are 820 nM and 8.6 µM. Strikingly, the Lys585Glu mutant exhibits for its first Ca^{2+}-binding event an affinity comparable to CBD1, whereas the Glu454Lys mutant has attained the binding affinities of CBD2. However intriguingly, only the Glu454Lys mutant shows an endothermic reaction typical of electrostatic interactions, while Ca^{2+} binding in wild-type CBD1, CBD2, and the Lys585Glu mutant is exothermic. Taken together, the NMR, gel mobility shift, and ITC data demonstrate that point mutations Glu454Lys and Lys585Glu have very similar properties to CBD2 and CBD1, respectively, and in a way reverse the functionality of their wild-type domains.

Model of the Intact Exchanger

A key aspect of the two Ca^{2+}-binding domains is their relative orientation to each other in the regulatory exchanger loop. A first indication of the arrangement between the domains originates from the turn at Asp498 in the CBD1 structure that directs residues 501–509, constituting strand A in CBD2, into an antiparallel orientation with respect to strand G of CBD1. Comparison of the $[^1H,^{15}N]$-HSQC spectrum of CBD12 with $[^1H,^{15}N]$-HSQC spectra of the individual domains in the presence of Ca^{2+} reveals that the majority of the shifting resonances is observed for residues in the Ca^{2+}-binding regions of both domains as well as for residues 374–375, 379–382, 490–505, 529–531, 563–566, 598–603, 632–637, and 641. Taking these shifting resonances as ambiguous interaction restraints in combination with restrictions imposed by the orientation of Asp499-Ala502, we obtained a model for CBD12. This model displays an antiparallel arrangement of the Ca^{2+}-binding domains with an extensive network of interactions in the center between residues in and around the β-bulges in strand A of CBD1 and strand G in CBD2.

In order to interpret Ca^{2+}-binding and release events in the context of the intact NCX, we also constructed a hypothetical model for the entire exchanger

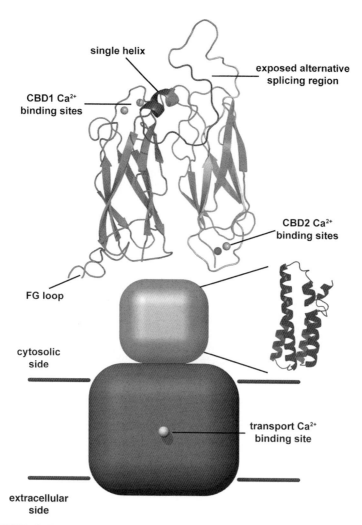

FIGURE 4. Hypothetical model of the intact exchanger, displaying the four domains, TM, CLD, CBD1, and CBD2 as well as the location of the Ca^{2+}-binding sites. (Figure adapted from Reference 8.)

with a simplified version displayed in FIGURE 4. Based on a homology we found with α-catenin,[10] we propose that the remainder of the regulatory loop (residues 217–370 and 651–705) forms a third domain, which we designate as CLD or catenin-like domain. Furthermore, as a consequence of sequence conservation with P-type ATPases,[11,12] we suggest that residues Val194, Val195, Val197, and Glu199 in NCX constitute a large part of the transport Ca^{2+}-binding site within the transmembrane domain (TM). In this model the Ca^{2+}-binding sites of CBD1 are approximately 90 Å away from the transport Ca^{2+}-binding site.

In contrast, the CBD2 Ca^{2+}-binding sites must be in close proximity to the CLD.

Gene Duplication

Members of the cation/Ca^{2+} exchanger superfamily[13] are characterized by the presence of two conserved regions located in the TM that form reentrant loops. These so-called α-repeat regions likely originated from an ancient gene duplication event.[14] Sequence and structural similarities between CBD1 and CBD2 now suggest that the entire NCX may have emerged from gene duplication.

CONCLUSIONS

Our study revealed the existence of three domains in the large intracellular loop of NCX. The two consecutive Ca^{2+}-binding domains, CBD1 and CBD2, are arranged in an antiparallel fashion and are connected via a third domain that we designated as CLD (FIG. 4) to the membrane part of the exchanger. CBD1 is the primary Ca^{2+} sensor in NCX and already detects small increases of cytosolic Ca^{2+}. In contrast, CBD2 undergoes comparably modest structural alterations and binds Ca^{2+} only at elevated Ca^{2+} concentrations. These two different sensitivity thresholds may enable NCX to function dynamically over a wide range of Ca^{2+} concentrations and permit high Ca^{2+} fluxes in excitable cells.

ACKNOWLEDGMENTS

We thank Stephan Hilge for professional help with the figures. This work was supported by a grant of the Netherlands Organisation for Scientific Research (NWO; grant 700-50-022). In addition, we gratefully acknowledge financial support from R.J. Nolte and J.C. van Hest.

REFERENCES

1. KANG, T.M. & D.W. HILGEMANN. 2004. Multiple transport modes of the cardiac Na^+/Ca^{2+} exchanger. Nature **427:** 544–548.
2. IWAMOTO, T. *et al.* 1999. Unique topology of the internal repeats in the cardiac Na^+/Ca^{2+} exchanger. FEBS Lett. **446:** 264–268.
3. NICOLL, D.A. *et al.* 1999. A new topological model of the cardiac sarcolemmal Na^+/Ca^{2+} exchanger. J. Biol. Chem. **274:** 910–917.

4. HILGEMANN, D.W., A. COLLINS & S. MATSUOKA. 1992. Steady-state and dynamic properties of cardiac sodium-calcium exchange. Secondary modulation by cytoplasmic calcium and ATP. J. Gen. Physiol. **100:** 933–961.

5. HILGEMANN, D.W. *et al.* 1992. Steady-state and dynamic properties of cardiac sodium-calcium exchange. Sodium-dependent inactivation. J. Gen. Physiol. **100:** 905–932.

6. LEVITSKY, D.O., D.A. NICOLL & K.D. PHILIPSON. 1994. Identification of the high affinity Ca^{2+}-binding domain of the cardiac Na^+/Ca^{2+} exchanger. J. Biol. Chem. **269:** 22847–22852.

7. OTTOLIA, M., K.D. PHILIPSON & S. JOHN. 2004. Conformational changes of the Ca^{2+} regulatory site of the Na^+/Ca^{2+} exchanger detected by FRET. Biophys. J. **87:** 899–906.

8. HILGE, M., J. AELEN & G.W. VUISTER. 2006. Ca^{2+} regulation in the Na^+/Ca^{2+} exchanger involves two markedly different Ca^{2+} sensors. Mol. Cell **22:** 15–25.

9. NICOLL, D.A. *et al.* 2006. The crystal structure of the primary Ca^{2+} sensor of the Na^+/Ca^{2+} exchanger reveals a novel Ca^{2+} binding motif. J. Biol. Chem. **281:** 21577–21581.

10. YANG, J. *et al.* 2001. Crystal structure of the M-fragment of α-catenin: implications for modulation of cell adhesion. EMBO J. **20:** 3645–3656.

11. NICOLL, D.A., S. LONGONI & K.D. PHILIPSON. 1990. Molecular cloning and functional expression of the cardiac sarcolemmal Na^+/Ca^{2+} exchanger. Science **250:** 562–565.

12. OGAWA, H. & C. TOYOSHIMA. 2002. Homology modeling of the cation binding sites of Na^+/K^+-ATPase. Proc. Natl. Acad. Sci. USA **99:** 15977–15982.

13. CAI, X. & J. LYTTON. 2004. The cation/Ca^{2+} exchanger superfamily: phylogenetic analysis and structural implications. Mol. Biol. Evol. **21:** 1692–1703.

14. NICOLL, D.A. *et al.* 1996. Mutation of amino acid residues in the putative transmembrane segments of the cardiac sarcolemmal Na^+/Ca^{2+} exchanger. J. Biol. Chem. **271:** 13385–13391.

Structure–Function Relationships of the NCKX2 Na$^+$/Ca^{2+}-K$^+$ Exchanger

Y. SHIBUKAWA, K. J. KANG, T. G. KINJO, R. T. SZERENCSEI,
H. F. ALTIMIMI, P. PRATIKHYA, R. J. WINKFEIN,
AND P. P. M. SCHNETKAMP

Department of Physiology and Biophysics, Faculty of Medicine, Hotchkiss Brain Institute, University of Calgary, Hospital Drive, N.W. Calgary, Alberta, T2N 4N1, Canada

ABSTRACT: K$^+$-dependent Na$^+$/Ca^{2+} exchangers (NCKX) have been shown to play important roles in physiological processes as diverse as phototransduction in rod photoreceptors, motor learning and memory in mice, and skin pigmentation in humans. Most structure–function studies on NCKX proteins have been carried out on the NCKX2 isoform, but sequence similarity suggests that the results obtained with the NCKX2 isoform are likely to apply to all NCKX1-5 members of the human *SLC24* gene family. Here we review our recent work on the NCKX2 protein concerning the topological arrangement of transmembrane segments carrying out cation transport, and concerning residues important for transport function and cation binding.

KEYWORDS: Na$^+$/Ca^{2+} exchange; NCKX; SLC24; calcium homeostasis; secondary transporters; photoreceptors

INTRODUCTION

K$^+$-dependent Na$^+$/Ca^{2+} exchange (NCKX) was first described in the late 1980s as the major Ca^{2+} extrusion pathway in the outer segments of vertebrate retinal rod photoreceptors and shown to operate at an electrogenic stoichiometry of 4Na$^+$:(1Ca^{2+} + 1K$^+$).[1–3] The requirement for and transport of K$^+$ associated with Na$^+$/Ca^{2+} exchange in rod photoreceptors constituted a clear distinction when compared with other Na$^+$/Ca^{2+} exchangers (NCX) studied extensively in the heart and in squid axons (reviewed in Ref. 4). Molecular cloning of cDNA of the dog heart NCX[5] and the bovine rod NCKX,[6] respectively, revealed that both represented the first member of two quite distinct gene families with surprisingly limited sequence similarity. The human *SLC8*

Address for correspondence: P. P. M. Schnetkamp, Department of Physiology and Biophysics, Faculty of Medicine, University of Calgary, 3330 Hospital Drive, N.W. Calgary, Alberta, T2N 4N1, Canada. Voice: 403-220-5448; fax: 403-283-8731.
pschnetk@ucalgary.ca

Ann. N.Y. Acad. Sci. 1099: 16–28 (2007). © 2007 New York Academy of Sciences.
doi: 10.1196/annals.1387.054

gene family of NCX proteins contains three members,[7] whereas the *SLC24* gene family of NCKX proteins comprises five members.[8] NCKX1 protein is located in high concentration only in the outer segments of bovine retinal rod photoreceptors, but not in other parts of bovine rod photoreceptors or in bovine cone photoreceptors.[9] For the other four NCKX isoforms our current knowledge is largely limited to distributions of NCKX transcripts. NCKX2 transcripts have been found throughout the rat brain[10] as well as in cone photoreceptors and ganglion cells of the human and chicken retina,[11,12] NCKX3 and NCKX4 transcripts are more ubiquitous and found in the brain, aorta, and various other tissues,[13,14] while NCKX5 transcripts have been found in skin and in the eye.[15] Although most previous studies on NCKX physiology concern the well-established role of NCKX1 in rod phototransduction,[16–18] exciting new roles have recently emerged for NCKX2 in synaptic Ca^{2+} homeostasis[19,20] and fish cone photoreceptor function,[21] and for NCKX5 in skin pigmentation in humans and fish.[15] In this article we will review recent progress in our studies on NCKX structure–function relationships.

NCKX cDNA ENCODE K^+-DEPENDENT Na^+/Ca^{2+} EXCHANGERS

The first NCKX1 cDNA was cloned from bovine retina and encoded a protein of 1216 residues.[6,22] Unfortunately, the bovine NCKX1 clone did not yield functional expression when transfected into various cell lines and this was not due to lack of protein expression.[23] By the late 1990s many NCKX-related sequences from both vertebrate and invertebrate species had emerged from cloning efforts and from various genomic sequencing projects. However, most NCKX sequences encoded proteins of 560–660 residues with the exception of all mammalian NCKX1, which contained 1100–1200 residues comparable to bovine NCKX1. Moreover, similarity between bovine NCKX1 and all other NCKX sequences was limited to two sets of hydrophobic segments totaling $\sim$300 residues, that is, one-fourth of the bovine NCKX1 sequence. We examined functional expression of NCKX1 and NCKX2 orthologs cloned from various mammals and chicken, and we also examined functional expression of NCKX paralogs cloned from *Caenorhabditis elegans* and *Drosophila*. Most were shown to encode functional K^+-dependent Na^+/Ca^{2+} exchangers with similar cation dependencies with the exception of the bovine and human NCKX1 clones, which did not yield functionally active proteins when expressed in cell lines.[11,24–27] Finally, when the two large hydrophilic loops were removed from the bovine NCKX1 sequence, the resultant NCKX1 deletion mutant showed K^+-dependent Na^+/Ca^{2+} exchange when expressed in insect High Five cells.[24] The observation of K^+-dependent Na^+/Ca^{2+} exchange by itself does not prove that the expressed NCKX proteins transport K^+ as had been shown previously for *in situ* NCKX1.[2,3] Taking advantage of the ability to grow large amounts of insect High Five cells stably expressing various NCKX proteins, we were able to show that several of the above NCKX proteins transport K^+ when expressed in these cells; moreover, the stoichiometry of the best expressing NCKX2 clones was $4Na^+:(1Ca^{2+} + 1K^+)$,[28] the same stoichiometry observed before for *in situ* NCKX1.[2]

A TOPOLOGICAL MODEL OF THE NCKX2
Na^+/Ca^{2+}-K^+ EXCHANGER

The results discussed in the previous section demonstrate that two sets of hydrophobic domains, presumably containing multiple transmembrane segments (TMS), are required and sufficient for Na^+/Ca^{2+}-K^+ exchange transport. Hydropathy analysis of NCKX sequences representing either different isoforms or representing the same isoform from different species yields a surprising range of predicted topological arrangements of the hydrophobic segments of NCKX proteins as assessed by web-based algorithms (e.g., Ref. 27). This clearly indicates the need for an experimental determination of the orientation of TMS and intra- or extracellular localization of the short connecting loops.

All NCKX sequences with the exception of *Drosophila* NCKX-X contain a hydrophobic segment within the first 50 residues from the N terminus. These hydrophobic segments are predicted to be either a cleavable signal peptide or a noncleaved signal anchor (http://cbs.dtu.dk/services/SignalP/). We examined signal peptide cleavage in the case of the NCKX1 and NCKX2 proteins expressed in cell lines by inserting epitope tags either before or after the predicted cleavage site. Signal peptide cleavage was observed by the absence of the tag inserted in front of the cleavage site but this only occurred in a fraction of the expressed NCKX proteins giving rise to two populations of NCKX proteins with different MW; moreover, the presence of the signal peptide was shown to be essential for correct targeting to the plasma membrane and only cleaved NCKX protein was found to be located in the surface membrane.[29,30]

We used insertion of N-glycosylation sites to determine which of the short connecting loops are exposed to the extracellular milieu; furthermore, cysteine-scanning mutagenesis was used to identify substituted cysteine residues accessible to the extracellular space. The results obtained are summarized in FIGURE 1 and yielded a consistent and unique topological model of the NCKX2 proteins.[31] A large number of single residue substitutions were made replacing the WT residue by cysteine, the mutant NCKX2 proteins were expressed in cultured insect High Five cells and inhibition of NCKX2 transport by pretreatment with the extracellular and membrane-impermeant sulfhydryl-modifying reagent MTSET was measured. K^+-dependent ^{45}Ca uptake was measured in Na^+-loaded High Five cells (so-called reverse exchange transport) as a quantitative indicator of (mutant) NCKX2 function. Pretreatment of WT NCKX2 with MTSET did not affect transport function indicating that none of the nine

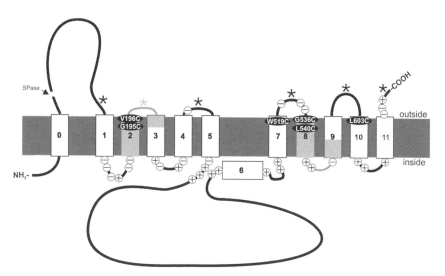

FIGURE 1. Topological model of NCKX2. Current topological model of the human NCKX2 protein is shown.[31] SPase signifies cleavage of the H0 signal peptide. The *gray bars* indicate the α1 (H2-H3) and α2 (H8-H9) repeats, respectively. Residues in black ovals indicate cysteine substitutions that are susceptible to the impermeant thiol reagent MTSET as described in the text. *Asterisks* indicate the positions of (inserted) *N*-glycosylation sites that resulted in glycosylation (glycosylation of the insert into the H2-H3 linker was considerably weaker than the inserts in the other linkers). Charged residues in the short connecting loops are indicated as well.

endogenous cysteine residues of the WT NCKX2 protein were accessible to the MTSET reagent, or, if they were accessible, modification by MTSET did not affect NCKX2 transport. In contrast, pretreatment with MTSET of cysteine substitutions of the residues indicated by the black ovals resulted in a strong inhibition of ^{45}Ca uptake in Na$^+$-loaded High Five cells expressing these mutant NCKX2 proteins, indicating that the residues involved were accessible to the extracellular space, and that these residues were likely to line an access channel to the cation binding sites as modification by the positively charged MTSET reagent impeded ^{45}Ca transport.[31]

N-glycosylation sites typically need to be placed ~25 residues away from the membrane surface to be glycosylated effectively.[32] As most of the predicted loops that connect the various TMS of the NCKX2 protein are very short, we used a 41 amino acid spacer to introduce a single *N*-glycosylation site separately into each of the connecting loops and examined glycosylation of the resultant mutant NCKX2 proteins by treatment with the glycosidase PNGase F.[31] Deglycosylation and the resultant increased mobility on SDS-PAGE were observed for the glycosylation site inserts of those loops marked with an asterisk (FIG. 1). Glycosylation indicates an extracellular localization of these loops. In one case, an endogenous *N*-glycosylation site (Asparagine111)

was present and located in the N-terminal hydrophilic loop (after signal peptide cleavage), which therefore is located in the extracellular space. Combined,
the results of the cysteine-scanning mutagenesis (residues indicated by black
ovals) and the insertion of glycosylation sites (asterisks) yield the complete
topological model of the NCKX2 protein illustrated in FIGURE 1. Demarcation
of transmembrane segments of the NCKX2 protein is also suggested by the frequent occurrence of pairs or clusters of charged residues, which are unlikely to
be embedded in the membrane. The position of these charged residues within
our topological model is indicated as well.

SCANNING MUTAGENESIS OF NCKX2: RESIDUES
IMPORTANT FOR CATION TRANSPORT

In order to select (blocks of) residues for mutagenesis experiments we first
carried out "selection through evolution," that is, we compared cation dependencies and K^+ transport of NCKX proteins from various species ranging from
mammals to worms as well as from a mutant NCKX1 protein from which the
two large hydrophilic loops were deleted.[24,26,33] Cation dependencies were
found to be very similar suggesting that domains/residues conserved between
all these NCKX proteins are responsible for cation binding and cation transport. This selection procedure narrows down the domains of interest to only
20% of the entire mammalian NCKX1 sequence. FIGURE 2 illustrates the alignment between the sequences of human NCKX2 and a NCKX paralog cloned
from *Caenorhabditis elegans*. We carried out scanning mutagenesis on all the
residues indicated by asterisks and examined functional consequences for all
the mutant NCKX2 proteins by using the K^+-dependent ^{45}Ca uptake assay
with the cultured insect High Five cells system mentioned above.[34] The interpretation of the results was facilitated by the fact that protein expression
levels were very similar for all the mutant NCKX2 proteins expressed in High
Five cells. In addition, the many residue substitutions in the NCKX2 protein
studied did not affect targeting of the mutant NCKX2 proteins to the plasma
membrane as judged by the observation that signal peptide cleavage was not
affected in any of the mutants and confirmed by surface biotinylation.

Twenty residues were identified for which substitutions resulted in a greater
than 70% reduction of the V_{max} of ^{45}Ca uptake via reverse exchange.[34] Two
substitutions were made for each residue, in most cases to alanine or cysteine,
while for the acidic residues charge-conservative and size-conservative substitutions were made (e.g., Glu to Asp or Gln). FIGURE 3 illustrates the position
of these 20 residues superimposed on our topological model of the NCKX2
protein. The residues are concentrated in the α1 and α2 repeats indicative of
the importance of these areas for NCKX-mediated exchange transport. Most of
the residues important for NCKX-mediated transport fall into three categories:
glycine residues, hydroxyl-containing residues (Ser, Thr), and acidic residues

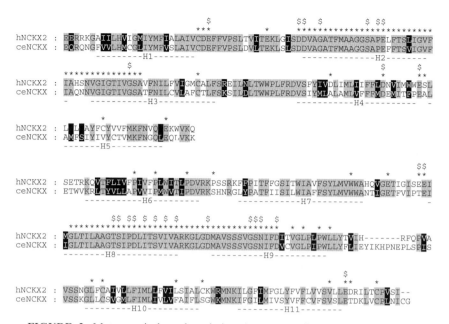

FIGURE 2. Mutagenesis through evolution. A sequence alignment is shown of the hydrophobic segments of the human NCKX2 and a NCKX paralog cloned from *Caenorhabditis elegans*. Asterisks indicate residues that were included in the scanning mutagenesis and were analyzed for changes in maximal transport capacity.[34] $ indicates residue substitutions analyzed for changes in apparent cation dissociation constants.[35,36]

(Asp, Glu). Glycine residues (in particular Gly176 and Gly210 for which even the conservative substitution to Ala caused a >85% loss of activity) may indicate swivel points that permit significant conformational changes or movement between two different domains of the NCKX proteins. Significant conformational changes are implicit in the alternating access model of NCKX transport function. Acidic residues are almost certainly required for high affinity Ca^{2+} binding by the NCKX protein and four critical acidic residues are found within the TMS of the NCKX2 protein. Complete conservation among all members of the NCX and NCKX gene families is observed for Glu188 and Asp548, while Asp258 and Asp575 are conserved in all five members of the NCKX gene family, but not in members of the NCX gene family. The five serine–threonine residues found to be important for NCKX2 function are conserved in all members of the NCX and NCKX gene families and we have suggested these residues play an important role in gating exchange fluxes and preventing uncoupled cation fluxes via the NCKX proteins.[34] For eight residues we found that highly conservative substitutions that merely lengthened the side chain by a methyl group (e.g., Ser to Thr, Gly to Ala, Asp to Glu) abolished transport function. This may suggest that side chains of these residues line a tight cation

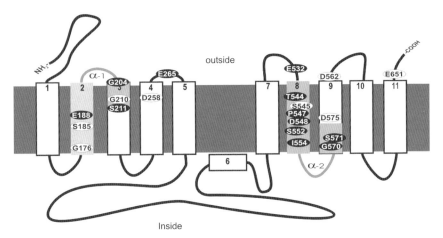

FIGURE 3. NCKX2 residues important for exchange function. Residue substitutions that resulted in a greater than 70% loss of exchanger function[34] are superimposed on our topological model of the NCKX2 protein. Conservative residue substitutions that increased the side chain length by just one -CH_2- and resulted in a greater than 70% inhibition of exchanger transport are indicated by the white ovals with black lettering.

binding pocket and that lengthening of the side chain does not leave enough space for Ca^{2+} to bind to this pocket (the position of these residues are indicated by the filled ovals in FIG. 3).

A FLUORESCENCE ASSAY FOR NCKX PROTEINS: RESIDUES IMPORTANT FOR Ca^{2+} AND K^+ BINDING

The ^{45}Ca uptake assay discussed above proved to be an efficient method to assess changes in V_{max} of mutant NCKX2 proteins. However, it proved more difficult to assess changes in apparent cation dissociation constants, in particular for mutant NCKX2 proteins with reduced maximal transport rates. We subsequently developed a simple fluorescence assay of NCKX function in transfected HEK293 cells based on the fluorescent Ca^{2+}-indicating dye Fluo-3.[35] The assay is illustrated in FIGURE 4. HEK293 cells are transfected with mutant or WT NCKX2 cDNA and loaded with the cell permeant form of the fluorescent Ca^{2+}-indicating dye Fluo-3. Like most NC(K)X assays ours measures the reverse mode of the exchanger that results in Ca^{2+} influx into the transfected cells. Na^+_i-dependent Ca^{2+} influx is measured as a Ca^{2+}- and K^+-dependent rise in Fluo-3 fluorescence indicative of a rise in intracellular-free Ca^{2+} concentration. The total amount of Fluo-3 loaded into the cells is obtained by addition of saponin, which permeabilizes the plasma membrane. Using this assay we identified nine residues for which substitutions resulted in shifts of the apparent dissociations constants for both Ca^{2+} and K^+,[35] and one residue

Fluo-3 Assay

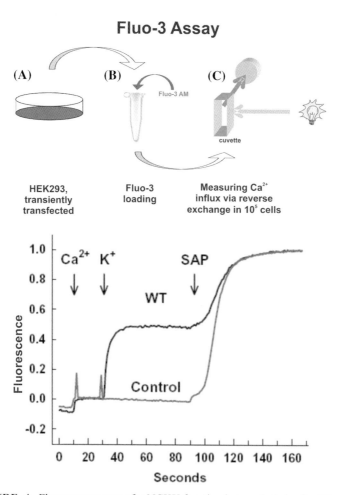

FIGURE 4. Fluorescence assay for NCKX function in transfected cells. Diagram that depicts our Fluo-3-based assay for NCKX function in transfected HEK293 cells. HEK293 cells are loaded using Fluo-3AM and washed to remove unhydrolyzed dye. Cells are subsequently placed in a cuvette filled with a buffered LiCl medium containing 0.1 mM EDTA. Addition of 0.35 mM $CaCl_2$ caused a small increase in fluorescence due to Fluo-3 leaked from the cells. Subsequent addition of 20 mM KCl caused a large increase in fluorescence, but only in cells transfected with NCKX2 cDNA. This increase indicates K^+-dependent Ca^{2+} influx via reverse exchange. Final addition of 0.01% saponin released all Fluo-3 from the cells and resulted in maximal fluorescence due to the high Ca^{2+} concentration. (For further details, see text and Ref. 35.)

for which substitutions resulted in an abolition of the K^+-dependence that characterizes all NCKX proteins[36] (the position of the residues is illustrated in FIG. 5). Previous *in situ* work on the bovine rod NCKX1 exchanger had shown that Na^+ and Ca^{2+} share a common binding site in view of competitive

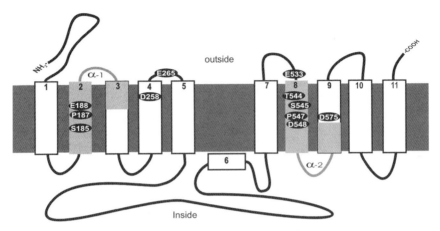

FIGURE 5. Residue substitutions that affect apparent Ca^{2+} and K^+ dissociation constants. Residue substitutions that result in changes in the apparent Ca^{2+} and K^+ dissociation constants[35,36] are superimposed (*black ovals*) on our topological model of NCKX2.

interactions between these two cations, while Ca^{2+} and K^+ do not compete for a common site, although K^+ has complex effects on the competitive interaction between Na^+ and Ca^{2+}.[37–39] The observation that residue substitutions studied so far invariably show a correlation between shifts in apparent Ca^{2+} affinities and shifts in apparent K^+ affinities is surprising. It suggests that Ca^{2+} and K^+ share a common translocation pathway rather than bind to separate sites. Little is known yet about residues that are important for Na^+ binding and transport. Further studies are required to evaluate whether certain NCKX residues could be important to the binding of all three substrate cations of the NCKX proteins.

SPATIAL ORGANIZATION OF RESIDUES
IMPORTANT FOR NCKX FUNCTION

From the previous overview it is clear that residues far apart in the linear NCKX sequence are important for cation transport and determine cation affinities. Some, but not all of these residues are concentrated in two clusters located in the α1 and α2 repeat, respectively. Again, most but not all of these residues are conserved between all members of the NCX and NCKX gene families as illustrated in FIGURE 6. These results suggest that the α1 and α2 repeat are in close apposition in the three-dimensional structure of the NCKX2 protein despite being far apart in the linear sequence. We used site-directed disulfide mapping (SDDM) to address proximity of residues important for cation binding and transport. We previously made a cysteine-free and functional NCKX2 mutant in which all nine cysteine residues present in WT NCKX2 were replaced by serine.[40] SDDM involves the cysteine substitution of two NCKX2

NCKX2 $_{179}$FMAAGG**SSAPE**LFTS $_{544}$**TS** I **PD**LI **TSV** I VA $_{570}$**GSN** I F**D** I TVGL

NCX1 $_{104}$LMAL**G SSAPE** I LLS $_{810}$**TSVPD**TFASKVAA $_{837}$**GSN**AV**N**VFLGI

FIGURE 6. Sequence alignment of conserved regions of the NCX and NCKX proteins. Members of the *SLC24* (NCKX) and *SLC8* (NCX) gene families share very limited sequence similarity confined to two short regions in α1 and α2 repeats, respectively. We identified Glu188, Asp548, and Asp575 as three residues critical in Ca^{2+} and K^+ binding to the NCKX2 protein.[35,36]

residues in different positions in the cysteine-free NCKX2 protein, one located in the α1 repeat and the other in the α2 repeat. Proximity of these two cysteine inserted residues is probed by copper–phenantroline-mediated disulfide bond formation, which requires a spacing not greater than ~3 Å. Using SDDM two key residues in the α1 repeat (Ser185 and Glu188) were shown to be in close proximity of key residues in the α2 repeat (Ser545, Asp548, and Ser552) and with Asp575 in H9.[30] Interestingly, Glu188Cys could form disulfide bonds with the three above cysteine replacements in the α2 repeat covering two full turns of an alpha helix or ~10 Å. To bring all these residues in sufficient prox-imity H2 and H8 may adopt helix–loop–helix motifs (found in several bacterial transporters), or H2 and H8 may move with respect to each other, perhaps rep-resenting the conformational change implicit in the alternating access model of transport. Consistent with a corkscrew movement of H2 with respect to H8, Glu188Cys could also form disulfide bonds with two cysteine replacements involving residues (Ile546 and Ile550) on the hydrophobic surface of H8 fac-ing away from the hydrophylic surface defined by residues Ser545, Asp548, and Ser552. Further experiments involving residues along the entire length of helices H2 and H8 will be needed to address this issue and such experiments are under way.

The above studies depend heavily on extensive mutagenesis of the NCKX2 cDNA and this raises the question whether the various mutant NCXK2 pro-teins represent the WT NCKX2 structure or improperly folded variants of this structure. This is particularly relevant for NCKX2 mutant proteins with greatly reduced transport function. It is generally thought that misfolded membrane proteins are recognized by the quality control apparatus of the ER and are not trafficked to the plasma membrane.[41] Heterologously expressed WT NCKX2 invariably shows a two band pattern when analyzed on SDS-PAGE; the lower MW form represents a subpopulation of the expressed NCKX2 protein from which N-terminal signal peptide is cleaved. Signal peptide cleavage is essen-tial for plasma membrane targeting and only the lower MW band represents protein present in the plasma membrane.[29] With very few exceptions we found that signal peptide cleavage was not affected in many mutant NCKX2 proteins studied. This suggests that plasma membrane targeting was not affected by the many residue substitutions studied and that these mutant proteins are likely to reflect properly folded NCKX2 protein.

CONCLUDING REMARKS

In this article we have reviewed biochemical and biophysical methods to obtain structural information on NCKX proteins and relate this to NCKX function. A high-resolution X ray crystal structure would greatly assist in understanding the NCKX, function. Unfortunately, crystallization of eukaryotic polytopic membrane proteins, such as NCKX, to obtain high-resolution structures has proven to be elusive.

ACKNOWLEDGMENT

This work was supported by an operating grant from the Canadian Institutes for Health Research to P.P.M.S. Y.S. is a recipient of a Grant-in-Aid (No. 18592050) for Scientific Research from the MEXT in Japan. P.P.M.S. is a scientist of the Alberta Heritage Foundation for Medical Research; T.G.K. is a recipient of a studentship from the Alberta Heritage Foundation for Medical Research, and K.J.K. and H.F.A. are recipients of a studentship from the Canadian Foundation Fighting Blindness.

REFERENCES

1. CERVETTO, L., L. LAGNADO, R.J. PERRY, et al. 1989. Extrusion of calcium from rod outer segments is driven by both sodium and potassium gradients. Nature 337: 740–743.
2. SCHNETKAMP, P.P.M., D.K. BASU & R.T. SZERENCSEI. 1989. Na-Ca exchange in the outer segments of bovine rod photoreceptors requires and transports potassium. Am. J. Physiol. (Cell Physiol.) 257: C153–C157.
3. SCHNETKAMP, P.P.M. 1989. Na-Ca or Na-Ca-K exchange in the outer segments of vertebrate rod photoreceptors. Prog. Biophys. Mol. Biol. 54: 1–29.
4. BLAUSTEIN, M.P. & W.J. LEDERER. 1999. Sodium/calcium exchange: its physiological implications. Physiol. Rev. 79: 763–854.
5. NICOLL, D.A., S. LONGONI & K.D. PHILIPSON. 1990. Molecular cloning and functional expression of the cardiac sarcolemmal Na^+-Ca^{2+} exchanger. Science 250: 562–565.
6. REILÄNDER, H., A. ACHILLES, U. FRIEDEL, et al. 1992. Primary structure and functional expression of the Na/Ca,K-exchanger from bovine rod photoreceptors. EMBO J. 11: 1689–1695.
7. QUEDNAU, B.D., D.A. NICOLL & K.D. PHILIPSON. 2004. The sodium/calcium exchanger family-SLC8. Eur. J. Physiol. 447: 543–548.
8. SCHNETKAMP, P.P.M. 2004. The SLC24 Na^+/Ca^{2+}-K^+ exchanger family: vision and beyond. Eur. J. Physiol. 447: 683–688.
9. KIM, T.S.Y., D.M. REID & R.S. MOLDAY. 1998. Structure-function relationships and localization of the Na/Ca-K exchanger in rod photoreceptors. J. Biol. Chem. 273: 16561–16567.

10. TSOI, M., K.-H. RHEE, D. BUNGARD, *et al.* 1998. Molecular cloning of a novel potassium-dependent sodium-calcium exchanger from rat brain. J. Biol. Chem. **273:** 4155–4162.

11. PRINSEN, C.F.M., R.T. SZERENCSEI & P.P.M. SCHNETKAMP. 2000. Molecular cloning and functional expression the potassium-dependent sodium-calcium exchanger from human and chicken retinal cone photoreceptors. J. Neurosci. **20:** 1424–1434.

12. PRINSEN, C.F.M., C.B. COOPER, R.T. SZERENCSEI, *et al.* 2002. The retinal rod and cone Na^+/Ca^{2+}-K^+ exchangers. Adv. Exp. Med. Biol. **514:** 237–251.

13. KRAEV, A., B.D. QUEDNAU, S. LEACH, *et al.* 2001. Molecular cloning of a third member of the potassium-dependent sodium-calcium exchanger gene family, NCKX3. J. Biol. Chem. **276:** 23161–23172.

14. LI, X.F., A.S. KRAEV & J. LYTTON. 2002. Molecular cloning of a fourth member of the potassium-dependent sodium-calcium exchanger gene family, NCKX4. J. Biol. Chem. **277:** 48410–48417.

15. LAMASON, R.L., M.A. MOHIDEEN, J.R. MEST, *et al.* 2005. SLC24A5, a putative cation exchanger, affects pigmentation in zebrafish and humans. Science **310:** 1782–1786.

16. KORENBROT, J.I. 1995. Ca^{2+} flux in retinal rod and cone outer segments. Cell Calcium **18:** 285–300.

17. SCHNETKAMP, P.P.M. 1995. Calcium homeostasis in vertebrate retinal rod outer segments. Cell Calcium **18:** 322–330.

18. FAIN, G.L., H.R. MATTHEWS, M.C. CORNWALL, *et al.* 2001. Adaptation in vertebrate photoreceptors. Physiol. Rev. **81:** 117–151.

19. LI, X.F., L. KIEDROWSKI, F. TREMBLAY, *et al.* 2006. Importance of K^+-dependent Na^+/Ca^{2+}-exchanger 2, NCKX2, in motor learning and memory. J. Biol. Chem. **281:** 39205–39216.

20. KIM, M.H., N. KOROGOD, R. SCHNEGGENBURGER, *et al.* 2005. Interplay between Na^+/Ca^{2+} exchangers and mitochondria in Ca^{2+} clearance at the calyx of Held. J. Neurosci. **25:** 6057–6065.

21. PAILLART, C., R.J. WINKFEIN, P.P.M. SCHNETKAMP, *et al.* 2007. Functional characterization and molecular cloning of the K^+-dependent Na^+/Ca^{2+} exchanger in intact retinal cone photoreceptors. J. Gen. Physiol. **129:** 1–16.

22. TUCKER, J.E., R.J. WINKFEIN, C.B. COOPER, *et al.* 1998. cDNA cloning of the human retinal rod Na/Ca+K exchanger: comparison with a revised bovine sequence. Invest. Ophthalmol. Vis. Sci. **39:** 435–440.

23. COOPER, C.B., R.J. WINKFEIN, R.T. SZERENCSEI, *et al.* 1999. cDNA-cloning and functional expression of the dolphin retinal rod Na-Ca+K exchanger NCKX1: comparison with the functionally silent bovine NCKX1. Biochemistry **38:** 6276–6283.

24. SZERENCSEI, R.T., J.E. TUCKER, C.B. COOPER, *et al.* 2000. Minimal domain requirement for cation transport by the potassium-dependent Na/Ca-K exchanger: comparison with an NCKX paralog from *Caenorhabditis elegans.* J. Biol. Chem. **275:** 669–676.

25. HAUG-COLLET, K., B. PEARSON, S. PARK, *et al.* 1999. Cloning and characterization of a potassium-dependent sodium/calcium exchanger in *Drosophila.* J. Cell Biol. **147:** 659–669.

26. SHENG, J.-Z., C.F.M. PRINSEN, R.B. CLARK, *et al.* 2000. Na^+-Ca^{2+}-K^+ currents measured in insect cells transfected with the retinal cone or rod Na^+-Ca^{2+}-K^+ exchanger cDNA. Biophys. J. **79:** 1945–1953.

27. WINKFEIN, R.J., B. PEARSON, R. WARD, *et al.* 2004. Molecular characterization, functional expression and tissue distribution of a second NCKX Na^+/Ca^{2+}-K^+ exchanger from *Drosophila.* Cell Calcium **36:** 147–155.
28. SZERENCSEI, R.T., C.F.M. PRINSEN & P.P.M. SCHNETKAMP. 2001. The stoichiometry of the retinal cone Na/Ca-K exchanger heterologously expressed in insect cells: comparison with the bovine heart Na/Ca exchanger. Biochemistry **40:** 6009–6015.
29. KANG, K.-J. & P.P.M. SCHNETKAMP. 2003. Signal sequence cleavage and plasma membrane targeting of the rod NCKX1 and cone NCKX2 Na^+/Ca^{2+}-K^+ exchangers. Biochemistry **42:** 9438–9445.
30. KINJO, T.G., K.-J. KANG, R.T. SZERENCSEI, *et al.* 2005. Site-directed disulfide mapping of residues contributing to the Ca^{2+} and K^+ binding pocket of the NCKX2 Na^+/Ca^{2+}-K^+ exchanger. Biochemistry **44:** 7787–7795.
31. KINJO, T.G., R.T. SZERENCSEI, R.J. WINKFEIN, *et al.* 2003. Topology of the retinal cone NCKX2 Na/Ca-K exchanger. Biochemistry **42:** 2485–2491.
32. POPOV, M., L.Y. TAM, J. LI, *et al.* 1997. Mapping the ends of transmembrane segments in a polytopic membrane protein. Scanning N-glycosylation mutagenesis of extracytosolic loops in the anion exchanger, band 3. J. Biol. Chem. **272:** 18325–18332.
33. SZERENCSEI, R.T., R.J. WINKFEIN, C.B. COOPER, *et al.* 2002. The Na/Ca-K exchanger gene family. Ann. N. Y. Acad. Sci. **976:** 41–52.
34. WINKFEIN, R.J., R.T. SZERENCSEI, T.G. KINJO, *et al.* 2003. Scanning mutagenesis of the alpha repeats and of the transmembrane acidic residues of the human retinal cone Na/Ca-K exchanger. Biochemistry **42:** 543–552.
35. KANG, K.-J., T.G. KINJO, R.T. SZERENCSEI, *et al.* 2005. Residues contributing to the Ca^{2+} and K^+ binding pocket of the NCKX2 Na^+/Ca^{2+}-K^+ exchanger. J. Biol. Chem. **280:** 6823–6833.
36. KANG, K.-J., Y. SHIBUKAWA, R.T. SZERENCSEI, *et al.* 2005. Substitution of a single residue, Asp^{575}, renders the NCKX2 K^+-dependent Na^+/Ca^{2+} exchanger independent of K^+. J. Biol. Chem. **280:** 6834–6839.
37. SCHNETKAMP, P.P.M. 1991. Optical measurements of Na-Ca-K exchange currents in intact outer segments isolated from bovine retinal rods. J. Gen. Physiol. **98:** 555–573.
38. SCHNETKAMP, P.P.M., X.B. LI, D.K. BASU, *et al.* 1991. Regulation of free cytosolic Ca^{2+} concentration in the outer segments of bovine retinal rods by Na-Ca-K exchange measured with Fluo-3. I. Efficiency of transport and interactions between cations. J. Biol. Chem. **266:** 22975–22982.
39. SCHNETKAMP, P.P.M., J.E. TUCKER & R.T. SZERENCSEI. 1995. Ca^{2+} influx into bovine retinal rod outer segments mediated by Na-Ca+K exchange. Am. J. Physiol. (Cell Physiol.) **269:** c1153–c1159.
40. KINJO, T.G., R.T. SZERENCSEI, R.J. WINKFEIN, *et al.* 2004. Role of cysteine residues in the NCKX2 Na^+/Ca^{2+}-K^+ exchanger: generation of a functional cysteine-free exchanger. Biochemistry **43:** 7940–7947.
41. ELLGAARD, L. & A. HELENIUS. 2003. Quality control in the endoplasmic reticulum. Nat. Rev. Mol. Cell Biol. **4:** 181–191.

Examining Ca^{2+} Extrusion of Na^+/Ca^{2+}-K^+ Exchangers

HAIDER F. ALTIMIMI AND PAUL P. M. SCHNETKAMP

Department of Physiology and Biophysics, Faculty of Medicine,
Hotchkiss Brain Institute, University of Calgary, N.W. Calgary, Alberta,
T2N 4N1, Canada

ABSTRACT: Na^+/Ca^{2+}-K^+ exchangers (NCKX) are plasma membrane transporters that are thought to mainly mediate Ca^{2+} extrusion (along with K^+) at the expense of the Na^+ electrochemical gradient. However, because they are bidirectional, most assays have relied on measuring their activity in the reverse (Ca^{2+} import) mode. Herein we describe a method to control intracellular ionic conditions, and examine the forward (Ca^{2+} extrusion) mode of exchange of NCKX2.

KEYWORDS: Na^+/Ca^{2+}-K^+ exchangers; NCKX; calcium extrusion; calcium sequestration

INTRODUCTION

Plasma membrane Na^+/Ca^{2+}-K^+ exchangers (NCKX) are bidirectional and electrogenic secondary transporters.[1] The transport of Ca^{2+} in NCKX is coupled to the symport of K^+ at a 1:1 stoichiometry and to antiport of Na^+ at a stoichiometry of 4 Na^+ for 1 Ca^{2+}. NCKX was first described in the retinal rod outer segment, where it mediates extrusion of Ca^{2+} that enters through the light sensitive cyclic nucleotide gated channel.[2,3] Since the isolation and cloning of the retinal rod outer segment exchanger (NCKX1),[4,5] a second isoform (NCKX2) was cloned by homology, and was found in many areas of rodent brain,[6] as well as human and chick retina.[7] The physiological role of NCKX2 is not as well established as that of NCKX1, but recent evidence indicates that it participates in Ca^{2+} clearance in some neurons.[8,9] We have previously reported on the residues responsible for Ca^{2+} and K^+ transport, but the residues that mediate Na^+ transport are as of yet unknown.[10] Hence, we set out to devise an assay that will allow us to control Na^+ concentrations, and thereby measure the affinity of NCKX2 for Na^+. Another objective was to examine the forward exchange (Ca^{2+} extrusion) mode of operation of NCKX2,

Address for correspondence: Paul P. M. Schnetkamp, Department of Physiology and Biophysics, Faculty of Medicine, University of Calgary, 3330 Hospital Dr., N.W. Calgary, Alberta, T2N 4N1, Canada. Voice: 403-220-5448; fax: 403-283-8731.
pschnetk@ucalgary.ca

Ann. N.Y. Acad. Sci. 1099: 29–33 (2007). © 2007 New York Academy of Sciences.
doi: 10.1196/annals.1387.056

since most studies on NCKX2 expressed in cell lines have only reported on reverse exchange (Ca^{2+} import) mode.

EXPERIMENTAL PROCEDURES

The details of the experimental procedures and assay for NCKX activity have been described elaborately elsewhere.[10,11] For the purposes of this study, we employed the channel forming alkali cation ionophore gramicidin to clamp Na^+ (and K^+) concentrations across the plasma membrane of HEK293 cells. The assay is based on the addition of various Na^+ concentrations to a suspension of NCKX2-transfected, fluo-3-loaded HEK293 cells placed in separate cuvettes containing 150 mM KCl and 250 μM free $CaCl_2$. To examine forward exchange mode of NCKX2, EDTA was added to chelate extracellular Ca^{2+} in the cuvette, thereby reversing the Ca^{2+} concentration gradient.

RESULTS AND DISCUSSION

The addition of 350 μM $CaCl_2$ to NCKX2-transfected HEK293 cells, previously loaded with fluo-3 and treated with 2 μM gramicidin, in suspension in a buffered medium of 150 mM KCl, 20 mM HEPES, 100 μM EDTA did not induce any changes in intracellular Ca^{2+} levels. Only upon addition of NaCl did the fluorescence rise sharply, as intracellular-free Ca^{2+} was elevated by NCKX2 operating in the reverse mode. The increase in intracellular-free Ca^{2+} reached a steady-state equilibrium plateau within 20 s, and remained constant throughout the data collection period (FIG. 1). These results show that HEK293 cells do not harbor an endogenous Na^+/Ca^{2+} exchange mechanism, because mock-transfected HEK293 cells did not respond to the same ionic conditions. They also demonstrate that the reverse exchange mode of operation of NCKX2 is absolutely dependent on Na^+ as substrate (on the intracellular side of the plasma membrane). The steady-state plateau level of intracellular-free Ca^{2+} also suggests that some equilibrium is established in these cells, between NCKX2 importing extracellular Ca^{2+} and other cellular Ca^{2+}-handling mechanisms.

To examine forward exchange mode of NCKX2, 1 mM EDTA was introduced into the cuvette to chelate extracellular Ca^{2+}, thereby reversing the concentration gradient for Ca^{2+}—the only relevant gradient for the exchanger, since we have effectively clamped both Na^+ and K^+ gradients with gramicidin. The resultant chemical gradient of high intracellular-free Ca^{2+} favors extrusion by NCKX2. The Ca^{2+} efflux signal is, in fact, much more rapid than the initial, NCKX2-mediated, Ca^{2+} influx; the exchanger effectively clears intracellular-free Ca^{2+} back to baseline levels in $\sim$10 s (FIG. 1 A).

Based on the evidence for involvement of other cellular Ca^{2+}-handling mechanisms in buffering NCKX2-mediated fluxes in Ca^{2+}, we used the ionophore A23187 to release any sequestered Ca^{2+}, after the addition of 1 mM EDTA and subsequent drop in intracellular-free Ca^{2+}. We found a large increase in fluorescence upon addition of 200 nM A23187 (FIG. 1 B); this signal was smaller when A23187 was added to the cuvette at earlier time points (data not shown), corroborating that NCKX2 continually mediates Ca^{2+} import

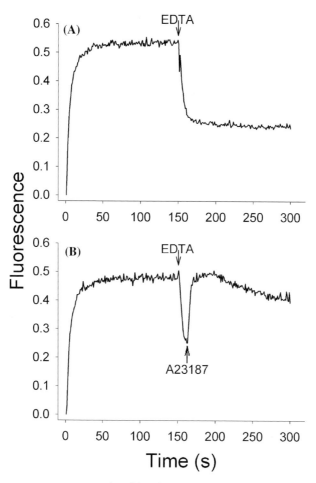

FIGURE 1. Assaying for $Na^+/Ca^{2+}-K^+$ exchange using gramicidin. (**A**) Fluo-3-loaded NCKX2-transfected HEK293 cells in a 50 μL suspension of buffered LiCl medium (in mM: 150 LiCl, 20 HEPES, 0.1 EDTA, 0.25 sulfinpyrazone) were placed in a cuvette containing 1950 μL buffered KCl medium (in mM: 150 KCl, 20 HEPES, 0.1 EDTA, 0.25 sulfinpyrazone, 0.5 DTT), and then 2 μM gramicidin added. After 3 min, 350 μM $CaCl_2$ was added, following which 75 mM NaCl was added to initiate reverse exchange (Ca^{2+} import) mode of NCKX2. At the arrow, 1 mM EDTA was added to chelate extracellular Ca^{2+}, and initiate forward exchange (Ca^{2+} extrusion) mode of NCKX2. (**B**) Same procedures were followed as in **A**; at the second arrow, 200 nM A23187 was added to the cuvette.

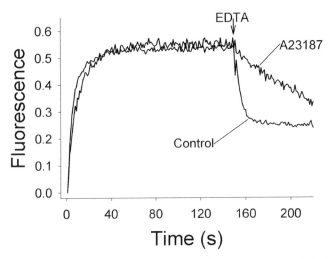

FIGURE 2. Same experimental procedures as in FIGURE 1 A, except 200 nM A23187 was added to the cuvette at the start of the experiment (at the same time as gramicidin; trace labeled A23187). Also shown for comparison is the trace from FIGURE 1 A (trace labeled Control).

while other intracellular mechanisms sequester Ca^{2+}. A similar observation was made in retinal rod outer segments, where NCKX1 in that case mediated continuous Ca^{2+} import when the Na^+ gradient was reversed, and intracellular Ca^{2+} sequestration in internal disks increased with prolonged Ca^{2+} loading.[12]

To isolate the activity of the exchanger, 200 nM A23187 was added to the fluo-3-loaded NCKX2-transfected HEK293 cells at the beginning of the trace (along with 2 μM gramicidin), and the assay was initiated by the addition of 350 μM $CaCl_2$, followed by 75 mM NaCl. After 150 s, the forward mode of NCKX2 transport was engaged by the addition of 1mM EDTA to the cuvette (FIG. 2). In this case, the efflux signal was markedly slower than in the case without the prior addition of A23187. This slow efflux signal signified that the exchanger was somehow inactivated and was not able to effectively extrude the intracellular Ca^{2+} load. It also points to the fact that intracellular Ca^{2+} measurements made in intact cells do not reflect the activity of the exchanger alone, but also reflect the contribution of several other cellular Ca^{2+} homeostatic mechanisms. For further analysis of inactivation of NCKX, and contribution of other Ca^{2+} homeostatic mechanisms, see Altimimi and Schnetkamp.[13]

ACKNOWLEDGMENT

This work was supported by an operating grant from the Canadian Institutes for Health Research to P.P.M.S. H.F.A. is recipient of a studentship from the

Canadian Foundation Fighting Blindness. P.P.M.S. is a scientist of the Alberta Heritage Foundation for Medical Research.

REFERENCES

1. SCHNETKAMP, P.P.M. 2004. The SLC24 Na^+/Ca^{2+}-K^+ exchanger family: vision and beyond. Pflügers Arch. **447:** 683–688.
2. CERVETTO, L., L. LAGNADO, R.J. PERRY, et al. 1989. Extrusion of calcium from rod outer segments is driven by both sodium and potassium gradients. Nature **337:** 740–743.
3. SCHNETKAMP, P.P.M., D.K. BASU & R.T. SZERENCSEI. 1989. Na^+-Ca^{2+} exchange in bovine rod outer segments requires and transports K^+. Am. J. Physiol. **257:** C153–C157.
4. COOK, N.J. & U.B. KAUPP. 1988. Solubilization, purification, and reconstitution of the sodium-calcium exchanger from bovine retinal rod outer segments. J. Biol. Chem. **263:** 11382–11388.
5. REILÄNDER, H., A. ACHILLES, U. FRIEDEL, et al. 1992. Primary structure and functional expression of the Na/Ca,K-exchanger from bovine rod photoreceptors. EMBO J. **11:** 1689–1695.
6. TSOI, M., K.H. RHEE, D. BUNGARD, et al. 1998. Molecular cloning of a novel potassium-dependent sodium-calcium exchanger from rat brain. J. Biol. Chem. **273:** 4155–4162.
7. PRINSEN, C.F., R.T. SZERENCSEI & P.P.M. SCHNETKAMP. 2000. Molecular cloning and functional expression of the potassium-dependent sodium-calcium exchanger from human and chicken retinal cone photoreceptors. J. Neurosci. **20:** 1424–1434.
8. LEE, S.H., M.H. KIM, K.H. PARK, et al. 2002. K^+-dependent Na^+/Ca^{2+} exchange is a major Ca^{2+} clearance mechanism in axon terminals of rat neurohypophysis. J. Neurosci. **22:** 6891–6899.
9. LI, X.F., L. KIEDROWSKI, F. TREMBLAY, et al. 2006. Importance of K^+-dependent Na^+/Ca^{2+}-exchanger 2, NCKX2, in motor learning and memory. J. Biol. Chem. **281:** 6273–6282.
10. KANG, K.J., T.G. KINJO, R.T. SZERENCSEI & P.P.M. SCHNETKAMP. 2005. Residues contributing to the Ca^{2+} and K^+ binding pocket of the NCKX2 Na^+/Ca^{2+}-K^+ exchanger. J. Biol. Chem. **280:** 6823–6833.
11. COOPER, C.B., R.T. SZERENCSEI & P.P.M. SCHNETKAMP. 2000. Spectrofluorometric detection of Na^+/Ca^{2+}-K^+ exchange. Methods Enzymol. **315:** 847–864.
12. SCHNETKAMP, P.P.M. & R.T. SZERENCSEI. 1993. Intracellular Ca^{2+} sequestration and release in intact bovine retinal rod outer segments. Role in inactivation of Na-Ca+K exchange. J. Biol. Chem. **268:** 12449–12457.
13. ALTIMIMI H.F. & P.P.M. SCHNETKAMP. 2007. Na^+-dependent inactivation of the retinal cone/brain Na^+/Ca^{2+}-K^+ exchanger NCKX2. J. Biol. Chem. **282:** 3720–3729.

Topologic Investigation of the NCKX2 Na$^+$/Ca^{2+}-K$^+$ Exchanger α-Repeats

TASHI G. KINJO, ROBERT T. SZERENCSEI,
AND PAUL P. M. SCHNETKAMP

*Department of Physiology and Biophysics, Faculty of Medicine,
Hotchkiss Brain Institute, University of Calgary, N.W. Calgary,
Alberta, T2N 4N1, Canada*

ABSTRACT: Algorithms suggest that NCKX proteins consist of an N-terminal signal peptide and 11 transmembrane segments divided in two groups of 5 and 6, respectively, separated by a large cytoplasmic loop. This predicted topology places the NCKX α-repeats with the same orientation in the plasma membrane. Using thiol-specific drug treatment and site-directed disulfide mapping, we have investigated the orientation of the NCKX2 α-repeats. Our results suggest that the NCKX2 α-repeats have an antiparallel orientation in the plasma membrane. In addition, these experiments suggest that the α-repeats are found in close proximity in the mature configuration of the protein.

KEYWORDS: Na/Ca exchange; calcium homeostasis; photoreceptors; membrane topology; secondary transporters

INTRODUCTION

The Na$^+$/Ca^{2+}-K$^+$ exchanger (NCKX) uses both the Na$^+$ and K$^+$ gradients to drive the extrusion of cytoplasmic Ca^{2+}. Physiologically, NCKX gene products play important roles in visual light adaptation, learning and memory, and skin pigmentation.[1-3] Products of the NCKX gene family are thought to consist of an N-terminal signal peptide followed by a set of five putative membrane-spanning helices, a large cytoplasmic loop, and a further set of five membrane-spanning helices, respectively.[4,5] Using glycosylation site insertion NCKX2 mutants, we have localized the various NCKX2 putative helix connecting loops to the extracellular space (FIG. 1 A).[5] The localization of these connecting loops has resulted in a complete topologic model of NCKX2 (FIG. 1 B). In this model, the α-repeat regions of NCKX2 are placed in

Address for correspondence: Paul P. M. Schnetkamp, Department of Physiology and Biophysics, Faculty of Medicine, University of Calgary, 3330 Hospital Drive, N.W. Calgary, Alberta, T2N 4N1, Canada. Voice: 403-220-5448; fax: 403-283-8731.

pschnetk@ucalgary.ca

Ann. N.Y. Acad. Sci. 1099: 34–39 (2007). © 2007 New York Academy of Sciences.
doi: 10.1196/annals.1387.055

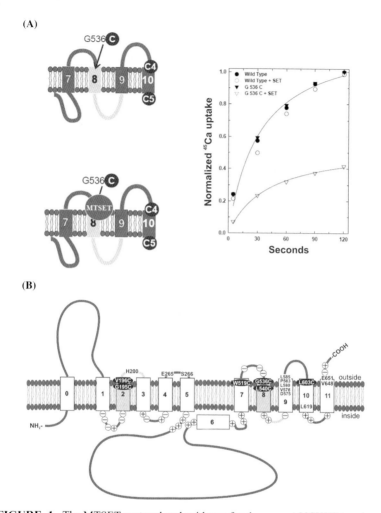

FIGURE 1. The MTSET protocol and evidence for the current NCKX2 topology. (**A**) schematic diagram of the MTSET method is shown in the left panel. The functional correlate is shown in the right panel. NCKX2 residues were replaced by cysteine residues (e.g., G536C). The function of the cysteine mutant was then assessed, using a $^{45}Ca^{2+}$ uptake assay, both pre (*black symbols*) and post (*white symbols*) treatment with extracellular MT-SET. Function was normalized to the untreated control WT NCKX2 and G536C mutant, respectively. Function of the G536C mutant was ~30% of that of wild-type NCKX2. MT-SET treatment of the wild-type NCKX2 did not result in a significant reduction in function (compare white with black circles). In contrast, treatment of the G536C mutant with MT-SET (compare white with black triangles) resulted in the inhibition of NCKX2 function suggesting an extracellular location for the G536 residue. (**B**) The current topology model of NCKX2 is presented along with a summary of MTSET experiments. Cysteine replacement of those residues in black background was significantly inhibited (>40%) by MTSET, while others resulted in only modest inhibition (<20%). (Reprinted from Reference 5 with permission from the American Chemical Society.)

antiparallel orientation in the plasma membrane. Unfortunately, the NCKX2 glycosylation site insertion mutants displayed greatly reduced functional activity. Resolving the orientation of the NCKX α-repeats is important in understanding the mechanism of NCKX ion transport as these regions house the residues found to be critical for cation binding.[6] In addition, the α-repeats contain the only sequence elements conserved between the NCKX and the Na^+/Ca^{2+} exchanger (NCX) gene families.[7]

We have investigated the relative orientation of the NCKX2 α-repeats using two methods: thiol-specific drug susceptibility of cysteine NCKX2 mutants, and site-directed disulfide mapping. The first method involves the creation of NCKX2 cysteine insertion mutants and then accessing their function pre and post treatment with a cysteine-modifying reagent. The second technique, site-directed disulfide mapping, involves the insertion of pairs of cysteine residues into a cysteine-free NCKX2 background and then assaying for cysteine proximity through the induction of disulfide bonds. Disulfide bond formation is detected as a shift in apparent molecular weight when the protein is subject to gel electrophoresis under nonreducing conditions. Our lab has recently created a functional cysteine-free NCKX2 mutant for use in these studies.[8] The results from these experiments suggest an opposite orientation for the NCKX2 α-repeats.

MATERIALS AND METHODS

A more detailed account of the procedures used has been described previously.[5,9] High Five cells expressing the indicated mutants were Na^+ loaded and then placed in $K^+/^{45}Ca^{2+}$ media. At various time points, aliquots were taken and scintillation counted as a measure of reverse NCKX2 function. When indicated, methanethiosulfonate ethyltrimethylammonio (MTSET) treatment was carried out prior to the Na^+ loading step. For site-directed disulfide mapping, double cysteine mutants were expressed in HEK293 cells and oxidized with copper–phenanthroline; next, cells were lysed and the lysate was subjected to gel electrophoresis in the presence or absence of dithiothreitol (DTT).

RESULTS AND DISCUSSION

FIGURE 1 illustrates the MTSET technique used for the investigation of NCKX2 topology. Using this method, cysteine residues are inserted into the wild-type NCKX2 background. These mutants are then expressed in High Five cells and assessed for reverse NCKX function using a $^{45}Ca^{2+}$ uptake assay. Mutants exhibiting significant NCKX2 function are then treated extracellularly with the membrane impermeant MTSET reagent. Inhibition of

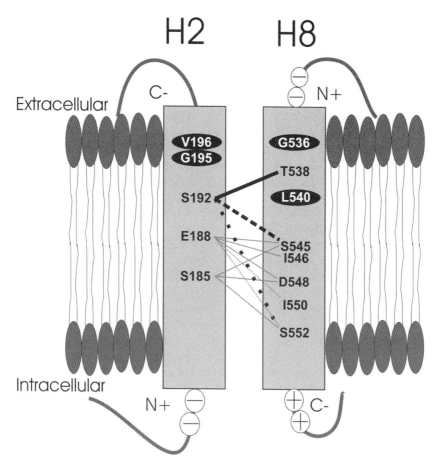

FIGURE 2. Cross-linking evidence for antiparallel NCKX2 α-repeat orientation. Schematic illustrating the cross-linking results of the NCKX2 α-repeats is shown. Pairs of cysteines found to exhibit high levels of copper-induced cross-linking are indicated by *solid lines*. Double cysteine mutants that exhibited lower levels of cross-linking are shown with *dotted lines*. The cysteine pairs involving S192C are shown with heavy black lines, while lighter lines indicated additional cross-linking pairs.[9]

NCKX2 mutant function by MTSET then suggests an extracellular location for the mutated residue. This procedure is illustrated for the G536C in FIGURE 1. As shown, MTSET treatment of the wild-type NCKX2 did not result in the inhibition of NCKX2 function. This suggests that endogenous NCKX2 cysteine residues were either not accessible to the MTSET reagent or were not located in a functionally significant area of the exchanger. FIGURE 1 B illustrates a summary of the NCKX2 cysteine mutants that were found to be inhibited by MTSET along with the current NCKX2 topologic model. As shown, several mutants were found to be significantly inhibited ($>40\%$) by MTSET treatment (black background). Our lab has recently shown that the residues forming the NCKX2 cation binding site are located in the α-repeats.[6] Four out of the six residues that were found to be highly sensitive to MTSET are located in the α-repeat regions. MTSET sensitivity of G195C and V196C mutants at the C terminus of H2 (α-1 repeat), and MTSET inhibition of mutants G536C and L540C at the N terminus of H8 (α-2 repeat) suggest that these helices have opposite orientations in the plasma membrane.

Site-directed disulfide mapping was used to investigate this α-repeat orientation. Pairs of cysteines were introduced into the cysteine-free NCKX2 sequence with one cysteine placed at the C-terminal end of H2 (S192C) and the other placed in H8. Three such double cysteine mutants were made, each with the second cysteine of the pair being placed progressively deeper into the membrane away from the N-terminal putative extracellular end of H8. These double cysteine mutants were; S192C/T538C, S192C/S545C, and S192C/S552C (FIG. 2). Following heterologous expression, oxidation and the electrophoretic separation of these mutant proteins under nonreducing conditions, an interesting pattern emerged. The strongest level of cross-linking was observed for the S192C/T538C mutant, while significantly lower cross-linking efficiency was observed for the S192C/S545C mutant. Little-to-no cross-linking was observed for the S192C/S552C mutant. These results support an antiparallel NCKX2 α-repeat orientation. In addition, our lab has found that the critical residues involved in NCKX2 cation binding (E188 and D548) are located in close proximity in three-dimensional space (FIG. 2). Collectively, these results suggest that the α-repeats of NCKX2 come together in opposite orientation to form the critical NCKX functional unit.

ACKNOWLEDGMENT

This work was supported by an operating grant from the Canadian Institutes for Health Research to P.P.M.S. T.G.K. is a recipient of a studentship from the Alberta Heritage Foundation for Medical Research and P.P.M.S. is a scientist of the Alberta Heritage Foundation for Medical Research.

REFERENCES

1. SCHNETKAMP, P. 2004. The SLC24 Na^+/Ca^{2+}-K^+ exchanger family: vision and beyond Pflügers Archiv European. J. Physiol. **447:** 683–688.
2. LI, X.F., L. KIEDROWSKI, F. TREMBLAY, et al. 2006. Importance of K^+-dependent Na^+/Ca^{2+}-exchanger 2, NCKX2, in motor learning and memory. J. Biol. Chem. **281:** 6273–6282.

3. LAMASON, R.L., M.A. MOHDEEN, J.R. MEST, *et al.* 2005. SLC24A5, a putative cation exchanger, affects pigmentation in zebrafish and humans. Science **310:** 1782–1786.

4. NICOLL, D.A., M.I.C.H. OTTOLIA & K.D. PHILIPSON. 2002. Toward a topological model of the NCX1 exchanger. Ann. N. Y. Acad. Sci. **976:** 11–18.

5. KINJO, T.G., R.T. SZERENCSEI, R.J. WINKFEIN, *et al.* 2003. Topology of the retinal cone NCKX2 Na/Ca-K exchanger. Biochemistry **42:** 2485–2491.

6. KANG, K.J., T.G. KINJO, R.T. SZERENCSEI & P.P.M. SCHNETKAMP. 2005. Residues contributing to the Ca^{2+} and K^+ binding pocket of the NCKX2 Na^+/Ca^{2+}-K^+ exchanger. J. Biol. Chem. **280:** 6823–6833.

7. REILÄNDER, H., A. ACHILLES, U. FRIEDEL, *et al.* 1992. Primary structure and functional expression of the Na/Ca,K-exchanger from bovine rod photoreceptors. EMBO J. **11:** 1689–1695.

8. KINJO, T.G., R.T. SZERENCSEI, R.J. WINKFEIN & P.P.M. SCHNETKAMP. 2004. Role of cysteine residues in the NCKX2 Na^+/Ca^{2+}-K^+ exchanger: generation of a functional cysteine-free exchanger. Biochemistry **43:** 7940–7947.

9. KINJO, T.G., K.J. KANG, R.T. SZERENCSEI, *et al.* 2005. Site-directed disulfide mapping of residues contributing to the Ca^{2+} and K^+ binding pocket of the NCKX2 Na^+/Ca^{2+}-K^+ exchanger. Biochemistry **44:** 7787–7795.

Transmembrane Segments I, II, and VI of the Canine Cardiac Na$^+$/Ca^{2+} Exchanger Are in Proximity

XIAOYAN REN, DEBORA A. NICOLL, AND KENNETH D. PHILIPSON

Department of Physiology and Cardiovascular Research Laboratories,
University of California at Los Angeles, Los Angeles, California 90095, USA

ABSTRACT: A helix-packing model for the NCX1 sodium calcium exchanger is presented based on cross-linking between introduced cysteine residues.

KEYWORDS: NCX1; cross-linking; helix packing

The cardiac Na$^+$/Ca^{2+} exchanger 1 (NCX1) is a membrane protein that extrudes Ca^{2+} from cells using the energy of the Na$^+$ gradient and is a key protein in regulating intracellular Ca^{2+} and contractility. Based on the current topological model, NCX1 consists of nine transmembrane segments (TMS). The N-terminal five TMS are separated from the C-terminal four TMS by a large intracellular loop. Cysteine 768 is modeled to be in TMS 6 close to the intracellular surface.

In this study, exchanger cDNAs were subcloned into pIB/V5-His vector (Invitrogen, Carlsbad, CA) and transfected into High Five Insect cells using Cellfectin (Invitrogen). Cholesterol-cyclodextrin complex (CHMβCD)[1] of 0.1 mM was added to the transfected cells to increase exchanger protein expression. Cells were harvested 24- to 48-h post transfection. The sodium-dependent calcium uptake into High Five Insect cells was assayed as described.[2] Cross-linking experiments were carried out using active NCX1 mutants. Intact cells were washed and cross-linking was carried out at 20°C by adding oxidative reagent (CuSO$_4$/ phenanthroline), thiol-specific homobifunctional cross-linker, or methanethiosulfonate (MTS) cross-linkers (Toronto Research Chemicals, Toronto, Ontario, Canada) to the intact cell suspension. The final concentrations of reagents were 1 mM CuSO$_4$, 3 mM phenanthroline, 0.5 mM

Address for correspondence: Dr. Kenneth D. Philipson, Department of Physiology and Cardiovascular Research Laboratory, MRL 3-645, David Geffen School of Medicine at UCLA, Los Angeles, CA 90095-1760. Voice: 310-825-7679; fax: 310-206-5777.
KPhilipson@mednet.ucla.edu

Ann. N.Y. Acad. Sci. 1099: 40–42 (2007). © 2007 New York Academy of Sciences.
doi: 10.1196/annals.1387.069

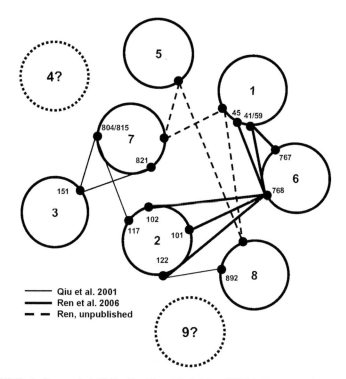

FIGURE 1. Expanded Helix-Packing Model for NCX1. Transmembrane segments are represented as circles. Cross-links that have been observed by Qiu *et al.*[2] (p. 194) are shown as thin, solid lines. Cross-links that have observed by Ren *et al.*[3] (p. 22808) are shown as thick, solid lines. Cross-links that have not previously been reported are shown as dashed lines.

o-PDM or *p*-PDM, 0.5 mM 1,3-propanediyl bismethanethiosulfonate (M3M), or 0.5 mM 1,6-hexanediyl bismethanethiosulfonate (M6M).

The proximity of TMS 6 to TMS 1 and 2 was examined. Insect High Five cells were transfected with cDNAs encoding mutant NCX1 proteins. Each mutant contained cysteine 768 and an introduced cysteine in TMS 1 or 2. Cross-linking between cysteines was determined after reaction with thiol-specific cross-linkers containing spacer arms of 6.5, 7.7, 10.4, or 12 Å. The data indicate that residues in TMS 1 and 2 are close to cysteine 768 in TMS 6. The intramolecular cross-linked product migrated with lower mobility than native NCX1 in SDS gels. Cysteine 768 cross-linked with residues at both ends of TMS 1 and 2, and is likely located toward the middle of TMS 6. Based on these results and others (unpublished), we present an expanded helix-packing model for NCX1 (FIG. 1).

REFERENCES

1. GIMPL, G. *et al*. 1995. Expression of the human oxytocin receptor in baculovirus-infected insect cells: high-affinity binding is induced by a cholesterol-cyclodextrin complex. Biochemistry **34:** 13794–13801.
2. QIU, Z., D.A. NICOLL & K.D. PHILIPSON. 2001. Helix packing of functionally important regions of the cardiac Na^+-Ca^{2+} exchanger. J. Biol. Chem. **276:** 194–199.
3. REN, X., D.A. NICOLL & K.D. PHILIPSON. 2006. Helix packing of the cardiac Na^+-Ca^{2+} exchanger: proximity of transmembrane segments 1, 2, and 6. J. Biol. Chem. **281:** 22808–22814.

Effect of Ca^{2+} on Protein Kinase A-Mediated Phosphorylation of a Specific Serine Residue in an Expressed Peptide Containing the Ca^{2+}-Regulatory Domain of Scallop Muscle Na$^+$/Ca^{2+} Exchanger

C. RYAN,[a] G. SHAW,[b] AND P. M. D. HARDWICKE[b]

[a]*Department of Physiology and Cardiovascular Research Laboratories, University of California, Los Angeles, School of Medicine, Los Angeles, California 90095-1760, USA*

[b]*Department of Biochemistry and Molecular Biology, Southern Illinois University, School of Medicine, Carbondale, Illinois 62901, USA*

ABSTRACT: Sequencing of the scallop muscle Na$^+$/Ca^{2+} exchanger revealed three consensus sequences for phosphorylation by PK-A in the large cytoplasmic loop (R^{363}KLTG, R^{379}RASV, and R^{618}RGSV). Site-directed mutagenesis of the expressed Glu^{384}-Ser^{713} segment of the f loop identified Ser^{621} as a residue phosphorylated by PK-A. The R^{618}RGSV sequence is located at the junction of the mutually exclusive exon and exon 9, a site where many alternatively spliced variants of vertebrate NCX1 and NCX3 are generated. Phosphorylation of Ser^{621} by PK-A in the isolated Glu^{384}-Ser^{713} peptide was blocked under conditions where Ca^{2+} was bound.

KEYWORDS: Na$^+$/Ca^{2+} exchanger; protein kinase A; phosphorylation; scallop

INTRODUCTION

Variant amino acid sequences are generated from *ncx1* by alternative splicing at the 3′-end of exon 2, where two mutually exclusive exons, 3 and 4 (or A and B) are followed by various combinations of four small cassette exons, 5, 6, 7, 8 (or C, D, E, and F) that precede exon 9.[1] This segment of the f loop

Address for correspondence: Peter Hardwicke, Department of Biochemistry and Molecular Biology, Mail Code 4413, Southern Illinois University, Carbondale, IL 62901-4413. Voice: 618-453-6469; fax: 618-453-6440.

phardwicke@siumed.edu

Ann. N.Y. Acad. Sci. 1099: 43–52 (2007). © 2007 New York Academy of Sciences.
doi: 10.1196/annals.1387.038

corresponds to the region represented by K^{601}-S^{677} of NCX1.1. This lies between the β-2 repeat and the highly conserved C-terminal part of the f loop and is located within the CBD2 Ca^{2+}-regulatory subdomain.[2] It is also the location of major differences between NCX1 and two other types of exchanger, NCX2 and NCX3, the products of different genes on different chromosomes in mammals. Overall, NCX1, NCX2, and NCX3 have very similar properties, and differ only in certain aspects of their regulation.[3] One of the two mutually exclusive exons, 3 (A), is preferentially present in the NCX1 isoforms located in excitable tissues.[4] The location of the alternatively spliced region in the f loop suggests that the different isoforms provide a basis for tissue-specific regulation of exchanger activity. In the three-dimensional structure of the enzyme, the part of the CBD2 region into which protein segments coded by the cassette exons are inserted lies close to Ca^{2+}-binding sites in the CBD1 subdomain.[2]

A potential phosphorylation site for cAMP-dependent protein kinase A (PK-A) is present in the f loop of NCX1 in the R^{385}KAVS sequence, just N-terminal to the Ca^{2+}-regulatory region.[5] Although none of the alternative splicing events observed in vertebrate exchangers inserts or deletes a PK-A site, activation of NCX1 and NCX3 by agents activating PK-A has been reported,[6,7] an effect that appears to be associated in some way with the presence of the A (3) exon. Agents that activate PK-A cause NCX1.1 expressed in öocytes to become phosphorylated,[8] and it may be associated with PK-A in the plasma membrane.[9]

Many studies have shown that ATP increases the affinity of the transport sites for cytoplasmic Ca^{2+} and extracellular Na^+ on squid axonal, optic lobe, and stellate ganglion Na^+/Ca^{2+} exchangers, and there is strong evidence that protein kinase activity may be involved in mediating these effects.[10–12] In previous studies,[13] the Na^+/Ca^{2+} exchanger from the adductor muscle of the bivalve mollusk, *Placopecten magellanicus* was found to be a substrate for PK-A. In the study presented here, a serine residue at the junction of the mutually exclusive exon and exon 9 was identified as a residue phosphorylated by PK-A in the intact membrane-bound scallop NCX.

MATERIALS AND METHODS

Sea scallops (*Placopecten magellanicus*) were obtained from the Marine Biology Laboratory, Woods Hole, MA.

Molecular Biology

Total RNA was extracted from freshly excised scallop cross-striated (phasic) muscle and Superscript II RT from GIBCO BRL (Baltimore, MD) used to synthesize first-strand cDNA, as described previously.[14] Overlapping PCR products were made using homologous primers, and cloned into vectors pCR2.1 pET-30 and pET35b (+) for sequencing. Data were extended into the untranslated regions by 3′- and 5′ -RACE, and the total sequence checked.

The PCR product corresponding to the Glu^{384}-Ser^{714} segment of the scallop Na^+/Ca^{2+} was cloned into pET-30 Xa/LIC and pET35b (+) vectors (Novagen, Madison, WI), and these then used to transform competent *E. coli* BL21 (DE3). When needed, the isolated Na^+/Ca^{2+} exchanger-specific peptide was cleaved from the pET35b (+) Glu^{354}-Ser^{714} fusion protein by digestion with factor Xa. The Quik-Change site-directed mutagenesis kit of Stratagene (La Jolla, CA) was used to change Ser^{621} to a cysteine. Conversion of triplet TCA of Ser^{621} to TGC was confirmed by sequencing of the modified insert in the pET-30 vector.

Circular Dichroism (CD) Measurements

The fractional saturation of the sites (θ) was assumed to be directly proportional to the change in CD signal at 218 nm relative to the maximal change,

$$\theta = (\Delta\varepsilon_{218} - \Delta\varepsilon_{218,min})/(\Delta\varepsilon_{218,max} - \Delta\varepsilon_{218,min}).$$

Electrophoresis

Discontinuous SDS gels were run in the Laemmli (Tris-glycine),[15] with 0.1 mM sodium thioglycholate present in the sample and cathode buffers. Treatment of pET-Glu^{384}-Ser^{714} and S621C pET-30 Glu^{384}-Ser^{714} with PK-A was carried out essentially as described in Reference 13.

RESULTS

Identification of a Serine Residue in CBD2 as the Site of a PK-A-Mediated Phosphorylation of Membrane-Bound Scallop Na^+/Ca^{2+} Exchanger

Sequencing demonstrated the presence of three consensus sequences for PK-A phosphorylation in the f loop: $R^{363}KLTG$ (PK1), $R^{371}RASV$ (PK2), and $K^{618}RGSV$ (PK3) (AY567834 GenBank). The relative positions of the PK-A consensus phosphorylation sequences in the putative structure of the scallop exchanger are shown in FIGURE 1. The 16 kDa soluble tryptic fragment (T_1–T_2) corresponds closely to CBD1, while the soluble 19 kDa peptide (T_2–T_3), which is phosphorylated by PK-A, approximates CBD2.[2,13]

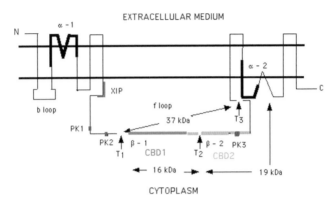

FIGURE 1. Schematic diagram of the Scallop NCX based on the putative structure of canine NCX1.1. [16]

PK1 and PK2 could be excluded as candidates for any site phosphorylated by PK-A in the intact membrane-bound scallop exchanger that is found in the soluble 37 kDa (T_1–T_3) tryptic fragment,[13] since they lie outside that segment of the f loop. The third, K^{618}RGSV, is located in the structural domains defined by the 37 kDa (T_1–T_3) peptide, which corresponds approximately to the segment of the f loop containing CBD1 and CBD2, and the 19 kDa (T_2–T_3) soluble tryptic fragment, which corresponds approximately to CBD2. K^{618}RGSV was therefore a plausible site for the PK-A-mediated phosphorylation previously reported in the native exchanger.

The scallop K^{618}RGSV sequence corresponds to K^{599}KPTG in the squid exchanger,[12] which also bridges the region coded by the cassette exons of vertebrate exchangers, and has in exactly the same way its first three residues coded by exon 3 and its last two by exon 9. R^{363}KLTG (PK1), is equivalent to R^{344}KLTG in the squid neuronal exchanger NCX-SQ1,[12] but is absent in vertebrate isoforms. R^{371}RASV (PK2) corresponds to R^{385}KAVS in vertebrate NCX1,[5] while NCX-SQ1 lacks this sequence.

Thus, as shown in TABLE 1, the scallop exchanger has in its large cytoplasmic domain one potential PK-A site in common with NCX1 (R^{371}RASV); and two in common with NCX-SQ1: R^{363}KLTG and K^{618}RGSV.

K^{618}RGSV, which lies within CBD2, has its first three residues coded by the last 9 nucleotides at the extreme 3′-end of the mutually exclusive exon, and

TABLE 1. Consensus sequences for PK-A in scallop NCX, NCX-SQ1, and NCX1

	PK1	PK2	PK3
Scallop NCX	R^{363}KLTG	R^{371}RASV	K^{618}RGSV
Squid (NCX-SQ1)	R^{344}KLTG	ABSENT	K^{599}KPTG
NCX1	ABSENT	R^{372}KAVS	ABSENT

its last two residues coded by the first six nucleotides of exon 9. Therefore, the PK3 consensus PK-A phosphorylation site sequence straddles the junction between the mutually exclusive exon and exon 9, and so bridges exactly the region where alternative splicing occurs in vertebrate exchanger isoforms.

Further characterization of the Ca^{2+}-regulatory domain and its relationship to a site phosphorylated by PK-A in the membrane-bound scallop Na^+/Ca^{2+} exchanger made use of soluble fusion proteins containing the Glu^{384}-Ser^{714} segment of the f loop expressed in *E. coli* using the Novagen pET-30 Xa/LIC and pET-35b (+) Xa/LIC vectors. The pET30 fusion protein consisted of a 5 kDa N-terminal domain bearing a $(His)_6$ tag and a 37 kDa C-terminal domain containing Glu^{384}-Ser^{714} (see FIG. 2). Glu^{384}-Ser^{714} corresponded closely to the soluble 37 kDa "T_1–T_3" tryptic fragment (Asp^{392}-possibly Lys^{728}), formed from native exchanger in the presence of Ca^{2+}. The isolated Glu^{384}-Ser^{714} peptide was prepared from the fusion proteins by digestion with Factor Xa.

Samples containing 1 mM Ca^{2+} and 10 mM EGTA showed essentially single bands, but of different mobility on Laemmli SDS gels. The sample applied containing Ca^{2+} had a mobility close to that expected for the SDS complex of a globular protein of 42 kDa, but the sample containing EGTA displayed an anomalously low mobility, corresponding to that of the SDS complex of a protein of ~ 53 kDa, suggesting an extended conformation. Mobility shifts in the SDS complexes of acidic Ca^{2+}-binding proteins, such as calmodulin and calcineurin in the presence of Ca^{2+}, are well documented,[17] and a similar effect has also been observed with a fusion protein containing the Ca^{2+}-regulatory domain of NCX1.[18] Although SDS is frequently referred to as a "denaturant," it

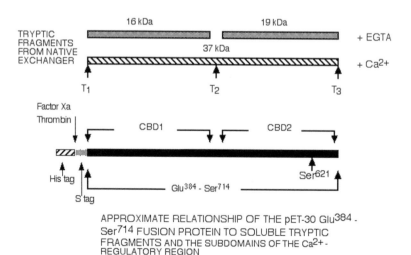

APPROXIMATE RELATIONSHIP OF THE pET-30 Glu^{384}-
Ser^{714} FUSION PROTEIN TO SOLUBLE TRYPTIC
FRAGMENTS AND THE SUBDOMAINS OF THE Ca^{2+}-
REGULATORY REGION

FIGURE 2. Approximate relationship of the pET-30 Glu^{384}-Ser^{714} fusion protein to soluble tryptic fragments and Ca^{2+}-regulatory subdomains of the scallop NCX.

is rather a refolding agent that can modify, stabilize, and even induce secondary structure.[19]

Identification of Ser[621] as the Site Phosphorylated by PK-A in Glu[384]-Ser[714] and the 37 and 19 kDa Tryptic Fragments

Treatment of the pET-30 Glu[384]-Ser[714] fusion protein or the isolated Glu[384]-Ser[714] peptide with PK-A and [γ-^{32}P]-ATP in the presence of EGTA led to their phosphorylation (FIG. 3), consistent with the previously reported presence of ^{32}P in the 19 and 37 kDa soluble tryptic fragments released from phosphorylated native exchanger.[13] When the codon for Ser[621] was replaced by that for cysteine, no phosphorylation of Glu[384]-Ser[714] by PK-A was observed (FIG. 3). This result indicated that Ser[621] was the site labeled with ^{32}P in the soluble

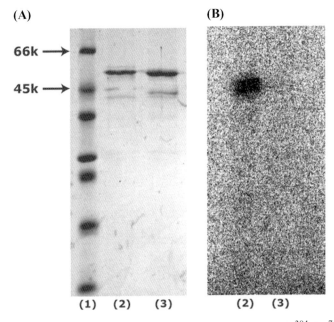

FIGURE 3. Effect of treatment of the wild-type pET-30 Glu[384]-Ser[714] fusion protein and the S621C mutant pET-30 Glu[384]-Ser[714] fusion protein with PK-A. Samples were run on a 12.5% Laemmli gel. Since the samples were in a medium containing EGTA for phosphorylation,[13] the protein–SDS complexes of the fusion proteins migrated with the anomalously low mobility typical of samples applied in EGTA. (There is no PK-A consensus sequence in the His tag of the fusion protein.) **(A)** Coomassie blue stained gel. Lane 1, Markers; Lane 2, 4 μg wild-type (wt) pET-30 Glu[384]-Ser[714] fusion protein; Lane 3, 6 μg S621C mutant pET-30 Glu[384]-Ser[714] fusion protein.**(B)** Phosphoimage of the gel shown in FIGURE 3 A. Lane 2, wt pET-30 Glu[384]-Ser[714] fusion protein; Lane 3, S621C mutant pET-30 Glu[384]-Ser[714] fusion protein. The wild-type fusion protein was labeled with ^{32}P, but the S621C mutant was not.

19 and 37 kDa tryptic fragments released from intact membrane-bound native scallop exchanger after phosphorylation with PK-A.

CD Measurements Confirm that a Large Change in Secondary Structure Occurs When Ca^{2+} Is Bound to Glu^{384}-Ser^{714} of Scallop NCX

Far UV CD spectra showed that the pET-30 fusion protein was rich in β-strand, as expected from the recent NMR structural studies of NCX.[2] There was a large increase in secondary structure content when Ca^{2+} was bound with high affinity to two sites in a positively cooperative fashion, in keeping with the major structural change induced by Ca^{2+} that had been detected in earlier proteolysis experiments. If scallop NCX behaves similarly to NCX1, the changes in the CD signal would be due primarily to modification of the CBD1 domain, since CBD2 in NCX1 does not unfold appreciably when its Ca^{2+} sites are emptied,[2] so that binding data based on changes in secondary structure would essentially reflect properties of CBD1 alone. Such a model is consistent with nonlinear least means squares fitting of the data using Kaleidograph, which showed that the whole data set best fit the equation,

$$\text{fractional saturation } \theta = [Ca^{2+}]^n / ([Ca^{2+}]^n + K_{0.5}^n),$$

when n was $= 1.67$, and $K_{0.5} = 0.67$ μM, with a correlation coefficient of 0.98978, implying two positively interacting sites, as expected for saturation of the two Ca^{2+}-binding sites of the CBD1 domain present in the N-terminal half of Glu^{384}-Ser.[7142]

Binding of Ca^{2+} to the N-Terminal Domain of Glu^{384}-Ser^{714} Blocks Phosphorylation of Ser^{621} by PK-A

In the presence of EGTA, isolated Glu^{384}-Ser^{714} peptide made from fusion protein by Factor Xa digestion was also phosphorylated with PK-A and $[\gamma\text{-}^{32}P]$-ATP in EGTA (FIG. 4). However, under conditions where the Ca^{2+}-regulatory binding sites were occupied, Ser^{621} was no longer accessible to PK-A.

DISCUSSION

Glu^{384}-Ser^{714} and the 37 kDa tryptic fragment contain the CBD1 and CBD2 Ca^{2+}-regulatory domains, the T_2 tryptic cleavage site whose accessibility is affected by Ca^{2+}, and the PK3 consensus PK-A sequence containing Ser^{621} straddling the junction of the mutually exclusive exon and exon 9. Ser^{621} is located within the CBD2 Ca^{2+}-binding domain, lying in the primary sequence approximately a quarter of the way between the major acidic segment D^{597}HEEYEKNE and the C-terminal end of CBD2. The location of PK3 in

(A)   **(B)**

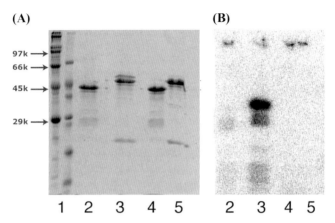

FIGURE 4. Effect of Ca^{2+}-regulatory domain on the accessibility of Ser^{621} to PK-A. When treatment with PK-A was carried out in the presence of EGTA, the peptide was strongly labeled with ^{32}P, but when exposure to the kinase was in the presence of Ca^{2+} concentrations sufficient to saturate the regulatory sites, only traces of ^{32}P could be detected. Although there are no reports of direct inhibition of PK-A by Ca^{2+}, to ensure that this was not a factor in these results, recombinant *Xenopus* histone H3 (expressed in *E. coli*) was treated with PK-A and $[\gamma\text{-}^{32}P]$-ATP under the same conditions as the Glu^{384}-Ser^{714} peptide, when no evidence was found for any direct effect of Ca^{2+} on the activity of the kinase. **(A)** Coomassie Blue Stained Gel. Lane 1, Markers; Lane 2, 5 μg Glu^{384}-Ser^{714} peptide after treatment with PK-A and $[\gamma\text{-}^{32}P]$-ATP in the presence of 0.66 mM Ca^{2+}; Lane 3, 5 μg Glu^{384}-Ser^{714} peptide after treatment with PK-A and $[\gamma\text{-}^{32}P]$-ATP in the presence of 0.59 mM $CaCl_2$ plus 11.76 mM EGTANa; Lane 4, 5 μg Glu^{384}-Ser^{714} peptide after exposure to $[\gamma\text{-}^{32}P]$-ATP under identical conditions in the presence of Ca^{2+}, but in the absence of PK-A; Lane 5, 5 μg Glu^{384}-Ser^{714} peptide after exposure to $[\gamma\text{-}^{32}P]$-ATP under identical conditions in the presence of EGTA, but in the absence of PK-A, **(B)** Phosphoimage of the gel shown in FIGURE 4 A.

the scallop and squid exchangers, which both lack all the cassette exons, coincides precisely with the site in vertebrate exchangers where cassette exon shuffling generates a large number of alternatively spliced isoforms in vertebrate exchangers. Introduction of the negatively charged phosphoryl group on the side chain of Ser^{621} could influence the binding of Ca^{2+} to CBD2 through modification of the local electrostatic field, or act to modify the conformation of the polypeptide chain in this region so that the Ca^{2+}-binding characteristics of the Ca^{2+}-regulatory domain is altered. Phosphorylation of the molluscan enzymes at PK3 by PK-A may modify regulation of Na^+/Ca^{2+} exchange in a way similar to the effect of some of the insertions/deletions of segments of peptide produced by alternative splicing in vertebrates. The location of PK3 may form part of the structural basis for integration of Ca^{2+} and cAMP signals by the molluscan exchangers.

The structural basis for inhibition of PK-A phosphorylation of Glu^{384}-Ser^{714}, by Ca^{2+} is likely to be related to the Ca^{2+}-dependent structural changes detected by trypsin digestion and CD measurements, and may reflect a reorganization of the Ca^{2+}-regulatory domain that alters accessibility of $K^{618}RGSV$ to PK-A, for example, by blockage of the site by the Ca^{2+}-bound CBD1 subdomain. This would be consistent with the condensation of the f loop that occurs on binding Ca^{2+}, an effect that leads to protection of the T_2 tryptic cleavage site.

ACKNOWLEDGMENT

This work was supported by a grant from the Central Research Committee, School of Medicine, Southern Illinois University.

REFERENCES

1. KOFUJI, P., W.J. LEDERER & D.H. SCHULZE. 1994. Mutually exclusive and cassette exons underlie alternatively spliced isoforms of the Na^+/Ca^{2+} exchanger. J. Biol. Chem. **269:** 5145–5149.
2. HILGE, M., J. AELEN & G.W. VULSTER. 2006. Ca^{2+}-regulation in the Na^+/Ca^{2+} exchanger involves two markedly different Ca^{2+} sensors. Mol. Cell **22:** 15–25.
3. LINCK, B., Z. QIU, Z. HE, *et al.* 1998. Functional comparison of the three isoforms of the Na^+/Ca^{2+} exchanger (NCX1, NCX2, NCX3). Am. J. Physiol. **274:** (Cell Physiol. 43) C415–C423.
4. DUNN, J., C.L. ELIAS, H.D. LE, *et al.* 2002. The molecular determinants of ionic regulatory differences between brain and kidney Na^+/Ca^{2+} exchanger (NCX1) isoforms. J. Biol. Chem. **277:** 33957–33962.
5. NICOLL, D.A., S. LONGONI & K.D. PHILIPSON. 1990. Molecular cloning and functional expression of the cardiac sarcolemmal Na^+-Ca^{2+} Exchanger. Science **250:** 562–564.
6. HE, S., A. RUKNUDIN, L.L. BAMBRICK, *et al.* 1998. Isoform-specific regulation of the Na^+/Ca^{2+} exchange in rat astrocytes and neurons by PKA. J. Neurosci. **18:** 4833–4841.
7. HE, L.-P., L. CLEEMAN, N.M. SOLDATOV & M. MORAD. 2003. Molecular determinants of cAMP-mediated regulation of the Na^+-Ca^{2+} exchanger expressed in human cell lines. J. Physiol. (Lond.) **548:** 677–689.
8. RUKNUDIN, A., S. HE, W.J. LEDERER & D.H. SCHULZE. 2000. Functional differences between the cardiac and renal isoforms of the rat Na^+/Ca^{2+} NCX1 expressed in *Xenopus* Oocytes. J. Physiol. (Lond.) **529:** 599–610.
9. SCHULZE, D.H., M. MUQHAL, W.J. LEDERER & A.M. RUKNUDIN. 2003. Sodium/calcium exchanger (NCX1) macromolecular complex. J. Biol. Chem. **278:** 28849–28855.
10. DIPOLO, R. & L. BEAUGÉ. 1994. Effects of vanadate on MgATP stimulation of Na-Ca exchange support kinase-phosphatase modulation in squid axons. Am. J. Physiol. **266:** C1382–C1391.
11. BERBERIAN, G., C. ASTEGGIANO, C. PHAM, *et al.* 2002. MgATP and phosphoinositides activate Na(+)/Ca(2+) exchange in bovine brain vesicles. Comparison with other Na(+)/(Ca2+) exchangers. Pflugers Arch. **444:** 677–684.

12. HE, Z., Q. TONG, B. QUEDAU, *et al.* 1998. Cloning, expression, and characterization of the squid Na^+-Ca^{2+} exchanger (NCX-SQI). J. Gen. Physiol. **111:** 857–873.
13. CHEN, M., Z. ZHANG, M.-A. BOATENG-TAWIAH & P.M.D. HARDWICKE. 2000. A Ca^{2+}-dependent tryptic cleavage site and a protein kinase a phosphorylation site are present in the Ca^{2+} regulatory domain of scallop muscle Na^+-Ca^{2+} exchanger. J. Biol. Chem. **275:** 22961–22968.
14. SHI, X., M. CHEN, P. HUVOS & P.M.D. HARDWICKE. 1998. Amino acid sequence of a Ca^{2+}-transporting ATPase from the sarcoplasmic reticulum of the cross-striated part of the adductor muscle of the deep-sea scallop: comparison to SERCA enzymes of other animals. Comp. Biochem. Physiol. Part B **120:** 359–374.
15. LAEMMLI, U.K. 1970. Cleavage of structural proteins during the assembly of the head of bacteriophage T4. Nature **227:** 680–685.
16. REN, X., D.A. NICOLL & K.D. PHILIPSON. 2006. Helix packing of the cardiac Na^+-Ca^{2+} exchanger. J. Biol. Chem. **281:** 22808–22814.
17. KLEE, C.B., T.H. CROUCH & M.H. KRINKS. 1979. Calcineurin: a calcium- and calmodulin-binding protein of the nervous system. Proc. Natl. Acad. Sci. USA **76:** 6270–6273.
18. LEVITSKY, D.O., D.A. NICOLL & K.D. PHILIPSON. 1994. Identification of the high affinity Ca^{2+}-binding domain of the cardiac Na^+-Ca^{2+} exchanger. J. Biol. Chem. **269:** 22847–22852.
19. MONTSERRET, R., M.J. MCLEISH, A. BÖCKMANN, *et al.* 2000. Involvement of electrostatic interactions in the mechanism of peptide folding induced by sodium dodecyl sulfate binding. Biochemistry **39:** 8362–8373.

Regulation of the Cardiac Na^+/Ca^{2+} Exchanger by Calcineurin and Protein Kinase C

MUNEKAZU SHIGEKAWA,[a] YUKI KATANOSAKA,[b,c]
AND SHIGEO WAKABAYASHI[b]

[a]Department of Human Life Sciences, Senri-Kinran University, Suita, Osaka
565-0873, Japan

[b]Department of Molecular Physiology, National Cardiovascular Center
Research Institute, Fujishiro-dai 5-7-1, Suita, Osaka 565-8565, Japan

[c]Department of Cardiovascular Physiology, Okayama University, Graduate
School of Medicine, Dentistry and Pharmaceutical Sciences, 2-5-1 Shikata-cho,
700-8588 Okayama, Japan

ABSTRACT: Na^+/Ca^{2+} exchanger (NCX) activity is markedly inhibited in hypertrophic neonatal rat cardiomyocytes subjected to chronic phenylephrine treatment. This inhibition is reversed partially and independently by acute inhibition of calcineurin and protein kinase C (PKC) activities. Similar NCX inhibition occurs in CCL39 cells expressing cloned wildtype NCX1, when they are infected with adenoviral vectors carrying activated calcineurin A and then treated acutely with phorbol myristoyl acetate or protein phosphatase-1 inhibitors. The data obtained with these cells suggest that calcineurin activity, PKCα-mediated NCX1 phosphorylation, and the central loop of NCX1 (possibly its β1 repeat) are required for the observed NCX inhibition. We observe partial inhibition of NCX activity independent of NCX1 phosphorylation when CCL39 cells are infected with activated calcineurin A but not further treated with phorbol myristoyl acetate or phosphatase inhibitors. Calcineurin thus appears to downregulate NCX activity via two independent mechanisms, one involving NCX1 phosphorylation and the other not involving NCX1 phosphorylation. These data indicate the existence of a novel regulatory mechanism for NCX1 involving calcineurin and PKC, which may be important in cardiac pathology.

KEYWORDS: Na^+/Ca^{2+} exchanger; phosphorylation; calcineurin; protein kinase Cα; cardiac hypertrophy

Address for correspondence: Dr. Munekazu Shigekawa, Department of Human Life Sciences, Senri-Kinran University, Fujishiro-dai 5-25-1, Suita, Osaka 565-0873, Japan. Voice: 81-6-6872-7846; fax: 81-6-6872-7872.
shigekaw@ri.ncvc.go.jp

Ann. N.Y. Acad. Sci. 1099: 53–63 (2007). © 2007 New York Academy of Sciences.
doi: 10.1196/annals.1387.059

INTRODUCTION

The primary function of Na^+/Ca^{2+} exchanger 1 (NCX1) in the heart is extrusion of Ca^{2+} from cardiomyocytes during relaxation and diastole, which balances Ca^{2+} entry via L-type Ca^{2+} channels during cardiac excitation.[1] The expression level of NCX1 is often elevated in the hearts of patients with heart failure and animal models of cardiac hypertrophy and heart failure.[2–5] In some of these animal models, elevated NCX1 expression is accompanied by enhanced NCX activity.[4,5] However, whether increased NCX expression invariably leads to enhanced NCX function is not clear, since NCX current density was reported to be downregulated in the hypertrophic hearts subjected to experimental myocardial infarction or pressure overload.[6,7] Transport activity of NCX1 is known to be influenced by a variety of factors, including hormones and growth factors, intracellular and extracellular cations, and protein and lipid phosphorylation.[8] Many of these factors could produce altered regulatory influence under pathological conditions, which may account for altered NCX activity in the diseased hearts, although this point has poorly been addressed in previous studies.

The role of protein phosphorylation in the NCX regulation has not been well defined. Cardiac NCX1 was shown to be modestly stimulated by acute treatment with protein kinase C (PKC) activators, such as phorbol myristoyl acetate (PMA) and $G\alpha q$-coupled receptor agonists.[9,10] The effects of PKA activators on NCX1 function have also been studied, but the results reported so far are conflicting.[11] On the other hand, some protein phosphatase inhibitors, such as calyculin A, reportedly inhibit NCX1 activity.[12] In contrast to these data, measurements of NCX1 currents using excised giant cardiac or oocyte membrane patches have never produced data supporting the involvement of protein kinases in NCX regulation.[13]

To further study the role of protein phosphorylation in NCX regulation, we undertook a search for regulatory proteins interacting with the central cytoplasmic loop of NCX1 by using a yeast two-hybrid screen. From this search and subsequent analysis, we identified a novel regulatory mechanism for cardiac NCX1 involving calcineurin and PKC in hypertrophic cardiomyocytes subjected to prolonged phenylephrine (PE) pretreatment. This mechanism is capable of markedly inhibiting NCX1 activity, which may account for the downregulation of NCX activity sometimes seen in hypertrophied hearts.

METHODS

The methods and experimental conditions employed have mostly been described elsewhere.[14] Measurement of ^{32}P-label incorporation into NCX1 proteins was performed essentially as described previously.[9] Briefly, confluent CCL39 cells stably expressing the wild-type NCX1 in 100-mm dishes were labeled for 5 h at 37°C in a phosphate-free, serum-free medium containing ^{32}P orthophosphate (10 MBq/mL). NCX1 proteins were immunoprecipitated from lysates of these cells with an anti-NCX1 antibody, and the immunoprecipitated materials were subjected to SDS-PAGE on a 8.5% gel. ^{32}P-labeled proteins on the gel were visualized by a Bioimage analyzer (Fuji Film Co., Tokyo, Japan). The proteins on the same gel were blotted to Immobilon membranes and visualized by immunostaining and the ECL detection system.

RESULTS AND DISCUSSION

Direct Association of NCX1 with Calcineurin A (CnA)

To isolate regulatory proteins interacting with NCX1, we performed the yeast two-hybrid screen of a human brain cDNA library using various segments of the large central loop of NCX1 as bait. We found that a ~100 amino acid C-terminal tail of CnAβ containing the autoinhibitory domain binds to a.a.407-478 of NCX1, known as the β1 repeat, which presumably constitutes part of the allosteric Ca^{2+} regulatory site of the exchanger (FIG. 1 A). We then examined whether calcineurin interacts with NCX1 and its isoforms (NCX2 and NCX3) at the protein level. Anti-pan CnA antibody was able to co-precipitate proteins reactive with antibody to each NCX isoform from lysates of rat brain and heart (FIG. 1 B). Thus, CnA was physically associated with NCX isoforms, consistent with the fact that the β1 repeat sequence is conserved in all these isoforms. It should be noted that anti-pan CnA immunoprecipitated single major proteins recognized by anti-CnAβ, indicating that the antibody predominantly precipitated CnAβ.

We also examined the interaction of NCX1 with CnA using 120-day-old normal and BIO14.6 hamster hearts. The BIO14.6 hamster is an animal model of cardiomyopathy and muscular dystrophy, which is caused by genetic deficiency of δ-sarcoglycan.[15] We found that association of calcineurin with NCX1 is significantly enhanced in BIO14.6 compared with normal hearts.[14]

Enhanced Expression and Depressed Activity of NCX1 in Hypertrophic Cardiomyocytes Subjected to Chronic PE Treatment or Activated CnA Infection

To examine the possible effect of calcineurin on NCX activity, we used neonatal rat cardiomyocytes pretreated with 10 μM PE for 72 h as an *in vitro* hypertrophic model. These myocytes exhibited a prominent increase in cell size with enhanced sarcomere organization and enhanced expression of ANP, as observed previously.[16,17] Immunoblot analysis revealed that NCX1 expression was markedly upregulated during the PE treatment, although CnA expression remained unchanged (FIG. 2, upper panel). By immunostaining, we observed significant overlapping localization of NCX1 and CnA in the sarcolemma of PE-treated myocytes.[14] Importantly, the rate of Na^+_i-dependent $^{45}Ca^{2+}$ uptake was markedly decreased in myocytes after 72-h PE treatment, although it was modestly increased in myocytes subjected to 24-h PE treatment (FIG. 2, lower panel).

To obtain insight into the underlying mechanism for the PE-induced NCX inhibition, we examined the effect of inhibition of calcineurin or PKC activity on the uptake rate, since these enzymes are reportedly activated in chronically PE-treated myocytes.[18] When myocytes were treated acutely with 1.0 μM FK506, a calcineurin inhibitor, at the end of the 72-h PE treatment, the uptake rate increased about twofold (FIG. 3). A similar result was obtained with another calcineurin inhibitor cyclosporin A (10 μM) (data not shown). Incubation with PMA during the last 24 h of the 72-h PE treatment also resulted in an increase in the uptake rate comparable to that seen at 1.0 μM FK506. This effect of PMA (PKC downregulation) was mimicked by a 30-min treatment with PKC inhibitors calphostin C (0.3 μM), GF109203X (50 nM), or chelerythrine (1 μM) (FIG. 3 and data not shown). These data suggest that calcineurin and PKC activities inhibited the uptake rate to similar extents in the PE-treated myocytes. Since they exert opposite effects on protein phosphorylation, calcineurin and PKC appear to modulate NCX function by different mechanisms involving distinct substrate proteins.

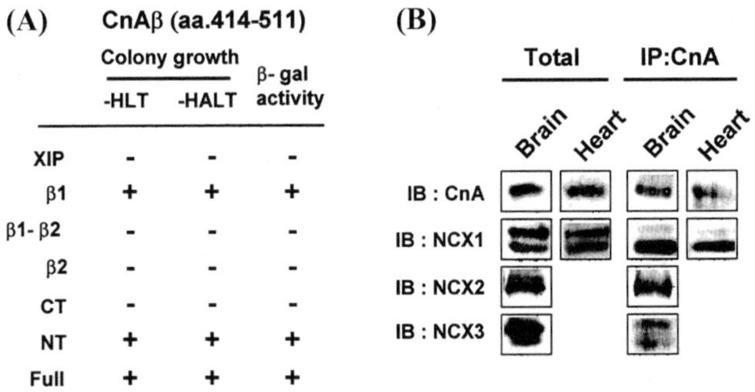

FIGURE 1. Identification of CnAβ as a NCX-binding protein. (**A**) We examined the interaction of NCX1 segments with the CnAβ tail (aa.415–511) using the yeast two-hybrid screen by one-on-one transformations and selection by colony growth in -HLT or -HALT medium and β-galactosidase assay in -HALT medium. We used DNA fragments corresponding to the following dog NCX1 segments as bait: XIP (aa. 250–406), β1 (aa.407–478), β1-β2 (aa.479–538), β2 (aa.539–613), CT (aa.614–796), NT (aa.250–613), and Full (aa.250–796). (**B**) Co-immunoprecipitation of CnA with NCX isoforms. Lysates from rat brain and heart (*total*) and materials immunoprecipitated with anti-pan CnA (*IP:CnA*) were subjected to immunoblot (*IB*) assays with antibodies to the indicated proteins.

Like PE treatment, chronic overexpression of recombinant-activated CnA in cardiomyocytes reportedly causes myocyte hypertrophy.[19] We found that in hypertrophic cardiomyocytes subjected for 48 h to adenovirus-mediated transfection of activated CnA, NCX1 expression was elevated about twofold, with the rate of Na^+_i-dependent $^{45}Ca^{2+}$ uptake being depressed by 70% relative to the uptake rate in nontransfected myocytes (data not shown). Such inhibition of NCX activity was partially reversed by calphostin C, indicating that overexpression of calcineurin activity causes PKC-mediated NCX inhibition in cardiomyocytes.

Effects of Activation of Calcineurin and PKC on NCX Activity in CCL39 Cells Expressing Cloned Wild-Type or Mutant NCX1

We asked whether NCX inhibition by calcineurin and PKC seen in hypertrophic cardiomyocytes can be reproduced in the fibroblastic cell line CCL39 cells expressing cloned wild-type NCX1. In the NCX1-expressing CCL39 cells infected with activated CnA, the rate of Na_i^+-dependent $^{45}Ca^{2+}$ uptake was 40% lower compared to control cells in which activated CnA expression

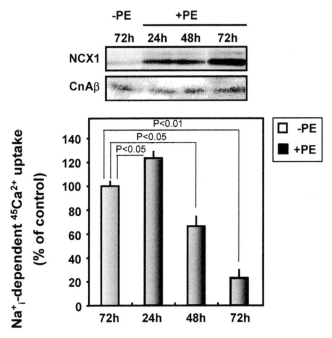

FIGURE 2. Time-dependent changes in the expression of NCX1 and CnAβ and the rate of Na_i^+-dependent $^{45}Ca^{2+}$ uptake (average $\pm$ SD; $n = 3$) in cardiomyocytes during treatment with 0 or 10 μM PE.

had been prevented by addition of 1 μM doxycycline (DOX) (FIG. 4 A). Such uptake inhibition was completely reversed by 1 μM FK506, but was further promoted by 0.3 μM PMA. Thus both calcineurin and PKC activities are capable of causing significant NCX inhibition in CCL39 cells as well. Furthermore, since PMA causes a mild increase in NCX activity in CCL39 cells not infected with activated CnA otherwise under comparable conditions,[9,20] we conclude that prior expression of calcineurin activity is somehow required for the occurrence of PKC-mediated NCX inhibition. In contrast to these data, the uptake rate in cells expressing an NCX1 mutant lacking most of its central loop (NCX1Δ246–672) was not affected by either activated CnA, FK506, or PMA (FIG. 4 A). These results strongly suggest that it is the functional state of the exchanger that is altered by calcineurin and PKC activities and that the central loop of NCX1 is required for the modulation of NCX activity.

We then studied whether endogenous calcineurin interacts with the β1 repeat of NCX1 in CCL39 cells not infected with activated CnA. We added thapsigargin to cells expressing the wild-type NCX1 or a β1 repeat deletion mutant (NCX1Δ407-478) 30 min before the uptake measurement to induce a low but

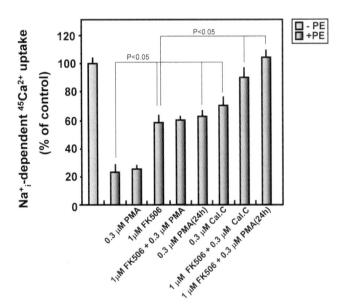

FIGURE 3. Effects of FK506, PKC modulators, and their combinations on NCX activity of PE-treated cardiomyocytes. Myocytes were incubated with the indicated agents during the last 30 min of a 72-h treatment with 0 or 10 μM PE, and then the rates of Na_i^+-dependent $^{45}Ca^{2+}$ uptake were measured. In some series, myocytes were incubated with 0.3 μM PMA during the last 24 h of PE treatment (*PMA(24h)*). *Cal.C*; calphostin C. Data are averages ± SD ($n = 3$).

sustained $[Ca^{2+}]_i$ increase by reducing Ca^{2+}-buffering capacity of the endo-plasmic reticulum, and thus activate endogenous calcineurin (FIG. 4 B). The uptake rate increased to $156 \pm 2\%$ ($n = 9, P < 0.01$ versus control) of the control value when1 μM FK506 was added to wild-type NCX1-expressing cells, whereas the same agent produced little effect in NCX1Δ407–478-expressing cells (FIG. 4 B). The uptake rate increased to $122 \pm 2\%$ ($n = 9, P < 0.05$ versus control) of the control value when FK506 was added to the wild-type NCX1-expressing cells not treated with thapsigargin. These data are consistent with the view that the β1 repeat of NCX1 is necessary for the endogenous calcineurin to regulate NCX activity.

Effects of Calcineurin and PKC on NCX1 Phosphorylation

We examined the effects of calcineurin and PKC on the phosphorylation status of NCX1 protein in CCL39 cells incubated for 5 h in a medium containing carrier-free ^{32}P-orthophosphate. In the wild-type NCX1-expressing cells, activated CnA infection did not induce significant ^{32}P-label incorporation into NCX1, although it decreased the uptake rate by about 40% as described above (FIG. 5 A). Subsequent short PMA treatment resulted in significant ^{32}P-label incorporation into NCX1 and further reduction of the uptake rate. We obtained very similar results when the PP1 inhibitor tautomycin (6 μM) was used in place of PMA (FIG. 5 A). Another PP1 inhibitor calyculin A (1 μM) also produced similar results (data not shown). These effects of PMA and PP1 inhibitors were hardly seen when calphostin C was added together with each agent. Thus, PMA and PP1 inhibitors caused simultaneous occurrence of PKC-mediated phosphorylation of NCX1 and downregulation of its activity.

We tested the effect of dominant-negative PKCα on ^{32}P-label incorporation and uptake inhibition in the wild-type NCX1-expressing cells infected with activated CnA (FIG. 5 B). PMA or tautomycin induced little ^{32}P-label incorporation and uptake inhibition when cells had been infected with dominant-negative PKCα (FIG. 5 B), although activated CnA-induced 40% reduction of uptake rate occurred normally under these conditions. Thus, PKCα activity is required for the PMA- or PP1 inhibitor-induced phosphorylation of NCX1, suggesting that PKCα and PP1 may act at the same phosphorylation site(s) in NCX1.

Regulatory Mechanism of Cardiac NCX1 by Calcineurin and PKC

We showed that NCX activity is markedly depressed in hypertrophic neonatal rat cardiomyocytes subjected to chronic PE treatment or activated CnA infection. Similar NCX inhibition can be reproduced in CCL39 fibroblastic cells expressing cloned wild-type NCX1, when they are infected with activated CnA and subsequently treated with PMA. The results described here indicate that both calcineurin and PKC activities contribute to the NCX inhibition. Furthermore, the data obtained with CCL39 cells suggest that the hypertrophic phenotype change seen in PE-treated or activated CnA-infected cardiomyocytes is not essential for the occurrence of the NCX inhibition.

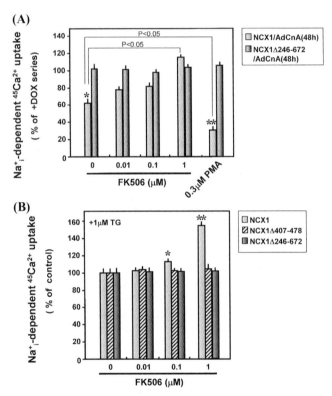

FIGURE 4. Effect of FK506 or PMA on NCX activity of CCL39 cells expressing NCX1 variants and/or activated CnA. (**A**) Cells expressing the wild-type NCX1 or NCX1Δ246-672 were infected with activated CnA (*AdCnA*) for 48 h. They were then incubated with indicated concentrations of FK506 or PMA for 30 min, and the rates of Na$_i^+$-dependent ^{45}Ca^{2+} uptake were measured. The uptake rate in cells infected with activated CnA in the presence of 1 μM doxycycline (DOX) (+DOX/AdCnA) was taken as 100%. (**B**) Cells expressing wild-type NCX1, NCX1Δ407-478, or NCX1Δ246-672 were incubated for 30 min with 1 μM thapsigargin (*TG*) and indicated concentrations of FK506, and the uptake rates were measured. The uptake rate with no FK506 was taken as 100% for each NCX variant. Data are averages $\pm$ SD ($n = 9$). *$P < 0.05$ versus control. **$P < 0.01$ versus control.

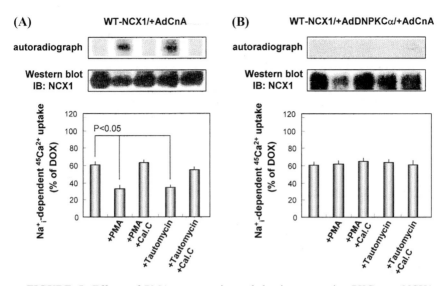

FIGURE 5. Effects of PMA, tautomycin, and dominant-negative PKCα on NCX1 phosphorylation and NCX activity. (**A**) CCL39 cells expressing the wild-type NCX1 were incubated with ^{32}P-orthophosphate during the last 5 h of a 48-h infection with activated CnA (*+AdCnA*). ^{32}P-label incorporation and the rate of Na$_i^+$-dependent ^{45}Ca^{2+} uptake were measured in the presence or absence of 0.3 μM PMA, 0.5 μM calphostin C, 6 μM tautomycin, or their combinations added 30 min before the end of the ^{32}P-labeling. (**B**) NCX1-expressing CCL39 cells were infected with adenoviral vectors carrying activated CnA and dominant-negative PKCα (*+AdDNPKCα*), and ^{32}P-label incorporation and the uptake rate were measured 48 h later otherwise as described in **A**. In panels **A** and **B**, the uptake rate in the presence of activated CnA and DOX was taken as 100%. Data are averages ± SD (*n* = 3).

We found that calcineurin binds to the β1 repeat of the central cytoplasmic loop of NCX1 and that the β1 repeat deletion from the exchanger leads to the loss of the ability of endogenous calcineurin to downregulate NCX activity in CCL39 cells (FIG. 4 B). These findings are consistent with the view that the NCX1 binding is necessary for endogenous calcineurin to regulate NCX function. It is noteworthy, however, that the activated CnA construct used here lacks its C-terminal tail and the ability to bind to the β1 repeat, yet it is able to regulate NCX activity (FIGS. 4 and 5). It thus appears that activated CnA, when overexpressed in cells, is able to regulate NCX activity without binding to NCX1.

PMA or PP1 inhibitors caused phosphorylation of NCX1 and the concomitant reduction of its activity in wild-type NCX1-expressing CCL39 cells infected with activated CnA (FIG. 5 A, B). Results obtained with dominant-negative PKCα suggest that NCX1 phosphorylation is required for the occurrence of NCX inhibition and that PKCα is involved in the PMA- or PP1 inhibitor-dependent NCX1 phosphorylation. Data obtained with PP1 inhibitors suggest that PP1 is the prominent phosphatase involved in the reversal of PKCα-dependent NCX1 phosphorylation. It is important to note that the occurrence of PMA-dependent NCX inhibition requires prior activation of calcineurin activity, since it was observed only after the infection of cells with activated CnA (FIG. 4 A). On the other hand, we observed partial inhibition of NCX activity independent of NCX1 phosphorylation in cells infected with activated CnA (FIG. 5 A, B). Calcineurin thus appears to cause downregulation of NCX activity through two independent mechanisms; one involving PKCα-mediated NCX1 phosphorylation and the other not involving NCX1 phosphorylation. In contrast to all these data, we previously observed that PMA causes mild stimulation of NCX activity in cells not infected with activated CnA.[9,20] The mechanism by which the presence or absence of activated calcineurin causes down- or upregulation of NCX activity by PMA is currently unclear and should be dealt with in future studies.

CONCLUDING REMARKS

Gαq/Gα11-mediated signaling plays an essential role in the development of hemodynamic overload-induced cardiac hypertrophy. We found that NCX activity is markedly depressed in hypertrophic neonatal rat cardiomyocytes subjected to chronic PE treatment. We identified a novel regulatory mechanism for cardiac NCX1 that involves its central cytoplasmic loop, calcineurin, and PKC. This mechanism might play an important role *in vivo* in the development of myocardial hypertrophy and subsequent contractile dysfunction, since NCX1 is an essential regulator of Ca^{2+} homeostasis in cardiomyocytes.

ACKNOWLEDGMENTS

This work was supported by the Special Coordination Funds from the Ministry of Education, Culture, Sports, Science and Technology, and a research grant from the Uehara Foundation. We thank Dr. Motoi Ohba of Showa university for a gift of *AdDNPKCα*.

REFERENCES

1. BRIDGE, J.H.B. *et al*. 1990. The relationship between charge movements associated with I_{Ca} and I_{Na-Ca} in cardiac myocytes. Science **248:** 376–378.
2. KENT, R.L. *et al*. 1993. Rapid expression of the Na^+-Ca^{2+} exchanger in response to cardiac pressure overload. Am. J. Physiol. **265:** H1024–H1029.
3. STUDER, R. *et al*. 1994. Gene expression of the cardiac Na^+-Ca^{2+} exchanger in end-stage human heart failure. Circ. Res. **75:** 443–453.

4. O'Rourke, B. *et al.* 1999. Mechanisms of altered excitation-contraction coupling in canine tachycardia-induced heart failure, I: experimental studies. Circ. Res. **84:** 562–570.

5. Ahmmed, G.U. *et al.* 2000. Changes in Ca^{2+} cycling proteins underlie cardiac action potential prolongation in a pressure-overload guinea pig model with cardiac hypertrophy and failure. Circ. Res. **86:** 558–570.

6. Wang, Z. *et al.* 2001. Na^+-Ca^{2+} exchanger remodeling in pressure overload cardiac hypertrophy. J. Biol. Chem. **276:** 17706–17711.

7. Quinn, F.R. *et al.* 2003. Myocardial infarction causes increased expression but decreased activity of the myocardial Na^+-Ca^{2+} exchanger in the rabbit. J. Physiol. **553:** 229–242.

8. Shigekawa, M. *et al.* 2001. Cardiac Na^+/Ca^{2+} exchange: molecular and pharmacological aspects. Circ. Res. **88:** 864–876.

9. Iwamoto, T. *et al.* 1996. Phosphorylation-dependent regulation of cardiac Na^+/Ca^{2+} exchanger via protein kinase C. J. Biol. Chem. **271:** 13609–13615.

10. Stengl, M. *et al.* 1998. Phenylephrine-induced stimulation of Na^+/Ca^{2+} exchange in rat ventricular myocytes. Cardiovasc. Res. **38:** 703–710.

11. Lin, X. *et al.* 2005. β-adrenergic stimulation does not activate Na^+/Ca^{2+} exchange current in guinea pig, mouse, and rat ventricular myocytes. Am. J. Physiol. **290:** C601–C608.

12. Condrescu, M. *et al.* 1999. Mode-specific inhibition of sodium–calcium exchange during protein phosphatase blockade. J. Biol. Chem. **274:** 33279–33286.

13. Collins, A. *et al.* 1992. The giant cardiac membrane patch method: stimulation of outward Na^+-Ca^{2+} exchange current by MgATP. J. Physiol. **454:** 27–57.

14. Katanosaka, Y. *et al.* 2005. Calcineurin inhibits Na^+/Ca^{2+} exchange in phenylephrine-treated hypertrophic cardiomyocytes. J. Biol. Chem. **280:** 5764–5772.

15. Nigro, V. *et al.* 1997. Identification of the Syrian hamster cardiomyopathy gene. Hum. Mol. Genet. **6:** 601–607.

16. Simpson, P. *et al.* 1982. Myocyte hypertrophy in neonatal rat heart cultures and its regulation by serum and by catecholamines. Circ. Res. **51:** 787–801.

17. Knowlton, K.U. *et al.* 1993. The alpha-adrenergic receptor subtype mediates biochemical, molecular, and morphologic features of cultured myocardial cell hypertrophy. J. Biol. Chem. **268:** 15374–15380.

18. Taigen, T. *et al.* 2000. Targeted inhibition of calcineurin prevents agonist-induced cardiomyocyte hypertrophy. Proc. Natl. Acad. Sci. USA **97:** 1196–1201.

19. De Windt, L.J. *et al.* 2000. Calcineurin-mediated hypertrophy protects cardiomyocytes from apoptosis *in vitro* and *in vivo*. An apoptosis-independent model of dilated heart failure. Circ. Res. **86:** 255–263.

20. Iwamoto, T. *et al.* 1998. Protein kinase C-dependent regulation of Na^+/Ca^{2+} exchanger isoforms NCX1 and NCX3 does not require their direct phosphorylation. Biochemistry **37:** 17230–17238.

New Modes of Exchanger Regulation

Physiological Implications

JOHN P. REEVES, MADALINA CONDRESCU, JASON URBANCZYK, AND OLGA CHERNYSH

Department of Pharmacology & Physiology, UMDNJ—Graduate School of Biomedical Sciences, Newark, New Jersey 07103, USA

ABSTRACT: Exchange activity is regulated principally by cytosolic Na^+, Ca^{2+}, and PIP2. However, the properties of these modes of regulation that have emerged from excised patch studies appear to be poorly suited to regulating exchange activity on a beat-to-beat basis. Here we summarize recent findings from our lab indicating that (*a*) allosteric activation by Ca^{2+} exhibits hysteresis, (*b*) elevated concentrations of cytosolic Na^+ induce a mode of activity that no longer requires regulatory Ca^{2+} activation, and (*c*) the requirement for PIP2 is reduced or eliminated after allosteric Ca^{2+} activation. Our results suggest that exchange activity in cardiac myocytes may be regulated by the time-integral of Ca^{2+} transients occurring over multiple beats.

KEYWORDS: phosphatidylinositol-4,5-*bis*phosphate (PIP2); hysteresis; allosteric calcium activation; fura-2

INTRODUCTION

Anyone studying the regulation of Na^+/Ca^{2+} exchange (NCX) activity in the heart must pause to reflect that there is little evidence that NCX activity is in fact regulated in functioning myocardial cells, and that, in any case, overexpression, knockout, and modeling studies suggest that alterations in NCX activity have only modest effects on excitation-contraction (EC) coupling (at least in mice). Hilgemann[1] has suggested that NCX activity might have little to do with beat-to-beat regulation of transmembrane Ca^{2+} traffic but may be important instead for "buffering" cytosolic Ca^{2+}, that is, shifting $[Ca^{2+}]_i$ toward the value specified by the thermodynamic driving forces for NCX activity. From this point of view, the efficiency of NCX in buffering Ca^{2+} would be increased as activity is upregulated, and vice versa, and this might have important consequences on EC coupling primarily when there is a change

Address for correspondence: John P. Reeves, Department of Pharmacology & Physiology, UMDNJ—Graduate School of Biomedical Sciences, 185 South Orange Avenue, P.O. Box 1709, Newark, NJ 07103. Voice: 973-972-3890; fax: 973-972-7950.

reeves@umdnj.edu

Ann. N.Y. Acad. Sci. 1099: 64–77 (2007). © 2007 New York Academy of Sciences.
doi: 10.1196/annals.1387.002

in some parameter that alters the thermodynamic driving forces for NCX (e.g., a change in action potential shape, beating frequency, $[Na^+]_i$, etc.).

It therefore makes teleological sense that NCX activity is regulated by the ions that comprise the major determinants of the NCX driving forces, that is, cytosolic Na^+ and Ca^{2+}. Increases in cytosolic Na^+ downregulate NCX activity (through the process of Na^+-dependent, or I_1, inactivation), whereas increases in cytosolic Ca^{2+} upregulate activity (through allosteric Ca^{2+} activation). There is a mutually antagonistic interplay between the two processes: I_1 inactivation increases the K_h for allosteric Ca^{2+} activation while elevations in $[Ca^{2+}]_i$ protect against I_1 inactivation. Phosphatidylinositol-4,5-*bis*phosphate (PIP2) plays an essential role in mediating these effects. This information has been amply documented over the years in the previous proceedings of this symposium.

This report presents information on some new aspects of NCX regulation that came from our studies of exchange activity in transfected Chinese hamster ovary (CHO) cells. We begin by arguing that Na^+-dependent inactivation and allosteric Ca^{2+} activation, as we understand these processes from *in vitro* studies, are remarkably ill suited to regulate NCX activity on a beat-to-beat basis. We will then review recent findings from our lab suggesting that both Ca^{2+}-dependent and Na^+-dependent modes of regulation act within a time frame that would preclude rapid changes of NCX activity within the space of a single cardiac cycle. We speculate that interactions between NCX and the cytoskeleton impose a constraint on allosteric Ca^{2+} activation, leading to hysteresis and the persistence of Ca^{2+} activation at low cytosolic Ca^{2+} concentrations. Finally, we suggest that NCX regulation takes place over multiple beats and is therefore well adapted to adjusting activity in response to the types of changes involving NCX driving forces described above.

PHYSIOLOGICAL REGULATION OF NCX ACTIVITY

The classical studies of NCX in squid giant axons revealed that activity was stimulated by adenosine 5′-triphosphate (ATP) and by cytosolic Ca^{2+}. ATP shifted the K_m values for $[Ca^{2+}]_i$ and for $[Na^+]_o$ to lower concentrations and had multiple effects on the interactions of Na^+, Ca^{2+}, and H^+ at the allosteric regulatory sites (recently reviewed in Ref. 2). In mammalian systems—most work has been done with cardiac myocytes or with the oocyte expression system—the major effect of ATP was to stimulate the synthesis of PIP2, an important positive regulator of NCX activity.[3] It was also reported that ATP (or PIP2) increased the affinity of the allosteric Ca^{2+} regulatory sites, and prolonged the relaxation of activity when regulatory Ca^{2+} was removed.[4,5] PIP2 is thought to interact with the so-called XIP region (XIP stands for eXchange Inhibitory Peptide) at the N-terminal portion of the hydrophilic regulatory domain. Current concepts of NCX regulation suggest that PIP2 binding disrupts an autoinhibitory interaction between XIP and another, unidentified docking site on the exchanger protein.[6] These effects do not appear to be absolutely specific for PIP2: other negatively charged phospholipids (e.g., phosphatidylserine) also stimulate NCX activity in a similar fashion.[7]

The role of PIP2 was first discerned in terms of its ability to protect against Na^+-dependent inactivation. In excised patches, Na^+-dependent inactivation is seen as a time-dependent decline in outward exchange currents following the application of cytosolic Na^+. This mode of inactivation is eliminated by high concentrations of cytosolic Ca^{2+}, or by treating the patches with ATP,[5] which acts by stimulating the synthesis of PIP2.[3] It seems unlikely that Na^+-dependent inactivation would be a significant mode of NCX regulation under normal physiological conditions because of the high concentrations of PIP2 normally present. One study demonstrated the existence of Na^+-dependent inactivation in intact cardiac myocytes,[8] but the cells that were used had been preincubated under conditions that induced blebbing, suggesting that PIP2 levels may have been low. Rather than regulating NCX activity under normal physiological conditions, Na^+-dependent inactivation might serve as a protective mechanism to shut off exchange activity under conditions such as ischemia; under ischemic conditions, the decline in ATP and PIP2 would inhibit reverse NCX activity so that the increase in cytosolic Na^+ would not overload the cell with Ca^{2+}.

At first sight, the teleological role of allosteric Ca^{2+} activation seems highly relevant to normal cell physiology, that is, the exchanger is turned "on" when needed to remove Ca^{2+} from the cell. However, it is not at all clear how the properties of allosteric Ca^{2+} activation could regulate Ca^{2+} efflux during the normal cardiac cycle. Thus, given the apparent K_h for allosteric Ca^{2+} activation (150–400 nM), the increase in $[Ca^{2+}]_i$ during the normal $[Ca^{2+}]_i$ transient is likely to bring about full, or nearly full, activation of NCX activity. Weber et al.[9] estimated that the submembrane $[Ca^{2+}]_i$ "seen" by NCX in cardiac myocytes was substantially higher than bulk $[Ca^{2+}]_i$ during the $[Ca^{2+}]_i$ transient (peak values of 3.5 and 1 μM, respectively). These results suggest that the dominant factor controlling NCX function during the peak of the $[Ca^{2+}]_i$ transient is the affinity of the translocation sites for Ca^{2+} ($K_m \sim 5$ μM) rather than allosteric Ca^{2+} activation ($K_h \sim 150$–400 nM).

To summarize, Na^+-dependent inactivation probably does not occur under normal physiological conditions and the rise in $[Ca^{2+}]_i$ during a typical action potential is sufficient to fully activate NCX activity. How then, can these modes of regulation serve to modulate NCX activity in beating myocytes? The situation is further complicated by recent findings indicating that once NCX is allosterically activated by Ca^{2+} it remains activated for an interval following the return of $[Ca^{2+}]_i$ to low values (persistent Ca^{2+} activation). These findings are described below.

HYSTERESIS OF ALLOSTERIC CALCIUM ACTIVATION

We recently reported that the K_h for allosteric Ca^{2+} activation of NCX in transfected CHO cells was ~300 nM.[10] For these studies we measured Ca^{2+} uptake by digital imaging of fura-2 loaded cells following the application of 0.1 mM Ca^{2+} in a Na^+-free medium. Strikingly, we found that the individual cells showed pronounced lag periods of variable duration before rapid Ca^{2+} uptake occurred. FIGURE 1 shows seven different cells from a single experiment in which more than 50 cells were monitored simultaneously. The cells display delays of increasing duration, including one cell that did not display Ca^{2+} uptake at any point. We reasoned that the low rate of Ca^{2+} uptake during the lag phase occurred because the cytosolic Ca^{2+} concentration of the cells (<50 nM) was initially too low to activate more than a small fraction of NCX activity. As $[Ca^{2+}]_i$ gradually increased during the lag phase, more exchangers became activated and the resulting positive feedback rapidly accelerated Ca^{2+} influx at the end of the lag phase. This interpretation was supported by the behavior of cells expressing a mutant of NCX, $\Delta(241–680)$, in which the allosteric Ca^{2+} binding sites were deleted so that the mutant did not require allosteric Ca^{2+} activation; these cells showed immediate Ca^{2+} uptake, without

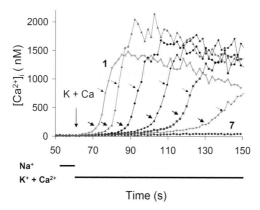

FIGURE 1. Exchange-mediated Ca^{2+} influx for seven individual cells. Transfected CHO cells expressing NCX1.1 were loaded with fura-2 and treated with ATP + Tg in Na-PSS + 0.3 mM EGTA to release Ca^{2+} from internal stores 10 min prior to beginning the experiment. At $t = 60$ s, K-PSS containing 0.1 mM $CaCl_2$ was applied to the cells. The seven cells were selected from a cover slip in which 60 cells were monitored simultaneously by digital imaging. The fura-2 signal was calibrated as described.[10] Upper arrow for each trace: data point corresponding to the maximal rate of Ca^{2+} uptake. Lower arrow (bold) for each trace: data point corresponding to half the maximal rate; $[Ca^{2+}]_i$ at this data point provides an estimate of the K_h for allosteric Ca^{2+} activation. Na-PSS—140 mM NaCl plus 5 mM KCl; K-PSS—140 mM KCl; each solution also contained 10 mM glucose, 1 mM $MgCl_2$, and 20 mM Mops/Tris, pH 7.4 (room temperature). This figure is reproduced from Ref. 10 (p. 621) with permission from Rockefeller University Press.

a lag phase, when reverse NCX activity was initiated. For the cells expressing NCX1.1, we computed the K_h value of 300 nM by determining, for each cell, the $[Ca^{2+}]_i$ concentration at which the rate of Ca^{2+} influx was half-maximal (bold arrows in FIG. 1).[10]

This value was in excellent agreement with the K_h values obtained in excised patches (200–600 nM),[5] but was much higher than the 50 nM K_h value we had obtained several years earlier using the same cell system.[11] In the earlier study, we had increased $[Ca^{2+}]_i$ using ionomycin and initiated NCX activity at various points as $[Ca^{2+}]_i$ declined, but in the experiments shown in FIGURE 1, we measured NCX activity as $[Ca^{2+}]_i$ increased. We believe that the difference in the two values reflects an intrinsic hysteresis in allosteric Ca^{2+} activation. That is, it takes more Ca^{2+} to activate NCX activity than to maintain the activated state once it has formed.

The practical effect of hysteresis is seen in the phenomenon of persistent Ca^{2+} activation, which is illustrated by the data in FIGURE 2. Here, Ca^{2+} was released from the endoplasmic reticulum by the application of the purinergic agonist ATP and thapsigargin (Tg), a specific inhibitor of the sarco (endo)plasmic reticulum Ca^{2+}-ATPase. The resulting increase in $[Ca^{2+}]_i$ activated the forward mode of NCX, and the clearing of Ca^{2+} from the cytosol by NCX sharply attenuated the $[Ca^{2+}]_i$ transient in comparison to nontransfected cells (not shown). Remarkably, when reverse-mode NCX activity was initiated shortly after $[Ca^{2+}]_i$ had returned to initial levels (<100 nM), Ca^{2+} uptake began immediately, without a lag period, indicating that the exchangers had retained their activated state despite the low $[Ca^{2+}]_i$. We had initially suggested that persistent Ca^{2+} activation might reflect the persistence of locally elevated submembrane concentrations of Ca^{2+}. However, further investigations provided no evidence supporting the existence of such a long-lasting elevation in submembrane Ca^{2+}. Thus, while this hypothesis cannot be completely ruled out, we now feel that hysteresis provides a more realistic explanation, that is, the effective K_h is lower following Ca^{2+} activation than before NCX is activated.

Hysteretical systems often involve positive feedback as a means of generating switch-like behavior leading to bistable states. It is easy to see how positive feedback occurs during reverse mode NCX activity, where increasing cytosolic Ca^{2+} stimulates activity through allosteric Ca^{2+} activation, leading to an increased rate of Ca^{2+} influx, as already described in connection with FIGURE 1. During forward mode activity, however, increasing allosteric Ca^{2+} activation would accelerate Ca^{2+} efflux, more likely leading to graded behavior and the formation of states with partially activated NCX activity. Nevertheless, hysteresis still seems to occur in the setting of forward mode activity, as illustrated by the results in FIGURE 3.[12] In this experiment, we applied 1 mM $CaCl_2$ to cells in which Ca^{2+} stores had previously been depleted by the addition of ATP + Tg, thereby activating store-operated Ca^{2+} channels (SOCs). As Ca^{2+} entered the cell through SOCs, the increase in $[Ca^{2+}]_i$ was at first similar in both transfected and nontransfected cells, suggesting that NCX was initially

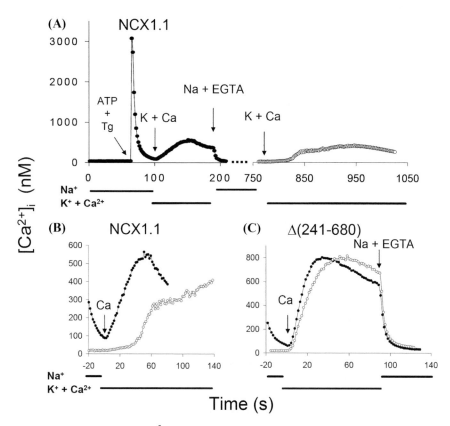

FIGURE 2. Persistent Ca^{2+} activation. (**A**) Cells expressing NCX1.1 were incubated in Na-PSS + 0.3 mM EGTA (see legend to FIG. 1) and 100 μM ATP plus 2 μM Tg was applied at 60 s; at 100 s, K-PSS + 0.1 mM CaCl$_2$ was applied. Na-PSS + 0.3 mM EGTA was again applied at 180 s and K-PSS + 0.1 mM Ca^{2+} was applied a second time at 780 s. (**B**) The traces in panel A for the two periods following application of K-PSS + 0.1 mM CaCl$_2$ are superimposed; note the absence of a lag phase when Ca^{2+} was applied shortly after ATP + Tg (filled circles). (**C**) Results of an indentical experiment to that shown in Panels A and B except that cells expressing the constitutive mutant, Δ(241–680), were used. Traces represent the average of >50 cells monitored simultaneously. This figure is reproduced from Ref. 10 (p. 621) with permission of Rockefeller University Press.

inactive at these [Ca^{2+}]$_i$ values. However, in the cells expressing NCX1.1, [Ca^{2+}]$_i$ rose to a peak and then declined to a steady state value as NCX was activated and began to carry out Ca^{2+} efflux. In contrast, when similar experiments were conducted with the constitutive Δ(241–680) mutant that does not require allosteric Ca^{2+} activation, no overshoot was observed and [Ca^{2+}]$_i$ increased monotonically to a steady state value (FIG. 3 B). In both cell types, the rise in [Ca^{2+}]$_i$ was much less than for nontransfected CHO cells, due to

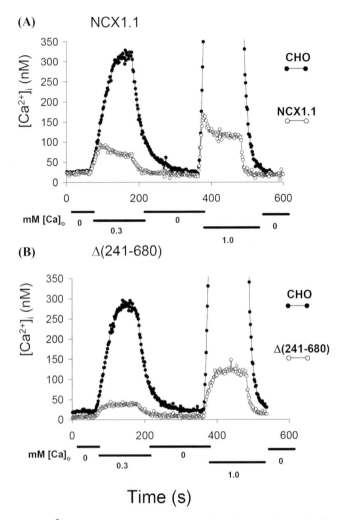

FIGURE 3. Ca^{2+} efflux by NCX. (**A**) Nontransfected CHO cells and cells expressing NCX1.1 were cultured on the same cover slip and pretreated with 100 μM ATP + 2 μM Tg to release Ca^{2+} from internal stores and activate SOCs; the cells were incubated in Na-PSS + 0.3 mM EGTA. Na-PSS + 0.3 mM CaCl$_2$ was applied (60 s) to initiate Ca^{2+} influx, followed after 2 min (180 s) by Na-PSS + EGTA and (at 360 s) by Na-PSS + 1 mM CaCl$_2$. Nontransfected CHO cells and cells expressing NCX1.1 were identified after concluding the experiment by assaying for NCX activity.[12] (**B**) An identical experiment to that described in Panel A except that cells expressing the constitutive Δ(241–680) mutant were used instead of cells expressing NCX1.1. Traces represent the average of 20–30 cells of each type. See legend to FIGURE 1 for the composition of solutions. This figure is reproduced from Ref. 12 (pp. 287, C797) with permission of the American Physiological Society.

the Ca^{2+} efflux activity of NCX. The overshoot is consistent with the hystereti-cal nature of allosteric Ca^{2+} activation, that is, the initial activation of NCX occurs at a higher Ca^{2+} concentration that is subsequently required to maintain activity. Moreover, the activation of forward mode NCX activity leads to per-sistent Ca^{2+} activation following the return of $[Ca^{2+}]_i$ to low values, another indication of hysteretical behavior (data not shown).[12] Thus, hysteresis of al-losteric Ca^{2+} activation appears to occur for both the Ca^{2+} influx and Ca^{2+} efflux modes of NCX activity.

The term hysteresis, derived from the Greek root *husterein* (to come late), implies that there may be a lag or constraint in the activation process itself. The mechanism underlying this constraint is unknown, but one factor might be interactions of NCX with the cytoskeleton. We have shown that the sur-face distribution of NCX is determined by the distribution of F-actin, that is, NCX protein was concentrated in regions that have a high F-actin content.[13] Indeed, in some cells, NCX formed linear arrays that coincided with under-lying stress fibers. When F-actin was disrupted with cytochalasin D (Cyto D), patches of fragmented F-actin were formed that appeared to "drag" NCX protein along with them. Colocalization of $\Delta(241–680)$ protein with F-actin was not observed, suggesting that the interactions with F-actin are mediated by the exchanger's hydrophilic domain. The report by Condrescu and Reeves[13] should be consulted for details.

Cyto D, and other agents that reduce F-actin turnover, increased the lag period for activation of NCX (FIG. 4 A), indicating that some aspect of the ac-tivation process had been impaired. Cyto D did not inhibit NCX activity *per se*, because *increased* activity in Cyto D-treated cells was seen for the constitutive $\Delta(241–680)$ mutant (FIG. 4 B), and also for the wild-type exchanger during persistent Ca^{2+} activation (not shown). The increased activity in the latter cases was due an increase in the surface concentration of NCX protein in the Cyto D-treated cells, possibly reflecting an inhibition of endocytosis by Cyto D. One interpretation of these results is that the interactions between F-actin and the regulatory domain of the exchanger impede the conformational transitions associated with allosteric Ca^{2+} activation; it might therefore require higher Ca^{2+} concentrations to overcome this constraint than would subsequently be required to maintain activity. Cyto D treatment inhibits actin turnover[14] and we suggest that greater F-actin stability might promote stronger interactions with NCX, thereby increasing the constraint on allosteric Ca^{2+} activation. However, this idea has not yet been tested.

A preliminary report indicated that NCX activity in cardiac myocytes also displayed persistent Ca^{2+} activation,[15] although detailed studies have not yet been carried out in these cells. In transfected CHO cells, persistent Ca^{2+} acti-vation decays over 60–90 s at room temperature and 15–30 s at 37°C. These time scales are obviously much longer than the normal cardiac cycle. The life-time of persistent Ca^{2+} activation depends upon the rate and extent at which $[Ca^{2+}]_i$ is reduced following a Ca^{2+} transient. Thus, in transfected CHO cells,

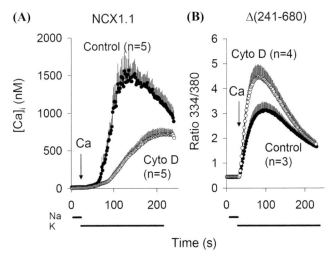

FIGURE 4. Effect of Cyto D treatment on reverse mode NCX activity. Cells expressing either NCX1.1 (**A**) or the constitutive Δ(241–680) mutant (**B**) were loaded with fura-2 for 30 min with or without 5 μM Cyto D present. The cells were treated with ATP + Tg (see FIG. 1) 10 min prior to beginning recordings. K-PSS + 0.1 mM CaCl$_2$ was applied at 30 s for each trace. Traces represent the means (+SEM bars) of the indicated number of cover slips; more than 50 cells were monitored simultaneously for each cover slip. This figure is reproduced from Ref. 13 (pp. 290, C691) with permission of the American Physiological Society.

the decay of persistent Ca^{2+} activation following an ATP-induced Ca^{2+} transient was slower in the presence of the SERCA inhibitor Tg than when Tg was omitted[10]; this undoubtedly reflects the ability of SERCA to quickly remove Ca^{2+} from the cytosol during the decay of the Ca^{2+} transient. This behavior suggests that the K$_h$ for deactivation of NCX activity (~50 nM?) is much lower than the K$_h$ for activation (~300 nM). The lifetime for persistent Ca^{2+} activation in cardiac myocytes following a normal Ca^{2+} transient is unknown, but if it were to extend beyond the length of the cardiac cycle, NCX would consequently remain activated at all times in beating myocytes. In this case, it would be difficult to see how allosteric Ca^{2+} activation could regulate NCX activity in a meaningful way.

This conundrum could be resolved in a particularly attractive manner if only a fraction of available exchangers were activated during any one cardiac cycle, perhaps (speculatively) because of the constraints on allosteric Ca^{2+} activation imposed by the cytoskeleton. If this were the case, a steady state population of active exchangers would develop over multiple beats reflecting the time dependence of activation on the one hand, and the relaxation of the activated state during diastole on the other. In this way, NCX activity could be adjusted up or down to reflect the needs of the cell and would be expected to increase in a graded fashion, for example, with increasing stimulation frequency. However, nearly all the available evidence suggests that allosteric Ca^{2+} activation is a very rapid process, occurring in <200 ms in excised patches[5] and in <50 ms in intact myocytes.[15,16] One study, however, reported that inward NCX currents in an excised patch were activated with a time constant of 0.6 s when $[Ca^{2+}]_i$ was rapidly increased by flash photolysis.[17] Further studies are required to resolve this important issue.

CYTOSOLIC SODIUM AND ALLOSTERIC CALCIUM ACTIVATION

Elevations in cytosolic $[Na^+]$ lower the reversal potential for NCX, thereby impeding forward mode activity. At higher Na^+ concentrations, NCX may actually reverse direction and bring Ca^{2+} into the cell. To protect against catastrophic Ca^{2+} overload in such a situation (e.g., during cardiac ischemia), elevations in $[Na^+]_i$ downregulate NCX activity through the process of Na^+-dependent inactivation.[18] The latter process, however, would only be expected to occur when ATP levels fall and membrane PIP2 levels are reduced. Recent results from our lab[19] suggest that in transfected CHO cells, with normal levels of ATP and PIP2, elevations in $[Na^+]$ actually upregulate activity by inducing a mode of activity that does not require allosteric Ca^{2+} activation (constitutive mode). The following paragraphs describe these new findings.

Mutation of the acidic residues within the regulatory Ca^{2+} binding sites of NCX leads to an increase in the K_h for allosteric Ca^{2+} activation. Matsuoka *et al.*[20] described one such mutant, D447V, which exhibited an apparent K_h of 1.8 μM, much higher than the 300 nM K_h of the wild-type exchanger. We expressed this mutant in CHO cells and, predictably, found little or no activity under physiological conditions. However, when the cells were treated with ouabain to elevate cytosolic $[Na^+]$, we unexpectedly saw strong Ca^{2+} influx when extracellular Na^+ was removed. The most striking aspect of these results, shown in FIGURE 5, was that Ca^{2+} uptake began without the lag period associated with allosteric Ca^{2+} activation, even though the initial $[Ca^{2+}]_i$ was ~25 nM, that is, very far below the expected K_h for allosteric Ca^{2+} activation for this mutant. The absence of a lag period was similar to the behavior of the constitutive mutant, $\Delta(241\text{--}680)$. This suggested that the D447V mutant was operating constitutively—that is, it did not require allosteric Ca^{2+} activation under the conditions of the experiment. Additional experiments in which $[Na^+]_i$ was "clamped" with gramicidin verified that NCX activity was essentially absent at 5 mM Na^+, but progressively increased as $[Na^+]_i$ was increased; lag periods were not observed at any Na^+ concentration. Similar experiments conducted with the wild-type NCX showed a progressive shortening of the lag periods as $[Na^+]_i$ increased; the lag periods became nearly undetectable at Na^+ concentrations of 40 mM or more.

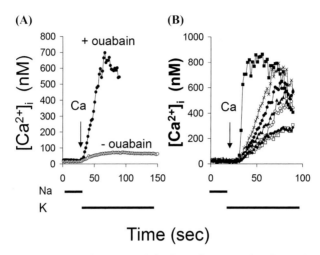

FIGURE 5. Reverse mode NCX activity by cells expressing the D447V mutant of NCX. (**A**) The cells were pretreated with ATP + Tg, and incubated in Na-PSS + 0.3 mM EGTA, with or without 1 mM ouabain, for 10 min prior to beginning recordings. K-PSS + 0.1 mM $CaCl_2$ was applied at 30 s, as indicated. Traces shown represent the average of >50 cells monitored simultaneously. (**B**) Several individual cells from the ouabain-treated cover slip in panel A. Note the absence of lag periods for any of the cells (cf. FIG. 1). Reproduced with permission of Blackwell Publishing from Ref. 19.

The activity of the D447V mutant depended critically upon PIP2. High concentrations of cytosolic Ca^{2+} stimulate phospholipase C activity, causing a rapid loss of PIP2 from membranes.[21] We adopted a 2-pulse protocol to test whether PIP2 depletion would inhibit the activity of the D447V mutant. We used reverse NCX activity of the mutant at 40 mM Na^+ to increase $[Ca^{2+}]_i$ for a period of 60 s; Ca^{2+} was then removed and when $[Ca^{2+}]_i$ had returned to initial values, a second pulse of Ca^{2+} was applied. As shown in FIGURE 6 A, no activity was observed if the second pulse of Ca^{2+} was applied shortly after the first. Activity was restored, however, if additional recovery time was provided at low $[Ca^{2+}]_i$. We followed membrane PIP2 levels using green fluorescent protein (GFP) fused to the plekstrin homology domain of phospholipase Cδ1, and found that PIP2 fell rapidly during the first period of elevated $[Ca^{2+}]_i$ and gradually recovered when low $[Ca^{2+}]_i$ was restored. Thus, the inhibition of D447V activity during the second pulse appeared to be due to the loss of PIP2. No inhibition of activity was observed in similar experiments conducted with the constitutive $\Delta(241–680)$ mutant (FIG. 6 B) or, strikingly, with cells expressing the wild-type exchanger (FIG. 6 C). We concluded that the loss of membrane PIP2 induced Na^+-dependent inactivation of the D447V mutant, leading to inhibition of activity, but that the wild-type exchanger was protected by the allosteric Ca^{2+} activation occurring during the first pulse and its persistence during the interpulse interval. This interpretation is consistent with

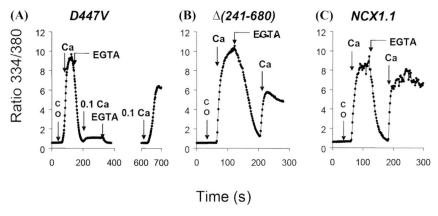

Time (s)

FIGURE 6. PIP2 depletion inhibits NCX activity of the D447V mutant, but not that of the $\Delta(241$–680) mutant or the wild-type exchanger. Cells expressing either D447V **(A)**, $\Delta(241$–680) **(B)** or NCX1.1 **(C)** were pretreated with ATP + Tg and gramicidin in 40/100 Na/K-PSS (40 mM NaCl + 100 mM KCl). At 30 s Cl-CCP (5 μM) and oligomycin (1 μg/mL) were applied (arrow labeled CO) to prevent Ca^{2+} uptake by mitochondria. At 60 s, 0.1 mM $CaCl_2$ in 40/100 Na/K-PSS containing Cl-CCP was applied, followed by 0.3 mM EGTA in 40/100 Na/K-PSS (120 s) and, at 200 s, a second application of 0.1 $CaCl_2$ in 40/100 Na/K-PSS. For the experiment in panel A, EGTA was again applied at 360 s and 0.1 mM Ca^{2+} was applied at third time at 630 s. Experiments (not shown)[19] in which membrane PIP2 was monitored with a fusion protein linking GFP with the pleckstrin homology domain of phospholipase Cδ1, showed that the fusion protein was strongly localized to the plasma membrane prior to the addition of Ca^{2+}, but showed a diffuse cytosolic distribution after 60 s of Ca^{2+} uptake; a gradual return of the fusion protein to the plasma membrane was seen after EGTA addition. Reproduced with permission of Blackwell Publishing from Ref. 19.

data from excised patches showing that high concentrations of cytosolic Ca^{2+} protect against Na^+-dependent inactivation.[5] We conclude that allosteric Ca^{2+} activation reduces the dependence of NCX activity on PIP2.

SUMMARY AND CONCLUSIONS

The picture of NCX regulation that emerges from these results indicates that both PIP2 and cytosolic Ca^{2+} are major positive regulators of NCX activity, and that each serves a different purpose. PIP2 seems to regulate the response of NCX to cytosolic Na^+. Thus, PIP2 is required for the induction of the constitutive mode of NCX activity at elevated $[Na^+]_i$. Na^+-dependent inactivation is counteracted by PIP2 and, when PIP2 levels fall and cytosolic Na^+ increases during ischemic events, this may protect cells against Ca^{2+} overload. On the other hand, PIP2 does not seem to be required for allosteric activation by cytosolic Ca^{2+} or for persistent Ca^{2+} activation (FIG. 6 C). A corollary of this statement is that PIP2 may play little if any role in regulating the forward, Ca^{2+} efflux mode of NCX.

In beating cardiac myocytes, persistent Ca^{2+} activation and the constitutive mode of NCX operation would tend to maintain NCX activity throughout the cardiac cycle. In this case, there would be little opportunity for regulation of NCX activity to occur within the time frame of a single cardiac cycle. It is more likely that NCX regulates its activity over multiple beats, and responds to the time-integral of changes in cytosolic Ca^{2+}. For this to be an effective process, however, there must be some mechanism for limiting the extent of NCX activation that occurs during a single beat. We have suggested the interactions between NCX and the cytoskeleton may impose a constraint on allosteric Ca^{2+} activation, possibly leading to the observed hysteresis in this process. Future studies will be directed toward testing these ideas and determining whether they apply to NCX in cardiac myocytes.

ACKNOWLEDGMENTS

The work described in this report was supported by grants from the National Institutes of Health and the American Heart Association. Jason Urbanczyk was supported by NIH training grant 1 T32 HL069752; Olga Chernysh was supported by AHA grant 0555759T.

REFERENCES

1. HILGEMANN, D.W. 2004. New insights into the molecular and cellular workings of the cardiac Na^+/Ca^{2+} exchanger. Am. J. Physiol. Cell Physiol. **287:** C1167–C1172.
2. DiPOLO, R. & L. BEAUGE. 2006. Sodium/calcium exchanger: influence of metabolic regulation on ion carrier interactions. Physiol. Rev. **86:** 155–203.
3. HILGEMANN, D.W. & R. BALL. 1996. Regulation of cardiac Na^+,Ca^{2+} exchange and KATP potassium channels by PIP2. Science **273:** 956–959.
4. COLLINS, A., A.V. SOMLYO & D.W. HILGEMANN. 1992. The giant cardiac membrane patch method: stimulation of outward Na^+- Ca^{2+} exchange current by MgATP. J. Physiol. (Lond.) **454:** 27–57.
5. HILGEMANN, D.W., A. COLLINS & S. MATSUOKA. 1992. Steady-state and dynamic properties of cardiac sodium-calcium exchange. Secondary modulation by cytoplasmic calcium and ATP. J. Gen. Physiol. **100:** 933–961.
6. MATSUOKA, S., D.A. NICOLL, Z. HE & K.D. PHILIPSON. 1997. Regulation of cardiac Na^+-Ca^{2+} exchanger by the endogenous XIP region. J. Gen. Physiol. **109:** 273–286.
7. HILGEMANN, D.W. & A. COLLINS. 1992. Mechanism of cardiac Na^+-Ca^{2+} exchange current stimulation by MgATP: possible involvement of aminophospholipid translocase. J. Physiol. (Lond.) **454:** 59–82.

8. MATSUOKA, S. & D.W. HILGEMANN. 1994. Inactivation of outward Na^+-Ca^{2+} exchange current in guinea- pig ventricular myocytes. J. Physiol. (Lond.) **476:** 443–458.

9. WEBER, C.R., V. PIACENTINO III, *et al.* 2002. Na^+-Ca^{2+} exchange current and submembrane $[Ca^{2+}]$ during the cardiac action potential. Circ. Res. **90:** 182–189.

10. REEVES, J.P. & M. CONDRESCU. 2003. Allosteric activation of sodium-calcium exchange activity by calcium: persistence at low calcium concentrations. J. Gen. Physiol **122:** 621–639.

11. FANG, Y., M. CONDRESCU & J.P. REEVES. 1998. Regulation of Na^+/Ca^{2+} exchange activity by cytosolic Ca^{2+} in transfected Chinese hamster ovary cells. Am. J. Physiol. **275:** C50–C55.

12. CHERNYSH, O., M. CONDRESCU & J.P. REEVES. 2004. Calcium-dependent regulation of calcium efflux by the cardiac sodium/calcium exchanger. Am. J. Physiol. Cell Physiol. **287:** C797–C806.

13. CONDRESCU, M. & J.P. REEVES. 2006. Actin-dependent regulation of the cardiac Na^+/Ca^{2+} exchanger. Am. J. Physiol. Cell Physiol. **290:** C691–C701.

14. STAR, E.N., D.J. KWIATKOWSKI & V.N. MURTHY. 2002. Rapid turnover of actin in dendritic spines and its regulation by activity. Nat. Neurosci. **5:** 239–246.

15. WEBER, C.R., K.S. GINSBURG & D.M. BERS. 2005. Allosteric regulation of cardiac Na/Ca exchange by cytosolic Ca: Dynamics during [Ca]i changes in intact myocytes. Biophys. J. **88:** 136a.

16. GOMEZ, A.M., B. SCHWALLER, H. PORZIG, *et al.* 2002. Increased exchange current but normal Ca^{2+} transport via Na^+-Ca^{2+} exchange during cardiac hypertrophy after myocardial infarction. Circ. Res. **91:** 323–330.

17. KAPPL, M. & K. HARTUNG. 1996. Rapid charge translocation by the cardiac Na^+-Ca^{2+} exchanger after a Ca^{2+} concentration jump. Biophys. J. **71:** 2473–2485.

18. HILGEMANN, D.W., S. MATSUOKA, G.A. NAGEL & A. COLLINS. 1992. Steady-state and dynamic properties of cardiac sodium-calcium exchange. Sodium-dependent inactivation. J. Gen. Physiol. **100:** 905–932.

19. URBANCZYK, J., O. CHERNYSH, M. CONDRESCU & J.P. REEVES. 2006. Sodium-calcium exchange does not require allosteric calcium activation at high cytosolic sodium concentrations. J. Physiol. **575:** 693–705.

20. MATSUOKA, S., D.A. NICOLL, L.V. HRYSHKO, *et al.* 1995. Regulation of the cardiac Na^+-Ca^{2+} exchanger by Ca^{2+}. Mutational analysis of the Ca^{2+}-binding domain. J. Gen. Physiol. **105:** 403–420.

21. BALLA, T. 2006. Phosphoinositide-derived messengers in endocrine signaling. J. Endocrinol. **188:** 135–153.

Shedding Light on the Na^+/Ca^{2+} Exchanger

MICHELA OTTOLIA,[a] SCOTT JOHN,[b] YI XIE,[a] XIAOYAN REN,[a]
AND KENNETH D. PHILIPSON[a]

[a]Department of Physiology, Cardiovascular Research Laboratories, David
Geffen School of Medicine at UCLA, Los Angeles, California 90095-1760, USA

[b]Department of Medicine, Cardiovascular Research Laboratories, David Geffen
School of Medicine at UCLA, Los Angeles, California 90095-1760, USA

ABSTRACT: The Na^+/Ca^{2+} exchanger (NCX) regulates cardiac contractility by adjusting the amount of Ca^{2+} inside myocytes. NCX accom-
plishes this by using the electrochemical gradient of Na^+: during each
cycle three Na^+ ions enter the cell and one Ca^{2+} ion is extruded against its
gradient. In addition to being transported, cytoplasmic Na^+ and Ca^{2+}
ions also regulate exchanger activity. The physiological relevance and
molecular processes underlying ionic regulation remain unclear. Also
unresolved are the events that regulate NCX trafficking to the membrane and its oligomeric state. This is essential information to interpret
structure–function data. The full-length exchanger was fused to both
CFP and YFP, creating active fluorescent exchangers used in FRET experiments to assess both conformational changes associated with ionic
regulation and the oligomeric state of NCX. Electrophysiological characterization demonstrates that these constructs behave similarly to the
wild-type (WT) exchanger. We have been able for the first time to monitor conformational changes of the exchanger Ca^{2+}-binding site in vivo.
These studies provide a better understanding of the molecular properties
of the NCX.

KEYWORDS: sodium–calcium exchange; FRET; calcium regulation;
dimer

INTRODUCTION

Heart contractility is tightly regulated by Ca^{2+} levels in cardiac cells. Following depolarization, Ca^{2+} enters heart cells through L-type Ca^{2+} channels
triggering Ca^{2+} release from the sarcoplasmic reticulum thereby initiating contraction. To relax, myocytes must lower intracellular Ca^{2+} to its resting level.

Address for correspondence: Michela Ottolia, Cardiovascular Research Laboratories, MRL 3-645,
David Geffen School of Medicine at UCLA, Los Angeles, CA 90095-1760. Voice: 310-794-7103;
fax: 310-206-5777.
mottolia@mednet.ucla.edu

Ann. N.Y. Acad. Sci. 1099: 78–85 (2007). © 2007 New York Academy of Sciences.
doi: 10.1196/annals.1387.044

Two proteins accomplish this task: the sarcoplasmic reticular Ca^{2+} ATPase, which refills the intracellular stores and the sarcolemmal Na^+/Ca^{2+} exchanger (NCX), which extrudes Ca^{2+} by using the electrochemical gradient of Na^+.[1] For each Ca^{2+} extruded three Na^+ ions are transported into the cell. This results in an ionic current that can be measured with the giant patch technique.[2] Detailed studies combining molecular biological and electrophysiological techniques have demonstrated that Na^+ and Ca^{2+} also modulate the activity of NCX. An increase in internal $[Na^+]$ (25–100 mM) drives NCX into an inactive state (Na^+-dependent inactivation),[3] while cytoplasmic Ca^{2+} (0.2–1 μM) promotes NCX activity and also removes the Na^+-dependent inactivation.[4] The extent to which these ions set NCX activity and the relevance of regulation remain unclear. Among the problems is determining the concentrations of Na^+ and Ca^{2+} ions that NCX senses in the subsarcolemmal space. Other cytosolic factors, such as pH and ATP, also affect NCX regulation. Since NCX influences cardiac contractility, it is important to determine how regulatory ions influence its activity. Electrophysiological studies have addressed this question. However, the system presents technical limitations, such as the difficulty of selectively blocking all other ionic currents and problems in the use of intact myocytes.

Also unresolved is the oligomeric state of this protein. At present the protein is thought to exist as monomer; however, there is no evidence to support this hypothesis. Knowledge of the oligomeric state of the NCX is essential to interpret structure–function data.

Here we describe optical techniques that help us answer some of these important questions. Our approach consisted of linking the cardiac NCX to the cyan (CFP) and yellow (YFP) variants of green fluorescent protein (GFP). The fusion of NCX allows us to both monitor the exchanger in living cells and to use fluorescence resonance energy transfer (FRET) to monitor changes in the conformation of NCX during ionic regulation *in vivo*.

RESULTS

Ca²⁺ Regulation of the NCX

One of our goals is to characterize the conformational changes of NCX associated with the binding of Ca^{2+} to its regulatory site and to monitor the changes in living cells. We have published results[5] indicating that the Ca^{2+}-binding domain (CBD) of NCX, by itself, binds and responds to Ca^{2+} during excitation–contraction coupling. To show Ca^{2+} modulation of the CBD, we used a noninvasive FRET-based technique. A peptide consisting of residues 371 to 508 was fused at the N terminus to yellow (YFP) and at the C terminus to cyan (CFP) variants of green fluorescent protein (YFP-CBD-CFP)

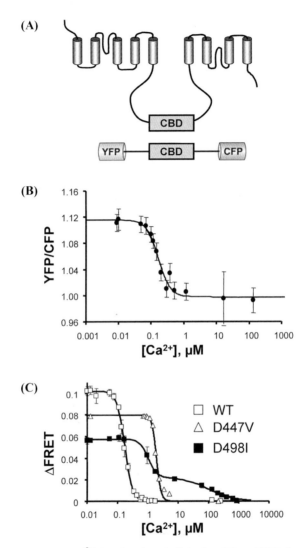

FIGURE 1. The NCX Ca^{2+}-binding domain linked to CFP and YFP detects changes in $[Ca^{2+}]$. (**A**) Schematic representation of NCX and YFP-CBD-CFP. CBD (Calcium Binding Domain) consists of residues 371 to 508 of the NCX large intracellular loop connecting transmembrane segments 5 and 6. (**B**) Changes in YFP/CFP ratio versus Ca^{2+} concentrations for YFP-CBD-CFP. (**C**) Mutations of important Asps in the CBD region shift the apparent affinity of YFP/CFP ratio values of YFP-CBD-CFP for Ca^{2+}.

(see FIG. 1 A). Changes in FRET, demonstrating changes in distance between fluorophores, were monitored as changes in the YFP/CFP emission ratio. During FRET, excitation of CFP induces YFP emission at the expense of CFP emission. Hence, the YFP/CFP fluorescence intensity ratio measures FRET efficiency. Using this technique we demonstrated that the distance between the two fluorophores increased with elevation of Ca^{2+}. The amount of Ca^{2+} needed to reduce FRET by 50% was ~200 nM (FIG. 1 B). Mutating Ca^{2+}-binding Asp residues within the CBD decreased Ca^{2+} affinity (D447V or D498I) (FIG. 1 C). When both residues were mutated, there was no Ca^{2+}-induced change in FRET (D447V-D498I; not shown). A limitation of these results is that YFP-CBD-CFP is a cytosolic protein and therefore may not represent movements that occur in the full-length NCX protein at its sarolemmal location. To overcome these problems we flanked the Ca^{2+}-binding site within the full-length NCX with CFP and YFP (NCX-YFP-CBD-CFP). Expression of NCX-YFP-CBD-CFP in both HEK cells and rat neonatal myocytes produced a marked fluorescence localized to the plasma membrane and a strong FRET signal. To determine whether insertion of CFP and YFP altered the transport properties of NCX, we studied the activity of the fusion protein using Na^+-gradient-dependent $^{45}Ca^{2+}$ uptake and the giant patch electrophysiological technique. $^{45}Ca^{2+}$ uptake reveals that this construct is as active as the wild-type (WT) exchanger when expressed in HEK cells. Giant patch experiments show that NCX-YFP-CBD-CFP has both Na^+- and Ca^{2+}-dependent regulation (see FIG. 2) with a Ca^{2+} affinity similar to WT NCX. These results indicate that NCX-YFP-CBD-CFP can be used to monitor conformational changes.

To monitor rapid conformational changes of the tagged exchanger in a membrane, we have developed a novel technique. We attached portions of *Xenopus* oocyte plasma membrane to coverslips with the cytoplasmic surface exposed to the bath by incubating devitellinized oocytes expressing NCX-YFP-CBD-CFP on polylysine-treated coverslips for ~10 min. The coverslip was then perfused with a jet of buffer to remove the oocyte except for the attached membrane. These patches offer several advantages: First, the membrane patches are wide (>30 μm) allowing fluorescent recordings from a large surface area, thereby optimizing the signal-to-noise ratio. Second, the background fluorescence generated by any protein localized to intracellular compartments is eliminated. Finally, rapid perfusion of the cytoplasmic surface of the plasma membrane is readily accomplished. This approach was originally developed for immunostaining experiments using *Xenopus* oocytes,[6] but to our knowledge, this is the first time it has been used to monitor changes in conformations of membrane proteins by FRET.

Changes in the YFP/CFP ratio were monitored upon rapid switching between solutions containing either 30 μM free Ca^{2+} or 10 mM EGTA in the presence of 100 mM Cs^+. The addition of Ca^{2+} evoked a decrease in the YFP/CFP fluorescence ratio and there was a return to the initial value upon removal of Ca^{2+}. As a control, CFP and YFP targeted to the plasma membrane were coexpressed in *Xenopus* oocytes and their response to changes in cytoplasmic Ca^{2+} concentration was analyzed using the plasma membrane sheets: in this case no change in FRET due to changes in Ca^{2+} concentrations was observed (not shown).

NCX

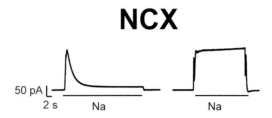

NCX-266CFP

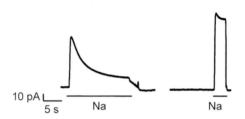

NCX-YFP-CBD-CFP

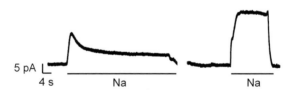

FIGURE 2. CFP/YFP NCX fusion proteins behave similarly to WT NCX. Outward currents recorded from oocytes expressing the WT exchanger (*top panel*), the full-length exchanger with either CFP inserted at position 266 (NCX-266CFP) (*middle panel*), or YFP and CFP inserted at positions 371 and 508 (NCX-YFP-CBD-CFP), respectively. NCX outward currents were elicited by application of cytoplasmic Na^+ in the presence of 0.5 (*left*) or 10 μM intracellular Ca^{2+} (*right*). Note that all constructs responded to a rise in the intracellular Ca^{2+} with the removal of Na^+-dependent inactivation (*left panel*).

Taken together, these experiments indicate that insertion of YFP and CFP into the NCX does not drastically disrupt the biophysical properties of NCX. We also demonstrate that the full-length exchanger with both CFP and YFP inserted changes conformation upon binding Ca^{2+} as indicated by the change in FRET. One future goal is to examine this conformational change *in vivo* by expressing NCX-YFP-CBD-CFP in rat neonatal myocytes to detect changes in FRET during excitation–contraction coupling.

In addition, we recently have developed the use of the zebrafish expression system to examine conformational changes of our fluorescent exchangers in intact contracting myocardium. These studies complement our other investigations and allow us to monitor changes in conformation due to increases in cytoplasmic Ca^{2+} in intact organisms.

Oligomeric State of the NCX

Interpreting structure–function data obtained by mutational analysis and optical experiments requires knowledge of the subunit composition of the exchanger. At present there are no reports on the oligomeric state of the NCX. To assess the subunit composition of the NCX, we monitored the trafficking of a functional NCX-YFP tagged exchanger to the plasma membrane in the presence of a fluorescent NCX–CFP mutant known to be retained in the intracellular compartments. If the exchanger is a multimeric protein, it is possible that one of the two exchangers will behave as a dominant form by altering the trafficking of the other protein. For this purpose we used the full-length exchanger with the fluorophores inserted at position 371 (NCX-371YFP). This construct is active and retains most of the biophysical properties of the untagged exchanger (not shown). However, the deletion of the last 7 amino acids from NCX-371CFP impairs the proper targeting of the tagged exchanger to the plasma membrane. FIGURE 3 shows that NCX-371CFP-Δ7 is retained within the endoplasmic reticulum.

Our experiments indicate that coexpression of this construct with a functional YFP-tagged exchanger prevented shuttling of the functional YFP-tagged exchanger to the plasma membrane (see FIG. 3) suggesting that dimerization (or higher order oligomerization) is important for the proper targeting of the protein to the plasma membrane.

To further investigate the multimeric status of the exchanger we monitored changes in FRET between two exchangers tagged at position 266 of the large cytopalsmic loop with CFP (NCX-266CFP) or YFP (NCX-266YFP). Insertion of CFP (or YFP) at this position does not alter the biophysical properties of the exchanger (see FIG. 2). The two fluorescent exchangers were coexpressed in *Xenopus* oocytes and plasma membrane sheets were isolated as described before. Upon addition of 30 μM cytoplasmic Ca^{2+} an increase in FRET was observed indicating that the distance between the two fluorophores (and therefore the distance between the two cytoplasmic loops) decreased. These results indicate that the large cytoplasmic loops of the two exchangers are in sufficiently close proximity to generate FRET and that the distance varies upon Ca^{2+} application.

Taken together these data suggest that the exchanger forms dimers (or higher order oligomers) in the plasma membrane and that the cytoplasmic portion of the multimeric complex undergoes Ca^{2+}-dependent conformational changes.

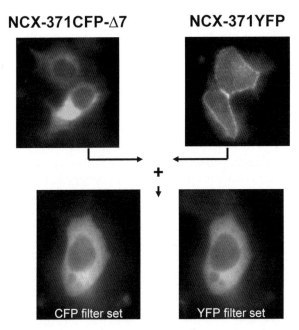

FIGURE 3. Oligomerization may be important for trafficking. NCX-371YFP is retained in the intracellular compartments when coexpressed with NCX-371CFP-Δ7. Shown are pictures of HEK cells transfected with the indicated constructs. Pictures were acquired using the appropriate filters sets for selectively imaging the fluorescence from CFP or YFP.

SUMMARY

Fluorescent-tagged exchangers are a useful tool to investigate the ionic regulation of NCX. Using these constructs, we observed, for the first time, transient conformational changes of the NCX upon the binding of Ca^{2+}. In addition, we have shown Ca^{2+}-dependent changes in the distance between exchangers indicating that the transporter forms a multimeric complex in the plasma membrane.

REFERENCES

1. PHILIPSON, K.D. & D.A. NICOLL. 2000. Sodium-calcium exchange: a molecular perspective. Ann. Rev. Physiol. **62:** 111–133.
2. HILGEMANN, D.W. 1990. Regulation and deregulation of cardiac Na^{+}-Ca^{2+} exchange in giant excised sarcolemmal membrane patches. Nature **344:** 242–245.
3. HILGEMANN, D.W., S. MATSUOKA, G.A. NAGEL & A. COLLINS. 1992. Steady-state and dynamic properties of cardiac sodium-calcium exchange. Sodium-dependent inactivation. J. Gen. Physiol. **100:** 905–932.

4. HILGEMANN, D.W., A. COLLINS & S. MATSUOKA. 1992. Steady-state and dynamic properties of cardiac sodium-calcium exchange. Secondary modulation by cytoplasmic calcium and ATP. J. Gen. Physiol. **100:** 933–961.

5. OTTOLIA, M., K.D. PHILIPSON, & S. JOHN. 2004. Conformational changes of the Ca^{2+} regulatory site of the Na^+-Ca^{2+} exchanger detected by FRET. Biophys. J. **87:** 899–906.

6. SINGER-LAHAT, D., N. DASCAL, L. MITTELMAN, *et al.* Imaging plasma membrane proteins in large membrane patches of Xenopus oocytes. Pflugers Arch. **440:** 627–633.

The Regulation of the Na/Ca Exchanger and Plasmalemmal Ca^{2+} ATPase by Other Proteins

ABDUL M. RUKNUDIN[a,b] AND EDWARD G. LAKATTA[a]

[a]Laboratory of Cardiovascular Science, Gerontology Research Center, National Institute on Aging, NIH, Baltimore, Maryland 21224, USA
[b]Department of Microbiology and Immunology, University of Maryland, School of Medicine, Baltimore, Maryland 21201, USA

ABSTRACT: Na/Ca exchanger (NCX) and plasma membrane Ca^{2+} ATPase are the Ca^{2+} efflux mechanisms known in mammalian cells. NCX is the main transporter to efflux intracellular Ca^{2+} in the heart. NCX protein contains nine putative transmembrane domains and a large intracellular loop joining two sets of the transmembrane domains. The intracellular loop regulates the activity of the NCX by interacting with other proteins and nonprotein factors, such as ions, PIP2. Several proteins that are associated with NCX have been identified recently. Similarly, plasmalemmal Ca^{2+} ATPase (PMCA) has 10 putative transmembrane domains, and the C-terminal intracellular region inhibits transporter activity. There are several proteins associated with PMCA, and the roles of the associated proteins of PMCA vary from specific localization to involving PMCA in signal transduction. Elucidation of structural and functional roles played by these associated proteins of NCX and PMCA will provide opportunities to develop drugs of potential therapeutic value.

KEYWORDS: calcium transport; phosphorylation; signal transduction

INTRODUCTION

The Na/Ca exchanger (NCX) is an important transporter in the Ca^{2+} homeostasis of most mammalian cells. The process of Na/Ca exchange was discovered about four decades ago, and the first member of NCX family of genes was cloned in 1990.[1–3] Following the discovery of the *NCX1* gene, two more genes, *NCX2* and *NCX3*, were cloned in mammals. The homologous genes of NCX1 have been identified in various species and microorganisms, and

Address for correspondence: Abdul M. Ruknudin, Ph.D., Laboratory of Cardiovascular Science, Gerontology Research Center, National Institute on Aging, NIH, Baltimore, MD 21224. Voice: 410-558-8270; fax: 410-558-8150.
aruknudi@umaryland.edu

Ann. N.Y. Acad. Sci. 1099: 86–102 (2007). © 2007 New York Academy of Sciences.
doi: 10.1196/annals.1387.045

they are now grouped under the superfamily, SLC8.[4] Among the members of the SLC8 family, the most researched member is NCX1. NCX1 is present in most cells, whereas NCX2 and NCX3 are present in skeletal and brain tissues. Interestingly, NCX1 is the only isoform present in the mammalian heart.[5,6] Numerous studies have identified the importance of NCX in the functioning of the heart and brain, and the NCX function is altered under pathological conditions in those organs.[7–12] This article summarizes the recent advances in the regulation of the NCX activity and another transporter, plasmalemmal Ca^{2+} ATPase (PMCA).

STRUCTURE OF NCX1

The Transmembrane Domains

The cloning of *NCX1* gene facilitated the studies of structure–function relationships of this important ion transporter. Early studies suggested that the NCX1 protein contained 11 putative transmembrane domains based on the hydropathy analysis of the protein sequence.[3,13] The transmembrane regions 1 to 5 were separated by a large intracellular loop from 6 to 11 transmembrane regions. The deletion of the large intracellular loop studies indicated that transmembrane regions were mainly responsible for the transport of the Na^+ and Ca^{2+} across the membrane.[14] This prompted further mutation studies in the transmembrane regions in an attempt to identify the residues responsible for the binding and transport of Na^+ and Ca^{2+}.[15] On the basis of these studies, the earlier concept of the 11 transmembrane domains underwent drastic revision, and it was suggested that the sixth transmembrane domain was intracellular and that the ninth forms a partial transmembrane segment. The current understanding is that NCX1 has nine putative transmembrane domains with a larger intracellular loop based on the cystine replacement studies (FIG. 1).[9,16,17] Schwarz and Benzer identified two conserved regions in the various NCX and their homologs, known as α1 and α2 regions.[18] The α1 region is localized between second and third transmembrane regions, and α2 is localized between eighth and ninth transmembrane regions. These highly conserved regions are considered to be responsible for the binding and transport of Na^+ and Ca^{2+}, and interestingly, the α1 region of NCX1 is oriented extracellularly in the membrane, while the α2 region is oriented intracellularly.[9,17] There are also conserved repeats in the intracellular loop referred to β1 and β2 repeats.[18]

Intracellular Loop

Elucidation of the regulatory mechanism of NCX1 has gradually progressed over time. Immediately after the cloning of NCX1, it was observed that more

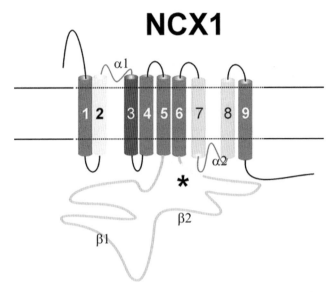

FIGURE 1. Topology of NCX1. The NCX1 protein contains two sets of putative trans-membrane domains separated by a large intracellular loop. The N-terminal set of TM contains five TM regions and the C-terminal set contains four TM regions. The conserved $\alpha1$ region is located between 2 and 3 TM regions whereas the $\alpha2$ region is localized between 7 and 8 TM regions. Asterisk denotes the alternatively spliced region. The repeats $\beta1$ and $\beta2$ regions are indicated in the intracellular loop. (Modified from Schulze *et al.* 2003.[29]) (In color in Annals online.)

than half of the protein forms an intracellular loop joining the proximal and distal transmembrane domains, and as mentioned earlier, the transmembrane regions were responsible for the transport of Na^+ and Ca^{2+}. However, any process without proper regulation is like a steam engine running without brakes. The NCX transports 3 Na^+ to 1 Ca^{2+} in opposite directions and the exchanger can either remove or bring Ca^{2+} into the cells depending on the conditions.[19] The electrochemical gradient of Na^+ across the membrane and the membrane potential are the determining factors for the activity and the direction of the flow of ions by the NCX. There are two levels of control in the NCX process. (*a*) The activity of the NCX is regulated by the interaction of the intracellular loop with (nonprotein) factors. (*b*) The intracellular loop also interacts with other proteins and these regulatory proteins modify the function of NCX activity (see below). Na^+ and Ca^{2+}, in addition to being transported, modulate the activity of the exchanger. Using giant patch technique, Hilgemann *et al.* showed that intracellular Na^+ inactivates the activity of the exchanger, and the inactivation process is referred to as I_1 and I_2. The inactivation I_2 is influenced by intracellular $[Ca^{2+}]$.[20–22] The giant patch and whole-cell patch electrophysiological studies showed that there was an absolute requirement of intracellular Ca^{2+} for the activation of the transport process of NCX, irrespective of the mode of exchange. The removal of the intracellular loop by deletion or by digestion with α-chymotrypsin eliminated the inhibition of the NCX activity, indicating that elements responsible for inactivation of NCX resided in the intracellular loop (FIG. 2). Interestingly, the activity is far higher after α-chymotrypsin treatment than the activity of the loopless NCX, suggesting that the digestion by enzyme could release the activity of the exchanger from unknown inhibitors associated with intracellular regions of NCX1. Moreover, the concept of absolute requirement for intracellular Ca^{2+} is questioned by a recent report that in the presence of higher intracellular $[Na^+]$, the NCX is constitutively active and there was no need for the requirement for intracellular Ca^{2+}.[23] There are other factors, such as pH, PIP2, ATP, and phosphorylation, which also modulate the activity of the

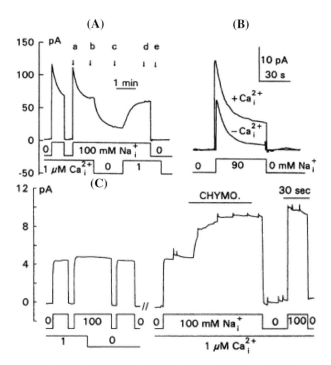

FIGURE 2. Regulation of NCX1 activity by intracellular loop in *Xenopus* oocytes expressing dog NCX1. (**A**) Current trace of giant patch at 0 mV of the wild-type NCX1. Applying 100 mM Na^+-evoked outward current in the presence of 1 μM Ca^{2+} and this current decays over time to steady state. (**B**) The steady-state current increased to maximum by applying α-chymotrypsin on the intracellular side of the giant patch. (**C**) A deletion mutant 240-679 (removing most of the intracellular loop) was expressed in *Xenopus* oocytes and the current measured using giant patch. The inactivation observed in A was not observed and α-chymotrysin treatment further increased the activity of NCX1. (Modified from Matsuoka *et al.,* 1993.[14])

NCX.[10,24–27] Moreover, the exchanger inhibitor peptide (XIP) completely inhibited exchange activity by interacting with the N-terminal end of the large intracellular loop, even though the physiological relevance of this action is not understood.[14,28]

REGULATORY PROTEINS OF NCX1

Another mechanism of modulation of the NCX has been recently discovered. A number of proteins have been identified to associate themselves with NCX1 and other NCX. These proteins have been shown to be associated with NCX1 on the intracellular side of the protein, most probably interacting with large intracellular loop. The common feature of these proteins is that they regulate the transport function of the NCX. The first sets of regulatory proteins, associated with NCX1 protein are kinases and phosphatases. Schulze *et al.* identified both catalytic and regulatory subunits of protein kinase A (PKA) and the catalytic subunit of protein kinase C (PKC) in association with the NCX1 protein in rat heart.[29] In the sarcolemma of the rat cardiomyocytes, two serine/thereonine phosphatases, PP1 and PP2A, are also found associated with NCX1 protein along with a scaffolding protein, mAKAP. The kinases and phosphatases that regulate the phosphorylation of NCX1 could be associated with it through mAKAP. Marx *et al.* have shown that PKA, PP1, and PP2A were associated with cardiac ryanodine receptor (RYR2) channels thorough Leucine–Zipper (LZ) domains.[30,31] There are several LZ domains in the NCX protein; two of these in the intracellular loop are shown in FIGURE 3. Additional experiments are required to show whether these regions are responsible for association with kinases and phosphatases with NCX1. Alternatively, PP1 also could associate directly with the NCX1 protein as it does with other proteins.[32,33] There is specific recognition sequence in the NCX1 protein (FIG. 4), but it is not clear how the PP1 is connected to NCX1. Another phosphatase, calcineurin (PP2B), has been shown to be associated with all three NCX and the β1 repeat of the intracellular loop region of NCX1 interacts with PP2B. In the cardiomyopathic hamster hearts, the amount of PP2B phosphatase is increased and inhibition of this phosphatase increased the NCX activity.[34]

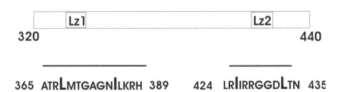

FIGURE 3. LZ motifs in NCX1 protein. Among the several LZ motifs identified in the NCX1 proteins, two of them in the intracellular loop are depicted. The numbers indicate the number of amino acid from N-terminal in the rat NCX1.

Putative PP1 binding motif

Rat NCX1   (774) YVMHFLTVFWKVLFAFV
Rat NCX2   (724) YVMHFLTVFWKVLFACL
Rat NCX3   (730) YVMHFLTVFWKVLFACV
Dros Dmel/Ncx (752) YVSHFVCLFWKVLFAFV
Squid NCX   (695) YIMHFVCLFWKVLFAFV

FIGURE 4. PP1-binding motifs in NCX. The conserved sequence denoting the putative-binding site for phosphatase, PP1, is present in all NCX as well as in squid NCX. The red color shows the motif. (Modified from Schulze *et al.*, 2003.[29]) (In color in Annals online.)

Whether PKA-dependent phosphorylation of the NCX1 occurs remains controversial. Some groups showed that the phosphorylation of cardiac NCX1 by β-adrenergic stimulation increased its activity while others did not observe this.[35,36] In smooth muscle cells, Iwamoto *et al.* showed the growth factor-induced modulation of NCX1 involves PKC-dependent phosphorylation. The cardiac alternatively spliced isoform of NCX1 also has been shown to be modulated by PKC, even though direct phosphorylation may not be occurring.[37,38] However, earlier studies using giant patches of NCX1 failed to show any modulation by either PKA or PKC catalytic enzymes.[20] Recently, Schulze *et al.* explained the controversy suggesting that the effects of phosphorylation could be seen or not seen based on the conditions that could influence either the kinases or phosphatases associated with NCX1 in heart[29] (FIG. 5 A) (see also Ruknudin *et al.* in this volume).

One of the recent significant developments in the regulation of NCX is the role phospholemman (PLM). PLM is a 72 amino acid protein, likely with a single transmembrane domain, abundantly expressed in heart.[39] Although this protein was cloned in 1991, the function of the protein in the heart was not clearly understood.[40,41] Recently, Bossuyt *et al.* and others reported that PLM was associated with Na/K ATPase in the heart, and the amount of expression changed during ischemia-induced heart failure.[42,43] Subsequently, Chueng's group discovered the PLM's association with NCX1, and has shown that PLM is an endogenous inhibitor of NCX activity in heart.[44,45] Later, Wang *et al.* identified the N-terminal region of the intracellular loop being the site of association with PLM.[46,47] In the early studies of PLM's association with Na/K ATPase, Fuller *et al.* showed the binding of phosphorylated PLM with another protein 14-3-3 *in vitro*.[43] The 14-3-3 protein is a dimer, and is known to be a central protein in the signal transduction of various pathways.[48] There are more than 200 proteins known to bind 14-3-3. Carafoli's group showed the 14-3-3 isoform ε binds to the PMCA and NCX, suggesting that NCX could be associated with a number of other signaling proteins through 14-3-3 protein.[49,50] One of them could be PKA-mediated effect of NCX1.[51,52]

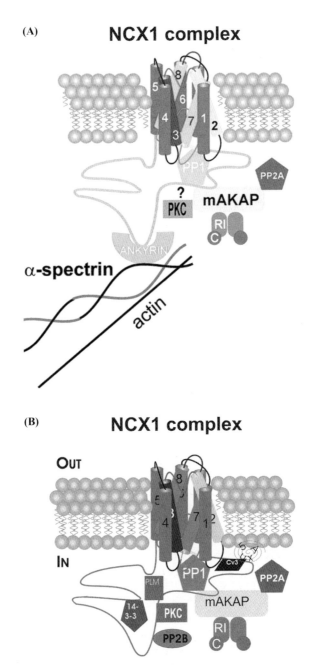

FIGURE 5. Cartoon showing the NCX and the associated proteins. (**A**) NCX macro-molecular complex with cytoskeletal proteins (Adopted from Schulze *et al.*, 2003[29]). (**B**) Recently identified protein associations with NCX protein. The cytoskeletal proteins are not shown for clarity. (In color in Annals online.)

Similar to PLM, sorcin is also a small phosphoprotein and it was thought to influence the activity of NCX as shown in the left ventricular dysfunction due to myocardial infarction in rabbit hearts.[53,54] Further experiments showing the physical interaction of sorcin with NCX are needed to verify this suggestion. NCX1 was found to be associated with transient receptor potential channels (TRPC3) in HEK293 cells, and the C-terminal end of the TRPC3 was involved in the interaction with NCX1. Both these proteins were shown to be functionally coupled.[55]

Other proteins, which are known to be associated with NCX regarding localization of NCX, are ankyrin, caveolin-3, and annexin-5 (FIG. 5 B).[56–58] It was shown that NCX1 is co-localized with ankyrin and this association could preferentially localize NCX1 in cardiomyocytes.[56,59] Mutation in ankyrin-B resulted in the long QT cardiac arrhythmia in mice and possibly in humans.[60] The caveolins localize proteins to the caveoli, the membrane structures found in the plasma membrane.[61] Also, caveolins are associated with the specialized regions in the membrane known as "lipid rafts" and the lipid rafts are the microdomains where specific signal transduction molecules are clustered.[62] Annexin-5 and caveolin-3 associate with NCX in the human heart, suggesting a role of these proteins in modulating the function of the NCX.[63] There are three annexin products, viz. A5, A6, and A7, which are abundant in the heart, and annexin A6-null mice exhibit an increase in myocytes contractility and more rapid removal of Ca^{2+} from the cytosol.[64] In human heart failure, the NCX1 protein forms subsarcolemmal complex with annexin-5 and caveolin-3 and *in vitro* experiments have demonstrated a role of intracellular loop in this complex formation. Camors *et al.* also suggested that the interaction of annexin-5 and NCX1 are stronger in failing hearts, and this could be responsible for altered Ca^{2+} handling in those hearts.[58]

Structure of the Intracellular Loop by NMR and Crystallography

This year, the structure of the major portions of the intracellular loop has been elucidated using multidimensional nuclear magnetic resonance (NMR) spectroscopy technique. This study revealed how regions of the catalytic Ca^{2+}-binding domains are organized in the intracellular loop of NCX1.[65] The investigators first solved the β1 and β2 repeats of the intracellular loop individually and then solved the combined structure of the major portion of the intracellular loop, including the alternatively spliced region. The classical immunoglobulin fold is found in each β1 and β2 regions. The β sandwich motif is formed by two parallel β sheets consisting of strands A-B-E and strands D-C-F-G. The intracellular loop is well defined by NMR spectroscopy, except the unstructured FG loop (FIG. 6 A). By looking at the structure of both β1 and β2 regions together, the NMR studies revealed a single helix joining the β1 and β2 regions. Interestingly, the Ca^{2+}-binding sites are about 180° to one another

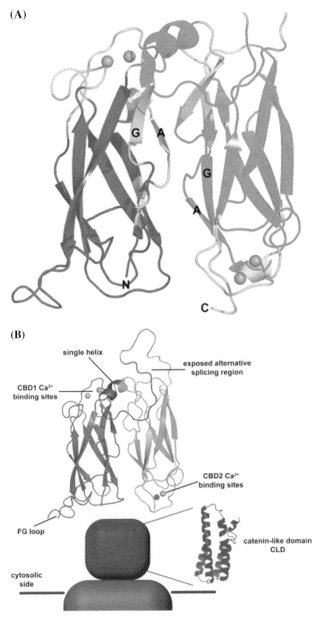

FIGURE 6. The structure of the intracellular loop of NCX1 as delineated by NMR spectroscopy. (**A**) Ribbon diagram of the interface between calcium-binding domains 1 and 2. The acidic residues that bind catalytic Ca^{2+} are seen 180° rotated around the vertical axis to one another. (**B**) N-terminal and C-terminal regions of the intracellular loop are denoted as catenin-like domains. Alternatively spliced region is also shown along with calcium-binding domains 1 and 2. (Modified from Hilge *et al.* 2006.[65]) (In color in Annals online.)

in the intracellular loop. Hilge *et al.* suggest that N-terminal and C-terminal ends of the intracellular loop form catenin-like domains on both ends to connect the whole of the intracellular loop with transmembrane regions (FIG. 6 A, B).[65] Using the crystallographic method Nicoll *et al.* also solved the structure of the β1 repeat of intracellular loop containing the Ca^{2+}-binding domain.[66] This study agreed with the findings of the NMR findings but it also identified 4 Ca^{2+} binding to the Ca^{2+}-binding region 1 (CAD1) of the β1 repeat, as opposed to 2 Ca^{2+} as suggested by NMR studies. The structural changes in the intracellular loop induced by associated proteins will be very interesting to study.

PMCA

PMCA is present ubiquitously in most cells and is shown to be important for Ca^{2+} homeostasis in these cells. PMCA represents a high affinity system for the expulsion of Ca^{2+} from the cell and is thought to be responsible for long-term setting and maintenance of intracellular Ca^{2+} levels. Four different isoforms of PMCA have been cloned so far and the expression of the isoforms 1, 2, and 4 has been shown in the myocardium. However, several experiments have demonstrated that PMCA contributes very little to excitation–contraction (EC) of cardiomyocytes.[67] Recent transgenic expression of PMCA 4 cDNA in the rat heart indicated that there was no significant change in the EC coupling in these hearts but the growth and differentiation of cardiomyocytes were highly influenced in the transgenic animals.[68,69]

Regulation of PMCA Activity

The PMCA protein has 10 putative TM domains and all four PMCA gene products undergo alternative splicing to produce a number of PMCA variants.[70] The variations occur in the intracellular regions between the second and third TM or in the C-terminal end of the protein (FIG. 7). Several PMCA isoforms and splice variants show distinct subcellular localizations.[71,72] Long before cloning the PMCA, it was known that C-terminal tail of the PMCA interacts with calmodulin.[73,74] When calmodulin is available, PMCA are inactive and remain in an autoinhibited state and the Ca^{2+} calmodulin binding deinhibit the pump, thereby increasing its apparent Ca^{2+} affinity and raising its V_{max}.[75] Phosphorylation by PKA, PKC, and tyrosine kinase have been shown to regulate the pump activity.[76] Moreover, the lipid environment plays an important role in the regulation of PMCA.[77]

Other Proteins Associated with PMCA

Among the various splice variants, the protein of PMCA 4b variant terminates in the conserved sequence—S(L/V)ETS(L/V) that matches the minimal-(T/S)XV* consensus motif for PDZ domain interaction.[78] The PDZ domains

are central in organizing protein complexes at the plasma membrane.[79] Kim
et al. showed that PMCA 4b isoform interacts with several membrane-
associated guanylate kinases (MAGUK) family members via PDZ domains,
and suggested that this interaction could influence the localization and incorporation into multiprotein complexes.[78] In this regard, another protein,
Ania3/Homer from Homer Family of scaffolding proteins, also associates with
every b-splice isoforms of PMCA.[80] DeMarco and Strechler identified PMCA
isoforms 2b and 4b associated with synapse-associated protein (SAP) through
the PDZ domains.[81] However, SAP102 was shown to be specifically associated
with PMCA 4b but not with PMCA 2b, and these specific associations with
SAP would help to recruit PMCA in specially localized areas in the plasma
membrane. Isoform-specific interaction of PMCA 2b was also seen with Na/H
exchanger regulatory factor-2 (NHERF-2).[82] Interestingly, Schuh *et al.* showed
association of PMCA 4b with calcium–calmodulin-dependent serine protein
kinase (CASK) in the brain and kidney.[83] The CASK contains a calmodulin-
dependent protein kinase-like domain followed by PDZ, SH3, and guanylate kinase-like domains. They also showed that functional PMCA influenced
the transcriptional activity. Other proteins, such as tumor suppression Ras-
associated factor 1 (RASSF1) and nitric oxide synthase, are also associated
with PMCA. Regarding the effect of PMCA on the transcription, calcineurin
is shown to be associated with PMCA 4b, and this in turn could regulate
NFAT pathway.[84] Recently, Williams *et al.* showed that PMCA forms a macromolecular complex with α1-syntrophin, neuronal nitric oxide synthase and
this complex could be attached to dystrophin and then to actin in the cells.[85]

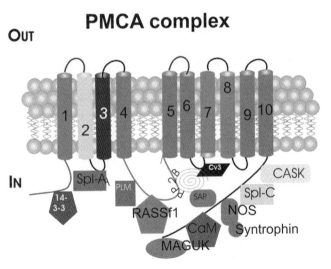

FIGURE 7. Cartoon of the PMCA with known associated proteins. The positions of
the many associated proteins are suggestive. (In color in Annals online.)

These studies indicate that several proteins associate with PMCA and some of these associations are isoform specific and regulate the function. In addition, the associated proteins of PMCA also help to recruit other proteins to form a local signal transduction complex.

CONCLUSION

During the last few years, exciting new developments have occurred in the field of NCX regulation. The high-resolution structure of the intracellular loop is now available. In addition to the regulation of the NCX transport by XIP, pH, PIP2 as well as Na^+ and Ca^{2+}, several proteins have been shown to be associated with NCX proteins, and these proteins modulate the activity of the NCX. Parallel to the recent developments in the NCX field (FIG. 5), several proteins that are associated with the PMCA pump (FIG. 7) have been identified. Both these transporters are regulated by their intracellular regions and by their associated proteins. It would be reasonable to assume that these transporters, in addition to their main role in Ca^{2+} homeostasis, could have other cellular functions. For example, NCX and PMCA might be involved in signal transduction related to β-adrenergic stimulation (NCX) or nitric oxide pathway (PMCA). Compared to unknown roles of associated proteins of NCX, the functional roles of many associated proteins of PMCA have been explored. Expansion and clarification of the role of these presently obscure functions of NCX-associated proteins will likely become a main focus of this field in the near future.

ACKNOWLEDGMENTS

The work was supported by the NIH intramural research programs (grants to E.G.L). A.M.R. is grateful for the University of Maryland, School of Medicine for permitting him to work at NIH.

REFERENCES

1. BAKER, P.F. & M.P. BLAUSTEIN. 1968. Sodium-dependent uptake of calcium by crab nerve. Biochim. Biophys. Acta 150: 167–170.
2. REUTER, H. & N. SEITZ. 1968. The dependence of calcium efflux from cardiac muscle on temperature and external ion composition. J. Physiol. 195: 451–470.
3. NICOLL, D.A., S. LONGONI & K.D. PHILIPSON. 1990. Molecular cloning and functional expression of the cardiac sarcolemmal Na(+)-Ca^{2+} exchanger. Science 250: 562–565.
4. QUEDNAU, B.D., D.A. NICOLL & K.D. PHILIPSON. 2004. The sodium/calcium exchanger family-SLC8. Pflugers Arch. 447: 543–548.

5. QUEDNAU, B.D., D.A. NICOLL & K.D. PHILIPSON. 1997. Tissue specificity and alternative splicing of the Na$^+$/Ca^{2+} exchanger isoforms NCX1, NCX2, and NCX3 in rat. Am. J. Physiol. **272**(4 Pt 1): C1250–C1261.
6. KOFUJI, P., R.W. HADLEY, R.S. KIEVAL, et al. 1992. Expression of the Na-Ca exchanger in diverse tissues: a study using the cloned human cardiac Na-Ca exchanger. Am. J. Physiol. **263**(6 Pt 1): C1241–C1249.
7. BLAUSTEIN, M.P. & W.J. LEDERER. 1999. Sodium/calcium exchange: its physiological implications. Physiol. Rev. **79:** 763–854.
8. PHILIPSON, K.D., D.A. NICOLL, S. MATSUOKA, et al. 1996. Molecular regulation of the Na(+)-Ca^{2+} exchanger. Ann. N. Y. Acad. Sci. **779:** 20–28.
9. PHILIPSON, K.D. & D.A. NICOLL. 2000. Sodium-calcium exchange: a molecular perspective 15. Annu. Rev. Physiol. **62:** 111–133.
10. DIPOLO, R. & L. BEAUGE. 2006. Sodium/calcium exchanger: influence of metabolic regulation on ion carrier interactions. Physiol. Rev. **86:** 155–203.
11. HOUSER, S.R. & E.G. LAKATTA. 1999. Function of the cardiac myocyte in the conundrum of end-stage, dilated human heart failure. Circulation **99:** 600–604.
12. BANO, D., K.W. YOUNG, C.J. GUERIN, et al. 2005. Cleavage of the plasma membrane Na$^+$/Ca^{2+} exchanger in excitotoxicity. Cell **120:** 275–285.
13. NICOLL, D.A., B.D. QUEDNAU, Z. QUI, et al. 1996. Cloning of a third mammalian Na$^+$-Ca^{2+} exchanger, NCX3. J. Biol. Chem. **271:** 24914–24921.
14. MATSUOKA, S., D.A. NICOLL, R.F. REILLY, et al. 1993. Initial localization of regulatory regions of the cardiac sarcolemmal Na(+)-Ca^{2+} exchanger. Proc. Natl. Acad. Sci. USA **90:** 3870–3874.
15. NICOLL, D.A., L.V. HRYSHKO, S. MATSUOKA, et al. 1996. Mutagenesis studies of the cardiac Na(+)-Ca^{2+} exchanger. Ann. N. Y. Acad. Sci. **779:** 86–92.
16. NICOLL, D.A., M. OTTOLIA, L. LU, et al. 1999. A new topological model of the cardiac sarcolemmal Na$^+$-Ca^{2+} exchanger. J. Biol. Chem. **274:** 910–917.
17. IWAMOTO, T., T.Y. NAKAMURA, Y. PAN, et al. 1999. Unique topology of the internal repeats in the cardiac Na$^+$/Ca^{2+} exchanger. FEBS Lett. **446:** 264–268.
18. SCHWARZ, E.M. & S. BENZER. 1997. Calx, a Na-Ca exchanger gene of *Drosophila melanogaster*. Proc. Natl. Acad. Sci. USA **94:** 10249–10254.
19. REEVES, J.P. 1998. Na$^+$/Ca^{2+} exchange and cellular Ca^{2+} homeostasis. J. Bioenerg. Biomembr. **30:** 151–160.
20. HILGEMANN, D.W., A. COLLINS & S. MATSUOKA. 1992. Steady-state and dynamic properties of cardiac sodium-calcium exchange. Secondary modulation by cytoplasmic calcium and ATP. J. Gen. Physiol. **100:** 933–961.
21. HILGEMANN, D.W., S. MATSUOKA, G.A. NAGEL & A. COLLINS. 1992. Steady-state and dynamic properties of cardiac sodium-calcium exchange. Sodium-dependent inactivation. J. Gen. Physiol. **100:** 905–932.
22. MATSUOKA, S. & D.W. HILGEMANN. 1994. Inactivation of outward Na(+)-Ca^{2+} exchange current in guinea-pig ventricular myocytes. J. Physiol. (Lond.) **476:** 443–458.
23. URBANCZYK, J., O. CHERNYSH, M. CONDRESCU & J.P. REEVES. 2006. Sodium-calcium exchange does not require allosteric calcium activation at high cytosolic sodium concentrations. J. Physiol. **575**(Pt 3): 693–705.
24. DOERING, A.E., D.A. EISNER & W.J. LEDERER. 1996. Cardiac Na-Ca exchange and pH. Ann. N. Y. Acad. Sci. **779:** 182–198.
25. HILGEMANN, D.W. & R. BALL. 1996. Regulation of cardiac Na$^+$,Ca^{2+} exchange and KATP potassium channels by PIP2. Science **273:** 956–959.

26. HE, S., A. RUKNUDIN, L.L. BAMBRICK, *et al*. 1998. Isoform-specific regulation of the Na$^+$/Ca^{2+} exchanger in rat astrocytes and neurons by PKA. J. Neurosci. **18:** 4833–4841.
27. IWAMOTO, T., Y. PAN, S. WAKABAYASHI, *et al*. 1996. Phosphorylation-dependent regulation of cardiac Na$^+$/Ca^{2+} exchanger via protein kinase C. J. Biol. Chem. **271:** 13609–13615.
28. LI, Z., D.A. NICOLL, A. COLLINS, *et al*. 1991. Identification of a peptide inhibitor of the cardiac sarcolemmal Na(+)-Ca^{2+} exchanger. J. Biol. Chem. **266:** 1014–1020.
29. SCHULZE, D.H., M. MUQHAL, W.J. LEDERER & A.M. RUKNUDIN. 2003. Sodium/calcium exchanger (NCX1) macromolecular complex. J. Biol. Chem. **278:** 28849–28855.
30. MARX, S.O., S. REIKEN, Y. HISAMATSU, *et al*. 2000. PKA phosphorylation dissociates FKBP12.6 from the calcium release channel (ryanodine receptor): defective regulation in failing hearts. Cell **101:** 365–376.
31. MARX, S.O., S. REIKEN, Y. HISAMATSU, *et al*. 2001. Phosphorylation-dependent regulation of ryanodine receptors: a novel role for leucine/isoleucine zippers. J. Cell Biol. **153:** 699–708.
32. DARMAN, R.B., A. FLEMMER & B. FORBUSH. 2001. Modulation of ion transport by direct targeting of protein phosphatase type 1 to the Na-K-Cl cotransporter. J. Biol. Chem. **276:** 34359–34362.
33. EGLOFF, M.P., D.F. JOHNSON, G. MOORHEAD, *et al*. 1997. Structural basis for the recognition of regulatory subunits by the catalytic subunit of protein phosphatase 1. EMBO J. **16:** 1876–1887.
34. KATANOSAKA, Y., Y. IWATA, Y. KOBAYASHI, *et al*. 2005. Calcineurin inhibits Na$^+$/Ca^{2+} exchange in phenylephrine-treated hypertrophic cardiomyocytes. J. Biol. Chem. **280:** 5764–5772.
35. WEI, S.K., A. RUKNUDIN, S.U. HANLON, *et al*. 2003. Protein kinase A hyperphosphorylation increases basal current but decreases beta-adrenergic responsiveness of the sarcolemmal Na$^+$-Ca^{2+} exchanger in failing pig myocytes. Circ. Res. **92:** 897–903.
36. GINSBURG, K.S. & D.M. BERS. 2005. Isoproterenol does not enhance Ca-dependent Na/Ca exchange current in intact rabbit ventricular myocytes. J. Mol. Cell Cardiol. **39:** 972–981.
37. IWAMOTO, T., S. WAKABAYASHI, T. IMAGAWA & M. SHIGEKAWA. 1998. Na$^+$/Ca^{2+} exchanger overexpression impairs calcium signaling in fibroblasts: inhibition of the [Ca^{2+}] increase at the cell periphery and retardation of cell adhesion. Eur. J. Cell. Biol. **76:** 228–236.
38. IWAMOTO, T., S. WAKABAYASHI & M. SHIGEKAWA. 1995. Growth factor-induced phosphorylation and activation of aortic smooth muscle Na/Ca exchanger. J. Biol. Chem. **270:** 8996–9001.
39. CRAMBERT, G., M. FUZESI, H. GARTY, *et al*. 2002. Phospholemman (FXYD1) associates with Na,K-ATPase and regulates its transport properties. Proc. Natl. Acad. Sci. USA **99:** 11476–11481.
40. PALMER, C.J., B.T. SCOTT & L.R. JONES. 1991. Purification and complete sequence determination of the major plasma membrane substrate for cAMP-dependent protein kinase and protein kinase C in myocardium. J. Biol. Chem. **266:** 11126–11130.
41. DAVIS, C.E., M.K. PATEL, J.R. MILLER, *et al*. 2004. Effects of phospholemman expression on swelling-activated ion currents and volume regulation in embryonic kidney cells. Neurochem. Res. **29:** 177–187.

42. BOSSUYT, J., S. DESPA, J.L. MARTIN & D.M. BERS. 2006. Phospholemman phospho-
    rylation alters its fluorescence resonance energy transfer with the Na/K-ATPase
    pump. J. Biol. Chem. **281:** 32765–32773.
43. FULLER, W., P. EATON, J.R. BELL & M.J. SHATTOCK. 2004. Ischemia-induced
    phosphorylation of phospholemman directly activates rat cardiac Na/K-ATPase.
    FASEB J. **18:** 197–199.
44. AHLERS, B.A., X.Q. ZHANG, J.R. MOORMAN, et al. 2005. Identification of an en-
    dogenous inhibitor of the cardiac Na^+/Ca^{2+} exchanger, phospholemman. J. Biol.
    Chem. **280:** 19875–19882.
45. SONG, J., X.Q. ZHANG, B.A. AHLERS, et al. 2005. Serine 68 of phospho-
    lemman is critical in modulation of contractility, $[Ca^{2+}]i$ transients and
    Na^+/Ca^{2+} exchange in adult rat cardiac myocytes. Am. J. Physiol. Heart Circ.
    Physiol. **288:** 42342–42354.
46. WANG, J., X.Q. ZHANG, B.A. AHLERS, et al. 2006. Cytoplasmic tail of phospho-
    lemman interacts with the intracellular loop of the cardiac Na^+/Ca^{2+} exchanger.
    J. Biol. Chem. **281:** 32004–32014.
47. ZHANG, X.Q., B.A. AHLERS, A.L. TUCKER, et al. 2006. Phospholemman inhibition
    of the cardiac Na^+/Ca^{2+} exchanger. Role of phosphorylation. J. Biol. Chem.
    281: 7784–7792.
48. AITKEN, A. 2006. 14-3-3 proteins: a historic overview. Semin. Cancer Biol. **16:**
    162–172.
49. PULINA, M.V., R. RIZZUTO, M. BRINI & E. CARAFOLI. 2006. Inhibitory interaction
    of the plasma membrane Na^+/Ca^{2+} exchangers with the 14-3-3 proteins. J. Biol.
    Chem. **281:** 19645–19654.
50. RIMESSI, A., L. COLETTO, P. PINTON, et al. 2005. Inhibitory interaction of the 14-
    3-3{epsilon} protein with isoform 4 of the plasma membrane Ca(2+)-ATPase
    pump. J. Biol. Chem. **280:** 37195–37203.
51. CHOE, C.U., E. SCHULZE-BAHR, A. NEU, et al. 2006. C-terminal HERG (LQT2)
    mutations disrupt IKr channel regulation through 14-3-3{epsilon}. Hum. Mol.
    Genet. **15:** 2888–2902.
52. ALLOUIS, M., B.F. LE, R. WILDERS, et al. 2006. 14-3-3 is a regulator of the cardiac
    voltage-gated sodium channel Nav1.5. Circ. Res. **98:** 1538–1546.
53. SEIDLER, T., S.L. MILLER, C.M. LOUGHREY, et al. 2003. Effects of adenovirus-
    mediated sorcin overexpression on excitation-contraction coupling in isolated
    rabbit cardiomyocytes. Circ. Res. **93:** 132–139.
54. SMITH, G.L., E.E. ELLIOTT, S. KETTLEWELL, et al. 2006. Na(+)/Ca(2+) exchanger
    expression and function in a rabbit model of myocardial infarction. J. Cardiovasc.
    Electrophysiol. **17**(Suppl 1): S57–S63.
55. ROSKER, C., A. GRAZIANI, M. LUKAS, et al. 2004. Ca(2+) signaling by TRPC3
    involves Na(+) entry and local coupling to the Na(+)/Ca(2+) exchanger. J.
    Biol. Chem. **279:** 13696–13704.
56. LI, Z.P., E.P. BURKE, J.S. FRANK, et al. 1993. The Cardiac Na^+-Ca^{2+} ex-
    changer binds to the cytoskeletal protein ankyrin. J. Biol. Chem. **268:** 11489–
    11491.
57. BOSSUYT, J., B.E. TAYLOR, M. JAMES-KRACKE & C.C. HALE. 2002. Evidence for
    cardiac sodium-calcium exchanger association with caveolin-3. FEBS Lett. **511:**
    113–117.
58. CAMORS, E., D. CHARUE, P. TROUVE, et al. 2006. Association of annexin A5 with
    Na^+/Ca^{2+} exchanger and caveolin-3 in non-failing and failing human heart. J.
    Mol. Cell Cardiol. **40:** 47–55.

59. CHEN, F., G. MOTTINO, V.Y. SHIN & J.S. FRANK. 1997. Subcellular distribution of ankyrin in developing rabbit heart—relationship to the Na^+-Ca^{2+} exchanger. J. Mol. Cell Cardiol. **29:** 2621–2629.

60. MOHLER, P.J., J.J. SCHOTT, A.O. GRAMOLINI, *et al*. 2003. Ankyrin-B mutation causes type 4 long-QT cardiac arrhythmia and sudden cardiac death. Nature **421:** 634–639.

61. LIU, L., K. MOHAMMADI, B. AYNAFSHAR, *et al*. 2003. Role of caveolae in signal-transducing function of cardiac Na+/K+-ATPase. Am. J. Physiol. Cell Physiol. **284:** C1550–C1560.

62. COHEN, A.W., R. HNASKO, W. SCHUBERT & M.P. LISANTI. 2004. Role of caveolae and caveolins in health and disease. Physiol. Rev. **84:** 1341–1379.

63. CAMORS, E., V. MONCEAU & D. CHARLEMAGNE. 2005. Annexins and Ca^{2+} handling in the heart. Cardiovasc. Res. **65:** 793–802.

64. SONG, G., S.E. HARDING, M.R. DUCHEN, *et al*. 2002. Altered mechanical properties and intracellular calcium signaling in cardiomyocytes from annexin 6 null-mutant mice. FASEB J. **16:** 622–624.

65. HILGE, M., J. AELEN & G.W. VUISTER. 2006. Ca^{2+} regulation in the Na^+/Ca^{2+} exchanger involves two markedly different Ca^{2+} sensors. Mol. Cell **22:** 15–25.

66. NICOLL, D.A., M.R. SAWAYA, S. KWON, *et al*. 2006. The crystal structure of the primary Ca^{2+} sensor of the Na^+/Ca^{2+} exchanger reveals a novel Ca^{2+} binding motif. J. Biol. Chem. **281:** 21577–21581.

67. BERS, D.M., J.W. BASSANI & R.A. BASSANI. 1996. Na-Ca exchange and Ca fluxes during contraction and relaxation in mammalian ventricular muscle. Ann. N. Y. Acad. Sci. **779:** 430–442.

68. HAMMES, A., S. OBERDORF-MAASS, T. ROTHER, *et al*. 1998. Overexpression of the sarcolemmal calcium pump in the myocardium of transgenic rats. Circ. Res. **83:** 877–888.

69. CARTWRIGHT, E.J., K. SCHUH & L. NEYSES. 2005. Calcium transport in cardiovascular health and disease–the sarcolemmal calcium pump enters the stage. J. Mol. Cell Cardiol. **39:** 403–406.

70. STREHLER, E.E. & D.A. ZACHARIAS. 2001. Role of alternative splicing in generating isoform diversity among plasma membrane calcium pumps. Physiol. Rev. **81:** 21–50.

71. REINHARDT, T.A., A.G. FILOTEO, J.T. PENNISTON & R.L. HORST. 2000. Ca(2+)-ATPase protein expression in mammary tissue. Am. J. Physiol. Cell. Physiol. **279:** C1595–C1602.

72. DUMONT, R.A., U. LINS, A.G. FILOTEO, *et al*. 2001. Plasma membrane Ca^{2+}-ATPase isoform 2a is the PMCA of hair bundles. J. Neurosci. **21:** 5066–5078.

73. JAMES, P., M. MAEDA, R. FISCHER, *et al*. 1988. Identification and primary structure of a calmodulin binding domain of the Ca^{2+} pump of human erythrocytes. J. Biol. Chem. **263:** 2905–2910.

74. ENYEDI, A., T. VORHERR, P. JAMES, *et al*. 1989. The calmodulin binding domain of the plasma membrane Ca^{2+} pump interacts both with calmodulin and with another part of the pump. J. Biol. Chem. **264:** 12313–12321.

75. CARIDE, A.J., N.L. ELWESS, A.K. VERMA, *et al*. 1999. The rate of activation by calmodulin of isoform 4 of the plasma membrane Ca(2+) pump is slow and is changed by alternative splicing. J. Biol. Chem. **274:** 35227–35232.

76. ENYEDI, A., N.L. ELWESS, A.G. FILOTEO, *et al*. 1997. Protein kinase C phosphorylates the "a" forms of plasma membrane Ca^{2+} pump isoforms 2 and 3 and prevents binding of calmodulin. J. Biol. Chem. **272:** 27525–27528.

77. PENNISTON, J.T. & A. ENYEDI. 1998. Comparison of ATP-powered Ca^{2+} pumps. *In* Ion Pumps. Bittar, E.E., ed. 249–274. JAI Press. Greenwich, CT.

78. KIM, E., S.J. DEMARCO, S.M. MARFATIA, *et al.* 1998. Plasma membrane Ca^{2+} ATPase isoform 4b binds to membrane-associated guanylate kinase (MAGUK) proteins via their PDZ (PSD-95/Dlg/ZO-1) domains. J. Biol. Chem. **273:** 1591–1595.

79. SCHUH, K., S. ULDRIJAN, M. TELKAMP, *et al.* 2001. The plasma membrane calmodulin-dependent calcium pump: a major regulator of nitric oxide synthase I. J. Cell Biol. **155:** 201–205.

80. SGAMBATO-FAURE, V., Y. XIONG, J.D. BERKE, *et al.* 2006. The Homer-1 protein Ania-3 interacts with the plasma membrane calcium pump. Biochem. Biophys. Res. Commun. **343:** 630–637.

81. DEMARCO, S.J. & E.E. STREHLER. 2001. Plasma membrane Ca^{2+}-ATPase isoforms 2b and 4b interact promiscuously and selectively with members of the membrane-associated guanylate kinase family of PDZ (PSD95/Dlg/ZO-1) domain-containing proteins. J. Biol. Chem. **276:** 21594–21600.

82. DEMARCO, S.J., M.C. CHICKA & E.E. STREHLER. 2002. Plasma membrane Ca^{2+} ATPase isoform 2b interacts preferentially with Na+/H+ exchanger regulatory factor 2 in apical plasma membranes. J. Biol. Chem. **277:** 10506–10511.

83. SCHUH, K., S. ULDRIJAN, S. GAMBARYAN, *et al.* 2003. Interaction of the plasma membrane Ca^{2+} pump 4b/CI with the Ca^{2+}/calmodulin-dependent membrane-associated kinase CASK. J. Biol. Chem. **278:** 9778–9783.

84. BUCH, M.H., A. PICKARD, A. RODRIGUEZ, *et al.* 2005. The sarcolemmal calcium pump inhibits the calcineurin/nuclear factor of activated T-cell pathway via interaction with the calcineurin A catalytic subunit. J. Biol. Chem. **280:** 29479–29487.

85. WILLIAMS, J.C., A.L. ARMESILLA, T.M. MOHAMED, *et al.* 2006. The sarcolemmal calcium pump, alpha-1 syntrophin, and neuronal nitric-oxide synthase are parts of a macromolecular protein complex. J. Biol. Chem. **281:** 23341–23348.

Phosphorylation and Other Conundrums of Na/Ca Exchanger, NCX1

ABDUL M. RUKNUDIN,[a–c] SHAO-KUI WEI,[d] MARK C. HAIGNEY,[d]
W.J. LEDERER,[b] AND DAN H. SCHULZE[a,b]

[a]Department of Microbiology and Immunology, University of Maryland,
Baltimore, Maryland 21201, USA

[b]Medical Biotechnology Center, University of Maryland, Biotechnology Institute,
Baltimore, Maryland 21201, USA

[c]Laboratory of Cardiovascular Science, Gerontology Research Center,
National Institute on Aging, Baltimore, Maryland 21224, USA

[d]Department of Medicine, Uniform Services University of Health Sciences,
Bethesda 20814, Maryland, USA

ABSTRACT: The Na^+/Ca^{2+} exchanger (NCX) is an important Ca^{2+} transport mechanism in virtually all cells in the body. There are three genes that control the expression of NCX in mammals. There are at least 16 alternatively spliced isoforms of NCX1 that target muscle and nerve and other tissues. Here we briefly discuss three remarkable regulatory issues or "conundrums" that involve the most prevalently expressed gene, *NCX1*. (1) How is NCX1 regulated by phosphorylation? We suggest that the macromolecular complex of NCX1 plays a critical role in the regulation of NCX. The role of the macromolecular complex and evidence supporting its existence and functional importance is presented. (2) Can there be transport block of a single "mode" of NCX1 transport by drugs or therapeutic agents? The simple answer is "no." A brief explanation is provided. (3) How can NCX1 knockout mice live? The answer is "by other compensatory regulatory mechanisms." These conundrums highlight important features in NCX1 and lay the foundation for new experiments to elucidate function and regulation of NCX1 and provide a context for investigations that seek to understand novel therapeutic agents.

KEYWORDS: protein kinase A; regulation; NCX1 knockout; mode-specific inhibition; NCX macromolecular complex; β-adrenergic stimulation

INTRODUCTION

Following the discovery of the sodium–calcium (Na^+/Ca^{2+}) exchanger (NCX) in 1967, many investigations examined its regulation and function.[1–5]

Address for correspondence: Abdul M. Ruknudin, Ph.D., Department of Microbiology and Immunology, 660 W. Redwood Street, Baltimore, MD 21201. Voice: 410-706-5180; fax: 410-706-2129.
aruknudi@umaryland.edu

Ann. N.Y. Acad. Sci. 1099: 103–118 (2007). © 2007 New York Academy of Sciences.
doi: 10.1196/annals.1387.036

This effort was greatly aided by its cloning in 1990 by Philipson and co-workers who reported on the primary cDNA sequence of NCX1 showing possible phosphorylation sites[6]; here we discuss several controversies including the effects of NCX phosphorylation. The cloning of NCX1 facilitated the study of its function and regulation by enabling the expression of NCX in cells where NCX is largely absent. We and others have cloned NCX1 from various mammals and from *Drosophila* and expressed these clones in heterologous expression systems to study their function.[7–12]

Alternative splicing of NCX1 enables the expression of tissue-specific isoforms, which may have distinctive physiological properties and regulation. The major isoforms of NCX1 in neurons and astrocytes were cloned and expressed in *Xenopus* oocytes.[13–15] The neuronal isoform was regulated by PKA pathway but the glial isoform was not.[15] By expressing the major heart isoform in *Xenopus* oocytes, we found that the stimulation of PKA phosphorylation increased the activity whereas the major kidney isoform (under similar conditions) did not show the increased activity after activation of PKA pathway (FIG. 1).[16] The rat NCX1 protein of heart isoform when expressed in *Xenopus* oocytes as well as the NCX1 expressed in rat heart could be phosphorylated *in vitro* by PKA enzyme (FIG. 2).[16,17]

NCX CONUNDRUMS

There are at least three issues related to NCX1 function that persist in the literature and are of continuing interest today. (1) What are the functional consequences, if any, of NCX1 phosphorylation? (2) Is "modal" block of the NCX1 possible? and (3) How do mice lacking the NCX1 survive?

Conundrum 1. NCX Phosphorylation

We have found that NCX1 is the key member of a macromolecular complex that includes kinases and phosphatases and putative targeting elements (see FIG. 8). At least some of the inconsistency of the reported findings may reflect the importance of this macromolecular organization of the NCX1 transporter. This is briefly discussed in this article.

Controversy of NCX1 Phosphorylation by PKA and PKC

Even though we showed that the effect of PKA phosphorylation both in *Xenopus* oocytes expressing cardiac isoform and rat ventricular cardiomyocytes under physiological conditions results in the increased NCX1 activity,[16,17] the same rat heart isoform when expressed in HEK293 cells did not respond to the stimulation of PKA pathway (FIG. 3). This raised the possibility that the action of PKA on NCX1 could be more complex than simple phosphorylation of the NCX1 protein by the cytoplasmic kinases. Investigators from some labs also reported that they could not observe the effect of PKA phosphorylation on the cardiac NCX1.[18–21] However, some other reports supported our findings that PKA could modulate the activity of cardiac NCX1.[22,23] An effective demonstration of phosphorylation of NCX1 by another kinase, PKC, was first shown by Iwamoto *et al.*[24] The effect of PKC was demonstrated in smooth muscle cells first and then on the cloned cardiac exchanger expressed in CCL39 cells as well as on the rat cardiomyocytes.[24,25] The activity of NCX1 was shown to be increased by PKC activation. Using biochemical techniques, Iwamoto *et al.* found that the NCX1 protein was phosphorylated upon incubation with phorbol ester (PMA) to stimulate the PKC.[24] Interestingly, the basal phosphorylation of NCX1 protein was high and PMA treatment increased the phosphorylated moieties of the protein (FIG. 4). However, when all the consensus sites of the PKC phosphorylation were mutated in the NCX1, the protein of NCX1 still showed phosphorylation after PMA treatment[26] indicating that either the PKC effect is not ultimately affecting the NCX1 protein or that there could be nonconsensus sites for PKC to act on the NCX1 protein. So, the effect of PKA and PKC remained as an enigma and based on unknown conditions of the experiments, sometimes the effects of PKC and PKA could be detected but in other circumstances they were not always observed.

The Macromolecular Complex of NCX1

The phosphorylation conundrum of NCX1 could be explained by the hypothesis that kinases and phosphatases could be physically associated with the NCX1 protein as it has been shown with other membrane proteins like ryanodine receptor channels (RyR), β-adrenergic receptors (β-AR), or L-type Ca channels (LTCC).[27,28] Under conditions that promote kinase activity, the effects of phosphorylation could be observed. However, under conditions that promote the activity of the phosphatases, the effect of phosphorylation might be difficult to detect. To test the macromolecular complex hypothesis and investigate whether or not other proteins are associated with NCX1 protein, we immunoprecipitated NCX1 protein from rat cardiomyocytes. This protein was then subjected to polyacrylamide gel electrophoresis under reducing conditions to separate possible components.[29] By immunoblotting with antibodies specific for PKA, the separated NCX1 immunoprecipitate showed that it was associated with the catalytic subunit of PKA heteromultimer and also the regulatory subunit (FIG. 5). Using a pan antibody to PKC, we detected PKC in the immunoprecipitate of NCX1. Moreover, the serine/threonine phosphatases, PP1 and PP2 subunits, were also identified in the complex immunoprecipitated by the NCX1 antibody (FIG. 6). The scaffolding protein, mAKAP was part of

(A)   **AD isoform of NCX1 activity enhanced by PKA activation**

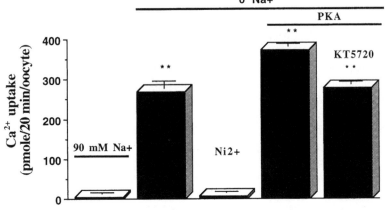

(B)   **BD isoform of NCX1 activity not enhanced by PKA activation**

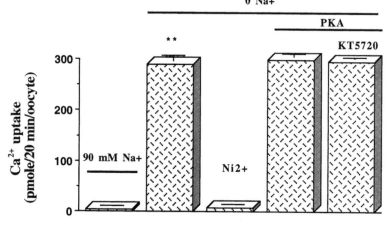

(C)

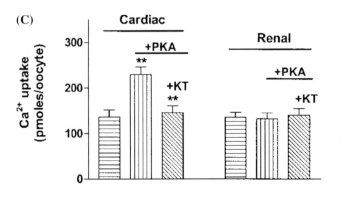

the complex too. Thus the macromolecular complex of NCX1 consisted of at least two kinases and two phosphatases with a scaffolding protein associated with it (FIG. 6).[17] Detection of these components of NCX1 complex was first obtained with no detergent in the extraction buffer (although the immunowash buffer did contain the detergents). Using extraction buffer with Triton X-100 detergent, we observed the association of mAKAP with NCX1 protein (FIG. 7). This is the first demonstration of the regulatory proteins interacting with NCX1 protein (FIG. 8). Recently, another phosphatase, calcineurin (PP2B) has been shown to associate with NCX1 in cardiomyocytes.[30]

Regulation of NCX1 by β-AR in Normal and Heart Failure Hearts

The effect of PKA on NCX1 could be physiologically relevant since sympathetic stimulation of β-AR in the heart has been shown to activate the PKA pathway in a relatively complex manner, phosphorylating important Ca^{2+} regulatory proteins, such as LTCC, RyR, and phospholamban (PLB).[31] During contraction–relaxation cycle of cardiomyocytes, the extracellular Ca^{2+} enters through LTCC upon depolarization, and releases the stored Ca^{2+} from sarcoplasmic reticulum (SR) (through RyR activation) producing the cardiac $[Ca^{2+}]_i$ transient that underlies contraction. During relaxation, about 80% of the cytoplasmic Ca^{2+} is taken up by SR Ca ATPase (SERCA) into SR stores and the remaining Ca^{2+} is extruded by NCX1. Together SERCA and NCX1 are primarily responsible for the reduction of the cytoplasmic $[Ca^{2+}]$ back to diastolic levels.[32] It is known that β-adrenergic stimulation phosphorylates LTCC, RYR, and PLB.[31] With all of the biochemical components in place, it is reasonable to hypothesize that NCX1 could be regulated by PKA phosphorylation as well. However, attempts to demonstrate the effect of β-AR stimulation showed mixed results. Perchenet *et al.* showed the increased Ni^{2+}-sensitive current in guinea pig cardiomyocytes

←——

FIGURE 1. Effect of PKA phosphorylation on alternatively spliced isoforms of rat NCX1 expressed in *Xenopus* oocytes. (**A**)Neuronal isoform (AD) was expressed in *Xenopus* oocytes. $^{45}Ca^{2+}$ uptake was measured in oocytes in the presence and absence of external Na^+. The uptake of $^{45}Ca^{2+}$ in the absence of external Na^+ was blocked by Ni^{2+}. After incubating oocytes with IBMX, dibutryl cAMP, and forskolin, there was an increase in the amount of $^{45}Ca^{2+}$ uptake and this increase could be prevented by including PKA inhibitor, KT5920. $**P < 0.001$ n = 5 or more. (**B**) Astrocytic isoform (BD) was expressed in *Xenopus* oocytes and the measurements and experimental conditions were same as in **A**. $**P < 0.001$ n = 5 or more. (Modified from He *et al.*, 1998[15].) (**C**) The cardiac isoform (ACDEF) and the renal/astrocytic (BD) were expressed in *Xenopus* oocytes. The experiments and analysis were same as in **A** except that the $^{45}Ca^{2+}$ uptake in the presence of 90 mM Na^+ and Ni^{2+} were not shown. $**P < 0.001$ n = 5 or more. Modified from Ruknudin *et al.*, 2000[16] with permission.

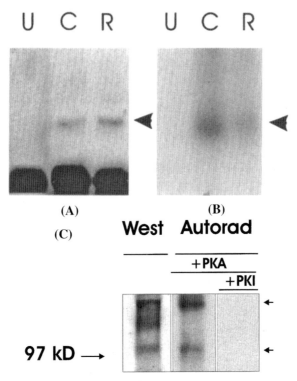

FIGURE 2. The cardiac NCX1 of rat is phosphorylated *in vitro*. (**A**) The proteins from oocytes expressing cardiac (C) and renal (R) isoforms were identified using antibodies to NCX1 protein in Western blot. The uninjected oocytes (U) did not show the NCX1 protein. (**B**) The cardiac isoform shows higher amount of phosphorylation than the renal isoform in the autoradiograph. (Modified from Ruknudin *et al.*, 2000[16].) (**C**) NCX1 protein from rat ventricle was extracted and identified in the Western blot (West) as 160, 140, and 120 kDa bands using antibodies to NCX1. Upon *in vitro* phosphorylation, the bands at 160 and 120 kDa showed phosphorylation (Autorad). The PKA inhibitor, PKI prevented the phosphorylation. (Modified from Schulze *et al.*, 2003[17].)

after β-AR stimulation.[22] We showed that the β-AR stimulation could increase NCX1 activity in pig cardiomyocytes (FIG. 9).[33] However, Ginsburg *et al.* could not find the increase in activity of NCX1 in rabbit cardiomyocytes using both perforated patch and ruptured patch whole-cell clamp methods.[34] Lin *et al.* did not detect any increased activity of NCX1 in mouse, rat, and guinea pig cardiomyocytes.[35] However, they observed increased Ni^{2+}-sensitive current in guinea pig cardiomyocytes after β-AR stimulation and this current was attributed to CFTR current. In the pig cardiomyocytes, neither removal of extracellular Cl^- nor application of niflumic acid and glibenclamide affected

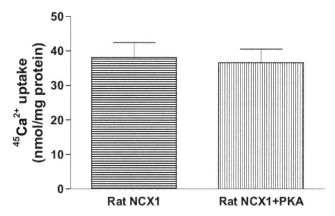

FIGURE 3. Effect of PKA phosphorylation on the cardiac NCX1 expressed in HEK293 cells. The activity of the expressed cardiac NCX1 was measured by $^{45}Ca^{2+}$ uptake by reverse mode of NCX in HEK293 cells transiently transfected with cDNA of cardiac isoform of NCX1. Incubating the cells with IBMX, dibutryl cAMP, and forskolin did not increase NCX1 activity in HEK293 cells (+PKA) as it was observed in *Xenopus* oocytes (see FIG. 1).

the Ni^{2+}-sensitive current. Moreover, removal of extracellular Ca^{2+} reduced the current substantially (FIG. 10). Taken together, the Ni^{2+}-sensitive current measured before and after the application of β-AR-stimulating isoproterenol represented mostly NCX1 current in pig cardiomyocytes. We also showed that there was attenuation of β-AR-stimulation effect on NCX1 current using tachycardia-induced heart failure model in pig and it was due to hyperphosphorylation of NCX1 protein in heart failure pig (FIG. 11).[33] The NCX1 activity was reduced in the ischemia-induced left ventricular dysfunction in rabbits even though the amount of NCX1 protein was higher than control suggesting that the reduced phosphorylation of NCX1 protein could be the underlying factor.[36]

Phosphorylation by Other Kinases

Regarding other kinases, calmodulin-kinase II (CaMKII) is shown to regulate LTCC, RYR, and SERCA-PLB in the heart.[37] NCX1 protein shows consensus sites for CamKII phosphorylation. It has been shown by pharmacological experiments that CamKII could modulate the function of NCX1.[38] The guinea pig ventricular myocytes showed indirect evidence of the tyrosine kinase activity on the NCX1. The tyrosine kinase inhibitor, tyrphostin A23, induced a current in guinea pig ventricular myocytes, which is sensitive to the application of Ni^{2+} and removal of Na^+ or Ca^{2+} in the external solution.[39] Recently,

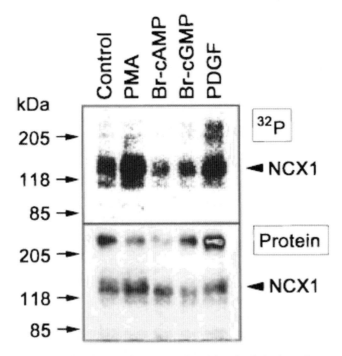

FIGURE 4. Phosphorylation of NCX1 induced by phorbol 12-myristate 13-acetate (PMA), 8-Br-cAMP, 8-Br-cGMP, and platelet-derived growth factor-BB (PDGF). The cloned dog cardiac NCX1 was expressed in CCL39 cells and the proteins were immuno-precipitated using NCX1 antibodies. The NCX1 protein was phosphorylated in the basal state and phosphorylation was increased by PMA and PDGF. (Modified from Iwamoto et al., 1998 [26].)

NCX1 appears to regulate the Ca^{2+} store in microglia and the NCX1 activity is influenced by tyrosine kinase.[40] In conclusion, PKA, PKC, CamKII, and tyrosine kinases are shown to regulate NCX1 but more experiments are needed to explain the conditions under which these kinases regulate the NCX1.

Conundrum 2: Is Block of a Single Transport "Mode" of NCX1 Possible?

Continuing reports on "block" of a single "mode" of NCX1 transport as a physiological and/or pharmacological mechanism are surprising.[41] The hypothesis central to this discussion is that a drug may only (or dominantly) affect one mode (e.g., the "reverse" [also called the "Ca^{2+} influx"] mode of NCX transport that produces net influx of Ca^{2+} on NCX1) without affecting the other mode (e.g., the "normal" or Ca^{2+}-efflux mode that produces net

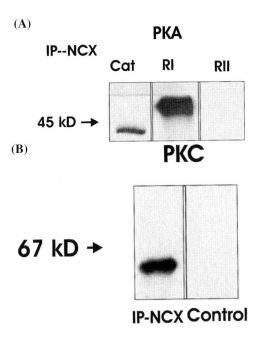

FIGURE 5. Identification of kinase components with NCX1. (**A**) The NCX1 protein immunoprecipitated from rat ventricle was immunoblotted with antibodies to PKA catalytic and regulatory subunits. The antibodies to PKA identified catalytic and RI subunits from the NCX1 immunoprecipitate. (**B**) The pan PKC antibodies showed PKC being associated with NCX1 protein and the omission of primary antibodies did not show any band. (Modified from Schulze *et al.*, 2003[17].)

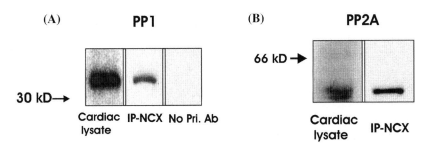

FIGURE 6. Characterization of phosphatases in the NCX1 immunoprecipitate. Immunoblotting of the NCX1 antibody-immunoprecipitated proteins with antibodies to PP1 (**A**) and PP2A showed that these phosphatases were immunoprecipitated by NCX1 antibody along with NCX1 protein (**B**). The presence of these phosphatases was confirmed in the cardiac lysates. (Modified from Schulze *et al.*, 2003[17].)

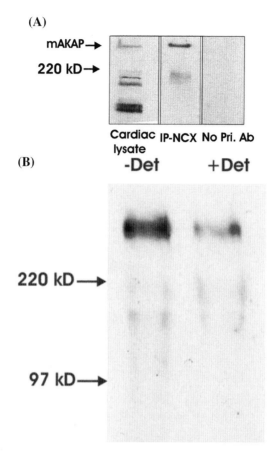

FIGURE 7. Immunoblotting to detect mAKAP in different isolation buffers. (**A**) Immunoblotting of the NCX1 –antibody-precipitated proteins with antibody to mAKAP showed that the scaffolding protein mAKAP was also associated with NCX1. (Modified from Schulze et al., 2003.[17]) (**B**) Using 0.1% Triton X-100 detergent in the extraction buffer (to dissolve the lipid membrane) only reduced the amount of mAKAP in the NCX1 antibody immunoprecipitate but mAKAP was still seen with NCX1 protein.

efflux of Ca^{2+} on NCX1). By this hypothesized mechanism, an NCX-specific drug could selectively stop net Ca^{2+} influx without affecting net Ca^{2+} efflux. However, this hypothesis conflicts with thermodynamic requirements. Consequently, whenever invoked, single-mode block must be carefully examined. Briefly, the reason one cannot block "one mode" without affecting the other is that both modes are just specific manifestations of the net overall transport of NCX. As an NCX protein transports Na^+ and Ca^{2+}, it undergoes a series of reactions that must be repeated for the next set of Na^+ and Ca^{2+} ions

NCX1 complex

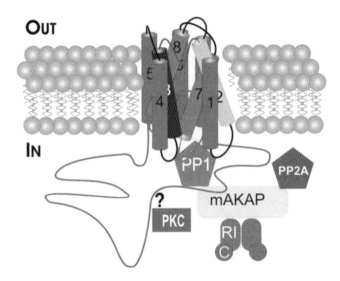

FIGURE 8. A cartoon of the macromolecular complex of NCX1 and associated proteins regulating the phosphorylation. All of the depicted proteins have been experimentally shown to be present in the NCX1 complex. Their positions are only suggested for discussion purposes. (In color in Annals online.)

to be transported by that protein. Each of the steps is, in principle, reversible. Blocking a rate-limiting step in the enzyme's transport sequence leads to a reduction of the turnover rate of the enzyme. This is associated with a redistribution of the populations of the enzyme in its various reaction states (in the transport reaction sequence). Thus "single-mode block" of NCX as described in the literature does not and cannot occur. (See review and discussion by Noble and Blaustein in this issue and a discussion of transport by NCX in a more comprehensive review[3].) It should be noted that isotopic flux, however, may be changed by drugs independent of the overall transport rate of NCX when the reaction involved is not the rate-limiting reaction. Such non-rate-limiting reactions do not need to reflect the overall function of the transport cycle and may reflect a subset of steps within the transport cycle.[3] While there can be a sidedness (inside, outside, or within the membrane) of drug action, the drug does not change the overall driving force for the transport reaction unless it leads to the change in stoichiometry. Even if that were to occur, the subsequent transport reactions must still follow the laws of thermodynamics.

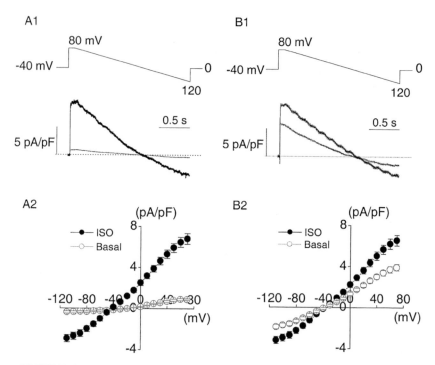

FIGURE 9. Isoproterenol treatment increased NCX current in pig cardiomyocytes. A1. A typical trace of Ni-sensitive NCX current from control pig cardiomyocytes was shown below the ramp voltage protocol used to record the current. A2. Averaged current in current–voltage relationship from 18 cells. B1. A representatively attenuated response of the heart failure cardiomyocyte to isoproterenol. B2. Averaged current from 37 cells from heart failure pigs. (Modified from Wei *et al.*, 2003[33] with permission.)

Conundrum 3. How Can NCX1 KO Mice Survive?

The observation that NCX1 knockout mice live is surprising.[42–44] This finding suggests that, in the extreme, there are compensatory mechanisms that can be activated to control cellular Ca^{2+} transport. These same transport mechanisms must be carefully assessed whenever NCX function is investigated. See other reports in this volume.

CONCLUSIONS

We have presented three problems that are central to investigations of Na^+/Ca^{2+} exchangers: (1) What are the actions, if any, of NCX phosphorylation on NCX function? The evidence suggests that the system is complicated

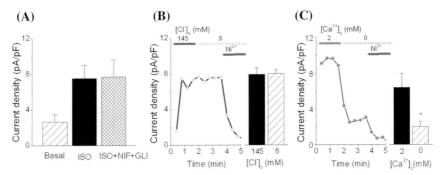

FIGURE 10. The NCX current measured from pig cardiomyocytes not blocked by inhibitors of CFTR or Cl^- currents and directly affected by external $[Ca^{2+}]$. (**A**) The average Ni-sensitive NCX current under basal condition, after isoproterenol treatment and with niflumic acid and glibenclamide. (**B**) Removal of external Cl^- did not influence the Ni-sensitive NCX current. (**C**) Removal of external Ca^{2+} reduced substantial outward current consistent with NCX current. $*P < 0.05$. (Modified from Wei *et al.*, 2006[45] with permission.) (In color in Annals online.)

and the overall findings are inconsistent and better experiments are still needed to address this question. (2) Can reverse mode and forward mode of NCX be inhibited independently of each other? There are theoretical and practical reasons why the answer is no. (3) How do NCX1 null mice survive? More experiments

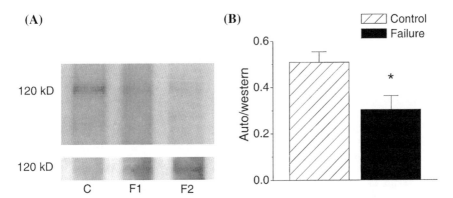

FIGURE 11. NCX protein is hyperphosphorylated in the heart failure. (**A**) Back phosphorylation of the immunoprecipitated NCX protein showed that the protein from control (C) pig heart could be significantly less phosphorylated at basal condition whereas the protein from heart failure (F1 and F2) showed a higher state of phosphorylation at basal state. (**B**) The amount of back phosphorylation of NCX protein in pig control cardiomyocytes is more when compared to NCX protein in the cardiomyocytes of heart failure pigs ($n = 5$). $*P < 0.05$. (Modified from Wei *et al.*, 2003[33] with permission.)

are needed to determine exactly what compensatory mechanisms are activated in this mouse model and how this occurs.

REFERENCES

1. REUTER, H. & N. SEITZ. 1968. The dependence of calcium efflux from cardiac muscle on temperature and external ion composition. J. Physiol. **195:** 451–470.

2. BAKER, P.F. & M.P. BLAUSTEIN. 1968. Sodium-dependent uptake of calcium by crab nerve. Biochim. Biophys. Acta **150:** 167–170.

3. BLAUSTEIN, M.P. & W.J. LEDERER. 1999. Sodium/calcium exchange: its physiological implications. Physiol. Rev. **79:** 763–854.

4. PHILIPSON, K.D., D.A. NICOLL, S. MATSUOKA, et al. 1996. Molecular regulation of the Na(+)-Ca^{2+} exchanger. Ann. N. Y. Acad. Sci. **779:** 20–28.

5. DIPOLO, R. & L. BEAUGE. 2006. Sodium/calcium exchanger: influence of metabolic regulation on ion carrier interactions. Physiol. Rev. **86:** 155–203.

6. NICOLL, D.A., S. LONGONI & K.D. PHILIPSON. 1990. Molecular cloning and functional expression of the cardiac sarcolemmal Na(+)-Ca^{2+} exchanger. Science **250:** 562–565.

7. HRYSHKO, L.V., S. MATSUOKA, D.A. NICOLL, et al. 1996. Anomalous regulation of the Drosophila Na(+)-Ca^{2+} exchanger by Ca^{2+}. J. Gen. Physiol. **108:** 67–74.

8. RUKNUDIN, A., C. VALDIVIA, P. KOFUJI, et al. 1997. Na^+/Ca^{2+} exchanger in Drosophila: cloning, expression, and transport differences. Am. J. Physiol. **273**(1 Pt 1): C257–C265.

9. EGGER, M., A. RUKNUDIN, P. LIPP, et al. 1999. Functional expression of the human cardiac Na^+/Ca^{2+} exchanger in Sf9 cells: rapid and specific Ni^{2+} transport. Cell Calcium **25:** 9–17.

10. KOFUJI, P., R.W. HADLEY, R.S. KIEVAL, et al. 1992. Expression of the Na-Ca exchanger in diverse tissues: a study using the cloned human cardiac Na-Ca exchanger. Am. J. Physiol. **263**(6 Pt 1): C1241–C1249.

11. REILLY, R.F. & C.A. SHUGRUE. 1992. cDNA cloning of a renal Na(+)-Ca^{2+} exchanger. Am. J. Physiol. **262**(6 Pt 2): F1105–F1109.

12. RUKNUDIN, A. & D.H. SCHULZE. 2002. Phosphorylation of the Na^+/Ca^{2+} exchangers by PKA. Ann. N. Y. Acad. Sci. **976:** 209–213.

13. KOFUJI, P., W.J. LEDERER & D.H. SCHULZE. 1994. Mutually exclusive and cassette exons underlie alternatively spliced isoforms of the Na/Ca exchanger. J. Biol. Chem. **269:** 5145–5149.

14. SCHULZE, D.H., P. KOFUJI, C. VALDIVIA, et al. 1996. Alternative splicing of the Na(+)-Ca^{2+} exchanger gene, NCX1. Ann. N. Y. Acad. Sci. **779:** 46–57.

15. HE, S., A. RUKNUDIN, L.L. BAMBRICK, et al. 1998. Isoform-specific regulation of the Na^+/Ca^{2+} exchanger in rat astrocytes and neurons by PKA. J. Neurosci. **18:** 4833–4841.

16. RUKNUDIN, A., S. HE, W.J. LEDERER & D.H. SCHULZE. 2000. Functional differences between cardiac and renal isoforms of the rat Na^+-Ca^{2+} exchanger NCX1 expressed in *Xenopus* oocytes. J. Physiol. **529**(Pt 3): 599–610.

17. SCHULZE, D.H., M. MUQHAL, W.J. LEDERER & A.M. RUKNUDIN. 2003. Sodium/calcium exchanger (NCX1) macromolecular complex. J. Biol. Chem. **278:** 28849–28855.

18. CONDRESCU, M., J.P. GARDNER, G. CHERNAYA, *et al.* 1995. ATP-dependent regulation of sodium-calcium exchange in Chinese hamster ovary cells transfected with the bovine cardiac sodium-calcium exchanger. J. Biol. Chem. **270:** 9137–9146.

19. HILGEMANN, D.W., A. COLLINS & S. MATSUOKA. 1992. Steady-state and dynamic properties of cardiac sodium-calcium exchange. Secondary modulation by cytoplasmic calcium and ATP. J. Gen. Physiol. **100:** 933–961.

20. HILGEMANN, D.W., S. MATSUOKA, G.A. NAGEL & A. COLLINS. 1992. Steady-state and dynamic properties of cardiac sodium-calcium exchange. Sodium-dependent inactivation. J. Gen. Physiol. **100:** 905–932.

21. MATSUOKA, S. & D.W. HILGEMANN. 1992. Steady-state and dynamic properties of cardiac sodium-calcium exchange. Ion and voltage dependencies of the transport cycle. J. Gen. Physiol. **100:** 963–1001.

22. PERCHENET, L., A.K. HINDE, K.C. PATEL, *et al.* 2000. Stimulation of Na/Ca exchange by the beta-adrenergic/protein kinase A pathway in guinea-pig ventricular myocytes at 37 degrees C. Pflugers Arch. **439:** 822–828.

23. LINCK, B., Z. QIU, Z. HE, *et al.* 1998. Functional comparison of the three isoforms of the Na^+/Ca^{2+} exchanger (NCX1, NCX2, NCX3). Am. J. Physiol. **274** (2 Pt 1): C415–C423.

24. IWAMOTO, T., Y. PAN, S. WAKABAYASHI, *et al.* 1996. Phosphorylation-dependent regulation of cardiac Na^+/Ca^{2+} exchanger via protein kinase C. J. Biol. Chem. **271:** 13609–13615.

25. IWAMOTO, T., S. WAKABAYASHI & M. SHIGEKAWA. 1995. Growth factor-induced phosphorylation and activation of aortic smooth muscle Na/Ca exchanger. J. Biol. Chem. **270:** 8996–9001.

26. IWAMOTO, T., Y. PAN, T.Y. NAKAMURA, *et al.* 1998. Protein kinase C-dependent regulation of Na^+/Ca^{2+} exchanger isoforms NCX1 and NCX3 does not require their direct phosphorylation. Biochemistry **37:** 17230–17238.

27. DAVARE, M.A., V. AVDONIN, D.D. HALL, *et al.* 2001. A beta2 adrenergic receptor signaling complex assembled with the Ca^{2+} channel Cav1.2 1. Science **293:** 98–101.

28. MARX, S.O., S. REIKEN, Y. HISAMATSU, *et al.* 2000. PKA phosphorylation dissociates FKBP12.6 from the calcium release channel (ryanodine receptor): defective regulation in failing hearts. Cell **101:** 365–376.

29. WALKER, J.M., Ed. 2002. *The Protein Protocols Handbook.* 2nd ed. Humana Press. Totowa, NJ.

30. KATANOSAKA, Y., Y. IWATA, Y. KOBAYASHI, *et al.* 2004. Calcineurin inhibits Na^+/Ca^{2+} exchange in phenylephrine-treated hypertrophic cardiomyocytes. J. Biol. Chem. **280:** 5764–5772.

31. BERS, D.M. 2001. Excitation-Contraction Coupling and Cardiac Force, 2nd ed. Kluwer Academic Publishers. Dordrecht, the Netherlands.

32. BERS, D.M., J.W. BASSANI & R.A. BASSANI. 1996. Na-Ca exchange and Ca fluxes during contraction and relaxation in mammalian ventricular muscle. Ann. N. Y. Acad. Sci. **779:** 430–442.

33. WEI, S.K., A. RUKNUDIN, S.U. HANLON, *et al.* 2003. Protein kinase A hyperphosphorylation increases basal current but decreases beta-adrenergic responsiveness of the sarcolemmal Na^+-Ca^{2+} exchanger in failing pig myocytes. Circ. Res. **92:** 897–903.

34. GINSBURG, K.S. & D.M. BERS. 2005. Isoproterenol does not enhance Ca-dependent Na/Ca exchange current in intact rabbit ventricular myocytes. J. Mol. Cell. Cardiol. **39**(6): 972–981.

35. LIN, X., H. JO, Y. SAKAKIBARA, *et al.* 2006. Beta-adrenergic stimulation does not activate Na^+/Ca^{2+} exchange current in guinea pig, mouse, and rat ventricular myocytes. Am. J. Physiol. Cell. Physiol. **290:** C601–C608.
36. QUINN, F.R., S. CURRIE, A.M. DUNCAN, *et al.* 2003. Myocardial infarction causes increased expression but decreased activity of the myocardial Na^+-Ca^{2+} exchanger in the rabbit. J. Physiol. **553**(Pt 1): 229–242.
37. BERS, D.M.. 2002. Cardiac excitation-contraction coupling 6. Nature **415:** 198–205.
38. ISOSAKI, M., N. MINAMI & T. NAKASHIMA. 1994. Pharmacological evidence for regulation of $Na(+)$-Ca^{++} exchange by Ca^{++}/calmodulin-dependent protein kinase in isolated bovine adrenal medullary cells. J. Pharmacol. Exp. Ther. **270:** 104–110.
39. MISSAN, S. & T.F. MCDONALD. 2004. Cardiac Na^+-Ca^{2+} exchanger current induced by tyrphostin tyrosine kinase inhibitors. Br. J. Pharmacol. **143:** 943–951.
40. MATSUDA, T., T. NAGANO, M. TAKEMURA & A. BABA. 2006. Topics on the $Na(+)/Ca(2+)$ exchanger: responses of $Na(+)/Ca(2+)$ exchanger to interferon-gamma and nitric oxide in cultured microglia. J. Pharmacol. Sci. **102:** 22–26.
41. AMRAN, M.S., N. HOMMA & K. HASHIMOTO. 2003. Pharmacology of KB-R7943: a Na^+-Ca^{2+} exchange inhibitor. Cardiovasc. Drug Rev. **21:** 255–276.
42. HENDERSON, S.A., J.I. GOLDHABER, J.M. SO, *et al.* 2004. Functional adult myocardium in the absence of Na^+-Ca^{2+} exchange: cardiac-specific knockout of NCX1. Circ. Res. **95:** 604–611.
43. POTT, C., X. REN & D.X. TRAN, *et al.* 2006. Mechanism of shortened action potential duration in Na^+-Ca^{2+} exchanger knockout mice. Am. J. Physiol. Cell. Physiol. doi:10.1152/AJPCELL.00177.2006-EPUB.
44. POTT, C., K.D. PHILIPSON & J.I. GOLDHABER. 2005. Excitation-contraction coupling in Na^+-Ca^{2+} exchanger knockout mice: reduced transsarcolemmal Ca^{2+} flux. Circ. Res. **97:** 1288–1295.
45. WEI, S.K., A. RUKNUDIN, M. SHOU, J. M. MC CURLEY, S.V. HANLON, E. ELGIN, D.H. SCHULZE, and M.C. HAIGNEY. 2006. Muscarinic modulation of the Sodium–Calcium exchanger in heart failure. Circulation. In press.

Regulation of Cardiac Na^+/Ca^{2+} Exchanger by Phospholemman

JOSEPH Y. CHEUNG,[a] LAWRENCE I. ROTHBLUM,[b]
J. RANDALL MOORMAN,[c] AMY L. TUCKER,[c] JIANLIANG SONG,[a]
BELINDA A. AHLERS,[a] LOIS L. CARL,[a] JuFANG WANG,[a]
AND XUE-QIAN ZHANG[a]

[a]*Department of Cellular and Molecular Physiology, Pennsylvania State
University College of Medicine, Hershey, Pennsylvania 17033, USA*

[b]*Weis Center for Research, Geisinger Clinic, Danville, Pennsylvania 17822, USA*

[c]*Department of Internal Medicine (Cardiovascular Division), University of
Virginia Health Sciences Center, Charlottesville, Virginia 22908, USA*

ABSTRACT: Phospholemman (PLM) is the first sequenced member of
the FXYD family of regulators of ion transport. The mature protein
has 72 amino acids and consists of an extracellular N terminus containing the signature FXYD motif, a single transmembrane (TM) domain, and a cytoplasmic C-terminal domain containing four potential
sites for phosphorylation. PLM and other members of the FXYD family are known to regulate Na^+-K^+-ATPase. Using adenovirus-mediated
gene transfer into adult rat cardiac myocytes, we showed that changes in
contractility and intracellular Ca^{2+} homeostasis associated with PLM
overexpression or downregulation are not consistent with the effects expected from inhibition of Na^+-K^+-ATPase by PLM. Additional studies
with heterologous expression of PLM and cardiac Na^+/Ca^{2+} exchanger
1 (NCX1) in HEK293 cells and cardiac myocytes isolated from PLM-
deficient mice demonstrated by co-localization, co-immunoprecipitation,
and electrophysiological and radioactive tracer uptake techniques that
PLM associates with NCX1 in the sarcolemma and transverse tubules
and that PLM inhibits NCX1, independent of its effects on Na^+-K^+-
ATPase. Mutational analysis indicates that the cytoplasmic domain of
PLM is required for its regulation of NCX1. In addition, experiments
using phosphomimetic and phospho-deficient PLM mutants, as well as
activators of protein kinases A and C, indicate that PLM phosphorylated at serine68 is the active form that inhibits NCX1. This is in sharp
contrast to the finding that the unphosphorylated PLM form inhibits
Na^+-K^+-ATPase. We conclude that PLM regulates cardiac contractility
by modulating the activities of NCX and Na^+-K^+-ATPase.

Address for correspondence: Joseph Y. Cheung, M.D., Ph.D., Capizzi Professor of Medicine,
Director, Division of Nephrology Thomas Jefferson University 833 Chestnut Street, Suite 700 Philadel-
phia, PA 19107. Voice: 215-503-6830; fax: 215-503-4099.
  joseph.cheung@jefferson.edu

Ann. N.Y. Acad. Sci. 1099: 119–134 (2007). © 2007 New York Academy of Sciences.
doi: 10.1196/annals.1387.004

KEYWORDS: Na$^+$/Ca^{2+} exchanger; FXYD proteins; phospholemman; cardiac myocytes; phosphorylation; Na$^+$-K$^+$-ATPase; ischemic cardiomyopathy

INTRODUCTION

Phospholemman (PLM), a 72-amino acid integral membrane phosphoprotein with a single transmembrane (TM) domain, is a major substrate for protein kinases A and C in heart and skeletal muscle.[1] It belongs to the FXYD gene family of small membrane protein regulator of ion transport.[2] Early work based on PLM overexpression in *Xenopus oocytes* suggests that PLM is a hyperpolarization-activated anion-selective channel.[3] When reconstituted in lipid bilayers, PLM forms a channel that is highly selective for taurine[4] and is thought to be involved in regulation of cell volume in noncardiac tissues.[5]

In 2000, Sweadner and Rael[2] described a new gene family of regulators of ion transport. There are seven known members of this FXYD family, plus a shark homolog of PLM termed *PLMS*. FXYD proteins have an extracellular N terminus containing the signature FXYD motif, a single membrane-spanning segment, and a cytoplasmic C terminus. As a family, FXYD proteins are found predominantly in tissues involved in solute and fluid transport (kidney, colon, mammary gland, pancreas, liver, lung, prostate, and placenta) or are electrically excitable (heart, skeletal, and neural tissues). With the exception of the γ-subunit of Na$^+$-K$^+$-ATPase (FXYD2), all other known members of the FXYD gene family have at least one serine or threonine within the cytoplasmic tail, indicating potential phosphorylation sites. In particular, PLM (FXYD1) is the only FXYD family member to have a consensus sequence for phosphorylation by protein kinase (PK) A (RRXS) and PKC (RXXSXR), and "never in mitosis" A kinase (FRXS/T). Physiologically, PLM is phosphorylated by PKA at serine68 and PKC at both serine63 and serine68. Phosphorylation of PLM in response to β-adrenergic stimulation parallels the positive inotropic effects.[6,7] In addition, PLM shares sequence similarity with phospholamban (PLB), a small protein in the sarcoplasmic reticulum (SR) membrane that regulates sarco(endo)plasmic reticulum Ca^{2+}-ATPase (SERCA2) transport activity (RSSIRRLST69 in PLM and RSAIRRAST17 in PLB). When phosphorylated by PKA at serine16, PLB dissociates from SERCA2 and thereby removes its inhibitory effects on Ca^{2+} transport.[8]

PLM and other members of the FXYD gene family including FXYD2,[9] channel inducing factor (CHIF; FXYD4),[10] FXYD7,[11] and PLMS,[12] associate with and regulate the activity of the α-subunits of Na$^+$-K$^+$-ATPase. When coexpressed with α- and β-subunits of Na$^+$-K$^+$-ATPase in *Xenopus oocytes*, PLM modulates Na$^+$-K$^+$-ATPase activity, primarily by decreasing K$_m$ for Na$^+$ and K$^+$ without affecting V$_{max}$.[13] In the heart, PLM co-immunoprecipitates with α-subunits of Na$^+$-K$^+$-ATPase[13,14] and regulates its activity either by modulating

K_m for Na^{+15} or V_{max}.[14] Mutational analysis suggests that FXYD proteins (FXYD2, 4, and 7) interact with TM segment 9 of the Na^+-K^+-ATPase.[16] Co-immunoprecipitation and covalent cross-linking studies demonstrate the TM segment of PLM is close to TM2 segment of Na^+-K^+-ATPase.[17] Molecular modeling based on the Ca^{2+}-ATPase crystal structure in the E_1ATP-bound conformation, suggests that the single TM segment of FXYD proteins docks into the groove between TM segments 2, 6, and 9 of the α-subunit of the Na^+-K^+-ATPase.[17] Phosphorylation of PLM[15,18] and PLMS[12] by PKA relieves its inhibition of Na^+-K^+-ATPase. Phosphorylated PLMS dissociates from shark Na^+-K^+-ATPase.[12] Whether phosphorylation of PLM or other FXYD proteins causes them to dissociate from the α-subunit of Na^+-K^+-ATPase is not clear.[18]

In the heart, inhibition of Na^+-K^+-ATPase activity by PLM is expected to raise intracellular Na^+ concentration ($[Na^+]_i$), thereby decreasing the thermodynamic driving force for forward Na^+/Ca^{2+} exchange (Ca^{2+} efflux) and increasing the driving force for reverse Na^+/Ca^{2+} exchange (Ca^{2+} influx). Both these actions increase intracellular Ca^{2+} concentration ($[Ca^{2+}]_i$) and result in enhanced cardiac contractility. Indeed, inhibition of Na^+-K^+-ATPase with secondary effects on the cardiac Na^+/Ca^{2+} exchanger 1 (NCX1) has long been proposed to be the mechanism of action of positive inotropy of digitalis glycosides.[19]

Effects of Manipulation of PLM Levels on Myocyte Contractility and [Ca²⁺]ᵢ Homeostasis

Overexpression of PLM by adenovirus-mediated gene transfer results in 1.4- to 3.5-fold increase in PLM protein levels in cultured adult rat myocytes after 72 hours.[20,21] Importantly, protein levels of NCX1, SERCA2, calsequestrin, and $\alpha 1$- and $\alpha 2$-subunits of Na^+-K^+-ATPase remain unchanged in PLM-overexpressed myocytes.[14,20–22] Contractility measurements show that at 0.6 mM extracellular Ca^{2+} concentration ($[Ca^{2+}]_o$), PLM-overexpressed myocytes contracted more than control myocytes infected with adenovirus-expressing green fluorescent protein.[20] By contrast, at 1.8 and 5.0 mM $[Ca^{2+}]_o$, PLM-overexpressed myocytes shortened less than control myocytes[20] (FIG. 1). The amplitudes of $[Ca^{2+}]_i$ transients in PLM-overexpressed myocytes mirror those of contraction amplitudes, that is, higher at 0.6 but lower at 1.8 and 5.0 mM $[Ca^{2+}]_o$.[20] The pattern of reduced dynamic range of contractility in response to increasing $[Ca^{2+}]_o$ observed in PLM-overexpressed myocytes is inconsistent with the hypothesis that PLM exerts its effects on cardiac contractility by inhibiting Na^+-K^+-ATPase since enhanced contractility would be expected at all $[Ca^{2+}]_o$.[19] Rather, the contractile phenotype of PLM-overexpressed myocytes is reminiscent of that observed in myocytes in which NCX1 is downregulated.[23] Indeed, co-overexpression of PLM and NCX1 in

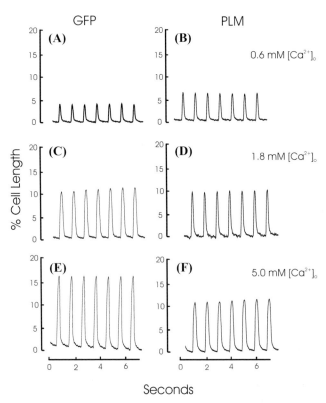

FIGURE 1. PLM overexpression alters contractile function in adult rat myocytes. Isolated myocytes are infected with recombinant adenovirus expressing either green fluorescent protein (GFP) or both GFP and PLM and then cultured for 72 h. For contraction studies, cultured myocytes are paced (1 Hz) to contract at $37°C$ and $[Ca^{2+}]o$ of 0.6, 1.8, and 5.0 mM. Shown are steady-state twitches from myocytes expressing either GFP (**A**, **C**, **E**) or PLM (**B**, **D**, **F**).

adult rat myocytes rescues the contractile abnormalities observed in myocytes overexpressing PLM alone.[21]

Downregulation of PLM by adenovirus-mediated antisense transfer in adult rat myocytes results in ~23% reduction in PLM after 72 h.[24] Expression of NCX1, SERCA2, α-subunit of Na^+-K^+-ATPase, and calsequestrin is not affected.[24] Contractility and $[Ca^{2+}]_i$ transient measurements demonstrate decreased amplitudes at 0.6 mM $[Ca^{2+}]_o$ but increased amplitudes at 5.0 mM $[Ca^{2+}]_o$ in PLM-downregulated myocytes.[24] Theoretically, decreased Na^+-K^+-ATPase inhibition due to lower levels of PLM would be expected to reduce contractility in PLM-downregulated myocytes. This is not observed, especially at high $[Ca^{2+}]_o$. The increased dynamic range of contractility in response to increasing $[Ca^{2+}]_o$ in PLM-downregulated myocytes, however, is similar to that observed in NCX1-overexpressed myocytes.[25]

A mouse line deficient in PLM is generated by replacing exons 3 to 5 of the PLM gene with LacZ and neomycin resistance genes[26] and backcrossed to a pure congenic C57BL/6 background. Absence of PLM does not affect protein levels of NCX1, SERCA2, calsequestrin, and α1-subunit of Na^+-K^+-ATPase.[27] Compared to their wild-type littermates, left ventricular myocytes isolated from PLM-null mice shorten less at 0.6 but more at 5.0 mM $[Ca^{2+}]_o$.[27] In PLM-null myocytes, observed changes in $[Ca^{2+}]_i$ transient amplitudes mirror those of contraction amplitudes.[27]

Phenotypically, overexpressing PLM in adult cardiac myocytes results in reduced dynamic range in contractility and $[Ca^{2+}]_i$ transients in response to increasing $[Ca^{2+}]_o$. Downregulating or knocking out PLM produces changes in the opposite direction. The observed phenotypic changes associated with PLM overexpression or downregulation/knockout cannot be accommodated by the hypothesis that PLM affects contractility by inhibiting Na^+-K^+-ATPase since enhanced contractility and $[Ca^{2+}]_i$ transient amplitudes would be expected at all $[Ca^{2+}]_o$.

Regulation of NCX1 activity by PLM can theoretically explain the differential effects of low versus high $[Ca^{2+}]_o$ on $[Ca^{2+}]_i$ transient and contraction amplitudes in PLM overexpressing[20] and downregulated[24] or knockout[27] myocytes. During diastole, the primary function of NCX1 is to extrude Ca^{2+}. This is consistent with our observations that overexpressing[25] or downregulating[23] NCX1 results in lower or higher diastolic $[Ca^{2+}]_i$ levels, respectively. Upon depolarization, when the membrane potential (E_m) exceeds the equilibrium potential of NCX1 (E_{NaCa}), Ca^{2+} influx is thermodynamically favored, although the amount and duration of Ca^{2+} influx via reverse Na^+/Ca^{2+} exchange during systole are at present unclear. At high $[Ca^{2+}]_o$, by increasing NCX1 activity with PLM downregulation or knockout without affecting other Ca^{2+} transport pathways, more Ca^{2+} enters via reverse NCX1 during systole. Increased Ca^{2+} influx results in the observed higher SR Ca^{2+} contents and larger $[Ca^{2+}]_i$ transient and contraction amplitudes in PLM knockout compared with wild-type myocytes.[27] The higher SR Ca^{2+} content leads to a corresponding increase in Ca^{2+} spark frequency, resulting in increased SR Ca^{2+} leak, which limits the otherwise inexorable increase in SR Ca^{2+} content. The increased SR Ca^{2+} leak, together with the Ca^{2+} that has entered via L-type Ca^{2+} channels, is extruded by NCX1 during diastole. In this way, the PLM-knockout myocyte stimulated at 5 mM $[Ca^{2+}]_o$ reaches a higher steady-state SR Ca^{2+} content compared to wild-type myocyte and maintains beat-to-beat Ca^{2+} balance. At 0.6 mM $[Ca^{2+}]_o$, the driving force for Ca^{2+} influx is much reduced during systole, but increased NCX1 activity (by PLM downregulation or knockout) will initially pump more Ca^{2+} out during diastole, thus partially depleting SR Ca^{2+} content until a new lower steady-state SR Ca^{2+} content is reached. Indeed, compared to control myocytes, PLM-downregulated or knockout myocytes have lower SR Ca^{2+} contents and $[Ca^{2+}]_i$ transient and contraction amplitudes.[24,27] The same paradigm can be applied to explain our observations on PLM-overexpressed myocytes. It is the unique ability of NCX1 to drive Ca^{2+} in and pump Ca^{2+} out during an excitation–contraction cycle that explains the differential effects of low versus high $[Ca^{2+}]_o$ on $[Ca^{2+}]_i$ transient and contraction amplitudes in myocytes in which NCX1 amounts[23,25] or activity[20,24,27] are altered.

PLM Co-Localizes with and Co-Immunoprecipitates Cardiac NCX

In native adult rat cardiac myocytes, PLM co-localizes with NCX1 to the sarcolemma and transverse tubules.[21] In solubilized cardiac sarcolemma, anti-PLM antibodies immunoprecipitate both PLM and NCX1[22,24,28] (FIG. 2). In HEK293 cells that are devoid of endogenous NCX1 and PLM and are electrically silent,[29] coexpression of exogenous PLM and NCX1 results in co-localization of these proteins to the plasma membrane.[28] Moreover, co-immunoprecipitation experiments demonstrate association of PLM with NCX1 in transfected HEK293 cells.[28]

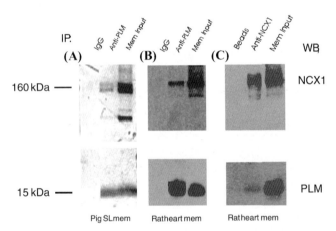

FIGURE 2. Association of PLM with NCX1 in native cardiac membranes. *Panel A.* Immunoprecipitates (IP) from 600 μg of solubilized pig sarcolemmal (SL) vesicles using μl of anti-PLM antibody or control IgG. NCX1 and PLM are identified by immunoblotting with R3F1 and anti-PLM antibodies, respectively. Solubilized SL vesicles (Mem Input) are derived from pig hearts expressing native levels of PLM and NCX1. *Panel B.* Immunoprecipitates from 2 mg of solubilized crude membrane preparations (Mem Input) from rat hearts using 5 μg of anti-PLM antibody or control IgG. NCX1 and PLM are identified by immunoblotting with R3F1 and anti-PLM antibodies, respectively. *Panel C.* Immunoprecipitates from 2 mg of solubilized crude membrane preparations (Mem Input) from rat hearts using 5 μl of anti-NCX1 antibody (R3F1). Control is using beads alone. NCX1 and PLM are identified by immunoblotting with R3F1 and anti-PLM antibodies, respectively. The antibodies used for immunoblots are indicated on the right and molecular mass markers (in kDa) are shown on the left.

PLM Regulates NCX1 Activity Independent of Its Effects on Na^+-K^+-ATPase

Cotransfection of PLM with NCX1 in HEK293 cells results in suppression of NCX1 current (I_{NaCa}) when compared to cells transfected with NCX1 alone.[28] The conditions chosen to measure I_{NaCa} are carefully designed to minimize contamination by L-type Ca^{2+} currents (verapamil), Na^+-K^+-ATPase currents (ouabain and absence of K^+), K^+ currents (absence of K^+ in extracellular and pipette solutions), and Cl^- currents (niflumic acid). In addition, E_m is held at the calculated E_{NaCa} for 5 min before onset of the voltage ramp. This precaution minimizes fluxes through NCX1 before the voltage ramp and thus allows $[Ca^{2+}]_i$ and $[Na^+]_i$ to equilibrate with those present in the pipette solution. FIGURE 3 shows that I_{NaCa} is undetectable in cells transfected with control

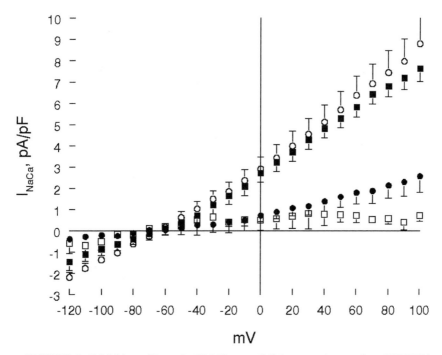

FIGURE 3. Inhibition of I_{NaCa} by PLM but not S68A mutant in transfected HEK293 cells. HEK293 cells are transfected with either pAdTrack (*open squares, n* = 3), pAdTrack$^+$NCX1 (*open circles, n* =11), PLM$^+$NCX1 (*filled circles, n* = 8), or S68A$^+$NCX1 (*filled squares, n* = 7). At 48 h post transfection, I_{NaCa} is measured at 5 mM $[Ca^{2+}]_o$ and 30°C with a descending–ascending voltage ramp protocol. Free $[Ca^{2+}]$ in the Ca^{2+} buffered pipette solution is 205 nM. Holding potential is at the calculated equilibrium potential of I_{NaCa} (−73 mV) under our experimental conditions. Ca^{2+}, Na^+-K^+-ATPase, Cl^-, and K^+ currents are blocked by appropriate inhibitors. Error bars are not shown if they fall within the boundaries of the symbols.

vector alone. In cells transfected with both NCX1 and PLM, I_{NaCa} is lower when compared to cells transfected with NCX1 alone. An important point is that PLM coexpression does not affect NCX1 protein levels.[28] In addition, the measured E_{NaCa} in cells transfected with NCX1 and PLM and in those transfected with NCX1 alone are similar at $\sim$ -60 mV, close to the theoretical E_{NaCa} of -73 mV. This observation suggests that the thermodynamic parameters ($[Ca^{2+}]_i$, $[Na^+]_i$, $[Ca^{2+}]_o$, and $[Na^+]_o$) that determine E_{NaCa}, and hence the driving force for I_{NaCa} (E_m-E_{NaCa}), are identical between cells transfected with NCX1 alone and those transfected with NCX1 and PLM. Thus the difference in I_{NaCa} between cells transfected with NCX1 alone and those transfected with NCX1 and PLM can be unambiguously assigned to the presence of PLM.

In adult rat cardiac myocytes, PLM overexpression by adenovirus-mediated gene transfer results in suppression of I_{NaCa}.[21,22] By contrast, downregulation of PLM by adenovirus-mediated antisense transfer results in significant increases in I_{NaCa}.[24] In addition, in cardiac myocytes isolated from adult PLM-deficient mouse, I_{NaCa} is significantly higher than that measured in myocytes isolated from wild-type littermates.[27]

Using a fundamentally different approach to measure NCX1 activity, we also demonstrate that Na^+-dependent $^{45}Ca^{2+}$ uptake is lower in HEK293 cells co-expressing PLM and NCX1 when compared to cells expressing NCX1 alone[28] (FIG. 4). Therefore, the results from three different model systems (adenovirus-mediated gene transfer, heterologous expression, and gene knockout) using two independent measurements of NCX1 activity (electrophysiology and radioactive tracer uptake) unequivocally demonstrate that inhibition of NCX1 by PLM is independent of its effects on Na^+-K^+-ATPase.

Endogenous PLM Is Partially Phosphorylated in Cardiac Myocytes

Incorporation of ^{32}P into PLM in intact guinea pig ventricles is enhanced $\sim$2.6-fold with isoproterenol treatment, suggesting wild-type PLM is partially phosphorylated in the unstimulated state.[7] Based on CP68 and C2 Ab, which are antibodies specific for phosphorylated (at serine68) and unphosphorylated PLM, respectively,[14,30] it has been estimated that $\sim$41% of PLM in adult rat myocytes[14] and $\sim$25% of PLM in guinea pig myocytes[18] are phosphorylated at serine68 under the basal state. Using another approach of comparing the effects of wild-type PLM and its serine68 and serine63 mutants on I_{NaCa} in adult rat myocytes, $\sim$46% of serine68 and $\sim$16% of serine63 are estimated to be phosphorylated in the resting state.[22] The results from these three fundamentally different experimental approaches strongly indicate that PLM is only partially phosphorylated in cardiac myocytes. In addition, overexpression of PLM does not grossly distort the relative level of serine68 phosphorylation in adult rat cardiac myocytes.[14]

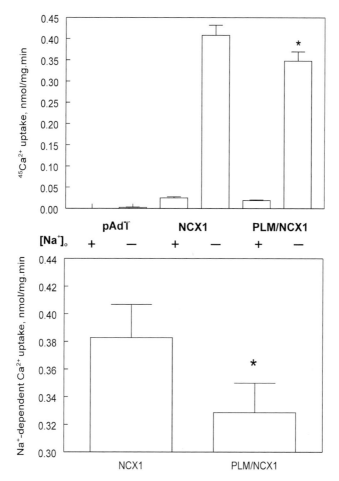

FIGURE 4. Inhibition of Na$^+$-dependent Ca^{2+} uptake by PLM in transfected HEK293 cells. Top: HEK293 cells transfected for 48 h with either pAdTrack (pAdT), pAdT+NCX1 (NCX1), or NCX1+PLM (PLM/NCX1) are loaded with Na$^+$ as described.[28] ^{45}Ca^{2+} uptake into Na$^+$-loaded cells is measured during the initial 30 s in normal balanced salt solution (+[Na$^+$]$_o$) or Na$^+$-free solution (−[Na$^+$]$_o$), both containing 1 mM ouabain. *$P < 0.035$, NCX1 (−[Na$^+$]$_o$) versus PLM/NCX1 (−[Na$^+$]$_o$). Bottom. Na$^+$-dependent ^{45}Ca^{2+} uptake is calculated by subtracting ^{45}Ca^{2+} uptake values obtained in the presence of Na$^+$ from those obtained in the absence of Na$^+$. *$P < 0.032$, NCX1 versus PLM/NCX1. Data represent the mean ± SE of four independent experiments.

PLM Phosphorylated at Serine68 Is the Active Form that Inhibits NCX1

In HEK293 cells expressing both NCX1 and PLM, forskolin treatment results in additional suppression of I$_{NaCa}$.[31] This observation suggests that PLM, when phosphorylated at serine68, inhibits cardiac Na$^+$/Ca^{2+} exchange in a heterologous expression system. The importance of phosphorylated serine68 in mediating the inhibition of I_{NaCa} is supported by the experimental results with serine68 mutants. Serine68-to-alanine (S68A) mutant, which cannot be phosphorylated, is ineffective in inhibiting I_{NaCa} in both transfected HEK293 cells[28,31] (FIG. 3) and rat cardiac myocytes.[22] Serine68-to-glutamic acid (S68E) mutant, which mimics 100% phosphorylation, results in additional suppression of I_{NaCa} when compared to wild-type PLM both in transfected HEK293 cells[31] and in adult rat myocytes.[22] These results, when taken together, suggest that PLM phosphorylated at serine68 is the active form that inhibits NCX1.

Activation of PKC by phorbol 12-myristate 13-acetate (PMA) results in a large increase in I_{NaCa} in HEK293 cells transfected with NCX1 alone.[31] In HEK293 cells coexpressing NCX1 and PLM, PMA treatment also results in enhancement of I_{NaCa}. The magnitude of I_{NaCa} increase, however, is much smaller in cells coexpressing NCX1 and PLM when compared with cells expressing NCX1 alone.[31] These results suggest that the stimulatory effects of PMA on NCX1 are attenuated by increased PLM phosphorylation at serine68 and serine63. The importance of serine63 in mediating inhibition of NCX1 by PLM is examined using a serine63-to-alanine (S63A) mutant. Whereas S68A mutant abolishes the inhibitory effects of PLM on NCX1, in HEK293 cells coexpressing S63A mutant and NCX1, forskolin treatment results in large reduction of I_{NaCa}, testifying to the primacy of serine68 phosphorylation in mediating the inhibitory effects of PLM on I_{NaCa}.[31]

In cardiac myocytes isolated from PLM-null mice, PMA treatment results in much larger increase in I_{NaCa} ($\sim$132%) when compared with that measured in wild-type myocytes ($\sim$91%).[31] Thus in intact cardiac myocytes, the direct stimulatory effects of PKC activators on NCX1 are somewhat opposed by an indirect inhibitory effect by increased phosphorylated PLM. The results from transfected HEK293 cells and PLM-null myocytes are consistent with the notion that phosphorylated PLM, specifically PLM phosphorylated at serine68, inhibits NCX1.

Effects of PLM on Cardiac Contractility Are Dependent on Its Ability to Inhibit NCX1

The S68A mutant of PLM (which cannot be phosphorylated) has no effects on I_{NaCa}, $[Ca^{2+}]_i$ transient and contractility amplitudes when introduced into adult rat myocytes.[22] By contrast, wild-type PLM, S68E (phosphomimetic), or S63A (intact serine68) mutants, inhibit I_{NaCa} and alter $[Ca^{2+}]_i$ transient and contractility amplitudes when compared to control myocytes.[22] These observations suggest that the ability of PLM to affect cardiac contractility is associated with its ability to regulate NCX1 function.

Inhibition of NCX1 by PLM Requires Interaction between Cytoplasmic Tail of PLM and Intracellular Loop of NCX1

The cardiac NCX consists of a N-terminal domain comprising the first five TM segments, a large cytoplasmic loop (residues 218–764), and a C-terminal domain consisting of the last four TM segments.[32] The single TM segment of PLM can theoretically interact with the TM domains of NCX1, similar to interaction between FXYD proteins and Na^+-K^+-ATPase.[17] Alternatively, in view of the importance of serine68 in regulation of NCX1 by PLM,[22,31] the cytoplasmic tail of PLM can interact with the large intracellular loop of NCX1. Two cytoplasmic tail truncation PLM mutants, TM43 (terminal 29 residues deleted) and TM65 (terminal 7 residues deleted), although correctly expressed in the sarcolemma of adult rat myocytes, do not affect myocyte shortening characteristics when compared to wild-type PLM.[22] This observation suggests that the cytoplasmic domain of PLM is important in mediating its effects on cardiac contractility and by inference, NCX1 activity.

Another inference is drawn from observations of the effects of S68A, S68E, and S63A PLM mutants on PMA-induced increases in I_{NaCa} in transfected HEK293 cells. In cells transfected with NCX1 and PLM, PMA induced a significant increase in I_{NaCa} amplitude.[31] Transfecting cells with PLM mutants in which serine68 is mutated (either S68A or S68E) results in total abolition of PMA's stimulatory effect on I_{NaCa}.[31] Preserving serine68 integrity restores the ability of PMA to significantly increase I_{NaCa} in cells transfected with NCX1 and PLM or S63A.[31] These results suggest that changes in conformation in PLM by mutating serine68 alters its interaction with NCX1, resulting in NCX1 not being accessible to PKC action perhaps due to steric hindrance. In this light, it is important to recall that PKC activation is associated with increased NCX1 phosphorylation at serine249, serine250, and serine357: all these serine residues lie in the intracellular loop.[33] Therefore, serine68 in the cytoplasmic tail of PLM hinders access of PKC to serine249, serine250, or serine357 in the intracellular loop of NCX1. In other words, these results are consistent with the hypothesis that the cytoplasmic tail of PLM interacts with the intracellular loop of NCX1.

Using GST pulldown assays, we recently demonstrated that the intracellular loop, but not the N- or C-terminal TM domains of NCX1, associates with PLM. Further analysis using protein constructs of GST fused to various segments of the intracellular loop of NCX1 indicates that PLM binds to residues 218–371 and 508–764 but not 371–508 (the calcium-binding domain of NCX1).[34] Split NCX consisting of N- or C-terminal domains with different lengths of the intracellular loop were coexpressed with PLM in HEK293 cells.[34] I_{NaCa} measurements indicate that PLM decreases NCX1 current only when the split exchangers contain residues 218–358 of the intracellular loop. Co-immunoprecipitation experiments with PLM and split exchangers suggest that PLM associates with the N-terminal of NCX1 only when it contains the intracellular loop segment spanning residues 218–358.[34] TM43, the cytoplasmic truncation mutant of PLM, fails to associate with NCX1 when coexpressed in HEK293 cells.[34] These observations suggest that PLM interacts with residues 218–358 of the intracellular loop of NCX1.

PLM: Regulator of Na^+-K^+-ATPase and NCX

To date, PLM is the only known FXYD member that regulates the activities of both Na^+-K^+-ATPase and NCX in the heart. Indeed, in cardiac sarcolemma, both α1- and α-2 subunits of Na^+-K^+-ATPase form distinct protein complexes with the NCX,[35] suggesting PLM, NCX, and Na^+-K^+-ATPase form a macro-molecular complex. There are significant differences, however, between the mechanisms by which PLM regulates the activities of these two important ion transporters. First, phosphorylation of PLM (at serine68) relieves its inhibition on Na^+-K^+-ATPase.[15,18] By contrast, it is the phosphorylated form of PLM that actively inhibits NCX.[22,31] Second, interaction between the single TM segment of PLM with TM2, 6, and 9 of α-subunit of Na^+-K^+-ATPase is important in the modulatory influences of PLM on the Na^+ pump.[16,17] By contrast, it is likely that interaction occurs between the cytoplasmic tail of PLM and the intracellular loop of NCX1.[34]

In the intact heart, β-adrenergic stimulation increases chronotropy leading to more frequent depolarizations and increased Na^+ entry. In addition, L-type Ca^{2+} current and SERCA2 activity are also increased in response to β-adrenergic stimulation, resulting in increased Ca^{2+} entry and SR Ca^{2+} loading. Increased SR Ca^{2+} content available for release largely accounts for the increased inotropy associated with β-adrenergic stimulation. To maintain steady-state Ca^{2+} balance, the increased myocyte Ca^{2+} entry must be balanced by enhanced Ca^{2+} efflux mediated by forward Na^+/Ca^{2+} exchange, thereby bringing more Na^+ into the myocyte. Therefore, enhanced Na^+-K^+-ATPase activity (by PLM phosphorylation) during β-adrenergic stimulation is necessary to prevent cellular Na^+ overload. On the other hand, unchecked stimulation of Na^+-K^+-ATPase will decrease $[Na^+]_i$, thereby increasing the thermodynamic driving force for forward Na^+/Ca^{2+} exchange (Ca^{2+} efflux), resulting in Ca^{2+} depletion. The resultant decreased inotropy is clearly not in the best interests of the organism under conditions of fight or flight. Our data suggest a coordinated paradigm in which PLM, upon phosphorylation at serine68 in response to β-adrenergic stimulation, enhances Na^+-K^+-ATPase (to minimize cellular Na^+ overload) but inhibits Na^+/Ca^{2+} exchange (to prevent cellular Ca^{2+} depletion) activities, thereby preserving chronotropic and inotropic responses.

PLM: Potential Involvement in Cardiac Pathology

In rat hearts subjected to coronary ligation, application of cDNA microarrays (containing 86 known genes and 989 unknown cDNAs) to analyze transcript levels indicates that PLM is 1 of only 19 genes to increase after myocardial infarction (MI).[36] Specifically, when compared with sham-operated rat ventricles, PLM expression is increased twofold as early as 3 days after MI and remains elevated for at least 2 weeks after MI. Subsequent work demonstrates that PLM protein levels increased 2.4- and 4-fold at 3 and 7 days post-MI, respectively, when compared to sham-operated rat hearts.[14] In post-MI rat hearts, both Na^+-K^+-ATPase[37] and Na^+/Ca^{2+} exchange[38] activities are depressed. Therefore, increased PLM expression may account for the depressed activities of both Na^+-K^+-ATPase and NCX observed in post-MI rat hearts.

In acute cardiac ischemia, Na^+-K^+-ATPase activity in the sarcolemma/particulate fraction is increased threefold compared with aerobic controls.[39] Interestingly, α-1 subunit of Na^+-K^+-ATPase is not phosphorylated during acute ischemia but PLM is.[39] The authors postulate that the increase in V_{max} of Na^+-K^+-ATPase observed during acute ischemia is due to phosphorylation of PLM with subsequent relief of inhibition of Na^+-K^+-ATPase.[39]

Two classes of drugs that clinically have been proven to be beneficial to patients with chronic ischemic cardiomyopathy are β-adrenergic antagonists and angiotensin-converting enzyme (ACE) inhibitors. It should be recalled that β-adrenergic receptor signals via PKA while angiotensin II receptor signals via PKC. Serine68 in PLM is phosphorylated by both PKA and PKC[6] and its phosphorylation governs the actions of PLM on Na^+-K^+-ATPase[15,18,38] and NCX.[31] In addition, expression of PLM is increased in experimental ischemic cardiomyopathy.[14,35] PLM may therefore be one of the therapeutic targets for β-adrenergic antagonists and ACE inhibitors.

CONCLUSION

Recent experimental evidence strongly supports that in addition to inhibiting Na^+-K^+-ATPase, PLM regulates cardiac NCX activity. Phosphorylation of PLM by β-adrenergic agonists inhibits NCX but relieves the inhibition of Na^+-K^+-ATPase. The consequences of Na^+-K^+-ATPase stimulation on the one hand and NCX inhibition on the other on cellular Ca^{2+} homeostasis and contractility are complex and difficult to predict or model. The stoichiometry of interaction between PLM and NCX and that between PLM and Na^+-K^+-ATPase is unknown. Structure–function studies suggest interaction between the TM domains of Na^+-K^+-ATPase with the TM domain of PLM. By contrast, the cytoplasmic tail of PLM interacts with the intracellular loop of the NCX. Much remains to be learned about the physiological relevance of regulation of NCX and Na^+-K^+-ATPase by PLM and the potential involvement of PLM in pathophysiological states.

ACKNOWLEDGMENTS

This work was supported in part by the National Institutes of Health Grants HL-58672 and HL-74854 (J.Y. Cheung), DK-46678 (J.Y. Cheung, co-investigator), GM-69841 and HL-77814 (L.I. Rothblum), HL-60074 (A.L. Tucker), HL-70548 and GM-64640 (J.R. Moorman); American Heart Association Pennsylvania Affiliate Grants-in-Aid 0265426U (X. Zhang), and 0355744U (J.Y. Cheung); American Heart Association Pennsylvania Affiliate Post-Doctoral Fellowship 0425319U (B.A. Ahlers); and by grants from the Geisinger Foundation (L.I. Rothblum and J.Y. Cheung).

REFERENCES

1. PALMER, C.J., B.T. SCOTT & L.R. JONES. 1991. Purification and complete sequence determination of the major plasma membrane substrate for cAMP-dependent protein kinase and protein kinase C in myocardium. J. Biol. Chem. **266:** 11126–11130.
2. SWEADNER, K.J. & E. RAEL. 2000. The FXYD gene family of small ion transport regulators or channels: cDNA sequence, protein signature sequence, and expression. Genomics **68:** 41–56.
3. MOORMAN, J.R. *et al.* 1995. Unitary anion currents through phospholemman channel molecules. Nature **377:** 737–740.
4. CHEN, Z. *et al.* 1998. Structural domains in phospholemman: a possible role for the carboxyl terminus in channel inactivation. Circ. Res. **82:** 367–374.
5. DAVIS, C.E. *et al.* 2004. Effects of phospholemman expression on swelling-activated ion currents and volume regulation in embryonic kidney cells. Neurochem. Res. **29:** 177–187.
6. WAALAS, S.I. *et al.* 1994. Protein kinase C and cyclic AMP-dependent protein kinase phosphorylate phospholemman, an insulin and adrenaline-regulated membrane phosphoprotein, at specific sites in the carboxy terminal domain. Biochem. J. **304(Pt 2):** 635–640.
7. PRESTI, C.F., L.R. JONES & J.P. LINDEMANN. 1985. Isoproterenol-induced phosphorylation of a 15-kilodalton sarcolemmal protein in intact myocardium. J. Biol. Chem. **260:** 3860–3867.
8. SIMMERMAN, H.K. & L.R. JONES. 1998. Phospholamban: protein structure, mechanism of action, and role in cardiac function. Physiol. Rev. **78:** 921–947.
9. THERIEN, A.G. *et al.* 1997. Tissue-specific distribution and modulatory role of the gamma subunit of the Na,K-ATPase. J. Biol. Chem. **272:** 32628–32634.
10. BEGUIN, P. *et al.* 2001. CHIF, a member of the FXYD protein family, is a regulator of Na,K-ATPase distinct from the gamma-subunit. EMBO J. **20:** 3993–4002.
11. BEGUIN, P. *et al.* 2002. FXYD7 is a brain-specific regulator of Na,K-ATPase alpha 1-beta isozymes. EMBO J. **21:** 3264–3273.

12. MAHMMOUD, Y.A., H. VORUM & F. CORNELIUS. 2000. Purification of a phospholemman-like protein from shark rectal glands. J. Biol. Chem. **275:** 35969–35977.

13. CRAMBERT, G. *et al.* 2002. Phospholemman (FXYD1) associates with Na,K-ATPase and regulates its transport properties. Proc. Natl. Acad. Sci. USA **99:** 11476–11481.

14. ZHANG, X.Q. *et al.* 2006. Phospholemman overexpression inhibits Na^+-K^+-ATPase in adult rat cardiac myocytes: relevance to decreased Na^+ pump activity in post-infarction myocytes. J. Appl. Physiol. **100:** 212–220.

15. DESPA, S. *et al.* 2005. Phospholemman-phosphorylation mediates the beta-adrenergic effects on Na/K pump function in cardiac myocytes. Circ. Res. **97:** 252–259.

16. LI, C. *et al.* 2004. Structural and functional interaction sites between Na,K-ATPase and FXYD proteins. J. Biol. Chem. **279:** 38895–38902.

17. LINDZEN, M. *et al.* 2006. Structural interactions between FXYD proteins and Na^+,K^+-ATPase: alpha/beta/FXYD subunit stoichiometry and cross-linking. J. Biol. Chem. **281:** 5947–5955.

18. SILVERMAN, B.Z. *et al.* 2005. Serine 68 phosphorylation of phospholemman: acute isoform-specific activation of cardiac Na/K ATPase. Cardiovasc. Res. **65:** 93–103.

19. GRUPP, I. *et al.* 1985. Regulation of sodium pump inhibition to positive inotropy at low concentrations of ouabain in rat heart muscle. J. Physiol. (Lond.) **360:** 149–160.

20. SONG, J. *et al.* 2002. Overexpression of phospholemman alter contractility and $[Ca^{2+}]_i$ transients in adult rat myocytes. Am. J. Physiol. Heart Circ. Physiol. **283:** H576–H583.

21. ZHANG, X.Q. *et al.* 2003. Phospholemman modulates Na^+/Ca^{2+} exchange in adult rat cardiac myocytes. Am. J. Physiol. Heart Circ. Physiol. **284:** H225–H233.

22. SONG, J. *et al.* 2005. Serine 68 of phospholemman is critical in modulation of contractility, $[Ca^{2+}]_i$ transients, and Na^+/Ca^{2+} exchange in adult rat cardiac myocytes. Am. J. Physiol. Heart Circ. Physiol. **288:** H2342–H2354.

23. TADROS, G.M. *et al.* 2002. Effects of Na^+/Ca^{2+} exchanger downregulation on contractility and $[Ca^{2+}]_i$ transients in adult rat myocytes. Am. J. Physiol. Heart Circ. Physiol. **283:** H1616–H1626.

24. MIRZA, M.A. *et al.* 2004. Effects of phospholemman downregulation on contractility and $[Ca^{2+}]_i$ transients in adult rat cardiac myocytes. Am. J. Physiol. Heart Circ. Physiol. **286:** H1322–H1330.

25. ZHANG, X.Q. *et al.* 2001. Overexpression of Na^+/Ca^{2+} exchanger alters contractility and SR Ca2+ content in adult rat myocytes. Am. J. Physiol. Heart Circ. Physiol. **281:** H2079–H2088.

26. JIA, L.G. *et al.* 2005. Hypertrophy, increased ejection fraction, and reduced Na-K-ATPase activity in phospholemman-deficient mice. Am. J. Physiol. Heart Circ. Physiol. **288:** H1982–H1988.

27. TUCKER, A.L. *et al.* 2006. Altered contractility and $[Ca^{2+}]_i$ homeostasis in phospholemman-deficient murine myocytes: role of Na^+/Ca^{2+} exchange. Am. J. Physiol. Heart Circ. Physiol. In press.

28. AHLERS, B.A. *et al.* 2005. Identification of an endogenous inhibitor of the cardiac Na^+/Ca^{2+} exchanger, phospholemman. J. Biol. Chem. **280:** 19875–19882.

29. DONG, H., J. DUNN & J. LYTTON. 2002. Stoichiometry of the cardiac Na$^+$/Ca^{2+} exchanger NCX1.1 measured in transfected HEK cells. Biophys. J. **82:** 1943–1952.
30. REMBOLD, C.M. *et al.* 2005. Serine 68 phospholemman phosphorylation during forskolin-induced swine carotid artery relaxation. J. Vasc. Res. **42:** 483–491.
31. ZHANG, X.Q. *et al.* 2006. Phospholemman inhibition of the cardiac Na$^+$/Ca^{2+} exchanger. Role of phosphorylation. J. Biol. Chem. **281:** 7784–7792.
32. PHILIPSON, K.D. & D.A. NICOLL. 2000. Sodium-calcium exchange: a molecular perspective. Annu. Rev. Physiol. **62:** 111–133.
33. IWAMOTO, T. *et al.* 1998. Protein kinase C-dependent regulation of Na$^+$/Ca^{2+} exchanger isoforms NCX1 and NCX3 does not require their direct phosphorylation. Biochemistry **37:** 17230–17238.
34. WANG, J. *et al.* 2006. Cytoplasmic tail of phospholemman interacts with the intracellular loop of the cardiac Na$^+$/Ca^{2+} exchanger. J. Biol. Chem. In press.
35. DOSTANIC, I. *et al.* 2004. The alpha 1 isoform of Na,K-ATPase regulates cardiac contractility and functionally interacts and co-localizes with the Na/Ca exchanger in heart. J. Biol. Chem. **279:** 54053–54061.
36. SEHL, P.D. *et al.* 2000. Application of cDNA microarrays in determining molecular phenotype in cardiac growth, development, and response to injury. Circulation **101:** 1990–1999.
37. DIXON, I., T. HATA & N. DHALLA. 1992. Sarcolemmal Na$^+$-K$^+$-ATPase activity in congestive heart failure due to myocardial infarction. Am. J. Physiol. Cell. Physiol. **262:** C664–C671.
38. ZHANG, X. *et al.* 1996. Na$^+$/Ca^{2+} exchange currents and SR Ca^{2+} contents in postinfarction myocytes. Am. J. Physiol. Cell. Physiol. **271:** C1800–C1807.
39. FULLER, W. *et al.* 2004. Ischemia-induced phosphorylation of phospholemman directly activates rat cardiac Na/K-ATPase. FASEB J. **18:** 197–199.

The Squid Preparation as a General Model for Ionic and Metabolic Na⁺/Ca²⁺ Exchange Interactions

Physiopathological Implications

R. DIPOLO[a,b] AND L. BEAUGÉ[b,c]

[a]Laboratorio de Fisiología Celular, Centro de Biofísica y Bioquímica, IVIC Apartado, 21827 Caracas 10020 A, Venezuela

[b]Marine Biological Laboratory, Woods Hole, Massachusetts 02543, USA

[c]Laboratorio de Biofísica, Instituto de Investigación Médica "Mercedes y Martín Ferreyra" Casilla de Correo 389, 5000 Córdoba, Argentina

ABSTRACT: We propose an integrated kinetic model for the squid nerve Na⁺/Ca²⁺ exchanger based on experimental evidences obtained in dialyzed axons. This model satisfactorily explains the interrelationship between ionic (Na^+_i–H^+_i–Ca^{2+}_i) and metabolic (ATP, phosphoarginine (PA)) regulation of the exchanger. Data in dialyzed axons show that the Ca_i-regulatory site located in the large intracellular loop plays a central role in the modulation by ATP by antagonizing the inhibitory Na^+_i–H^+_i synergism. We have used the Na_o/Na_i exchange mode to unequivocally measure the affinity of the Ca_i-regulatory site. This allowed us to separate Ca_i-regulatory from Ca_i-transport sites and to estimate their respective affinities. In this work we show for the first time that under conditions of saturation of the Ca_i-regulatory site (10 μM Ca^{2+}_i, pH_i 8.0), ATP have no effect on the Ca_i-transport site. In addition, we have expanded our equilibrium kinetic model of ionic and metabolic interactions to a complete exchange cycle (circular model). This model, in which the Ca_i-regulatory site plays a central role, accounts for the decrease in Na_i inactivation, at high pH_i, high Ca^{2+}_i, and MgATP. Furthermore, the model also predicts the net Ca^{2+} movements across the exchanger based on the exchanger complexes redistribution both during physiological and pathological conditions (ischemia).

KEYWORDS: sodium–calcium exchange; ionic and metabolic regulation; squid axons; cellular ischemia

Address for correspondence: Reinaldo DiPolo, Laboratorio de Fisiología Celular, Centro de Biofísica y Bioquímica, IVIC, Apartado 21827, Caracas 10020-A, Venezuela. Voice: 58-212-5041230; fax: 58-212-5041093.

dipolor@ivic.ve

Ann. N.Y. Acad. Sci. 1099: 135–151 (2007). © 2007 New York Academy of Sciences.
doi: 10.1196/annals.1387.049

INTRODUCTION

The Na^+/Ca^{2+} exchanger, a reversible carrier-mediated ion transport system, has been subjected to extensive investigation due to its implications in the physiology of several cellular processes including cardiac relaxation and intracellular calcium clearance during neurosecretion.[1-4] Furthermore, it has also been implicated in a number of physiopathological conditions, such as in the ischemia–reperfusion syndrome[5-7] and arterial hypertension.[2,3] This exchanger is highly regulated by transported and nontransported intracellular ions as well as by the metabolic state of the cell (ATP).[2,8] Most of the known regulatory process that controls the activity of the Na^+/Ca^{2+} exchanger occurs at the large intracellular loop since its deletion renders an exchanger without ionic and nucleotide (ATP) regulation (see FIG. 1).[9,10] Within that loop, a region between amino acids 218 and 238 is responsible for the Na_i-inactivation process and that from amino acid 371 to 508 comprises the Ca_i-regulatory binding domain.[11,12] Intracellular protons interact with these two ion regulatory sites promoting inhibition.[13,14] Finally, a putative site located on the loop is responsible for the upregulation of the exchanger by ATP. In contrast, upregulation by phosphoarginine (PA) does not require the integrity of that loop (see Berberian *et al.* in this volume).[15] In the past few years, our laboratories have developed an integrated kinetic model for ATP and PA modulation of the squid Na^+/Ca^{2+} exchanger on the basis of Na^+_i–H^+_i–Ca^{2+}_i interactions with the Ca_i-regulatory site and the Na^+_i and Ca^{2+}_i interactions with the

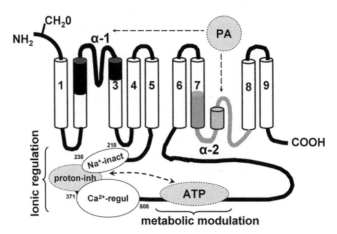

FIGURE 1. Schematic model of the Na^+/Ca^{2+} exchanger. The nine transmembrane segments comprise the two putative transport repeat regions: α-1 and α-2. Shown in the large cytoplasmic "loop" are the ionic and metabolic regulatory sites: the Na_i-regulatory region (Na_i inactivation or endogenous XIP region), the proton target region, the Ca_i-regulatory site, and the nucleotide (ATP) site. Regulation by guanidine phosphates (PA) is shown out of the intracellular "loop" (see text for detail).

transporting sites and the way they are affected by MgATP and PA.[15] The results described below show that by using the electroneutral Na_o/Na_i exchange reaction of the exchanger to measure the affinity of the exchanger for $Ca^{2+}{}_i$, it is possible to dissect the effects of transported and nontransported ligands on the Ca_i-regulatory site. In turn, this allows to establish a coherent steady-state model of ionic and metabolic interactions of the Na^+/Ca^{2+} exchanger.[8,15] Here we present experimental and simulation results conforming to a model where modulation by MgATP takes place only at the central Ca_i-regulatory site of the intracellular loop. Moreover, extensions of that model into a scheme representing the overall transport cycle predict the pre-steady-state and steady-state behavior of the exchanger and give valuable information concerning the net Ca^{2+} fluxes that occur through the exchanger under physiopathological circumstances.

Measurement of the Affinity of the Ca_i-Regulatory Site

The Na^+/Ca^{2+} exchanger can work basically in four different modes[2,16] (but see Ref. 17). Two of them electrogenics, Na_o/Ca_i (forward exchange) and Na_i/Ca_o (reverse exchange) and two electroneutral, Ca_o/Ca_i and Na_o/Na_i exchanges. A basic problem in measuring the affinity of the Ca_i-regulatory site for Ca^{2+} is that in these exchange modes (Na_o/Ca_i, Na_i/Ca_o, and Ca_o/Ca_i exchange) Ca^{2+} binds to intracellular and/or extracellular transport sites. Therefore, activation of the exchange by intracellular Ca^{2+} reflects mixed effects on the Ca_i-regulatory and Ca_i-transport sites. In squid axons, Na_o/Na_i exchange is both electroneutral and voltage independent.[18] Using intracellular dialyzed squid axons we have measured the Na_o/Na_i mode of the exchanger as well as its Ca_i-dependence under conditions in which no Ca^{2+} will be bound to either the internal or external transport sites.[15,19] FIGURE 2 A shows the measurement the Ca_i-dependent Na_o-dependent $^{22}Na^+$ efflux component in a dialyzed squid axon bathed in $0Ca^{2+}{}_o$. In the absence of intracellular Ca^{2+}, Na^+ efflux is less than 2 pmole/cm^2/s^1 and close to the expected leak value for Na^+ efflux.[19] Addition of 5 µM $Ca^{2+}{}_i$ markedly activates the efflux of Na^+ and subsequent removal of external Na^+ brings the efflux back to leak values. FIGURE 2 B shows the experimental protocol (right cartoon) and results of measurements of the affinity of the Ca_i-regulatory site (left figure). It is clear that under conditions of no Ca^{2+} transport (40 mM $Na^+{}_i$, 400 mM $Na^+{}_o$, pH$_i$ 7.3, and 2 mM ATP) and in the range from 0.1 to 10 µM $[Ca^{2+}]_i$, the Ca_i-dependent Na_o/Na_i exchange component saturates at about 10 µM with a K_d of about 0.6 µM $Ca^{2+}{}_i$. In conclusion, the Na_o/Na_i experiments provide an unequivocal method for estimation of the affinity of the Ca_i-regulatory site; this permits to study the way by which $Na^+{}_i$, $H^+{}_i$ and metabolism (ATP) modify the Ca_i-regulatory site affinity and is the basis for the integrated model of ionic and metabolic interactions discussed below.

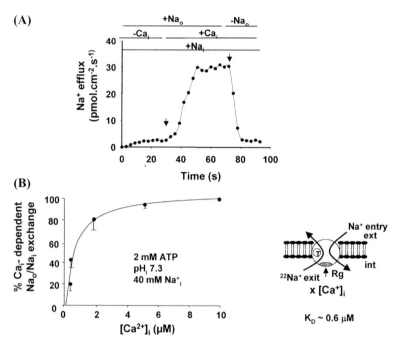

FIGURE 2. (**A**) Measurement of the affinity of the Ca_i-regulatory site using the Na_o/Na_i transport mode of the exchanger. The results are the summary of 10 different axons dialyzed with 2 mM ATP, 40 mM internal Na^+ at a physiological pH_i of 7.3. The *error bars* indicate SEM. The K_d for the affinity of the Ca_i-regulatory site is about 0.6 μM Ca^{2+}_i. (**B**) Cartoon of the experimental protocol in measuring the affinity of the Ca_i-regulatory site.

Na^+_i–H^+_i–Ca^{2+}_i Interactions in the Absence of Metabolic Regulation

FIGURE 3 A shows the magnitude (%) of the intracellular proton inhibition of forward (Na_o/Ca_i) exchange in a group of dialyzed squid axons in which intracellular Na^+, ATP, and PA were removed, at a constant $[Ca^{2+}]_i$ of 1 μM. A striking feature of these experiments is that intracellular protons still inhibit the exchanger. Under the above conditions, even at a physiological pH_i of 7.3, almost 80% of the activity of the exchanger is impaired, being released only at high pH_i. Using the experimental protocol depicted in FIGURE 2 B, it is possible to determine the affinity of the Ca_i-regulatory site over a wide range of pH_i. FIGURE 3 B shows the results of several experiments carried out in dialyzed squid axons in which the affinity of the Ca_i-regulatory site was determined at pH_i 6.9, 7.3, and 8.8 in the presence of 40 mM Na^+_i and no ATP. Decreasing the pH_i by only 0.4 units (from 7.3 to 6.9) causes more than 40 times reduction in the Ca^{2+} affinity while increasing the pH_i to 8.8 causes a dramatic increase in that affinity. The main conclusion from these experiments is that intracellular

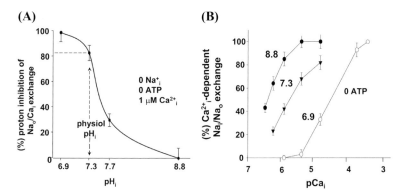

FIGURE 3. Proton inhibition of forward Na^+/Ca^{2+} exchange and its effect on the affinity of the Ca_i-regulatory site. (**A**) (%) proton inhibition of the Na_o/Ca_i exchange in the absence of intracellular Na^+ and ATP at a constant $[Ca^{2+}]_i$ of 1 μM. Notice the per se sharp proton inhibition around physiological pH_i. (**B**) (%) Ca^{2+}_i-dependent Na_o/Na_i exchange at pH_i 6.9 (*open circles*), 7.3 (*triangles*), and 8.8 (*fill circles*) in the absence of ATP, and at a physiological $[Na^+]_i$ of 40 mM. The measurements at 0.3 μM Ca^{2+}_i were made with BAPTA as a Ca^{2+} chelator. All other measurements were made with dibromo-BAPTA. The *error bars* indicate SEM.

protons, per se, markedly inhibit the Na^+/Ca^{2+} exchange activity, and that the target of this inhibition is the Ca_i-regulatory site. If intracellular protons are acting at the Ca_i-regulatory site, then one should expect that increasing $[Ca^{2+}]_i$ will relieve proton inhibition. FIGURE 4 shows the results of several experiments in which the activity of the Na^+/Ca^{2+} exchanger was measured as a function

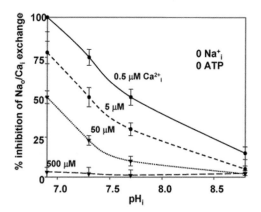

FIGURE 4. Effect of $[Ca^{2+}]_i$ on proton inhibition of Na_o/Ca_i exchange in the absence of both Na^+_i and ATP. Notice that Ca^{2+}_i antagonizes H^+_i inhibition since increasing Ca^{2+}_i from 0.5 μM to 500 μM drastically reduced proton inhibition. The *error bars* indicate SEM mean temperature 17.5°C.

of pH_i, at different $[Ca^{2+}]_i$, in dialyzed axons containing no Na^+_i nor ATP. The results clearly show that the H^+_i inhibition can be reverted by increasing the $[Ca^{2+}]_i$. At a physiological pH_i of 7.3, H^+_i causes a 75% inhibition of the forward Na^+/Ca^{2+} exchange activity at 0.5 μM Ca^{2+}_i, but less than 5% at 500 μM. The main conclusion from this section is that intracellular protons and Ca^{2+}_i compete either for the same site, or for the same form of the exchanger molecule, thus resulting in competition.[19]

After characterizing the H^+_i–Ca^{2+}_i interactions in dialyzed squid axons, we reexplored the original observation by Doering and Lederer of a synergistic interactions between H^+_i and Na^+_i.[13,14] This point is critical in the building of an integrated model between ionic and metabolic regulation of the Na^+/Ca^{2+} exchanger. FIGURE 5 A shows inhibition by Na^+_i of the forward (Na_o/Ca_i) exchange at acid pH_i (6.9), and at a physiological pH_i (7.3) in the absence of ATP and at a constant 1 μM $[Ca^{2+}]_i$. In the first part of the experiment (pH_i 6.9), addition of 25 mM Na^+_i causes almost 80% inhibition in the exchange activity; total inhibition was obtained with 50 mM Na^+_i. Removal of Na^+_i brings the exchanger to its original base line. At this point increasing intracellular pH by just 0.4 units (to pH_i 7.3) causes a large increase in the activity of the forward exchange. Under these conditions, increasing $[Na^+]_i$ of 25, 50, and 100 mM caused a 50%, 75%, and 100% inhibition, respectively. FIGURE 5 B summarizes the results of several axons in which the activity of the forward Na^+/Ca^{2+} exchanger was explored at acid, normal, and alkaline pH_i as a function of the $[Na^+]_i$. The K_i for Na^+_i inhibition changed from 92 mM at pH_i 8.8 to

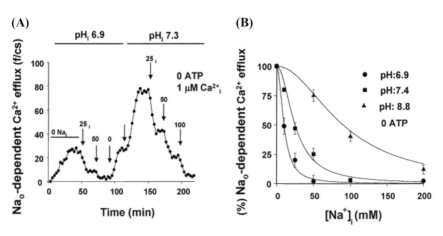

FIGURE 5. Effect of $[Na^+]_i$ on the Na_o/Ca_i exchange at pH_i of 6.9 and 7.3. (A) A representative dialyzed squid axon experiment in which the Na^+_o-dependent Ca efflux was measured at different $[Na^+]_i$, first at pH_i 6.9 and then at 7.3. Notice the strong inhibitory synergism between Na^+_i and H^+_i. (B) (%) Na^+_o-dependent Ca^{2+} efflux as a function of $[Na^+]_i$ in a series of dialyzed squid axons at three different pH_i: 6.9, 7.3, and 8.8. The K_i for Na^+_i decreases from 92 mM at pH_i 8.8 to 11 mM at pH_i 6.9. Notice the strong inhibition of Na^+_i at acidic pH_i. The *error bars* indicate SEM.

32 mM at pH_i 7.3 and 11 mM at pH_i 6.9. The conclusion of this second section is that, as it happens in the cardiac Na^+/Ca^{2+} exchanger,[13,14] the squid exchanger shows a the strong Na^+_i–H^+_i synergistic inhibition in its activity.[8,19]

Na^+_i–H^+_i–Ca^{2+}_i Interactions under Metabolically Regulated Conditions

The above experiments were carried out in the absence of ATP in order to look into Na^+_i–H^+_i–Ca^{2+}_i interactions in the absence of metabolic regulation. We now include MgATP in those interactions. In the first part of FIGURE 6 A an axon is dialyzed with acid pH_i (6.9) and no ATP. In this case, 25 mM Na^+_i causes 78% inhibition of the forward Na^+/Ca^{2+} exchange. In the second part, 2 mM MgATP were added. From the base line in $0Na^+_i$, MgATP causes a large increase in the exchange activity. Subsequent additions of 25, 50, 100, and 200 mM Na^+_i causes 28%, 33%, 57%, and 77% inhibition, respectively. FIGURE 6 B summarizes the result of several experiments on the effect of MgATP on Na_i inhibition at acid pH_i (6.9). The K_i for Na^+_i at acid pH_i (6.9) changes from 12 mM to 100 mM in the absence and presence of MgATP, respectively. The main conclusion from this experiment is that MgATP antagonizes Na^+_i –$^+H^+_i$ inhibition.

The obvious question is which is the target of MgATP: the Ca_i-regulatory site, the Ca_i-transport site, or both? The answer requires separating both Ca_i

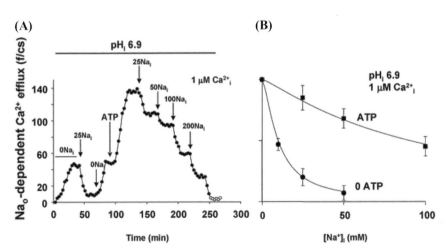

FIGURE 6. Effect of ATP on Na^+_i–H^+_i inhibition of Na_o/Ca_i exchange. (**A**) Intracellular Na^+ inhibition of Na^+_o-dependent Ca^{2+} efflux in a representative axon dialyzed with pH_i 6.9 in the absence and presence (3 mM) of ATP. (**B**) Summary of several dialyzed squid axon experiments in which inhibition of Na^+/Ca^{2+} exchange by $[Na^+]_i$ is explored at pH_i 6.9. Notice that in the absence of ATP, 25 mM internal Na^+ causes more than 70% inhibition of the forward exchange indicating that ATP protects against the inhibitory synergism between Na^+_i and H^+_i.

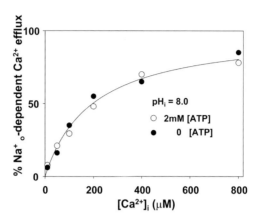

FIGURE 7. Lack of effect of ATP on Ca_i-transport site affinity in a dialyzed squid axon. Ordinate: (%) Na^+_o-dependent Ca^{2+} efflux. Abscissa: $[Ca^{2+}]_i$ in μM. In this experiment the Ca_i-regulatory site was saturated with 10 μM Ca^{2+}_i, 2 mM ATP at a pH_i of 8.0. Activation of the Ca_i-transport site was followed from 10 to 800 μM. Notice the absence of ATP effect on the affinity of the Ca_i-transport site. The $K_{0.5}$ for the internal Ca^{2+} transport site is close to 200 μM. Temperature 17.6°C.

sites. From the experiments of FIGURE 2 B where the Ca_i-dependence of the Na_o/Na_i exchange mode was measured, it is clear that at a physiological pH_i of 7.3 the Ca_i-regulatory site is practically saturated at 10 μM Ca^{2+}_i. On the other hand, intracellular alkalinization markedly increases the affinity of the Ca_i-regulatory site (FIG. 3 A). Therefore, at pH_i 8.0 and with 10 μM Ca^{2+}_i in the dialysis medium, the Ca_i-regulatory site will be expected to be completely saturated and therefore any increase in the activity of the exchanger upon raising $[Ca^{2+}]_i$ above 10 μM should reflect activation of the Ca_i-transport site. FIGURE 7 shows an experiment with an increase in $[Ca^+]_i$ from 10 up to 800 μM at pH_i 8.0 with and without MgATP. The Na_o-dependent Ca^{2+} efflux (forward exchange) is activated by $[Ca^{2+}]_i$ along a hyperbolic curve with a $K_{0.5}$ of about 200 μM Ca^{2+}_i. This represents the Ca_i activation of the intracellular calcium transport site and indicates that it is not affected by MgATP; thus, all MgATP effects on the Na^+/Ca^{2+} exchanger take place at the Ca_i-regulatory site on the intracellular regulatory loop. In these experiments there was no PA in the dialysis medium, otherwise the Ca_i-transport site affinity will be increased by a factor of 20.[15]

An Integrated Kinetic Model for Ionic–Metabolic Regulation of the Na^+/Ca^{2+} Exchanger: Physiological and Physiopathological Predictions

Pre-Steady-State Conditions

Early overall models of ionic regulation of the Na^+/Ca^{2+}: Na_i-dependent inactivation and H^+_i–Na^+_i synergism[14,20–22] did not take into account the

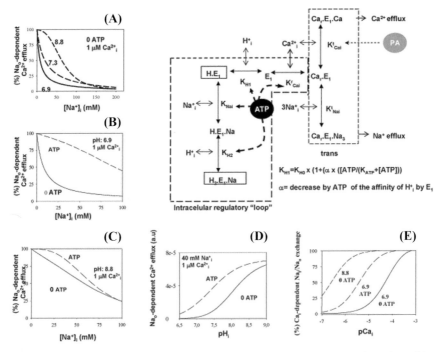

FIGURE 8. Integrated kinetic model showing ATP regulation of Na^+_i–H^+_i–Ca^{2+}_i interactions with the Ca_i-regulatory site and PA modulation of the Ca_i-transport site in the squid Na^+/Ca^{2+} exchanger. The *broken lines* represent both the intracellular regulatory "loop" where the Ca_i-regulatory site is located, and the transmembrane segments where the Ca_i-transport site is present. (**A–E**) Steady-state model simulations for the Na^+_i–H^+_i–Ca^{2+}_i and ATP interactions in the modulation of the squid Na^+/Ca^{2+} exchanger.

Ca_i-regulatory site. On the basis of the above experimental data, in the model proposed in FIGURE 8, H^+_i–Na^+_i–Ca^{2+}_i and ATP interactions are taken all together. The kinetic scheme we proposed includes H^+_i–Ca^{2+}_i competition and H^+_i–Na^+_i synergism as the basis for ionic regulation and MgATP modulation of these interactions.[8,19] In this case only intracellular ionic interactions and internally facing sites of the exchanger (E_1) at equilibrium were considered. The resulting forms are: E_1, the free carrier; $Ca_r.E1$, the carrier loaded with Ca^{2+} at the regulatory site; $C_r.E_1.Ca$, and $C_r.E_1.Na_3$ the carriers loaded with 1 Ca^{2+} or 3 Na^+ in the transporting sites. $H.E_1$, $H.E_1.Na$, and $H_2.E_1.Na$ are carriers binding H^+ and Na^+ at their inhibitory sites. Proton inhibition takes place also in the absence of Na^+_i; therefore, protons must bind first ($H.E_1$ complex). Because Na^+_i enhances H^+_i inhibition we propose that it binds to $H.E_1$ ($H.E_1.Na$ form), and in that way it allows the binding of a second proton forming the $H_2.E_1.Na$ dead end complex. In this way, inhibition of the exchanger by intracellular Na^+ will have two components: H^+_i–Na^+_i synergism on the regulatory site and Na^+_i–Ca^{2+}_i competition at the transporting sites; the second effect can be separated from the first by internal alkalinization (see FIG. 8 A). The experimental and simulation data show that H^+_i–Na^+_i synergism is the major component at low $[Na^+]_i$ whereas competition at the transport sites predominates at high $[Na^+]_i$. FIGURE 8 A and D is simulation of the Na^+/Ca^{2+} exchange activity during steady-state as a function of $[Na^+]_i$ and $[Ca^{2+}]_i$ at different pH_i in the absence of MgATP. All these and other results[19] led us to propose that upregulation of the squid Na^+/Ca^{2+} exchanger by MgATP occurs through a reduction in the apparent affinity for H^+_i and Na^+_i binding to their inhibitory sites. An "ATP regulatory region" is surrounded by dotted lines in FIGURE 8. The actual metabolic pathway is not known and does not appear in the scheme. This model predicts practically all Ca^{2+}–Na^+_i–H^+_i and ATP interactions observed in dialyzed squid axons. Four of them are illustrated here (FIG. 8 B, C, D, and E).

It also explored if our kinetic model of Na^+_i–H^+_i–Ca^{2+}_i–ATP interactions can also explain pre-steady-state Na^+_i inactivation, and the prevention of that inactivation by alkaline pH_i, ATP, and high $[Ca^{2+}]_i$ observed in other Na^+/Ca^{2+} exchangers. To that aim an overall cycle scheme was constructed by using unidirectional rate constants instead of the equilibrium constants used before. This model is described in FIGURE 9. The choice of the unidirectional rate constant was obviously arbitrary. All forward binding constants were taken equal to 1×10^8 /s/M. The backward rate constants were elected in order that the ratio forward/backward (or on/off) resulted in the equilibrium constants used by us before.[15] The values chosen for the translocation rate constants were 1×10^3/s for Na^+ efflux and influx and for Ca^{2+} efflux; because Ca^{2+} influx is rate limiting we took the corresponding translocation rate constant equal to 0.5×10^2/s.

The simulations show that this kinetic model indeed predicts the pre-steady-state Na^+_i inactivation and the prevention of that inactivation by alkaline pH, ATP, and Ca^{2+}_i. Obviously, similar results could also be obtained with other values for the unidirectional rate constants, but the important point here is that the model we proposed can explain experimental data obtained with Na^+/Ca^{2+} exchangers (FIG. 9 A, B, and C).

TABLE 1 summarizes the most relevant features of the three (I, II, and III) current models of Na^+/Ca^{2+} exchanger regulation. A crucial point concerning these models is that although they are based on experimental data obtained in cardiac cells (model I and II) and squid nerve (model III), there are many similarities including: Na_i inactivation, Ca_i regulation, H^+_i inhibition, and Na_i–H_i synergism.

Steady-State Conditions

Although the metabolic modulation of the exchanger has not been implicitly added in model I and II of TABLE 1, the experimental data indicate a similar

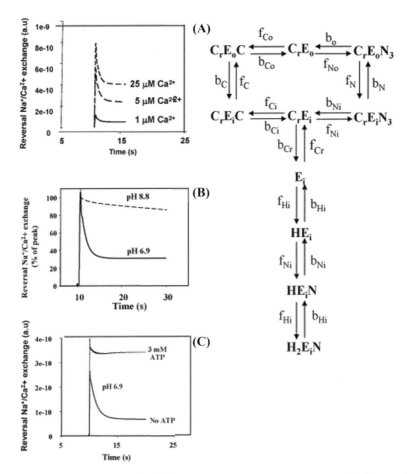

FIGURE 9. Complete Na^+/Ca^{2+} exchange cycle derived from the equilibrium model depicted in FIGURE 8. This circular model was used to simulate pre-steady-state conditions to explore if the mechanisms proposed for Na^+, H^+, Ca^{2+}, and MgATP interactions modifying the apparent of the $Ca^{2+}{}_i$-regulatory site could predict the $Na^+{}_i$-dependent inactivation observed in mammalian heart.[23] The choice of the unidirectional rate constant was obviously arbitrary and is explained in the text. The simulation results indicate that our kinetic model can also explain pre-steady-state $Na^+{}_i$ inactivation (**A**) prevention of that inactivation by increasing $[Ca^{2+}]_i$; (**B**) prevention of that inactivation by increasing pH_i; and (**C**) prevention of that inactivation in the presence of ATP_i.

kinetic consequence, that is, decrease in both Na_i-dependent inactivation and in the rate of entrance into inactivation induced by ATP.[8,14,20,21,23] An interesting feature experimentally found in both cardiac and squid and predicted by models I and III is that rising $[Na^+]_i$ at a constant $[Ca^{2+}]_i$ and $[Ca^{2+}]_o$ induces a monotonic stimulation of Ca^{2+} influx (reverse exchange) reaching a plateau level with no sign of a bell shape.[24] An interesting experimental observation

TABLE 1. Current kinetic models of NCX regulation

	Model I Na^+_i–Ca^{2+}_i modulation	Model II Na^+_i–H^+_i modulation	Model III Na^+_i–Ca^{2+}_i–H^+_i–ATP modulation	Reference
Experimental basis	Na_i-dependent fast inactivation (I_1)	H^+_i block in the absence of Na^+_i.	H^+_i block in the absence of Na^+_i.	(8,13,14,19,20,22)
	Na_i-independent Ca_i modulation (I_2)	Strong synergic inhibition by Na^+_i-H^+_i interactions	Decrease affinity of Ca^R_i site by H^+_i	(13,14,19,20)
		Protons do not compete with the Na^T_i site.	Strong synergic inhibition by Na^+_i-H^+_i interactions	(8,9,14,19)
	Occurs at $loop_i$	Occurs at $loop_i$	Protons do not compete with the Na^T_i site.	(8,14)
			ATP increase the Ca^R_i site affinity and relief Na^+_i-H^+_i synergism	(8,19)
			Occurs at $loop_i$	(15)
Ligands involved	Na^+_i–Ca^{2+}_i	H^+_i–Na^+_i	Na^+_i–H^+_i–Ca^{2+}_i–ATP	(8,14,19,20,22)
Model predictions	Pre- and steady-state Na_i inactivation	Partial inactivation of reverse NCX current in the absence of Na^+_i.	Pre- and steady-state Na_i inactivation	(13,14,20,21,22)
	Regulatory Ca^{2+}_i- dependence of reverse NCX current.	Increase proton block Na^+_i (synergism)	Relief of Na_i inhibition at alkaline pH_i.	(14,19,23)
			Increase proton block by Na^+_i (synergism)	(14,19)
	Monotonically increases in reverse NCX current by $[Na^+]_i$		Monotonically decrease in forward NCX by $[Na^+]_i$	(24)
			Monotonically increase in reverse NCX by $[Na^+]_i$	(22,24)
			Na^+_i–H^+_i inhibition relief by ATP	(8,19)

Ca^R_i: Ca_i-regulatory binding site. Na^T_i: Na_i-transport site. NCX: sodium-calcium exchanger. $Loop_i$: intracellular regulatory loop.

in squid axons concerning the effect of $[Na^+]_i$ on exchange functioning, is that forward exchange is monotonically inhibited, while reverse exchange is monotonically activated by $[Na^+]_i$. These findings, which are also experimentally found in the cardiac preparation,[22] are predicted by both models II and III. In the case of the squid, model simulations on the effect of $[Na^+]_i$ on both forward and reverse exchange are shown in FIGURE 10. At a physiological pH_i of 7.3 and in the presence of physiological concentrations of external Ca^{2+} and Na^+, Ca^{2+} efflux is inhibited by Na^+_i, being markedly dependent on the metabolic state of the cell. In the case of Na^+ exit (reverse and Na/Na exchange) although activation by Na^+_i is also markedly dependent on the metabolic state of the cell, internal Na^+ always activate.

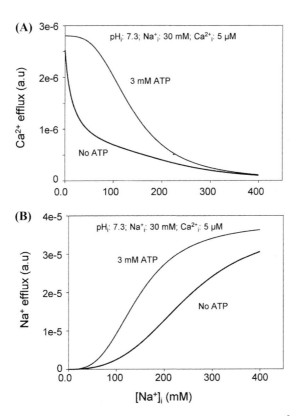

FIGURE 10. Model simulations on the effect of $[Na^+]_i$ on both **(A)** Ca^{2+} and **(B)** Na^+ exit by the Na^+/Ca^{2+} exchanger in the presence and absence of ATP_i and with physiological concentrations of external Na^+ and Ca^{2+}. Notice that at a physiological pH_i of 7.3 and in the presence of external Ca^{2+} and Na^+, Ca^{2+} efflux is inhibited by Na^+_i. This inhibition is markedly dependent on the metabolic state of the cell. In the case of Na^+ exit (reverse and Na/Na exchange) although activation by Na^+_i is also strongly dependent on the metabolic state of the cell, internal Na^+ always activate (see text for explanation).

Our simulations for steady state, which reproduce experimental data, are based on our model where proton inhibition of the Ca_i-regulatory site is enhanced by Na^+_i. As Na^+_i increases, the expected monotonic inhibition of Ca^{2+} efflux is developed but at the same time a monotonic stimulation of Na^+ efflux through the exchanger takes place. The explanation for these results is quite simple and depends on the redistribution of exchanger forms. Upon Na^+_i increases, the exchanger form with binding Ca^{2+} at the regulatory site $(Ca_r.E_1)$ will decrease, but, due to the Na^+–Ca^{2+} competition for the transporting sites, the form binding Ca^{2+} to intracellular transporting site $(Ca_r.E_1.Ca)$ will also be reduced and therefore Ca^{2+} efflux will eventually cease. The case is completely different for Na^+–Na^+ exchange. In this case, Na^+_i will have two opposite effects. On the one hand, it will promote proton displacement of Ca^{2+} from the regulatory site but, on the other, will compete with protons by driving the carrier to the form binding Na^+ to their intracellular transporting sites $(C_r.E_1.Na_3$ form). As a result of that competition, although the actual levels of fluxes will be lower at acidic pH, Na^+_i will always show monotonic activation of Na^+–Na^+ exchange. Actually, when simulations are taken to $[Na^+]_i$ concentrations as high as 4 M, the rates of Na^+–Na^+ exchange are the same at pH 6.9, 7.3, and 8.8. Actually, this mechanism is in some respect similar to that observed for the forward Ca^{2+} efflux mode in the absence of intracellular Na^+: by increasing $[Ca^{2+}]_i$ high enough proton inhibition is overcome not only by competition with protons but also by driving the carrier into the $Ca_r.E_1.Ca$ conformation. Experimentally and in simulations we observed that at 1 mM Ca^{2+}_i the forward exchange mode has identical values at the three pH just considered.[19]

Importance in Physiopathology

Our integrate model of ionic and metabolic regulation of the exchanger[8,15,19] considers variables (H^+_i, Na^+_i, Ca^{2+}_i, and ATP) most likely to change during physiopathological conditions, such as the ischemia–reperfusion syndrome.[6] We have used the steady-state model shown in FIGURE 9 to simulate situations largely resembling those occurring during a nerve ischemia by varying critical *CIS* values of $[Ca^{2+}]_i$, $[Na^+]_i$, pH_i, and ATP in the presence of physiological *TRANS* conditions of extracellular $[Na^+]$ and $[Ca^{2+}]$. TABLE 2 presents the predicted values of the forward and reverse Na^+/Ca^{2+} exchange fluxes as well as the Ca^{2+} efflux/influx exchange ratio under these situations. The relevant predictions can be summarized as follows: (*a*) With normal pHi, $[Na^+]_i$ and $[ATP]_i$ and in the presence of normal $[Na^+]_o$ and $[Ca^{2+}]_o$, the exchanger maintains a stable $[Ca^{2+}]_i$ around 0.1 μM (column I) with a balance of flux close to one (ratio Ca^{2+} efflux/Ca^{2+} influx = 0.97); (*b*) Rising the $[Ca^{2+}]_i$ 10 times (1 μM Ca^{2+}_i) without any change in the other variables leads to an efflux/influx ratio of 2.2 (column II) thus indicating that the exchanger

TABLE 2. Predicted effects of changes in intracellular proton, ATP, Na^+ and Ca^{2+} concentrations on the efflux and influx of Ca^{2+} through the squid nerve Na^+/Ca^{2+} exchanger

Column	I	II	III	IV	V	VI
pH_i	7.3	7.3	6.3	6.3	6.3	6.3
Ca^{2+}_i (μM)	0.1	1	1	1	1	3
Na^+_i (mM)	30	30	30	30	90	90
ATP (mM)	2	2	0.2	0.8	0.2	0.2
Ca^{2+} efflux (a.u.)	3.2×10^{-8}	6.8×10^{-7}	3.1×10^{-7}	3.8×10^{-7}	2.3×10^{-7}	1.1×10^{-6}
Ca^{2+} influx (a.u.)	3.3×0^{-8}	3.1×10^{-7}	2.1×10^{-7}	1.8×10^{-7}	5.8×10^{-7}	1.1×10^{-6}
Effl/Inf ratio	0.97	2.2	1.5	2.1	0.4	1.0

Extracellular Na^+ and Ca^{2+} concentrations were maintained constants at 440 mM and 5 mM, respectively. The K_m for ATP was taken as 200 μM.

works largely in the Ca^{2+} extrusion mode (forward exchange); (c) With 1 μM $[Ca^{2+}]_i$ and 30 mM $[Na^+]_i$ simultaneous acidification (pH_i: 6.3) and a fall in $[ATP]_i$ (0.2 mM) reduces the Ca^{2+} efflux/Ca^{2+} influx ratio to 1.5 (compare columns II and III). Nevertheless, this higher $[Ca^{2+}]_i$ is still able to cause net Ca^{2+} extrusion from the cell; (d) The "protective" effect of ATP is observed in column V in which a moderate increase in [ATP] from 0.2 to 0.8 mM (column III and column IV) increases Ca^{2+} efflux and decreases Ca^{2+} influx leading to a larger increase in net Ca^{2+} extrusion from the cell; (e) However, if with internal acidification and low ATP, the concentration of Na^+_i is raised threefold (90 mM; column V), there is a reduction in the efflux of Ca^{2+} together with an increase in Ca^{2+} influx leading to a net Ca^{2+} gain; and (f) Column VI illustrates a critical prediction of our steady-state model: at low pH_i (6.3), high Na^+_i (90 mM), and low ATP (0.2 mM), the exchanger can balance Ca^{2+} efflux/Ca^{2+} influx ratio at 3 μM Ca^{2+}_i, that is, 30 times the assumed basal $[Ca^{2+}]_i$ of the cell (0.1 μM in column I). Therefore, an alteration of critical intracellular variables renders an inefficient Na^+/Ca^{2+} exchanger. In other words, our overall kinetic model of ionic–metabolic interactions proposed here predicts that changes likely to occur in ischemic nerve cells will affect the efflux and influx of Ca^{2+} through the Na^+/Ca^{2+} exchanger in a way that will be deleterious for the cell due to an increase in the steady-state $[Ca^{2+}]_i$. The three critical variables are: pHi, $[Na^+]_i$, and [ATP]. It must be stressed that in TABLE 2 only changes in the Ca^{2+} fluxes through the Na^+/Ca^{2+} exchanger are taken into consideration. During ischemic conditions, other Ca^{2+} influx pathways might be in operation, which will exacerbate Ca^{2+} efflux/influx ratio unbalance in favor of net Ca^{2+}_i load and cell deterioration.

CONCLUSIONS

Our experiments with dialyzed squid axons provide new insight into the mechanism of regulation of the Na^+/Ca^{2+} exchanger. Using the

Ca^{2+}_i-dependence of the Na_o/Na_i exchange reaction as an unequivocal way to measure the affinity of the Ca_i-regulatory site we have been able to integrate the ionic and metabolic regulation of the exchanger in a kinetic model that takes into account the cross-interactions between ionic (Na^+_i-H^+_i-Ca^{2+}_i) and metabolic (ATP) modulations predicting both steady-state and pre-steady-state kinetics. Whereas the Na^+/Ca^{2+} exchange can function without MgATP, it markedly influences its activity. MgATP acts by antagonizing the H^+_i–Na^+_i synergistic inhibition and increasing the apparent affinity of the Ca^+_i-regulatory site, without affecting the transport site's affinities for Ca^{2+}_i or the maximal translocation rate. The central role of the Ca_i-regulatory site in exchange modulation is in line with the two recent findings: first, that the binding of Ca^{2+}_i to the regulatory site causes a marked conformational change in the exchange protein and second, that this regulatory site is working physiologically during beat-to-beat cardiac contractions.[25] Our integrated (ionic and metabolic variables) consecutive scheme (circular model), shows for the first time predictions on the relative importance of intracellular variables (Na^+_i–H^+_i–Ca^{2+}_i) on net Ca^{2+} movements through the Na^+/Ca^{2+} exchanger in nerve cells under normal *trans* conditions (Na^+_o and Ca^{2+}_o). These predictions might be valuable when analyzing pathological situations, such as nerve ischemia.

ACKNOWLEDGMENTS

This work was supported by grants from the U.S. National Science Foundation (MCB 0444598), FONACIT (S1-9900009046 and G-2001000637) Venezuela, and Foncyt (PICT-05-12397) and CONICET (PIP 5118) Argentina. We also thank Fundación Polar (Venezuela) and Fundación Interior Argentina for their constant support to science.

REFERENCES

1. BERS, D. 1991. Excitation-contraction coupling and cardiac contractile force. *In* Developments in Cardiac Vascular Medicine. Vol. 122. Kluwer Academic Publisher. Dordrecht/Boston/London.
2. BLAUSTEIN, M.P. & J.N. LEDERER. 1999. Sodium/calcium exchange: its physiological implication. Physiol. Rev. **79:** 763–854.
3. BLAUSTEIN, M.P. 1977. Sodium ions, calcium ions, blood pressure regulation, and hypertension: a reassessment and a hypothesis. Am. J. Physiol. **232:** C165–C173.
4. HURTADO, J., S. BORGES & M. WILSON. 2002. Na^+/Ca^{2+} exchanger controls the gain of Ca^{2+} amplifier in the dendrites of amacrine cells. J. Neurophysiol. **88:** 2765–2777.
5. MATSUDA, T., K. TAKUMA & A. BABA. 1997. Na^+-Ca^{2+} exchange; physiology and pharmacology. Jpn. J. Pharmacol. **74:** 1–20.
6. MURPHY, E., H.R. CROSS & C. STEENBERGEN. 2002. Is Na/Ca exchange during ischemia and reperfusion beneficial or detrimental? Ann. N. Y. Acad. Sci. **976:** 421–430.

7. NOBLE, D. 2002. Simulation of Na/Ca exchange activity during ischemia. Ann. N. Y. Acad. Sci. **976:** 431–437.
8. DIPOLO, R. & L. BEAUGE. 2006. Physiol. Rev. **86:** 155–203.
9. HILGEMANN, D.W. 1990. Regulation and deregulation of cardiac Na^+Ca^{2+} exchange in giant excised sarcolemmal membrane patches. Nature **344:** 242–245.
10. HE, Z., Q. TONG, B.D. QUEDNAU, et al. 1998. Cloning, expression, and characterization of the squid Na^+-Ca^{2+} exchanger (NCX-SQ1). J. Gen. Physiol. **111:** 857–873.
11. PHILIPSON, K.D. & D.A. NICOLL. 2000. Sodium-calcium exchange. A molecular perspective. Ann. Rev. Physiol. **62:** 111–133.
12. MATSUOKA, S., D.A. NICOLL, R.F. REILLY, et al. 1993. Initial localization of regulatory regions of the cardiac sarcolemmal Na^+-Ca^{2+} exchanger. Proc. Natl. Acad. Sci. USA **90:** 3870–3874.
13. DOERING, A.E. & W.J. LEDERER. 1993. The mechanism by which cytoplasmic protons inhibit the sodium-calcium exchanger in guinea pig heart cells. J. Physiol. (Lond.) **466:** 481–499.
14. DOERING, A.E. & W.J. LEDERER. 1994. The action of Na^+ as a cofactor in the inhibition by cytoplasmic protons of the cardiac Na^+-Ca^{2+} exchanger in the guinea pig. J. Physiol. (Lond.) **480:** 9–20.
15. DIPOLO, R., G. BERBERIÁN & L. BEAUGÉ. 2004. Phosphoarginine regulation of the squid nerve Na^+/Ca^{2+} exchanger: metabolic pathway and exchanger-ligand interactions different from those seen with ATP. J. Physiol. (Lond.) **554:** 387–401.
16. DIPOLO, R. & L. BEAUGÉ. 1990. Calcium transport in exitable cells. In Intracellular Calcium Regulation. Bronner, F., Ed.: 381–413. John Wiley & Sons Inc. New York.
17. KANG, T.M. & D.W. HILGEMANN. 2004. Multiple transport modes of the cardiac Na^+/Ca^{2+} exchanger. Nature **427:** 544–548.
18. DIPOLO, R. & L. BEAUGÉ. 1990. Asymmetrical properties of the Na-Ca exchanger in voltage clamped, internally dialyzed squid axons under symmetrical ionic conditions. J. Gen. Physiol. **85:** 819–835.
19. DIPOLO, R. & L. BEAUGÉ. 2002. MgATP counteracts intracellular proton inhibition of the sodium-calcium exchanger in dialyzed squid axons. J. Physiol. (Lond.) **539:** 791–803.
20. HILGEMANN, D.W. 1996. The cardiac Na-Ca exchanger in giant membrane patches. Ann. N. Y. Acad. Sci. **779:** 136–158.
21. MATSUOKA, S. & D.W. HILGEMANN. 1994. Inactivation of outward Na^+-Ca^{2+} exchange current in guinea pig ventricular myocytes. J. Physiol. (Lond.) **476:** 443–458.
22. FUJIOKA, Y., K. HIROE & S. MATSUOKA. 2000. Regulation kinetics of Na^+-Ca^{2+} exchange current in guinea-pig ventricular myocytes. J. Physiol. (Lond.) **529:** 611–623.
23. HILGEMANN, D., S. MATSUOKA, A. NAGEL & A. COLLINS. 1992. Steady-state and dynamic properties of cardiac sodium-calcium exchange. Sodium-dependent inactivation. J. Gen. Physiol. **100:** 905–932.
24. DIPOLO, R. & L. BEAUGE. 1991. Regulation of Na-Ca exchange. An overview. Ann. N. Y. Acad. Sci. **639:** 100–111.
25. OTTOLIA, M., K.D. PHILIPSON & S. JOHN. 2004. Conformational changes of the Ca^{2+} regulatory site of the Na^+-Ca^{2+} exchanger detected by FRET. Biophys. J. **87:** 899–906.

Some Biochemical Properties of the Upregulation of the Squid Nerve Na$^+$/Ca^{2+} Exchanger by MgATP and Phosphoarginine

GRACIELA BERBERIÁN,[a,b] REINALDO DiPOLO,[b,c]
AND LUIS BEAUGÉ[a,b]

[a]*Laboratorio de Biofísica, Instituto de Investigación Médica "Mercedes y Martín Ferreyra" (INIMEC-CONICET), 5000 Córdoba, Argentina*

[b]*Marine Biological Laboratory, Woods Hole, 02543 Massachusetts, USA*

[c]*Laboratorio de Fisiología Celular, Centro de Biofísica y Bioquímica, IVIC, Caracas1020-A, Venezuela*

ABSTRACT: In squid nerve MgATP upregulation of Na$^+$/Ca^{2+} exchange requires a soluble cytosolic regulatory protein (SCRP) of about 13 kDa; phosphoarginine (PA) stimulation does not. MgATP-γ-S mimics MgATP. When a 30-10-kDa cytosolic fraction is exposed to 0.5 mM [^{32}P]ATP in the same solution used for transport assays, and in the presence of native membrane vesicles, a 13-kDa and a 25-kDa band become phosphorylated. Membrane vesicles alone do not show these phosphorylated bands and heat denaturation of the cytosolic fraction prevents phosphorylation. Moreover, staurosporine, a general inhibitor of kinases, does not affect MgATP + SCRP stimulation of the exchanger or the phosphorylation of the 13 kDa but prevents phosphorylation of the 25-kDa cytosolic band. The 30-10-kDa fraction phosphorylated in the presence of staurosporine stimulates Na$^+$/Ca^{2+} exchange in vesicles in the absence of ATP but with Mg^{2+} in the medium. The 30-10-kDa fraction is not phosphorylated by PA. In membrane vesicles two protein bands, at about 60 kDa and 70 kDa identified as the low molecular weight neurofilament (NF), are phosphorylated by PA, but not by MgATP. This phosphorylation is specific for PA, insensitive to staurosporine (similar to the PA-stimulated fluxes), and labile. In addition, co-immunoprecipitation was observed between the NF and the exchanger protein. Under the conditions of these experiments no phosphorylation of the exchanger is detected, either with MgATP or PA.

Address for correspondence: Luis Beaugé, Instituto de Investigación Médica "Mercedes y Martín Ferreyra" (INIMEC-CONICET), Casilla de Correo 389, 5000 Córdoba, Argentina. Voice: 54-351-468-1465; fax: 54-351-469-5163.

lbeauge@immf.uncor.edu

Ann. N.Y. Acad. Sci. 1099: 152–165 (2007). © 2007 New York Academy of Sciences.
doi: 10.1196/annals.1387.009

KEYWORDS: Na$^+$/Ca^{2+} exchange; metabolic regulation; ATP; phospho-arginine; squid axons

INTRODUCTION

In dialyzed squid giant axons MgATP induces a substantial stimulation of the Na$^+$/Ca^{2+} exchanger.[1,2] On the other hand, initial experiments with isolated membrane vesicles from squid nerve failed to show this stimulation. One possibility was that a critical factor related to MgATP effect is lost during vesicle preparation. The first evidence supporting this idea came from the progressive run down of MgATP stimulation when axons were subjected to prolonged dialysis with highly porous capillaries (about 18 kDa molecular weights cut off). Partial purification of the squid axoplasm and brain led to the isolation of about 13-kDa soluble cytosolic protein or proteins that restored the MgATP effect lost after prolonged dialysis.[3,4] In addition, squid nerve membrane vesicles incubated with this protein displayed a substantial ATP stimulation of the exchanger. For this reason we called it soluble cytosolic regulatory protein, or SCRP.[3,4] On the other hand, no stimulation of the exchanger was observed with this protein/s alone or in the presence of nonhydrolyzable ATP analogs.[4] In mammalian heart and nerve, phosphoinositides, and particularly PtdIns-4,5-P2 are involved in MgATP stimulation of the Na$^+$/Ca^{2+} exchanger.[5–8] Conversely, in squid nerve vesicles, MgATP + SCRP activation occurs without any significant change in the patterns of phosphoinositides (PtdIns- 4-P and PtdIns-4,5-P2) formation[9]; also, in dialyzed squid axons the injection of PtdIns-PLC at concentrations high enough to reduce the PtdIns levels, or the injection of the PtdIns- 4,5-P2-specific antibody has no effect on MgATP modulation of the exchanger.[9]

A second metabolic upregulator of the squid Na$^+$/Ca^{2+} exchanger is phosphoarginine (PA).[10,11] Its metabolic pathway looks different from that of ATP[11,12] since: (*a*) Activation of the exchanger by PA is effective even when the MgATP effect is blocked by CrATP; (*b*) full PA stimulation is still observed at saturating ATP concentrations; and (*c*) PA does not require cytosolic factor since it activates in axons after prolonged dialysis and also in isolated squid nerve membrane vesicles in the absence of the SCRP. On the other hand, PA modulation seems also related to a phosphorylation process because, as it happens with MgATP, it is blocked by exogenous alkaline phosphatases.[2] The effects of MgATP and PA are also different on the kinetics of the exchanger: unlike MgATP, PA regulation does not affect the H^+_i, Na^+_i, and Ca$^{2+}_i$ interactions with the intracellular regulatory sites, but increases the affinity of the intracellular transport sites, preferentially for Ca$^{2+}_i$ (about 20-fold) over Na^+_i (50%).[11,12]

Although the complete mechanisms by which MgATP and PA upregulate the squid Na$^+$/Ca^{2+} exchanger are not yet known, this work provides important new information leading to their understanding.

METHODS

Preparation of a Cytosolic Fraction Containing the SCRP

Initially, the SCRP was obtained from a cytosol filtrate from a 30-kDa cutoff filter passed through a FPLC Superdex-75 column (Amersham Biosciences, Piscataway, NJ). The eluent fraction that promoted MgATP stimulation of the exchanger biological activity coincided with that where the cytochrome C marker (FW 12.9 kDa) came out.[3,4] The amount of protein obtained was exceedingly small, usually detected only with silver stain. Still, it showed three bands near 19 kDa, 13 kDa, and 9 kDa. In the experiments reported here we used the 30-kDa cytosolic fraction partially purified. When that fraction was passed through anionic (Dowex 1×8–400) and cationic (Dowex 50×8–400) exchange columns, activity was recovered only in the anionic column, indicating a protein with a net positive charge. Therefore, through out this work we used the eluent of an anionic column retained by a 10-kDa cutoff filter (30-10-kDa fraction).

Preparation of Nerve Membrane Vesicles, Phosphorylation, and Transport Studies

Squid nerve membrane vesicles were prepared from optic nerves as described previously[4]; about 40% of the vesicles are inside out allowing the study of intracellular exchanger–ligand interactions. The method for transport assays is also described elsewhere.[4] The phosphorylating conditions are indicated in the legends to figures. The patterns of protein phosphorylation were followed by a Storm 840 image analyzer (Molecular Dynamics Inc., Sunnyvale, CA) and a Scion PC software (NIH, bethesda, MD). Other technical details are given in the legends to the figures.

Synthesis of [^{32}P]-Labeled PA

The procedure for biochemical synthesis of [^{32}P]PA is shown in detail elsewhere.[13] The only point worth mentioning here is that the synthesized [^{32}P]PA is fully competent, free of contaminants, and with a specific activity of 1-2 $\times$ 10^3 cpm/pmol.

RESULTS AND DISCUSSION

Stimulation by ATP-γ-S of Na$^+$/Ca^{2+} Exchange Fluxes in Squid Nerve Membrane Vesicles

In dialyzed squid axons ATP-γ-S mimics ATP; however, the effects of this analoge had not been investigated in squid nerve membrane vesicles. If

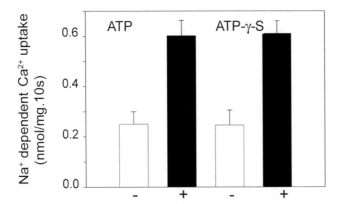

FIGURE 1. Stimulation by ATP-γ-S of the Na$^+$/Ca^{2+} exchange fluxes in membrane vesicles from squid nerve. Comparison between 1 mM ATP and 1 mM ATP-γ-S in the upregulation of the Na$^+$-dependent Ca^{2+} uptake in membrane vesicles of squid optic nerve. Vesicles were prepared as indicated in Methods and were loaded with 300 mM NaCl, 20 mM MOPS-Tris (pH 7.4 at 20°C), and 0.1 mM EDTA.[4] Influx of ^{45}Ca^{2+} were measured during 10 sc, incubating the vesicles at 20°C in the following solutions: 300 mM or 30 mM NaCl, 1 mM free Mg^{2+}, 20 mM MOPS-Tris (pH 7.3 at 20°C), 0.150 mM EGTA, 0.2 mM vanadate, and 1 mM Ca^{2+} without or with 1 mM ATP or 1 mM ATP-γ- S . Extravesicular NaCl was replaced with iso-osmolar concentrations of NMG-Cl. In all cases the cytosolic fraction of 10- 30 kDa obtained from squid optic ganglia (see "Methods" section) was presented. In the absence of the 10-30-kDa fraction the ATP and the ATP-γ-S upregulation fails to appear (results not shown here). The bars represent the Mean ± SE of triplicate determinations.

these vesicles are indeed a reliable preparation to follow transport–biochemical reactions ATP-γ-S should be able to stimulate the exchanger and only in the presence of the SCRP. One experiment to test this point is described in FIGURE 1. It is clear that there is no difference between ATP and its hydrolysable analoge; that is, these membrane vesicles are adequate to our aims. Not shown here is the lack of ATP-γ-S effect in the absence of SCRP. In a way, this result marks another difference with the mammalian cardiac and nerve exchangers. In the membrane vesicles from the mammalian heart[6] and brain,[8] MgATP upregulation requires the simultaneous presence of vanadate, whereas in dialyzed squid axons[14] and squid nerve vesicles (not shown here) it does not.

ATP and PA Phosphorylation of Membrane Proteins in Squid Nerve Membrane Vesicles

Phosphorylation of squid nerve membrane vesicles was studied by using [^{32}P]P-γ-ATP and [^{32}P]PA under conditions where both compounds stimulate the exchange fluxes.[12] One of these experiments, in 8% SDS-PAGE, is in FIGURE 2. With MgATP (*Panel A*, Lane 4), abundant phosphorylated bands were observed, including one around the molecular weight corresponding to

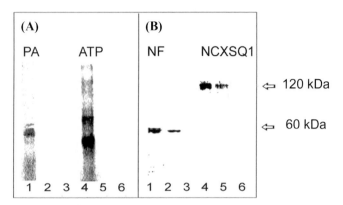

FIGURE 2. Phosphorylation of squid nerve membrane vesicles by ATP and PA. For ATP phosphorylation squid nerve membrane vesicles (15 µg) were incubated for 10 sc, in the presence of 0.5 mM [^{32}P]-γ-ATP, 0.7 µM Ca^{2+}, 1 mM Mg^{2+}, 0.2 mM vanadate, 260 mM NMG.Cl, 30 mM NaCl, at pH 7.3 (20°C) and 5 µL of a solution containing the 30-10-kDa cytosolic fraction (1.4 mg/mL total protein). [^{32}P]PA phosphorylation was performed under similar conditions without ATP and with 2 mM [^{32}P]PA. Phosphorylation was stopped by 5% TCA for total proteins or by dilution adding detergents for immunoprecipitation. Proteins were separated by 8% SDS-PAGE and electroblotting on PVDF membrane. *Panel A* shows phosphoproteins visualized by phosphoimage. Lanes 1 and 4, membrane vesicles; Lane 2, immunoprecipitate with anti- NF Ab; Lanes 3 and 6, negative control using preimmune serum in the immunoprecipitation step. Lane 5, immunoprecipitate with anti-NCXSQ1 Ab . *Panel B* is a Western blot with the same PVDF membrane use for phosphoimage. The small NF and the Na$^+$/Ca^{2+} exchanger proteins in the blots were identified with their respective specific antibodies: Lanes 1 to 3, anti-NF Ab; Lanes 4 to 6, anti-NCXSQ1 Ab. Lanes 3 qnd 6, negative controls.

the Na$^+$/Ca^{2+} exchanger. Nevertheless (*Panel A*, Lane 5), Na$^+$/Ca^{2+} exchanger immunoprecipitated with a specific antibody (kindly provided by Dr. K. Philipson) does not show any [^{32}P]Pi incorporation. The lack of phosphorylation is not due to the loss of exchanger protein because this protein is detected in the immunoblots (*Panel B*, Lanes 4 and 5). With [^{32}P]PA we used millimolar concentrations. In must be noted that the phosphorylation patterns of MgATP and PA are different. Two main bans, at about 60 kDa and 70 kDa, which appear with PA (*Panel A*, Lane 1) are not seen with MgATP (*Panel A*, Lane 4).

Amino acid sequencing by mass spectrometry of the PA bands (W.M. Keck Biomedical Mass Spectrometry Laboratory, University of Virginia, VA0) reveals a mix of proteins: the most abundant corresponds to a low molecular weight (60-70 kDa) neurofilament (small NF); the others, in much lower quantity, are tubuline A and B. The small NF was verified in Western blots with specific antibodies (kindly provided by Dr. H. C. Pant from the Neurochemistry Laboratory at NIH). It is interesting that while MgATP does not phosphorylate the 60-70 NF, it does so to the 200–220 kDa large NF.[15] An

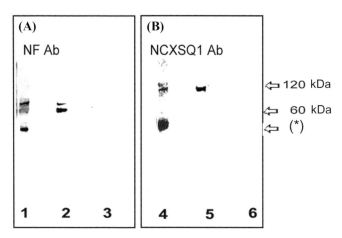

FIGURE 3. Western blot of the co-immune precipitation of the squid nerve NF and Na^+/Ca^{2+} exchanger proteins. Membrane vesicles of squid optic nerve were treated with buffer RIPA and the immunoprecipitation was performed using the anti-NF antibody.[15] The immunoprecipitate was then run in SDS-PAGE, transblotting on PVDF, and revealed using anti-NF antibody (*left, panel* **A**). After striping, the same PVDF membrane was exposed to anti-NCXSQ1 antibody (*right, panel* **B**). The samples are: Lanes 1 and 4: immunoprecipitate with NF-Ab; Lanes 2 and 5, positive controls by using membrane vesicles and Lanes 3 and 6, negative controls (the first Ab in the immunoprecipitation step was replaced by no-immune rabbit serum). The *arrows* indicate the band of Na/Ca exchange (120 kDa), the band of NF (60 kDa), and the band of immunoglobulin derived from the immunoprecipitate (∗). (Unpublished results of D. Raimunda.)

additional observation was that the PA-phosphorylated NF is quite labile, losing the $[^{32}P]Pi$ upon stopping the reaction by dilution; that might be the reason for not detecting $[^{32}P]Pi$ incorporation in the immunoprecipitate (*Panel A*, Lane 2) while the NF protein is detected in the Western blots (*Panel B*, Lane 1 and 2). So far, we have found no data in the literature on the properties of the small NF other that in the squid is associated with plasma membrane and cytoskeleton components.[15] However, FIGURE 3 shows a novel important observation: the small NF does co-immunoprecipitate with the Na^+/Ca^{2+} exchanger protein. Yet, when the NCXSQ1 was immunoprecipitated with its specific antibody the NF was not detected in the Western blot. This could be explained if (*a*) not all NF and NCXSQ1 form a complex, and (*b*) the NCXSQ1 Ab attaches to the free exchanger protein and, for steric reasons and/or conformational changes, does not do so to the NF–NCXSQ1 complex.

The different transport kinetic effects and phosphorylation patterns due to MgATP and PA, and the fact that upregulation of the exchanger by PA does not require the SCRP[11,12] agree with the existence of two metabolic pathways for stimulation of the squid nerve Na^+/Ca^{2+} exchanger by these ligands.[10–12] Based on the labile PA phosphorylation of the small NF, it could be argued that

MgATP phosphorylation of the exchanger may be lost after the hard conditions of immunoprecipitation. Nonetheless, MgATP phosphorylation patterns
of total membrane proteins not subjected to immunoprecipitation are exactly
the same with or without the SCRP (not shown here). This would favor a lack
of exchanger phosphorylation by MgATP in the squid.

Phosphorylation of the 30-10-kDa Cytosolic Fraction

FIGURE 4 A is a gradient gel stained with Coomassie blue showing that
the 30-10-kDa cytosolic fraction (Lane 1) contains several proteins, two of
them conspicuous, around 25 kDa and 13 kDa. FIGURE 4 B is a phosphoimage
of a gradient gel with samples from phosphorylation experiments in which
the 30-10-kDa cytosolic fraction and nerve membrane vesicles (alone or in
combination) were exposed for 10 s at room temperature to 0.5 mM $[^{32}P]P-\gamma$-
ATP in the same solution used for Na^+/Ca^{2+} transport experiments.

Lane 1 corresponds to the 30-10-kDa fraction alone, Lane 2 to a mixture of
native cytosolic fraction and membrane vesicles, Lane 3 to native membrane
vesicles alone, and Lane 4 native membrane vesicles mixed with heat denatured

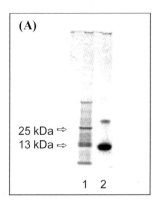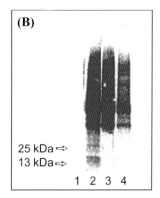

FIGURE 4. Protein content pattern and protein phosphorylation of the 30-10-kDa cytosolic fraction. (**A**) Protein pattern: Aliquots of the 30-10-kDa fraction were run in a gradient gel (8–16 % NOVEX; Invitrogen Corp., carlsbad, CA) and stained with Coomassie blue.
Note that in Lane 1 there are two major bands with molecular weights around 25 kDa and
13 kDa, respectively. Lane 2 is cytochrome C (main band at 12.9 kDa. The top band in this
Lane is a dimmer usually seen at high concentrations). The *arrows* indicate the molecular
weight markers. (**B**) Protein phosphorylation: The following components of squid nerve
were incubated for 10 s at 20°C with 0.5 mM $[^{32}P]P-\gamma$-ATP, 0.7 μM Ca^{2+}, 1 mM Mg^{2+},
200 μM vanadate, 260 mM NMG.Cl, and 30 mM NaCl: Lane 1: 30-10-kDa native fraction
alone. Lane 2: 30-10-kDa native fraction plus membrane vesicles of squid optic nerve. Lane
3: native membranes of squid optic ganglia alone. Lane 4: native membranes of squid optic
ganglia plus heat inactivated (10 min at 70°C) 30-10-kDa cytosolic fraction. When present,
the total protein of nerve vesicles was 20 μg per tube.

cytosolic fraction. Note that only in the presence of membrane vesicles does the cytosolic fraction become phosphorylated, and phosphorylation occurs just in the two major bands around 25 kDa and 13 kDa. This means that the responsible kinase/s for its phosphorylation is/are located in the plasma membrane of the cells. Other important results include: (*a*) membrane vesicles alone do not show phosphorylation in the 30-10-kDa bands zone and (*b*) heat denaturation of the cytosolic fraction prevents phosphorylation. This coincides with the inability of this fraction to promote MgATP stimulation of the exchanger after similar heat denaturation.[3,4]

Effects of Staurosporine on MgATP Phosphorylation of the 30-10-kDa Cytosolic Fraction and on PA Phosphorylation of the Small NF

Staurosporine, particularly at high concentrations, is a general inhibitor of kinases.[16] In dialyzed axons, MgATP and PA stimulations of the Na^+/Ca^{2+} exchanger are insensitive to a large variety of kinases and phosphatase inhibitors, including staurosporine up to 100 μM concentrations.[2,4] On those bases, we investigated the effects of staurosporine on MgATP and PA phosphorylation and transport stimulation in nerve vesicles. Protein phosphorylation of the whole membrane vesicles by MgATP was also examined. These results are in FIGURES 5 and 6. FIGURE 5 A shows that 50 μM staurosporine does not affect the ATP + SCRP stimulation of Na^+/Ca^{2+} exchange fluxes. On the other hand, in the presence of 1, 10, and 50 μM staurosporine (FIG. 5 B)

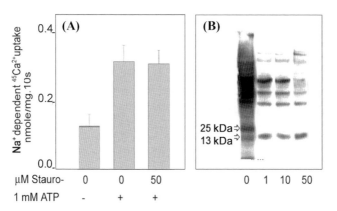

FIGURE 5. Effect of staurosporine on the ATP plus 30-10-kDa cytosolic fraction effects. *Left panel* (**A**): The stimulation MgATP plus the 30-10-kDa cytosolic fraction is unimpaired by the presence of 50 μM staurosporine. *Right panel* (**B**): MgATP phosphorylation of the 13-kDa protein from the squid cytosolic 10-30- kDa fraction is not affected by staurosporine up to 50 μM. On the other hand, the 25-kDa band does not incorporate Pi in the presence of the kinases inhibitor. Transport and phosphorylations assays were carried out as described in the legends to FIGURES 1 and 4.

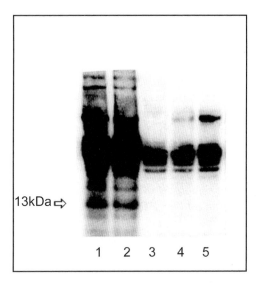

FIGURE 6. Stability of the phosphoproteins in the 30-10-kDa cytosolic fraction phosphorylated from MgATP. Phosphorylation was carried out as described in the legend to FIGURE 4. Phosphorylation was stopped by 5 × Laemmli sample buffer and the samples incubated for 10 min at 30°C in the following reaction mixtures: Lane 1: 0.1 M Mops-Tris (pH 7.6); Lane 2: 0.1 M acetate (pH 5.2); Lane 3: 0.6 M hydroxylamine in acetate (pH 5.2); Lane 4: 0.1 N HCl (pH 1); Lane 5: 0.1 N NaOH (pH 13). The figure is a phosphoimage of gradient (8–16%) SDS-PAGE.

[^{32}P]Pi incorporation in the 25-kDa band is abolished whereas phosphorylation of the 13-kDa band remains unaffected. This gives strong support to the idea that the 13-kDa band is indeed involved in MgATP stimulation of the exchanger.

In these experiments, after phosphorylation was stopped, the vesicles and the 30-10-kDa cytosolic fraction were not separated; that is, they went together into the gel and the membrane proteins appear on top. By looking at membrane proteins we can see that staurosporine eliminates most of their MgATP-dependent phosphorylation. This may provide an excellent tool for further analysis of protein phosphorylation related to stimulation of the exchanger.

Stability of the Phosphorylated 30-10-kDa Fraction

Several amino acid residues can incorporate a phosphate group through the action of kinases. One way to estimate which of them might be involved is to study the stability of the phosphoproteins in different media. FIGURE 6 describes one experiment where this was tested for the 30-10-kDa cytosolic protein phosphorylated from MgATP.

After stopping the reaction aliquots were incubated for 10 min at 30°C in solutions of variable composition. FIGURE 6 shows that the 13-kDa band remains unaltered in MOPS-Tris, pH 7.4 (Lane 1), decreases slightly in Na-acetate, pH 5.3 (Lane 2), and completely disappears in hydoxylamine Na-acetate (Lane 3), NaOH (Lane 4), and HCl (Lane 5). This indicates that the amino acid residues that incorporate phosphoryl groups could be tyrosine, aspartic acid, and/or glutamine but not serine, threonine, or histidine.[17]

Specificity, Stability, and Lack of Staurosporine Effect in PA Phosphorylation of the Small NF Protein in Squid Optic Nerve Membrane Vesicles

FIGURE 7 shows experiments on specificity and stability of PA phosphorylation of the small NF protein. *Panel A* (*left*) shows that this phosphorylation is indeed specific for PA since it is attenuated by nonradioactive PA (Lane 2) but not by ATP (Lane 3), AMP-PCP (Lane 4), or Pi (Lane 5).

Panel B (*right*) in the figure shows that the phosphorylated 60-70 NF is stable in acetate pH 5.4 (Lane 1), acetate pH 5.4 plus hydroxylamine (Lane 2), and 0.1 N HCl (Lane 3) but is labile at pH 7.6 (Lane 4) and in 0.1 N NaOH (Lane 5). These characteristics agree with serine and/or threonine residues but not with tyrosine, aspartic, glutamic acid, or histidine.[17] These characteristics are different from those seen in the MgATP phosphorylation of the 30-10-kDa cytosolic fraction.

FIGURE 8 illustrates the correlative lack of effect of 100 μM staurosporine on PA stimulation of Na^+/Ca^{2+} fluxes and of $[^{32}P]Pi$ incorporation from $[^{32}P]PA$

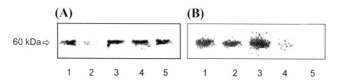

(A)　　　　　**(B)**

60 kDa ⇨

　　1　2　3　4　5　　1　2　3　4　5

FIGURE 7. Specificity and stability of $[^{32}P]PA$ phosphorylation of the 60-70-kDa NF protein present in squid nerve membrane vesicles. Aliquots of 50 μg total squid nerve vesicles protein were incubated for 10 s at 20° C in the usual medium for $[^{45}Ca]Ca^{2+}$ uptake through the Na^+/Ca^{2+} exchanger (1 μM Ca^{2+}, 2 mM free Mg^{2+}, 30 mM NaCl, 260 mM NMG.Cl, 0.1 mM Vanadate, pH 7.3) with 2 mM $[^{32}P]PA$ (500 cpm/pmol). (**A**) (*left panel*) is the phosphoimage of a SDS-PAGE of one experiment on specificity. In addition to 2 mM $[^{32}P]PA$ the following ligands were present: Lane 1: None; Lane 2: 10 mM cold PA; Lane 3: 10 mM ATP; Lane 4: 10 mM AMP-PCP; and Lane 5: 10 mM Pi. (**B**) (*right panel*) refers to a stability experiment. Here, phosphorylation was stopped by 5 × Laemmli sample buffer and the samples incubated for 10 min at 30°C in the following reaction mixtures: Lane 1: 0.1 M acetate (pH 5.2); Lane 2: 0.6 M hydroxylamine in acetate (pH 5.2); Lane 3: 0.1 N HCl (pH 1); Lane 4: 0.1 M Mops-Tris (pH 7.6); and Lane 5: 0.1 N NaOH (pH 13). The *arrow* indicates the position of the 60-kDa molecular weight marker.

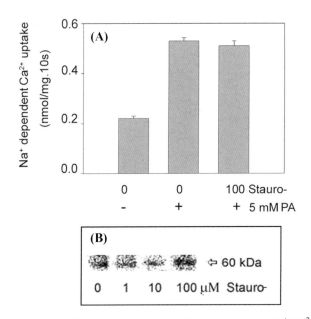

FIGURE 8. Lack of effect of staurosporine on PA stimulation of Na^+/Ca^{2+} exchange fluxes and on phosphorylation of the small NF protein. (**A**)(*top panel*): The stimulation by PA of the exchange fluxes is not affected by 100 μM staurosporine. (**B**) (*bottom panel*): Similarly, Pi incorporation from $[^{32}P]PA$ into the small NF protein is unaffected by staurosporine up to 100 μM concentration. Transport and phosphorylation assays were carried out as described above.

in squid nerve membrane vesicles. This concurs with the staurosporine insensitivity of PA stimulation of the Na^+/Ca^{2+} exchanger in dialyzed squid axons.[11]

Stimulation of Na^+/Ca^{2+} Exchanger by the Phosphorylated Cytosolic Factor

Experiments not described here showed that the incorporation of $[^{32}P]Pi$ in the 13-kDa cytosolic band of 30-10 kDa increased when phosphorylation was stopped with the addition of 5 mM EDTA and a mixture of phosphatases inhibitors (0.8 mM Vanadate, 0.5 mM glycerol-phosphate, and 10 μM okadaic acid). Under these conditions, after the cytosolic fraction was separated from the membrane vesicles by centrifugation (30 min at 25 spi in an Beckman Airfuge; Beckman Coulter, Inc., Fullerton, CA) the 13-kDa band remained fully phosphorylated for at least 36 h (the longest time tested) at –70° C, 4° C, and room temperature. These results provided us with the opportunity to investigate if the phosphorylated factor, without ATP, could stimulate Na^+/Ca^{2+} exchanger in squid nerve membrane vesicles. Phosphorylation was done as in FIGURE 2,

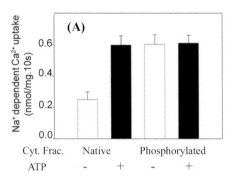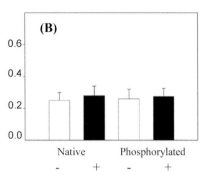

FIGURE 9. Effect of native and phosphorylated 30-10-kDa cytosolic factor on the Na^+-dependent Ca^{2+} uptake in squid nerve vesicles in the absence and presence of ATP and Mg^{2+} ions. Phosphorylation of cytosolic proteins was done as in FIGURE 2 in the presence of 50 µM staurosporine and stopped by adding EDTA and phosphatase inhibitors. The cytosolic phosphoproteins were separated from the vesicles by 30-min centrifugation at 25 psi in a Beckman Airfuge. The uptake solution contained 1µM [^{45}Ca]Ca^{2+}, 0.15 mM EGTA, 0.15 mM Vanadate, 30 or 300 mM NaCl, 20 or 200 mM Mops-NMG (pH.7.3), 70 mM NMG.Cl. The following contaminants from the phosphorylation reaction were present: 187 µM EDTA, 5 µM ATP, 45 µM Mg^{2+}, and 50 µM Vanadate. *Bars* are the means ± SD of triplicate determinations. (**A**)*Left panel*: 1 mM ionized Mg^{2+}; (**B**)*right panel*: no Mg^{2+} added. The state of the 30-10-kDa cytosolic fraction and the presence of 1 mM ATP is indicated in the figure.

in the presence of 50 µM staurosporine, and stopped by adding EDTA and the indicated phosphatase inhibitors.

The results, illustrated in FIGURE 9, show that the phosphorylated 30-10-kDa cytosolic fraction can stimulate Na^+/Ca^{2+} exchange fluxes in the absence of ATP (*left, panel A*). Interesting, as it happens with ATP, that stimulation does not take place in the absence of millimolar [Mg^{2+}] (*right, panel B*). This may indicate that a transphosphorylation takes place (from the factor to another structure) or that the binding of the phosphorylated protein to its target is Mg^{2+}-dependent. On the other hand, it opens up a series of important experimental approaches to continue investigating this system.

CONCLUSIONS

In this work we provide novel information regarding biochemical properties of MgATP and PA upregulation of the squid nerve Na^+/Ca^{2+} exchanger. Regarding MgATP, the data verified that the soluble cytosolic regulatory factor required for the nucleotide effect has indeed a molecular weight around 13 kDa and becomes phosphorylated by one or more kinase/s present in the plasma membrane. Furthermore, when phosphorylated, and only in the presence of millimolar ionized Mg^{2+} concentrations, this protein/s can stimulate the exchanger in the absence of ATP. At present efforts are directed to the identification, cloning, and expression of this factor. Regarding PA upregulation, we have shown that the small NF protein is phosphorylated by PA and also that its immunoprecipitate contains the exchanger protein. However, despite these interesting results, we cannot assert at this stage what role, if any, the small NF plays in the phosphagen stimulation of the exchanger. On the other hand, the fact that PA is able to phosphorylate a protein related to membrane and cytoskeleton structures opens a relevant field for research on the metabolic function of this phosphagen, which goes beyond the regulation of the Na^+/Ca^{2+} exchanger.

ACKNOWLEDGMENTS

This work was supported by grants from the US-NSF (MCB 0444598), FONCYT (PICT-05-12397), and CONICET (PIP 5118), Argentina and Fonacit (S1-9900009046 and G-2001000637) Venezuela.

REFERENCES

1. BLAUSTEIN, M.P. & J.W. LEDERER. 1999. Sodium/calcium exchange: its physiological implications. Physiol. Rev. **79:** 763–854.
2. DIPOLO, R. & L. BEAUGÉ. 2006. Sodium/calcium exchanger: influence of metabolic regulation on ion carrier interactions. Physiol. Rev. **86:** 155–203.
3. BEAUGÉ, L., D. DELGADO, H. ROJAS, G. BERBERIÁN & R. DIPOLO 1996. A nerve cytosolic factor is required for MgATP stimulation of a Na^+ gradient-dependent Ca^{2+} uptake in plasma membrane vesicles from squid optic nerve. Ann. N Y Acad. Sci. **779:** 208–216.
4. DIPOLO, R., G. BERBERIÁN, D. DELGADO, H. ROJAS & L. BEAUGÉ 1997. A novel 13 kDa cytoplasmic soluble protein is required for the nucleotide (MgATP) modulation of the Na/Ca exchange in squid nerve fibers. FEBS Lett. **401:** 6–10.
5. HILGEMANN, D.W. 1997. Cytoplasmic ATP-dependent regulation of ion transporters and channels: mechanisms and messengers. Annu. Rev. Physiol. **59:** 193–220.
6. BERBERIAN, G., C. HIDALGO, R. DIPOLO & L. BEAUGE. 1998. ATP stimulation of Na^+/Ca^{2+} exchange in cardiac sarcolemmal vesicles. Am. J. Physiol. **274:** C724–C733.
7. ASTEGGIANO, C., G. BERBERIÁN & L. BEAUGÉ. 2001. Phosphatidylinositol-4,5-biphosphate bound to bovine cardiac Na/Ca exchanger displays a MgATP regulation similar to that of the exchange fluxes. Eur. J. Biochem. **268:** 437–442.
8. BERBERIÁN, G., C. ASTEGGIANO, C. PHAM, *et al.* 2002. MgATP and phosphoinositides regulation of bovine brain Na^+/Ca^{2+} exchange. Comparison with other Na^+/Ca^{2+} exchangers. Pflugers Arch. **444:** 677–684.
9. DIPOLO, R., G. BERBERIÁN & L. BEAUGÉ. 2000. In squid nerves intracellular Mg2+ promotes deactivation of the ATP up-regulated Na^+/Ca^{2+} exchanger. Am. J. Physiol. **279:** C1631–C1639.

10. DiPOLO, R. & L. BEAUGÉ. 1995. Phosphoarginine stimulation of Na^+/Ca^{2+} exchange in squid axons. A new pathway for metabolic regulation? J. Physiol. **487:** 57–66.

11. DiPOLO, R. & L. BEAUGÉ. 1998. Differential up-regulation of Na^+/Ca^{2+} exchange by phosphoarginine and ATP in dialysed squid axons. J. Physiol. **507:** 737–747.

12. DiPOLO, R., G. BERBERIÁN & L. BEAUGÉ. 2004. Phosphoarginine regulation of the squid nerve Na^+/Ca^{2+} exchanger: metabolic pathway and exchanger-ligand interactions different from those seen with ATP. J. Physiol. **554:** 387–401.

13. BEAUGÉ, L., M. SIRAVEGNA & G. BERBERIÁN. 2000. Enzymatic synthesis of [32P]-labeled phosphoarginine. Anal. Biochem. **286:** 306–308.

14. DiPOLO, R. & L. BEAUGÉ. 1987. Characterization of the reverse Na/Ca exchange in squid axons and its modulation b y Ca_i and ATP: Ca_i-dependent Na_i-Ca_o and Na_i-Na_o exchanges modes. J. Gen. Physiol. **90:** 505–525.

15. JAFFE, H., P. SHARMA, P. GRANT & H.J. PANT. 2001. Characterization of the phosphorylation sites of the squid (Loligo pealei) high-molecular-weight neurofilament protein from giant axon axoplasm. J. Neurochem. **76:** 1022–1031.

16. REDDY, M.M. & P.M. QUINTON. 2004. Control of dynamic CFTR selectivity by glutamate and ATP in epithelial cells. Nature **423:** 756–760.

17. MUIMO, R., Z. HORNICKOVA, C.E. RIEMEN, *et al.* Histidine phosphorylation of annexin I in airway epithelia. J. Biol. Chem. **275:** 36632–36636.

Ionic Selectivity of NCKX2, NCKX3, and NCKX4 for Monovalent Cations at K^+-Binding Site

JU-YOUNG LEE, WON-KYUNG HO, AND SUK-HO LEE

Department of Physiology and National Research Laboratory for Cell Physiology, Seoul National University College of Medicine, Chongno-Ku, Yongon-Dong 28, Seoul, 110-799, Korea

ABSTRACT: To determine the ionic selectivity of K^+-sites in three members of $Na^+/Ca^{2+}+K^+$ exchanger (NCKX) family: NCKX2, NCKX3 and NCKX4, we compared the amplitudes of reverse mode NCKX current (I_{NCKX}) activated by K^+-substitutes (Rb^+, NH_4^+, Cs^+, or Li^+) relative to that by K^+ in an HEK293 cell overexpressing each of NCKX isoforms. In all three isoforms, monovalent cations activated I_{NCKX} with similar order of potency: $K^+ > Rb^+ > NH_4^+ > Cs^+ \gg Li^+$. However, the relative potency of Cs^+ and NH_4^+ for activating NCKX2 was significantly higher than those for NCKX3 and NCKX4, indicating that the selectivity of NCKX2 for K^+ is weaker than the other two isoforms.

KEYWORDS: NCKX; K^+-binding site; ionic selectivity

INTRODUCTION

Six members of $Na^+/Ca^{2+}+K^+$ exchanger (NCKX) family have been cloned, and their transcripts are expressed with wide tissue distribution.[1,2] With no specific inhibitor available, NCKX activity in native cells has been identified by demonstrating that the Na^+/Ca^{2+} exchange activity is dependent on intracellular K^+.[3,4] Despite the importance of ion selectivity of the K^+-binding site on NCKX protein for such studies, differences in the ionic selectivity at K^+-binding sites between NCKX isoforms have not been extensively studied. Here, we investigated whether various monovalent cations are able to substitute for K^+ on the side where Ca^{2+} is present in activation of NCKX2, NCKX3, and NCKX4.

Address for correspondence: Suk-Ho Lee, Ph.D., Department of Physiology and National Research Laboratory for Cell Physiology, Seoul National University College of Medicine, Chongno-Ku, Yongon-Dong 28, Seoul, 110-799, Korea. Voice. 82-2-740-8222; fax: 82-2-763-9667.
leesukho@snu.ac.kr

Ann. N.Y. Acad. Sci. 1099: 166–170 (2007). © 2007 New York Academy of Sciences.
doi: 10.1196/annals.1387.012

RESULTS AND DISCUSSION

We compared the ionic selectivity of K^+-sites in three members of the NCKX family: NCKX2, NCKX3, and NCKX4, each of which was overexpressed in HEK293 cells (cDNA: gift from Dr. J. Lytton). The reverse mode NCKX current (I_{NCKX}) was elicited by applying a bath solution containing both Ca^{2+} (1 mM) and K^+ (120 mM), while the cell was internally dialyzed with the high Na^+ pipette solution containing 10 mM BAPTA. In addition, we confirmed that the outward NCKX current was induced neither by K^+ nor by Ca^{2+} ion alone.

To determine the ionic selectivity of the K^+-transport site of NCKX2, we compared the amplitudes of I_{NCKX2} activated by K^+-substitutes (Rb^+, NH_4^+, Cs^+, or Li^+) relative to that by K^+ in the same cell. FIGURE 1 A shows a representative whole-cell recording of I_{NCKX2} at a holding potential of 0 mV induced by a series of monovalent cations together with 1 mM Ca^{2+}. The activation of I_{NCKX2} in the presence of each K^+-substitute was intervened by applying the control bathing solution containing LiCl and EGTA (0.5 mM). The I-V plot of I_{NCKX2} induced by each monovalent cation was obtained by subtracting the ramp currents in a control bathing solution from that in the presence of the monovalent cation plus Ca^{2+} (FIG. 1 A, right). We repeated similar experiments with HEK293 cells heterologously transfected with cDNA of NCKX3 or NCKX4.

We measured the amplitude of I_{NCKX} induced by bath application of K^+ + Ca^{2+} in every individual cell, and normalized the amplitude of I_{NCKX} induced by other monovalent cation with respect to that induced by K^+. The normalized amplitudes of I_{NCKX} induced by four different monovalent cations are summarized in FIGURE 2.

In general, monovalent cations activated I_{NCKX} with the following order of potency: $K^+ > Rb^+ > NH_4^+ > Cs^+ \gg Li^+$. In addition, organic cations (TMA^+ and TEA^+) induced no NCKX current (data not shown). The selectivity of NCKX2 for K^+ was weaker than the other two isoforms. The relative potency of Cs^+ and NH_4^+ for activating NCKX2 was significantly higher than those for NCKX3 and NCKX4. Moreover, the potency of Rb^+ for activating NCKX2 was comparable to that of K^+. These properties of NCKX2 seems to be in agreement with the selectivity data of NCKX1,[5] reminiscent of the high similarity in amino acid sequence of α-repeat regions between NCKX1 and NCKX2.[6]

It is generally accepted that the presence of monovalent cations on Ca^{2+}-side does not inhibit the activity of NCX except Na^+.[7] Since Li^+ on the Ca^{2+} side exhibits an activating effect for NCX, but not for NCKX,[8] intracellular dialysis with Li^+ or organic cations might be a good strategy for differentiating NCKX activity from NCX activity in native cells.

ACKNOWLEDGMENTS

This research was supported by a grant (M103KV010008-06K2201-00810) from Brain Research Center of the 21st Century Frontier Research Program

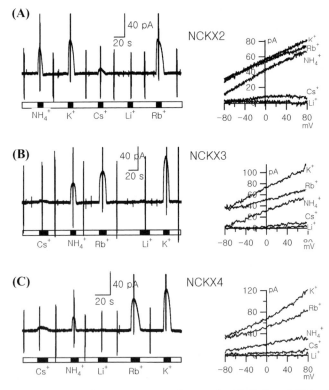

FIGURE 1. Representative current recordings at 0 mV (*left*) and I-V relationships (*right*) of the reverse mode NCKX currents recorded from HEK293 cells expressing either NCKX2 (**A**) or NCKX3 (**B**) or NCKX4 (**C**). To reverse the transmembrane Na^+ gradient, the HEK293 cells were internally dialyzed by a pipette solution containing 120 NaCl, 20 TEA-Cl, 10 BAPTA, 20 HEPES, 4 Mg ATP (in mM, pH 7.2 with NaOH). In control condition, cells were superfused with a Ca^{2+}-free bathing solution containing 120 LiCl, 0.5 EGTA, 1 $MgCl_2$, 20 TEA-Cl, 20 HEPES, 10 glucose (in mM, pH 7.4 with TMA-Cl; open bars). During periods indicated by the filled bars, 1 mM Ca^{2+} was bath applied together with different monovalent cations (120 mM, replacement of LiCl in the control bathing solution) indicated underneath the bars. The I-V relationships were determined as the difference in the current responses to a ramp pulse (-80 mV to $+80$ mV over 200 ms) before and during the bath application of different monovalent cations. Some of ramp artifacts are truncated for clarity. All experiments were performed at room temperature (22–24°C).

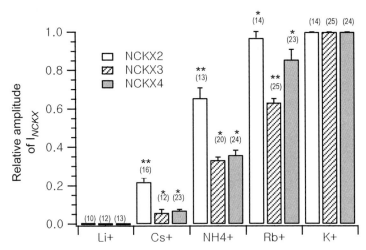

FIGURE 2. Effects of different monovalent cations on outward NCKX current. Mean values for relative amplitude of outward NCKX current activated by different monovalent cations are summarized. The amplitude of NCKX current in each ionic condition was normalized to that in the K^+-condition. The numbers in parentheses indicate number of cells tested. Error bar: SEM double asterisks (**) indicate that the specified isoform is different from the other two isoforms ($P < 0.01$) under the same cationic condition. Single asterisks (*) indicate that the specified isoform is different from only one other isoform ($P < 0.01$, Student's *t*-test).

and the grant for National Research Laboratory funded by the Ministry of Science and Technology, the Republic of Korea.

REFERENCES

1. SCHNETKAMP, P.P.M. 2004. The SLC24 Na^+/Ca^{2+}-K^+ exchanger family: vision and beyond. Pflugers Arch. **447:** 683–688.
2. CAI X., K. ZHANG & J. LYTTON. 2002. A novel topology and redox regulation of the rat brain K^+-dependent Na^+/Ca^{2+} exchanger, NCKX2. J. Biol. Chem. **277:** 48923–48930.
3. LEE S.H. *et al.* 2002. K^+-dependent Na^+/Ca^{2+} exchanger is a major Ca^{2+} clearance mechanism in axon terminals of rat neurohypophysis. J. Neurosci. **22:** 6891–6899.
4. KIM M.H. *et al.* 2005. Interplay between Na/Ca exchangers and mitochondria in Ca^{2+}-clearance at the calyx of Held. J. Neurosci. **25:** 6057–6065.
5. SCHNETKAMP, P.P.M. & R.T. SZERENCSEI. 1991. Effect of potassium ions and membrane potential on the Na-Ca-K exchanger in isolated intact bovine rod outer segments. J. Biol. Chem. **266:** 189–197.
6. TSOI M. *et al.* 1998. Molecular cloning of a novel potassium-dependent sodium-calcium exchanger for rat brain. J. Biol. Chem. **273:** 4155–4162.

7. UEHARA A. *et al.* 2004. Forefront of Na^+/Ca^{2+} exchanger studies: physiology and molecular biology of monovalent cation sensitivities in Na^+/Ca^{2+} exchangers. J. Pharmacol. Sci. **96:** 19–22.
8. IWAMOTO T. & M SHIGEKAWA. 1998. Differential inhibition of Na^+/Ca^{2+} exchanger isoforms by divalent cations and isothiourea derivative. Am. J. Physiol. **275:** C423–C430.

In Bovine Heart Na^+/Ca^{2+} Exchanger Maximal Ca^{2+}_i Affinity Requires Simultaneously High pH_i and PtdIns-4,5-P2 Binding to the Carrier

VELIA POSADA, LUIS BEAUGÉ, AND GRACIELA BERBERIÁN

Laboratorio de Biofísica, Instituto de Investigación Médica "Mercedes y Martín Ferreyra" (INIMEC-CONICET), 5000, Córdoba, Argentina

ABSTRACT: Na^+_i-dependent Ca^{2+} uptake, Na^+-dependent Ca^{2+} release, and PtdIns-4,5-P2 binding to Na^+/Ca^{2+} exchanger (NCX1) as a function of extravesicular (intracellular) $[Ca^{2+}]$ were measured. Alkalinization increases Ca^{2+}_i affinity and PtdIns-4,5-P2 bound to NCX1; these effects are abolished by pretreatment with PtdIns-PLC and are insensitive to MgATP. Acidification reduces Ca^{2+}_i affinity. MgATP reverts it only partially despite the fact that the PtdIns-4,5-P2 bound to NCX1 reaches the same levels as at pH 7.8. Extravesicular Na^+-stimulated and Ca^{2+}-dependent Ca^{2+} efflux indicate the Ca^{2+} regulatory site is involved. Therefore, to display maximal affinity to Ca^{2+}_i, PtdIns-4,5-P2 binding and deprotonation of NCX1 are simultaneously need.

KEYWORDS: Na^+/Ca^{2+} exchange; NCX1; pH and PtdIns-4,5-P2 modulation; Ca^{2+}_i affinity

MgATP and pH modulations of bovine heart Na^+/Ca^{2+} exchanger (NCX1)[1–3] were reexplored in sarcolemmal vesicles prepared by differential centrifugation[4] by following the intravesicular Na^+-dependent Ca^{2+} uptake (forward mode), and the extravesicular $Ca^{2+}+Na^+$-dependent Ca^{2+} release (reversal mode) as function of extravesicular (intracellular) $[Ca^{2+}]$.[4,5] For $[^{45}Ca]Ca^{2+}$ uptake vesicles loaded with 160 mM NaCl were incubated for 1 or 2 s, at 37°C in media containing low (16 mM) or high (160 mM) NaCl, 3 mM free Mg^{2+}, 0.4 mM vanadate, 0.15 mM EGTA, 20 mM Mops/Tris (pH 7.4, 6.8, or 7.8), several $[Ca^{2+}]$ (from 0.5 μM to 40 μM buffered with EGTA) with or without 1 mM ATP. Extravesicular Na^+ was replaced with iso-osmolar amounts of NMG-Cl. For $[^{45}Ca]Ca^{2+}$ efflux vesicles loaded with 10 μM $[^{45}Ca]CaCl_2]$

Address for correspondence: Graciela Berberián, Laboratorio de Biofísica, Instituto de Investigación Médica "Mercedes y Martín Ferreyra" (INIMEC-CONICET), Casilla de Correo 389, 5000 Córdoba, Argentina. Voice: 54-351-4681465; fax: 54-351-4695163.

gelso@immf.uncor.edu

Ann. N.Y. Acad. Sci. 1099: 171–174 (2007). © 2007 New York Academy of Sciences.
doi: 10.1196/annals.1387.010

TABLE 1. Effect of pH and MgATP on Ca^{2+}_i affinity of Na^+/Ca^{2+} exchanger from native and PI-PLC bovine heart sarcolemmal vesicles

pH	ATP 1mM	Km μM	Vm nmol/s mg	Km μM	Vm pmol/3s mg
		From $^{45}Ca^{2+}$ uptake		From $^{45}Ca^{2+}$ release	
7.4	−	1.33 ± 0.02	0.89 ± 0.02	1.73 ± 0.09	200 ± 9
7.4	+	0.24 ± 0.01	0.94 ± 0.04	ND	ND
6.8	−	55.8 ± 13.1	0.40 ± 0.03	13.9 ± 0.90	201 ± 14
6.8	+	10.5 ± 3.60	0.40 ± 0.02	3.75 ± 0.40	178 ± 5
7.8	−	0.24 ± 0.04	0.88 ± 0.02	0.17 ± 0.04	181 ± 9
7.8	+	0.19 ± 0.02	0.87 ± 0.01	ND	ND
7.8*	−	1.24 ± 0.48	0.88 ± 0.07	1.71 ± 0.11	190 ± 9.8

[^{45}]Ca^{2+} influx and [^{45}]Ca^{2+} efflux were assayed as function of $[Ca^{2+}]_i$ in media with different pHs and with or without MgATP as indicated. The experimental points (Mean ± SE of triplicates for Ca^{2+} uptake or quintuplicates for Ca^{2+} release) were fitted to Michaelian equations.

*Vesicles were preincubated for 4 min, at 37°C, with a 4 U/mL of PtdIns-PLC.

in 160 mM NMG-Cl, 20 mM Mops-Tris (pH 7.4 at 25 °C) were incubated for 3 s at 25°C, in 160 mM NaCl, 3 mM free Mg^{2+}, 0.4 mM vanadate, 0.15 mM EGTA, 20 mM Mops/Tris (pH 7.4, 6.8, or 7.8 at 37°C), with no Ca^{2+} (0.15 mM EGTA) or with variable $[Ca^{2+}]$ (using EGTA as Ca^{2+}) buffer with or without 1 mM ATP. The kinetic data were fitted to Michaelian functions using the SCoP program 3.5 (Scop, Simulation Resources, Inc., Redlands, CA). Under identical conditions the amounts of PtdIns-4,5-P2 bound to NCX1 were estimated by Western blot, in the first step with the anti- PtdIns-4,5-P2 antibody and, after stripping, with the antibody anti-NCX1. The band densities obtained with the anti-PtdIns-4,5-P2 Ab (in arbitrary units) were divided by those obtained from the anti-NCX1 Ab (in arbitrary units). Comparisons of different experiments were made by assigning a value of 1 to the ratio of densities in control conditions as described.[6] The results are Tables 1 and 2. (*a*) In native vesicles, alkalinization increases both the apparent affinity for Ca^{2+}_i and the amount of PtdIns-4,5-P2-bound NCX1 and these are unaffected by MgATP. Similar

TABLE 2. Effect of pH and 1 mM MgATP at 1 μM $[Ca^{2+}]_i$, on PtdIns-4,5-P2 bound to NCX1 and on Na^+_i-dependent $^{45}Ca^{2+}$ uptake and Ca^{2+}_i-dependent Ca^{2+} release in native and PtdIns-PLC-treated bovine heart sarcolemmal vesicles

pH	PtdIns-4,5-P2/NCX1 arbitrary units		Ca^{2+} uptake nmol/s mg		Ca^{2+} release pmol/3s mg	
	−ATP	+ATP	−ATP	+ATP	−ATP	+ATP
7.4	1	2.09 ± 0.11	0.41 ± 0.05	0.81 ± 0.11	47 ± 24	154 ± 23
6.8	0.98 ± 0.08	2.31 ± 0.10	0.08 ± 0.02	0.21 ± 0.04	15 ± 8	39 ± 10
7.8	2.11 ± 0.15	2.19 ± 0.08	0.81 ± 0.05	0.84 ± 0.05	160 ± 17	167 ± 28
7.8*	0.18 ± 0.04	0.41 ± 0.06	0.47 ± 0.07	ND	54 ± 19	ND

*Vesicles were preincubated for 4 min, at 37°C, with a 4 U/mL of PtdIns-PLC. See text for details.

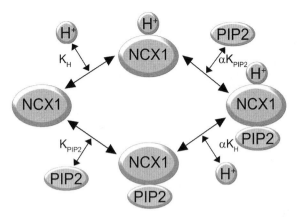

FIGURE 1. Scheme showing possible conversion equilibrium between free NCX1 and NCX1 complexed with H^+ and PtdIns-4,5-P2. K_H and K_{PIP2} are the dissociation constants for H+ and PtdIns-4,5-P2, respectively, and α is the cross reactivity coefficient. See text for details.

results were seen at pH 8.8 (not shown); that is, the relevant alkalinization is already maximal at pH 7.8. (*b*) Vesicles preincubated for 4 min, at 37°C, with 4 U/mL of PtdIns-PLC do not respond to an increase in pH, but become sensitive to alkalinization after addition of exogenous PtdIns-4,5-P2 or PtdIns plus MgATP. (*c*) Acidification markedly reduces the apparent affinity for $Ca^{2+}{}_i$. In Ca^{2+} uptake the V_{max} is reduced whereas in Ca^{2+} efflux remains unaltered. In both modes the pH effects are only partially reverted by MgATP although the PtdIns-4,5-P2 bound to NCX1 reaches the levels similar to those seen at pH 7.8. In addition, H^+ may influence the distribution, or the exposure, of PtdIns-4,5-P2 molecules in the vicinity of the exchanger since net synthesis of PtdIns-4,5-P2 was not influenced by pH (results not shown).

These data provide experimental evidences that PtdIns-4,5-P2 and pH regulation of NCX1 are related leading to changes in the affinity for $Ca^{2+}{}_i$. Irrespective of the intimate mechanism (changes in the protonation of PtdIns-4,5-P2, the carrier or both) we propose an equilibrium distribution between the four exchanger complexes (FIG. 1) where maximal $Ca^{2+}{}_i$ affinity of the regulatory site is displayed by the NCX1-PtdIns-4,5-P2 form and is reduced following the sequence NCX1 > H.NCX1-PtdIns-4,5-P2 > H.NCX1. In other words, there is cross reactivity where binding of H^+ is more difficult to NCX1.PIP2 than to NCX1 while binding of PtdIns-4,5-P2 is more difficult to H.NCX1 than NCX1 state (the α coefficient has a value higher than 1).

ACKNOWLEDGMENTS

This work was supported by grants from FONCYT (PICT-05-12397) and CONICET (PIP 5118), Argentina.

REFERENCES

1. BLAUSTEIN, M.P. & W.J. LEDERER. 1999. Sodium/calcium exchange: its physiological implications. Physiol. Rev. **79:** 763–854.
2. BERBERIAN, G., S.G. ROBERTS & L. BEAUGÉ. 2004.The phosphoinositides in the regulation of mammalian heart Na^+/Ca^{2+} exchanger. Curr. Top. Biochem. Res. **6:** 76–80.
3. DIPOLO, R. & L. BEAUGÉ. 2006.Sodium/Calcium exchanger. Influence of metabolic regulation on ion carrier interactions. Physiol. Rev. **86:** 155–203.
4. REEVES, J.P. 1985. The sarcolemmal sodium-calcium exchange system. Curr. Top. Membr. Transp. **25:** 77–119.
5. BERBERIÁN, G., C. HIDALGO, R. DIPOLO & L. BEAUGÉ. 1998. ATP stimulation of Na^+/Ca^{2+} exchange in cardiac sarcolemmal vesicles. Am. J. Physiol. **274:** 43, C724–C733.
6. ASTEGGIANO, C., G. BERBERIÁN & L. BEAUGÉ. 2001. Phosphatidylinositol-4,5-biphosphate bound to bovine cardiac Na^+/Ca^{2+} exchanger displays a MgATP regulation similar to that of the exchange fluxes. Eur. J. Biochem. **268:** 437–442.

NCX Current in the Murine Embryonic Heart

Development-Dependent Regulation by Na$^+$

MICHAEL REPPEL,[a,b] HANNES REUTER,[c] PHILIPP SASSE,[d]
JÜRGEN HESCHELER,[a] AND BERND K. FLEISCHMANN[d]

[a]*Institute of Neurophysiology, University of Cologne, D-50931 Cologne,
Germany*

[b]*Medizinische Klinik II, University of Schleswig-Holstein, Campus Lübeck,
23538 Lübeck, Germany*

[c]*Department of Internal Medicine III, University of Cologne, D-50931 Cologne,
Germany*

[d]*Institute of Physiology I, University of Bonn, Life and Brain Center,
53127 Bonn, Germany*

ABSTRACT: Due to its high functional expression, I_{NCX} may serve as an
important mechanism to ensure intracellular Ca^{2+} homeostasis, especially during the early embryonic stage. At the fetal stage I_{NCX} density is
significantly decreased but underlies regulatory processes, that is, regulation by Na$^+$.

KEYWORDS: NCX; cardiac development; electrophysiology; embryonic
heart; calcium

INTRODUCTION

NCX is strongly expressed in the adult heart and catalyzes the electrogenic
exchange of intracellular Ca^{2+} ions for extracellular Na$^+$ ions.[1] Its physiological role is to maintain low intracellular Ca^{2+} concentrations at rest and
to remove excess intracellular Ca^{2+} after the fast Ca^{2+}-induced Ca^{2+} release
(CICR) process during systole.[2]

In the embryonic heart[3] the close relationship between the Ca^{2+} intrusion
mechanisms of the sarcolemma and the extrusion mechanisms of the sarcoplasmatic reticulum, RyR2, serving as a prerequesite for CICR is lacking.

Address for correspondence: Michael Reppel, University of Cologne, Institute of Neurophysiology,
Robert-Koch-Str. 39, D-50931 Cologne, Germany. Voice: 0049-221-478-87099; fax: 0049-221-478-
3834.
akp72@uni-koeln.de

Ann. N.Y. Acad. Sci. 1099: 175–182 (2007). © 2007 New York Academy of Sciences.
doi: 10.1196/annals.1387.064

RyR2, as well as other important proteins involved in CICR and excitation–contraction (EC) coupling, for example, Ca^{2+} ATPase of the sarcoplasmatic reticulum (SERCA), phospholamban (Plb), and L-type Ca^{2+} channels, have been shown to be expressed only at low levels in the developing heart with a significant *increase* during proliferation.[4] In contrast, NCX was demonstrated to have higher expression levels in embryonic mouse hearts compared to adults.[5] An increased NCX-mediated Ca^{2+} influx and extrusion of Ca^{2+} is therefore thought to potentially serve as an alternative mechanism to maintain intracellular Ca^{2+} homeostasis in the developing heart.

Besides regulation of I_{NCX} by membrane potential, potentially being involved in pathophysiological conditions, for example, early afterdepolarization as a trigger for cardiac arrhythmia in adults,[6] NCX is also affected by Na^+ and Ca^{2+} itself. Most importantly, a subsarcolemmal microdomain has been proposed for Na^+ since Na^+ channels increase cytoplasmatic Na^+ only by approximately 0.1%, which is not sufficient to increase NCX activity.[7] We therefore aimed in this work to study I_{NCX} density during embryonic heart development and its regulation by Na^+.

METHODS

Murine embryonic cardiomyocytes were obtained from superovulated mice of the strain HIM:OF1 as described earlier.[8] Single atrial and ventricular cardiomyocytes were obtained using collagenase B (1 mg/mL, Roche, Mannheim, Germany). Cells were cultivated in Dulbecco's modified Eagle's medium (DMEM) supplemented with 20% fetal calf serum (FCS) (selected batches) on gelatine (0.1%)-coated coverslips and measured 24 to 48 h after dissociation.

I_{NCX} density was determined using the whole cell patch-clamp technique as described recently.[9–11] $[Ca^{2+}]_i$ was buffered to 150 nM with BAPTA (calculated using the Maxchelator program, D. Bers, Loyola University, Chicago, IL). The external solution (standard NCX solution) was K^+-free and contained (in mmol/L): Na^+ 135, Ca^{2+} 2, $MgCl$ 1, Glucose 10, HEPES 10, CsCl 10 (to block the inward rectifier K^+ current, I_{K1}, and the Na^+/K^+ pump), (in μmol/L): niflumic acid 100 (to block Ca^{2+}-activated Cl^- currents), Ouabain 10 (Na^+/K^+ pump inhibitor), and verapamil 10 (dihydropyridine antagonist), adjusted to pH 7.4 (CsOH). The internal solution contained (in mmol/L): CsCl 136, NaCl 10, aspartic acid 42, $MgCl$ 3, HEPES 5, tetraethylammonium (TEA) 20, MgATP 10, and 150 nM free $[Ca^{2+}]_i$, adjusted to pH 7.4 (CsOH) as described.[10] The holding potential was set to -30 mV to block T-type Ca^{2+} and Na^+ channels. Slow-ramp pulses were applied ($+60$ to -120 mV; 0.09 V/s) at 10-s intervals and current–voltage (I/V) relationships constructed.[11] Due to the lack of specific inhibitors of NCX, I_{NCX} was measured as the bidirectional Ni^{2+} (5 mM) sensitive current.

Results are given as means $\pm$ SEM. Where applicable, unpaired t-tests or one-way analysis of variance (ANOVA) were performed. Statistical significance was accepted when $P < 0.05$.

RESULTS

To study the functional role of NCX, we measured I_{NCX} in single ventricular-like cardiomyocytes of embryonic day E10.5 and E16.5, resembling early embryonic and fetal cardiomyocytes. All potentially contaminating currents like I_{K1}, Ca^{2+}-activated Cl^- currents and I_{Ca-L} as well as the Na^+/K^+ pump were blocked. In the presence of these blockers Ni^{2+} (5 mM) led to a strong and reversible decrease in a current at both negative and positive step potential and the applied voltage ramps from $+60$ to -120 mV evidenced an almost linear I/V relationship.

While the SERCA, RyR2, and phospholamban gene expression was reported to increase during embryonic heart development, we found that I_{NCX} strongly decreased (TABLE 1). FIGURE 1 summarizes the relative stage-related differences in I_{NCX} density at the Ca^{2+} outward (-120 mV) and Ca^{2+} inward mode ($+60$ mV) in embryonic cardiomyocytes of ventricular-like cells of E10.5 and E16.5. We observed a significant reduction to 55.7 $\pm$ 12.6% for the Ca^{2+} outward mode ($P < 0.01$) and to 48.9 $\pm$ 8% for the Ca^{2+} inward mode ($P < 0.01$) from E10.5 ($n = 14$) toward E16.5 ($n = 10$) similar to previous reports.[12]

As for Ca^{2+}[13] a subsarcolemmal microdomain has been proposed for Na^+.[7] To test for the basic electrophysiological parameter of ionic regulation, we applied 10 μM monensin, a Na^+ ionophore to E10.5 and E16.5 ventricular-like cardiomyocytes to increase bulk Na^+ (bulk Ca^{2+} hold at 150 nM). In clear contrast to unchanged I_{NCX} densities in EDS cardiomyocytes (*Ca^{2+} outward mode*: -4.05 ± 0.5 pA/pF, $n = 14$ (controls) versus -6.2 ± 1.3 pA/pF (Monensin), $n = 10$, $P > 0.05$; *Ca^{2+} inward mode*: 4.62 $\pm$ 0.4 pA/pF versus 5 $\pm$ 1.1 pA/pF, $P > 0.05$), we found a significant increase of I_{NCX} in LDS ventricular cardiomyocytes (*Ca^{2+} outward mode*: -2.25 ± 0.54 pA/pF, $n = 10$ (controls)

TABLE 1. I_{NCX} density in embryonic and fetal ventricular cardiomyocytes

Membrane potential	E10.5V (pA/pF)	E16.5V (pA/pF)	P value
-120 mV	-4.05 ± 0.5 $n = 14$	-2.25 ± 0.5 $n = 10$	$P < 0.01$
$+60$ mV	4.62 ± 0.4 $n = 14$	2.26 ± 0.4 $n = 10$	$P < 0.01$

I_{NCX} density of the Ca^{2+} outward (-120 mV) and the Ca^{2+} inward mode ($+60$ mV) at E10.5 and E16.5 (fetal stage) ventricular-like cardiomyocytes.

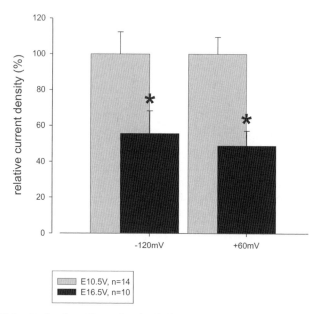

FIGURE 1. Reduction of I_{NCX} density during embryonic heart development. Asterisks indicate statistical significance.

versus -5.0 ± 1.1, $n = 15$ (Monensin), $P < 0.02$; Ca^{2+} *inward mode*: 2.3 $\pm$ 0.4 pA/pF versus 5.6 $\pm$ 1.2 pA/pF, $P < 0.03$), suggesting a high intrinsic activity of I_{NCX} at early stages of cardiac development or different biophysical properties of NCX. FIGURES 2 and 3 summarize the stage-dependent relative changes of I_{NCX} under control conditions and after application of monensin.

DISCUSSION

In this study we found high I_{NCX} density in early stages of murine heart development with a decrease toward fetal developmental stages. In the early embryonic heart the close spatial relationship between the VDCCs of the sarcolemma and the RyR2 of the sarcoplasmatic reticulum, is not fully developed.[3] Since the heart starts to beat already at day E8.5, enhanced sarcolemmal Ca^{2+} intrusion and extrusion due to the action of NCX, may play an important role for EC coupling.

During the course of end-stage heart failure embryonic gene profiles are reactivated and mRNA and protein levels of RyR2,[14] SERCA2a, and Plb[15] as well as L-type Ca^{2+} channels[16] show a significant decrease. This finding led to the assumption that the failing heart might potentially resemble key functional features of the early embryonic heart. In line with this hypothesis, NCX was demonstrated to have higher protein expression levels in embryonic

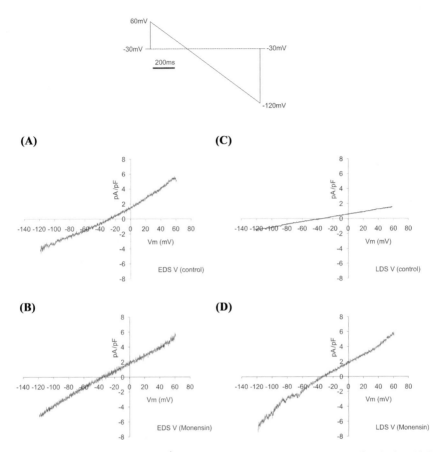

FIGURE 2. Effect of the Na$^+$ ionophore monensin (10 μM) on I_{NCX} density in E10.5 and E16.5 ventricular cardiomyocytes. Representative original traces in control cardiomyocytes derived from EDS (**A**) and LDS (**C**) hearts, and after monensin application (**B, D**).

mouse hearts compared to adults[5] and shows upregulation in the failing heart.[17] An increase of NCX function is therefore thought to potentially serve as an important compensating mechanism maintaining [Ca^{2+}]$_i$ homeostasis in the developing as well as in the failing heart. Accordingly, targeted deletion of NCX leads to embryonic death at around 11.0 days postcoitum (dpc) *in utero*,[18,19] whereas ventricular specific knockout at later developmental stages brings mice to adulthood and leads only to a 30% reduction of contractility.[20]

Interestingly, [Na$^+$]$_i$ is elevated in heart failure. In line with this observation at the earliest developmental stages a prominent Na$^+$ influx is present and serves as a prerequisite for normal cardiac function, that is, development of spontaneous excitability.[21] Besides Na$^+$ channels, Na$^+$ enters the cell also

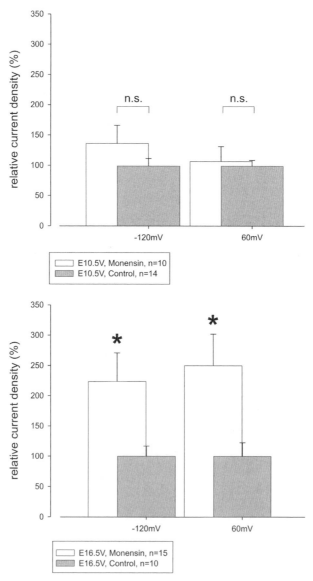

FIGURE 3. Statistical analysis of the monensin effect on I_{NCX} density in E10.5 and E16.5 ventricular cardiomyocytes. Asterisks indicate statistical significance.

through NCX and can also be extruded by NCX working in the reverse mode. As in embryonic cardiomyocytes increased I_{NCX} density and activation of NCX by Na^+ itself could lead to a compensation of increased Na^+ influx and subsarcolemmal $[Na^+]_i$, we found a prominent activation of the forward and reverse mode of NCX by the Na^+ ionophor monensin at fetal stages. However, elevated bulk $[Na^+]_i$ levels after application of monensin did not change I_{NCX} density in early embryonic ventricular cells. We assume that compartmentation might hinder local ion diffusion (fuzzy space) at early stages of murine heart development. The concept of a fuzzy space is supported by an earlier report from our group, where compartmentation of phosphodiesterases was found in early stage embryonic stem cell-derived cardiomyocytes,[22] by reports about subsarcolemmal microdomains described for Ca^{2+} and $Na^{+7,13}$ and compartmentalization of cAMP[23] in cardiomyocytes.

ACKNOWLEDGMENT

This study was supported by a grant from the "Deutsche Gesellschaft für Kardiologie" (Michael Reppel).

REFERENCES

1. HILGEMANN, D.W., A. COLLINS, D.P. CASH & G.A. NAGEL. 1991. Cardiac Na(+)-Ca2+ exchange system in giant membrane patches. Ann. N. Y. Acad. Sci. **639:** 126–139.
2. BARCENAS-RUIZ, L., D.J. BEUCKELMANN & W.G. WIER. 1987. Sodium-calcium exchange in heart: membrane currents and changes in [Ca2+]i. Science **238:** 1720–1722.
3. FABIATO, A. & F. FABIATO. 1978. Calcium-induced release of calcium from the sarcoplasmic reticulum of skinned cells from adult human, dog, cat, rabbit, rat, and frog hearts and from fetal and new-born rat ventricles. Ann. N. Y. Acad. Sci. **307:** 491–522.
4. LIU, W., K. YASUI, T. OPTHOF, *et al.* 2002. Developmental changes of Ca(2+) handling in mouse ventricular cells from early embryo to adulthood. Life Sci. **71:** 1279–1292.
5. REED, T.D., G.J. BABU, Y. JI, *et al.* 2000. The expression of SR calcium transport ATPase and the Na(+)/Ca(2+)exchanger are antithetically regulated during mouse cardiac development and in hypo/hyperthyroidism. J. Mol. Cell Cardiol. **32:** 453–464.
6. LEDERER, W.J. & R.W. TSIEN. 1976. Transient inward current underlying arrhythmogenic effects of cardiotonic steroids in Purkinje fibres. J. Physiol. **263:** 73–100.
7. CARMELIET, E. 1992. A fuzzy subsarcolemmal space for intracellular Na+ in cardiac cells? Cardiovasc. Res. **26:** 433–442.
8. FLEISCHMANN, M., W. BLOCH, E. KOLOSSOV, *et al.* 1998. Cardiac specific expression of the green fluorescent protein during early murine embryonic development. FEBS Lett. **440:** 370–376.
9. REPPEL, M., P. SASSE, R. PIEKORZ, *et al.* 2005. S100A1 enhances the L-type Ca2+ current in embryonic mouse and neonatal rat ventricular cardiomyocytes. J. Biol. Chem. **280:** 36019–36028.

10. ARTMAN, M., H. ICHIKAWA, M. AVKIRAN & W.A. COETZEE. 1995. Na+/Ca2+ exchange current density in cardiac myocytes from rabbits and guinea pigs during postnatal development. Am. J. Physiol. **268:** H1714–H1722.

11. KIMURA, J., A. NOMA & H. IRISAWA. 1986. Na-Ca exchange current in mammalian heart cells. Nature **319:** 596–597.

12. REPPEL, M., P. SASSE, D. MALAN, et al. 2007. Functional expression of the Na$^+$/Ca^{2+} exchanger in the embryonic mouse heart. J. Mol. Cell. Cardiol. **42:** 121–132.

13. CHENG, H., W.J. LEDERER & M.B. CANNELL. 1993. Calcium sparks: elementary events underlying excitation-contraction coupling in heart muscle. Science **262:** 740–744.

14. GO LO, MC MOSCHELLA, J. WATRAS, et al. 1995. Differential regulation of two types of intracellular calcium release channels during end-stage heart failure. J. Clin. Invest. **95:** 888–894.

15. FLESCH, M., R.H. SCHWINGER, P. SCHNABEL, et al. Sarcoplasmic reticulum Ca2+ATPase and phospholamban mRNA and protein levels in end-stage heart failure due to ischemic or dilated cardiomyopathy. J. Mol. Med. 1996 **74:** 321–332.

16. TAKAHASHI, T., P.D. ALLEN, R.V. LACRO, et al. 1992. Expression of dihydropyridine receptor (Ca2+ channel) and calsequestrin genes in the myocardium of patients with end-stage heart failure. J. Clin. Invest. **90:** 927–935.

17. SIPIDO, K.R., P.G. VOLDERS, M.A. VOS & F. VERDONCK. 2002. Altered Na/Ca exchange activity in cardiac hypertrophy and heart failure: a new target for therapy? Cardiovasc. Res. **53:** 782–805.

18. WAKIMOTO, K., K. KOBAYASHI, O. KURO, et al. 2000. Targeted disruption of Na+/Ca2+ exchanger gene leads to cardiomyocyte apoptosis and defects in heartbeat. J. Biol. Chem. **275:** 36991–36998.

19. KOUSHIK, S.V., J. WANG, R. ROGERS, et al. 2001. Targeted inactivation of the sodium-calcium exchanger (Ncx1) results in the lack of a heartbeat and abnormal myofibrillar organization. FASEB J. **15:** 1209–1211.

20. HENDERSON, S.A., J.I. GOLDHABER, J.M. SO, et al. 2004. Functional adult myocardium in the absence of Na+-Ca2+ exchange: cardiac-specific knockout of NCX1. Circ. Res. **95:** 604–611.

21. SATIN, J., I. HEHAT, O. CASPI, et al. 2004. Mechanism of spontaneous excitability in human embryonic stem cell derived cardiomyocytes. J. Physiol. **559:** 479–496.

22. ABI-GERGES, N., G..J. JI, Z.J. LU, et al. 2000. Functional expression and regulation of the hyperpolarization activated non selective cation current in embryonic stem cell-derived cardiomyocytes. J. Physiol. (Lond) **523:** 377–389.

23. BERS, D.M. & M.T. ZIOLO. 2001. When is cAMP not cAMP? Effects of compartmentalization. Circ. Res. **89:** 373–375.

Gender Differences in Na/Ca Exchanger Current and β-Adrenergic Responsiveness in Heart Failure in Pig Myocytes

SHAO-KUI WEI, JOHN M. MCCURLEY, STEPHEN U. HANLON, AND MARK C. P. HAIGNEY

Division of Cardiology, Department of Medicine, Uniformed Services University of the Health Sciences, Bethesda, Maryland 20814, USA

ABSTRACT: Clinical trials suggest females experience less heart failure (HF) progression, mortality, and arrhythmia frequency. HF increases Na/Ca exchanger (NCX) expression and activity contributing to both depressed contractility and ventricular arrhythmias, but whether gender modifies this effect is unknown. Left ventricular myocytes were isolated from control and from tachycardic pacing-induced failing swine hearts of both sexes. The Ni-sensitive NCX current (I_{NCX}) was measured in voltage clamp after blocking other channels. In control myocytes there is no difference in basal I_{NCX} and β-adrenergic responsiveness between male and female animals. HF greatly increased I_{NCX} and reduced β-adrenergic responsiveness in males compared to females, an effect that was eliminated by PP1. Diuretic therapy (furosemide, 1 mg/kg/day) further enhanced I_{NCX} and reduced β-adrenergic responsiveness in females and eliminated the gender difference. Gender-specific differences in calcium handling may contribute to improved survival of females in HF.

KEYWORDS: gender; sodium–calcium exchange; heart failure; phosphorylation

INTRODUCTION

The incidence of cardiovascular disease differs significantly between men and women. Clinical studies have reported that the progression of heart failure (HF) is slower in women than in their male counterparts, despite a similar extent of left ventricular dysfunction[1-6] and that the survival of women was

Address for correspondence: Mark C.P. Haigney, M.D., Division of Cardiology, Department of Medicine, Uniformed Services University of the Health Sciences, A3060, USUHS, 4301 Jones Bridge Road, Bethesda, MD 20814. Voice: 301-295-3826; fax: 301-295-3557.
mhaigney@usuhs.mil

Ann. N.Y. Acad. Sci. 1099: 183–189 (2007). © 2007 New York Academy of Sciences.
doi: 10.1196/annals.1387.026

significantly better in the nonischemic HF.[7] However, even the underlying mechanisms explaining this gender difference are still unclear.

Changes in cellular calcium handling have been implicated in contributing to the HF phenotype.[8] Several groups reported that Na/Ca exchanger (NCX) protein and/or activity was significantly increased in human[9,10] and animal HF.[11,12] Increased NCX activity could impair contractile function and promote unstable repolarization with early and/or delayed afterdepolarizations (EADs/DADs), especially in β-adrenergic stimulation.[13] However, it is unknown whether gender influences NCX activity in control or HF states. In addition, hormonal differences between males and females result in a favorable reduction in the effects of renin–angiotensin–aldosterone system activity in females.[14] We have recently reported that furosemide, a widely used loop diuretic, accelerates the development of systolic dysfunction in HF tachycardic-paced pigs.[15] Furosemide exposure was also associated with an increase in NCX activity compared to placebo. In the present study using the same model, we compared the magnitude of the NCX current (I_{NCX}) and its response to β-adrenergic stimulation (β-AR) between male and females control and HF pigs (with and without furosemide).

METHODS

Induction of heart failure and ventricular cardiomyocyte isolation and culture were carried out as described previously.[12] In brief, Yorkshire pigs of either sex between 7 and 10 weeks of age were paced at 200 bpm. All animals were randomized to treatment with furosemide (1 mg/kg/day IM) versus placebo and underwent periodic transthoracic echocardiography every 5 days. On development of severe left ventricular dysfunction (fractional shortening < 16%), animals were euthanized and myocytes were isolated by collagenase digestion, and the cells cultured overnight at 37°C in serum-free medium. Whole-cell recordings were obtained at 37°C using standard patch-clamp techniques. The external solution contained (mmol/L): NaCl 145, $MgCl_2$ 1, HEPES 5, $CaCl_2$ 2, CsCl 5, and glucose 10 (pH 7.4, adjusted with NaOH). Ouabain (0.02 mmol/L) and nifedipine (0.01 mmol/L) were added to the solution. The internal solution contained (mmol/L): CsCl 65, NaCl 20, Na_2ATP 5, $CaCl_2$ 6, $MgCl_2$ 4, HEPES 10, tetraethyl ammonium chloride (TEA) 20, EGTA 21, and ryanodine 0.05 (pH 7.2 adjusted with CsOH).

Membrane currents were elicited by using standard voltage ramp protocol. From a holding potential of −40 mV, a 100 ms step depolarization to +80 mV was followed by a descending voltage ramp (from +80 mV to −120 mV at 100 mV/s). I_{NCX} was measured as the Ni-sensitive current. Data are presented as the mean ± SEM. Continuous variables were compared by unpaired t-test. A P value of <0.05 was regarded as a statistically significant finding.

RESULTS

Gender Effects on NCX Current and β-AR
Responsiveness in Control Myocytes

Myocytes from males and females had similar peak outward and inward NCX currents as well as reversal potentials. Isoproterenol (ISO, 2 μM) significantly enhanced NCX current, the magnitude of increase being similar in myocytes from males and females. The mean population data (not shown) confirm that no significant difference in mean peak outward current exists between male and female controls ($P = 0.15$), while ISO responsiveness is almost identical with a robust increase in peak outward NCX current of 300%.

Gender Effects on NCX Current and β-AR
Responsiveness in HF Myocytes

FIGURE 1 shows a representative I_{NCX} (normalized to cell capacitance) in myocytes from male (1 A) or female (1 B) HF pigs in the presence or absence of a β-AR agonist (ISO, 2 μmol/L). In contrast to the control state, HF caused significant increase in basal NCX current and reduction of β-AR responsiveness in male animals. However, in myocytes from female HF animals, NCX current was only modestly increased (1 C), while β-AR responsiveness was preserved despite a similar extent of cardiac dysfunction (1 D). These results suggest that female HF animals experience a less severe remodeling of phenotype with respect to NCX current and its β-AR regulation, but the mechanism is unclear.

Comparison of the Response of NCX to Dephosphorylation
in Male and Female Myocytes in HF

The basal phosphorylation state of NCX reflects the net balance of activities of PKA and protein phosphatases. To investigate whether male HF animals have greater NCX phosphorylation than females in HF, protein phosphatase type 1 (PP1, 10 U/mL) was dialyzed through the intracellular solution to acutely dephosphorylate the NCX. FIGURE 2 A,B shows a representative I_{NCX} in myocytes from male or female HF pigs in the presence or absence of PP1, which significantly reduced NCX current to similar levels in both male and female HF myocytes. Since males had larger basal currents than female, the reduction of NCX current induced by PP1 was significantly greater in males than in females in HF. PP1 had no significant effect in control myocytes. These data support the hypothesis that NCX is more phosphorylated in males than in females with HF.

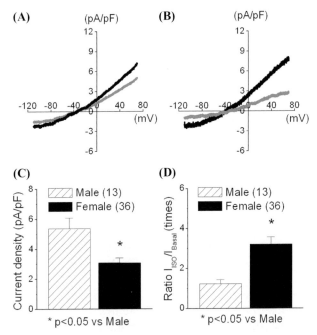

FIGURE 1. Gender difference in NCX current and β-adrenergic responsiveness in myocytes from HF pigs. (**A–B**). The representative I_{NCX} elicited by voltage ramp in myocytes from a male (**A**) or female (**B**) HF pig in the presence (black) or absence (gray) of ISO (2 μmol/L), showing myocytes from male HF pigs had significantly enhanced basal I_{NCX} but β-AR responsiveness was depressed compared to female pigs ($P < 0.05$, males versus females). (**C**). The average peak outward current density (at +70 mV) in myocytes from male or female pigs. Compared to control pigs, HF resulted in significantly increased I_{NCX} in male pigs ($P < 0.05$) but not in female pigs ($P = 0.15$). (**D**). The mean ratio of the peak outward I_{NCX} induced by ISO to basal current (ratio of I_{ISO}/I_{Basal}). Females retained significantly greater ISO responsiveness compared to males ($P < 0.05$).

Furosemide Effect on NCX Current and β-AR Responsiveness in Heart Failure

Basal NCX currents are significantly increased and β-AR responsiveness significantly blunted in both male and female furosemide-treated HF pigs ($P < 0.05$, FIG. 3 C,D). In contrast to the placebo-treated HF cells, furosemide further increased basal NCX current and reduced β-AR responsiveness in female animals but without additional effect on males. Furosemide therefore eliminated the gender-associated difference in NCX current and β-AR responsiveness.

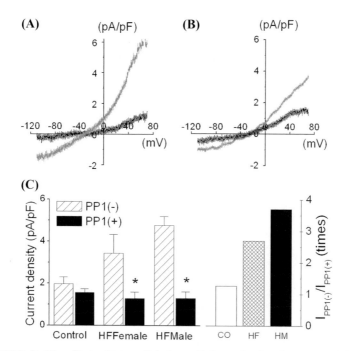

FIGURE 2. The effects of intracellularly applied protein phosphatase type 1 (PP1) on I_{NCX} in myocytes from different gender of HF pigs. (**A–B**). The representative I_{NCX} in HF myocytes from male (**A**) or female (**B**) pigs in the presence (black) or absence (gray) of PP1 (10 U/mL, dialyzed from internal solution); PP1 significantly depressed basal I_{NCX} in myocytes from both male or female HF pig. (**C**). The averaged peak outward current from control (CO), HF female (HF-F) and HF male (HF-M) pigs in the presence (black) or absence (shaded) of PP1 (left), and mean ratios of $I_{PP1(-)}/I_{PP1(+)}$ in control (CO), HF female (HF-F) and failing male (HF-M), showing that PP1 decreased I_{NCX} more in HF males than in HF females, suggesting that cells from HF males have a greater extent of phosphorylation of NCX compared with cells from HF females. There were no significant effects in cells from control pigs.

DISCUSSION

The significant findings of this study are: (*a*) in control pigs, males and females do not differ with respect to either basal NCX or ISO-stimulated NCX currents; (*b*) pacing-induced HF greatly increased basal NCX activity and decreased β-adrenergic responsiveness in male compared with that in female animals, and these differences seem to be largely due to greater tonic NCX phosphorylation in HF male pigs; and (*c*) furosemide, a loop diuretic drug that activates the RAAS, eliminated the gender difference in basal and ISO-stimulated NCX current.

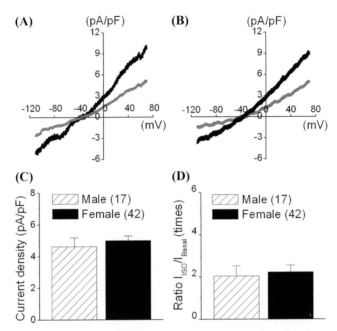

FIGURE 3. Gender differences in NCX current and β-adrenergic responsiveness in myocytes from HF pigs treated with furosemide. The representative I_{NCX} elicited by voltage ramp in myocytes from a male (**A**) or female (**B**) HF pig in the presence (*black*) or absence (*gray*) of ISO (2 μmol/L); no differences exist in either mean basal current (**C**) or β-AR responsiveness (**D**) between males and females. Compared to HF placebo group (FIG. 1), furosemide significantly increased basal I_{NCX} and depressed β-AR responsiveness in cells from female pigs but resulted in no additional effect in cells from male pigs.

Increased NCX activity could impair contractile function by reducing SR-Ca^{2+} content and promote unstable repolarization with EADs/DADs[15]; a smaller NCX current and largely preserved β-AR regulation observed in female failing myocytes might contribute to the better survival in females with HF. Further studies exploring the interaction between gender, the RAAS, and the expression/regulation of the NCX in HF need to be performed to determine the physiologic significance of these differences.

ACKNOWLEDGMENT

This work was supported in part by grants 0265463U-AHA (to S.-K.W.) and DOD C083QF (to J.M.M.).

REFERENCES

1. RARRET-CONNOR, E. 1997. Sex differences in coronary heart disease: why are women so superior? The 1995 Ancel Keys Lecture. Circulation **95**: 252–264.

2. ADAMS, K.F., C.A. SUETA, M. GHEORGHIADE, *et al.* 1999. Gender differences in survival in advanced heart failure. Insights from the FIRST study. Circulation **99**: 1816–1821.

3. HO, K., K. ANDERSON, W. KANNEL, *et al.* 1993. Survival after the onset of congestive heart failure in Framingham Heart Study subjects. Circulation **88**: 107–115.

4. ADAMS, K.F.J., S.H. DUNLAP, C.A. SUETA, *et al.* 1996. Relation between gender, etiology and survival in patients with symptomatic heart failure. J. Am. Coll. Cardiol. **287**: 1781–1788.

5. CROFT, J.B., W.H. GILES, R.A. POLLARD, *et al.* 1999. Heart failure survival among older adults in the United States: a poor prognosis for an emerging epidemic in the Medicare population. Arch. Intern. Med. **159**: 505–510.

6. SIMON, T., M. MARY-KRAUSE, C. FUNCK-BRENTANO, *et al.* 2001. Sex differences in the prognosis of congestive heart failure: results from the Cardiac Insufficiency Bisoprolol Study (CIBIS II). Circulation **103**: 375–380.

7. GHALI, J.K., H.J. KRAUSE-STEINRAUF, K.F. ADAMS, *et al.* 2003. Gender differences in advanced heart failure: insights from the BEST study. J. Am. Coll. Cardiol. **42**: 2128–2134.

8. O'ROURKE, B., D.A. KASS, G.F. TOMASELLI, *et al.* 1999. Mechanisms of altered excitation-contraction coupling in canine tachycardia-induced heart failure, I: experimental studies. Circ. Res. **84**: 562–570.

9. STUDER, R., H. REINECKE, J. BILGER, *et al.* 1994. Gene expression of the cardiac Na(+)-Ca2+ exchanger in end-stage human heart failure. Circ. Res. **75**: 443–453.

10. HASENFUSS, G., W. SCHILLINGER, S.E. LEHNART, *et al.* 1999. Relationship between Na^+-Ca^{2+}–Exchanger protein levels and diastolic function of failing human myocardium. Circulation **99**: 641–648.

11. POGWIZD, S.M., M. QI, W. YUAN, *et al.* 1999. Upregulation of Na(+)/Ca(2+) exchanger expression and function in an arrhythmogenic rabbit model of heart failure. Circ. Res. **85**: 1009–1019.

12. WEI, S.K., A. RUKNUDIN, S.U. HANLON, *et al.* 2003. PKA hyperphosphorylation increases basal current but decreases β-adrenergic responsiveness of the sarcolemmal Na^+/Ca^{2+} exchanger in failing pig myocytes. Circ. Res. **92**: 897–903.

13. SIPIDO, K.R., P.G. VOLDERS, M.A. VOS, *et al.* 2002. Altered Na/Ca exchange activity in cardiac hypertrophy and heart failure: a new target for therapy? Cardiovasc. Res. **53**: 782–805.

14. KANG, A.K., J.A. MILLER. 2003. Impact of gender on renal disease: the role of the renin angiotensin system. Clin. Invest. Med. **26**: 38–44.

15. MCCURLEY, J.M., S.U. HANLON, S.K. WEI, *et al.* 2004. Furosemide accelerates the progression of left ventricular dysfunction in experimental heart failure. J. Am. Coll. Cardiol. **44**: 1481–1487.

Roles of NCX and PMCA in Basolateral Calcium Export Associated with Mineralization Cycles and Cold Acclimation in Crayfish

M. G. WHEATLY, Y. GAO, L. M. STINER, D. R. WHALEN, M. NADE, F. VIGO, AND A. E. GOLSHANI

Biological Sciences, Wright State University, Dayton, Ohio 45435, USA

ABSTRACT: Basolateral Na^+/Ca^{2+} exchanger (NCX) and plasma membrane Ca^{2+} ATPase (PMCA) are the primary transmembrane proteins that export calcium (Ca^{2+}) from cells. In our lab we use a nonmammalian animal model, the freshwater crayfish, to study cellular Ca^{2+} regulation. Two experimental conditions are employed to effect Ca^{2+} dyshomeostasis: (*a*) in the postmolt stage of the crustacean molting cycle increased unidirectional Ca^{2+} influx associated with cuticular mineralization is accompanied by elevated basolateral Ca^{2+} export (compared with intermolt Ca balance); and (*b*) exposure of the poikilothermic crayfish to cold acclimation (4°C) causes influx of Ca^{2+} into cells, which is compensated by increased basolateral Ca^{2+} export (compared with exposure to 23°C). This study compares expression of both NCX and PMCA mRNA (real-time PCR) and protein (Western) in both epithelial (kidney) and nonepithelial tissue (tail muscle) during elevated basolateral Ca^{2+} export. Both experimental treatments produced increases in NCX and PMCA expression (mRNA and protein) in both tissues. Mineralization produced greater upregulation of mRNA in kidney than in tail, whereas cold acclimation yielded comparable increases in both tissues. Protein expression patterns were generally confirmatory of real-time PCR data although expression changes were less pronounced. Both experimental treatments appear to increase basolateral Ca^{2+} export.

KEYWORDS: NCX; PMCA; crayfish; mineralization cycle; cold acclimation

INTRODUCTION

Na^+/Ca^{2+} exchanger (NCX) and plasma membrane Ca^{2+} ATPase (PMCA) are the primary transmembrane proteins that export Ca^{2+} from the basolateral

Address for correspondence: Dr. M. G. Wheatly, Biological Sciences, Wright State University, 3640 Colonel Glenn Highway, Dayton, OH 45435-0001. Voice: 937-775-2611; fax: 937-775 3068.
michele.wheatly@wright.edu

Ann. N.Y. Acad. Sci. 1099: 190–192 (2007). © 2007 New York Academy of Sciences.
doi: 10.1196/annals.1387.022

membranes of living cells. In our laboratory we use a nonmammalian animal model, the freshwater crayfish, to study regulation of genes encoding Ca^{2+}-associated proteins. Two experimental conditions were employed to effect Ca^{2+} dyshomeostasis in epithelial tissue (kidney) and nonepithelial tissue (tail muscle): (*a*) Mineralization cycle: In the postmolt stage of the crustacean molting cycle, increased unidirectional Ca^{2+} influx is associated with elevated basolateral Ca^{2+} export (compared with intermolt Ca^{2+} balance). (*b*) Cold acclimation: Exposure of the poikilothermic crayfish to cold acclimation (4°C) precipitates Ca^{2+} influx into cells that is compensated by increased basolateral export (compared with exposure to 23°C).

Crayfish, *Procambarus clarkii,* were obtained from Carolina Biological Supply (Burlington, NC). Tissues (kidney and tail muscle) were harvested from (*a*) animals in postmolt (experimental) versus intermolt (control) or (*b*) maintained at 4°C (experimental) versus 23°C (control). mRNA for the genes of interest (NCX and PMCA) was quantified using real-time PCR and proteins were quantified through Western analysis using routine techniques published in our laboratory.[1,2]

MINERALIZATION CYCLES

Expression of mRNA for both NCX and PMCA increased in postmolt (compared with intermolt) in both tissues; however, quantitative increases were greater in kidney than in tail muscle (TABLE 1). In antennal gland the increase in PMCA expression was double the increase in NCX expression; in tail the increases were about the same for NCX and PMCA. Protein expression was also increased for both NCX and PMCA in both tissues; however, quantitative changes were about the same in both tissues (TABLE 2) and were generally lower than changes in mRNA.

COLD ACCLIMATION

Expression of mRNA for both NCX and PMCA increased in cold acclimation (4°C compared with 23°C) in both tissues; comparable increases were

TABLE 1. Quantitation of mRNA expression (fold changes relative to intermolt [mineralization cycles] or 23°C [cold acclimation]) determined via real-time PCR

Tissue	Gene	Intermolt	Postmolt	23°C	4°C
Kidney	NCX	1.00	11.9 ± 2.2	1.00	5.1 ± 0.7
	PMCA	1.00	22.2 ± 3.4	1.00	4.0 ± 0.7
Tail muscle	NCX	1.00	7.1 ± 1.8	1.00	3.3 ± 0.6
	PMCA	1.00	6.5 ± 2.6	1.00	2.1 ± 0.5

TABLE 2. Quantitation of protein expression (fold changes relative to intermolt [mineralization cycles] or 23°C [cold acclimation]) determined via Western analysis

Tissue	Gene	Intermolt	Postmolt	23°C	4°C
Kidney	NCX	1.00	1.6 ± 0.6	1.00	0.9 ± 0.4
	PMCA	1.00	2.2 ± 0.7	1.00	3.8 ± 0.9
Tail muscle	NCX	1.00	3.6 ± 1.3	1.00	1.6 ± 0.2
	PMCA	1.00	1.7 ± 0.2	1.00	5.7 ± 1.2

seen in both tissues (TABLE 1). In both tissues NCX mRNA expression somewhat exceeded PMCA expression. Protein expression was increased upon cold acclimation (except for kidney NCX) and there was a good correspondence with mRNA increases. In tail muscle PMCA protein expression exceeded NCX expression (TABLE 2).

TREATMENT COMPARISON

Both treatments employed to increase basolateral Ca^{2+} export were effective in increasing mRNA and protein expression for NCX and PMCA. Kidney showed elevated mRNA response (primarily with PMCA) in postmolt mineralization compared with tail muscle; cold acclimation produced comparable results in both tissues with NCX expression exceeding PMCA expression. Protein expression patterns were not so pronounced but generally confirmed the mRNA expression patterns. For cold acclimation PMCA expression was more pronounced than NCX expression and NCX expression in kidney was unchanged. Generally, there was better numeric correlation between mRNA and protein expression levels in cold acclimation than observed with mineralization treatment.

REFERENCES

1. GAO, Y., C.M. GILLEN & M.G. WHEATLY. 2006. Molecular characterization of the sarcoplasmic calcium-binding protein (SCP) from crayfish *Procambarus clarkii*. Comp. Biochem. Physiol. **144B:** 478–487.
2. GAO, Y. & M.G. WHEATLY. 2004. Characterization and expression of plasma membrane Ca^{2+} ATPase (PMCA3) in crayfish, *Procambarus clarkii,* antennal gland during molting. J. Exp. Biol. **207:** 2991–3002.

Modulation Pathways of NCX mRNA Stability

Involvement of RhoB

SACHIKO MAEDA,[a] ISAO MATSUOKA,[b] AND JUNKO KIMURA[a]

[a]Department of Pharmacology, Fukushima Medical University, Fukushima, 960-1295 Japan

[b]Laboratory of Pharmacology, Faculty of Pharmacy, Takasaki University of Health and Welfare, Gumma 370-0033, Japan

ABSTRACT: Cardiac Na^+/Ca^{2+} exchanger 1 (NCX1) expression levels change under various pathophysiological conditions. However, its mechanism is unknown. We found that fluvastatin, an HMG-CoA reductase inhibitor, decreased NCX1 mRNA and protein by inhibiting a small G protein, RhoB, in H9c2 cardiomyoblasts. Conversely, lysophosphatidylcholine (LPC) increased NCX1 mRNA and protein by activating RhoB. The effect of LPC was mediated by geranylgeranylation but not farnesylation of RhoB. Furthermore, we also detected that activation of RhoB increased NCX1 mRNA stability. Our results suggest that RhoB is involved in modulation of cardiac NCX1 mRNA expression.

KEYWORDS: H9c2 cells; Na^+/Ca^{2+} exchanger; small G proteins; isoprenoids

It has been demonstrated that cardiac Na^+/Ca^{2+} exchanger 1 (NCX1) expression changes under various pathological conditions. However, the mechanism regulating NCX1 gene expression is poorly understood. 3-Hydroxy-3-methylglutaryl CoA (HMG-CoA) reductase inhibitors, known as statins, have various cholesterol-independent "pleiotropic" effects including modulation of gene expression. We postulated that fluvastatin, one of the HMG-CoA reductase inhibitors, may modulate NCX1 mRNA expression. Therefore, we investigated the effect of fluvastatin on NCX1 mRNA levels in H9c2 cardiomyoblasts.

Semiquantitative reverse transcriptase polymerase chain reaction (RT-PCR) analyses revealed that fluvastatin decreased NCX1 mRNA levels in concentration- and time-dependent manners. Incubation of H9c2 cells with

Address for correspondence: Junko Kimura, Department of Pharmacology, Fukushima Medical University, Fukushima 960-1295, Japan. Voice: 81-24-547-1151; fax:81-24-548-0575.
jkimura@fmu.ac.jp

Ann. N.Y. Acad. Sci. 1099: 193–194 (2007). © 2007 New York Academy of Sciences.
doi: 10.1196/annals.1387.062

5 μM fluvastatin for 24 h significantly decreased NCX1 mRNA levels to about 60% of the control cells, accompanied by a parallel decrease in immunoreactive NCX1 protein. This effect of fluvastatin was due to the inhibition of HMG-CoA reductase, because fluvastatin failed to affect NCX mRNA levels in the presence of mevalonate or its downstream metabolites, such as farnesylpyrophosphate (FPP) and geranylgeranyl pyrophosphate (GGPP). The effect of fluvastatin was reproduced by expression of C3 toxin, the specific inhibitor of Rho family, indicating that fluvastatin inhibits the isoprenylation-dependent Rho signaling pathway, thereby decreasing NCX1 mRNA levels. Using siRNA against RhoA and RhoB, we identified that RhoB, but not RhoA, was involved in NCX1 mRNA expression. When transcription was blocked by 5,6-dichlorobenzimidazole riboside (DRB), the rate of decay of NCX1 mRNA level was accelerated by fluvastatin. These results suggest that a RhoB-mediated signaling pathway regulates cardiac NCX1 levels by modulating the NCX1 mRNA stability.[1]

Conversely, lysophosphatidylcholine (LPC), an activator of Rho-GTPase,[2,3] increased NCX1 mRNA and protein. This effect was sensitive to C3, indicating that RhoB is also involved in this effect. RhoB requires isoprenylation for its activation by either GGPP or FPP. Therefore we investigated which isoprenoid is involved in NCX1 increase by LPC. Incubation of H9c2 cells with fluvastatin for 24 h decreased NCX1 mRNA to about 60% of control. Under this condition, addition of GGPP or FPP restored NCX1 mRNA to the control level within 24 h. No significant difference was observed between GGPP and FPP. On the other hand, when LPC was applied with fluvastatin, NCX1 mRNA decreased by fluvastatin did not change. However, when LPC and GGPP were applied simultaneously, NCX1 mRNA was increased to a level significantly higher than the control. Unlike GGPP, FPP did not induce this increase. These results suggest that geranylgeranylation of RhoB, but not farnesylation, is involved in the effect of LPC increasing the NCX1 mRNA. This pathway of increasing mRNA stability might be involved in the altered levels of NCX1 expression under various cardiac pathophysiological conditions.

REFERENCES

1. MAEDA, S., I. MATSUOKA, T. IWAMOTO, et al. 2005. Down-regulation of Na^+/Ca^{2+} exchanger by fluvastatin in rat cardiomyoblast H9c2 cells: involvement of RhoB in Na^+/Ca^{2+} exchanger mRNA stability. Mol. Pharmacol. **68:** 414–420.
2. LI, L., I. MATSUOKA, Y. SUZUKI, et al. 2002. Inhibitory effect of fluvastatin on lysophosphatidylcholine-induced nonselective cation current in Guinea pig ventricular myocytes. Mol. Pharmacol. **62:** 602–607.
3. YOKOYAMA, K., T. ISHIBASHI, H. OHKAWARA, et al. 2002. HMG-CoA reductase inhibitors suppress intracellular calcium mobilization and membrane current induced by lysophosphatidylcholine in endothelial cells. Circulation **105:** 962–967.

Regulation of *Ncx1* Gene Expression in the Normal and Hypertrophic Heart

DONALD R. MENICK,[a] LUDIVINE RENAUD,[a] AVERY BUCHHOLZ,[a] JOACHIM G. MÜLLER,[a,b] HONGMING ZHOU,[c] CHRISTIANA S. KAPPLER,[a] STEVEN W. KUBALAK,[a] SIMON J. CONWAY,[c] AND LIN XU[a]

[a]*Division of Cardiology, Department of Medicine, Gazes Cardiac Research Institute, Medical University of South Carolina, Charleston, South Carolina, 29425, USA*

[b]*Department of Medicine, Fletcher Allen Health Care, University of Vermont, Burlington, Vermont 05401, USA*

[c]*Cardiovascular Development Group, Wells Center for Pediatric Research, Indiana University School of Medicine, Indianapolis, Indiana 46202, USA*

ABSTRACT: The Na^+/Ca^{2+} exchanger (NCX1) is crucial in the regulation of $[Ca^{2+}]_i$ in the cardiac myocyte. The exchanger is upregulated in cardiac hypertrophy, ischemia, and failure. This upregulation can have an effect on Ca^{2+} transients and possibly contribute to diastolic dysfunction and an increased risk of arrhythmias. Studies from both *in vivo* and *in vitro* model systems have provided an initial skeleton of the potential signaling pathways that regulate the exchanger during development, growth, and hypertrophy. The *Ncx1* gene is upregulated in response to α-adrenergic stimulation. We have shown that this is via p38α activation of transcription factors binding to the *Ncx1* promotor at the −80 CArG element. Interestingly, most of the elements, including the CArG element, which we have demonstrated to be important for regulation of *Ncx1* expression are in the proximal 184 bp of the promotor. Using a transgenic mouse, we have shown that the proximal 184 bp is sufficient for expression of reporter genes in adult cardiomyocytes and for the correct spatiotemporal pattern of *Ncx1* expression in development but not for upregulation in response to pressure overload.

KEYWORDS: Na^+/Ca^{2+} exchanger; NCX1; cardiac hypertrophy; regulation of gene expression; signal transduction; transgenic mice

INTRODUCTION

The heart is very dynamic and adapts to changes in hemodynamic demands by the enlargement of the ventricle via individual cardiomyocyte hypertrophy.

Address for Correspondence: Donald R. Menick Ph.D., Gazes Cardiac Research Institute, Medical University of South Carolina, 114 Doughty St., Charleston, SC 29425. Voice: 843-876-5045; fax: 843-792-4762.

menickd@musc.edu

Ann. N.Y. Acad. Sci. 1099: 195–203 (2007). © 2007 New York Academy of Sciences.
doi: 10.1196/annals.1387.058

The cardiac hypertrophic response is characterized by an increase in cardiac mass, protein content, and size of the individual cardiocyte. Initially, this results in increased myofilaments and improved contractile function. But, if the pathological stimulus is prolonged or sufficiently severe, and the increase in mass is insufficient to normalize ventricular wall stress, decompensated hypertrophy or heart failure will occur. These stimuli are often initiated by chronic hypertension, myocardial infarction, or ischemia associated with coronary artery disease. One of the features of the failing heart is a prolonged action potential and depressed contractility. In many models of heart disease and failure, the expression and activity of the Ca^{2+} sequestering sarcoplasmic reticulum (SR) Ca^{2+}-ATPase is decreased, and/or the activity and protein level of Na^+/Ca^{2+} exchanger (NCX1) is increased.[1–4] The exchanger catalyzes the electrogenic exchange of Ca^{2+} and Na^+ across the plasma membrane in either the Ca^{2+}-influx or Ca^{2+}-efflux mode. NCX1 is one of the essential regulators of Ca^{2+} homeostasis within cardiomyocytes and changes in its activity can affect contractility. The exchanger is upregulated at the transcriptional level in cardiac hypertrophy and failure.[1,5–8] Increased exchanger activity leads to increased Ca^{2+} extrusion and acts to preserve low diastolic Ca^{2+} levels, which may compensate in part for depressed SR Ca^{2+}-ATPase function. However, this adaptation has been demonstrated to have a number of deleterious consequences. First, in the failing heart, peak systolic Ca^{2+} is decreased because a higher percentage of $[Ca^{2+}]_i$ is being removed by NCX1, rather than returning to the SR.[9] In addition to the increase in NCX1 expression and activity, and reduced SR Ca^{2+}-ATPase function, Marx et al.[10] have shown that the ryanodine receptor is hyperphosphorylated in the failing heart. This results in increased overall Ca^{2+} release from the channel and has also been shown to contribute to a persistent unloading of SR Ca^{2+} stores.

Second, changes in Ncx1 expression and/or activity can contribute to aberrant contractile function. Because the exchanger is an electrogenic transporter, the increase in Ca^{2+} translocation across the plasma membrane results in greater risk of an inappropriately triggered depolarization (delayed afterdepolarization, DAD) before the relaxation cycle has completed.[3,11] This can cause arrhythmia and sudden death, a common cause of death from heart disease. Another potential problem arises from the lower levels of cytosolic Ca^{2+} in combination with prolonged action potential duration, which can promote reverse mode, Ca^{2+} in Na^+ out, activity. This would not only slow the rate of decay of the Ca^{2+} transient and relaxation, but also contribute to possible arrhythmogenesis and diastolic dysfunction.[12] Finally, in hearts that have been exposed to ischemia or hypoxia, reoxygenation can result in serious injury due to hypercontracture caused by spontaneous Ca^{2+} oscillations.[13] Several studies have shown that these oscillations are primarily elicited by reverse mode activity of the NCX1.[14–16] There is still a great deal more to be understood, but clearly, changes in exchanger expression are important to cardiac function. Therefore, we have begun to identify the molecular pathways and transcription factors and elements that are responsible for the upregulation of the exchanger gene in cardiac hypertrophy and failure.

There are multiple tissue-specific variants of *Ncx1* resulting from alternative promotor usage (H1, K1, and Br1) and alternative splicing.[17–19] Using transgenic models we demonstrated that the *1831Ncx1* H1 promotor was sufficient for driving the normal spatiotemporal pattern of NCX1 expression in cardiac development.[17] Importantly, at later fetal stages (E14 and older), and in the adult the *1831Ncx1* H1 promotor-driven reporter gene expression remains heart specific when the normal heart-restricted pattern of the endogenous NCX1 is no longer present.[20] Koban *et al.* showed cardiac-specific expression of reporter genes using either 2.1 kb or 2.6 kb rat *Ncx1* H1 promotor.[21] Expression was first detected at E7.5 and remained cardiac specific in the adult animal. Interestingly, here reporter gene expression was diminished with age. Importantly, we demonstrated that the feline *1831Ncx1* H1 promotor is sufficient for the upregulation of *Ncx1* in response to pressure overload in transgenic mice.[17]

p38 ACTIVITY IS IMPORTANT FOR MEDIATING *Ncx1* EXPRESSION

Several signaling pathways, including those activated by G protein–coupled receptor (GPCR) agonists, integrins, nonreceptor protein tyrosine kinases, protein kinase C, intracellular Ca^{2+}, and calcineurin are implicated in the initiation and maintenance of hypertrophy. The mitogen-activated protein kinases (MAPK), which transduce extracellular signals into the nucleus, are induced by hypertrophic stimuli. Importantly, their activation can initiate a hypertrophic response. p38 MAPK is of distinct interest since it has been shown to play a possible role in cardiocyte growth,[22] proapoptotic[23] or antiapoptotic[24] pathways. *In vitro* studies in neonatal cardiomyocytes show that activation of p38 induces expression of atrial natriuretic peptide[22,23,25] and inhibition of p38 reduced agonist-induced brain natriuretic peptide promotor activity.[26,27] Activation of p38 has been shown to mediate a negative inotropic effect, contribute to loss of contractility, and increase myocardial stiffness.[28] Andrews *et al.* have also shown that p38 can mediate the downregulation of SERCA2, which resulted in altered $[Ca^{2+}]_i$ regulation.[29] In addition to regulating SERCA, p38 also plays a role in mediating *Ncx1* expression.[30–32] We showed that the p38 α/β inhibitor SB202190 inhibits 50% of *Ncx1* upregulation mediated by α-adrenergic stimulation.[32] Overexpression of a dn p38 in adult cardiomyocytes also leads to a 50% decrease in α-adrenergic stimulation of *Ncx1*-promotor-driven reporter gene as well as endogenous *Ncx1* transcript. Our results clearly showed that p38 contributed to the α-adrenergic-mediated *Ncx1* upregulation. The MEK-1 and -2 inhibitor U0126 inhibits approximately 35% of the PE-stimulated *Ncx1* upregulation. Therefore, p38 and ERK-1 and -2 together mediate the majority of the α-adrenergic stimulation of the *Ncx1* promotor in adult cardiomyocytes (FIG. 1). In order to determine if activated p38 is sufficient for induction of the *Ncx1* promotor, we overexpressed the constitutively active forms of the p38 upstream kinases, MKK3 and MKK6. Overexpression of either MKK3 or MKK6 resulted in upregulation of the *Ncx1* promotor-driven reporter gene.[32] p38α, p38δ, and p38γ have been detected in the heart with p38α as the predominant isoform.[33] p38β has not been detected at the mRNA or protein level in human,[33] rat,[34] or mouse[28] heart. SB202190 completely prevented NCX1 upregulation in response to overexpression of MKK3bE or MKK6bE. p38α is the only p38 isoform activated by MKK3bE, which is also inhibited by SB202190. Therefore, activation of p38 is not only critical for a portion of the α-adrenergic-stimulated *Ncx1* upregulation, it is sufficient for *Ncx1* upregulation and it appears that this is mediated chiefly via p38α.

THE *Ncx1*-80 CArG ELEMENT MEDIATES p38-STIMULATED UPREGULATION OF THE EXCHANGER

The *Ncx1* promotor contains several consensus sequences for a number of potential DNA-binding factors, which we have shown to be important in

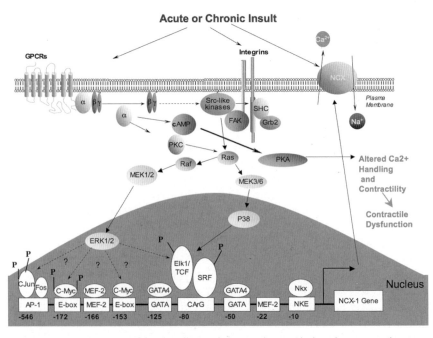

FIGURE 1. Diagram of the signaling pathways and transcription elements on the proximal *Ncx1* promotor important for the regulation of exchanger expression in the adult cardiocyte.

regulating *Ncx1* expression in neonatal[6] and adult cardiomyocytes. These include two GATA elements, two MEF2 elements, two E-Box elements, a CArG element, and a binding site for Nkx 2.5 (FIG. 1). These elements are also present in the human and rat and interestingly, the first 184 bp of the human[35] and feline[19] *Ncx1* H1 promotors have a 97% identity and the rat[36] and feline sequences 92%. In isolated adult cardiocytes the –123 GATA and –50 GATA element resulted in 35% and 29% of wild-type promotor activity respectively (FIG. 2). A point mutation disrupting the –80 CArG element resulted in luciferase activity of less than 20% of the control level.[37] In addition to the CArG and GATA elements being important to basal *Ncx1* expression in adult cardiomyocytes, the –80 CArG element is required for *Ncx1* upregulation in response to p38 activation (see FIG. 1).[32] Our study also showed that the –80 CArG element is responsible for part of the α-adrenergic-stimulated upregulation.

THE PROXIMAL 184 bp OF THE PROMOTOR IS SUFFICIENT FOR EXPRESSION BUT NOT FOR UPREGULATION IN RESPONSE TO PRESSURE OVERLOAD

The majority of the consensus sequences for DNA-binding factors, which are important in regulating *Ncx1* expression, are found in the proximal 184 bp

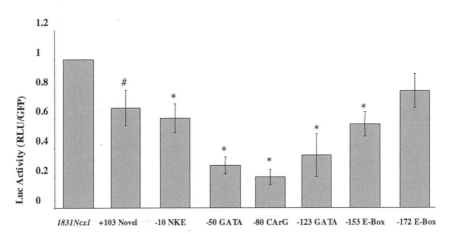

FIGURE 2. Effect of *Ncx1* promotor transcriptional element mutations on basal gene expression in adult feline cardiomyocytes. Adult feline cardiomyocytes were infected with the wild-type or mutant *Ncx1* promotor luciferase reporter adenoviruses (MOI 1.5). Cells were then incubated for 48 h and lysed in reporter buffer. Luciferase activity was determined relative to green fluorescent protein. All values are averages from four separate cell isolation experiments performed in triplicate. *$P < 0.0002$ when compared to wild-type promotor and #$P < 0.007$ when compared to wild-type promotor. (Figure taken with permission from Ref. 37.)

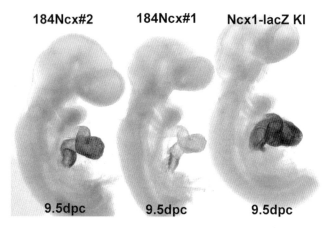

FIGURE 3. Temporal–spatial detection of endogenous $Ncx1^{lacZ}$ and *184Ncx1* β-galactosidase reporter activity within the mouse cardiovascular system. Both *184Ncx1* and endogenous $Ncx1^{lacZ}$ *lacZ* expression are present *in utero* throughout the E9.5 embryonic heart, and absent from the rest of the embryo and extraembryonic tissues. Reporter activity is restricted to the common ventricle and atria, but is absent from the aortic sac/truncus arteriosus and sinus venosus.

of the promotor. Our early work demonstrated that the *184Ncx1* promotor is sufficient for driving reporter expression in both neonatal and adult cardiomyocytes. Interestingly, the *184Ncx1* promotor is upregulated in response to α-adrenergic stimulation in neonatal cardiomyocytes but is recalcitrant to α-adrenergic stimulation in adult cardiomyocytes.[6,37] Subsequently, Xu *et al.* demonstrated that the *184Ncx1* promotor fragment could direct *LacZ* expression to the embryonic heart, with no expression in extracardiac tissues.[37] Expression is restricted to cardiomyocytes and is not observed in endothelial or endocardial cushion cells. The penitrance of expression varies between transgenic lines (FIG. 3). But even in lines where most cardiomyocytes express the reporter at 9.5 days (see line 1 FIG. 3) the expression pattern becomes patchy by E14.5 and diminished overall in the newborn[37] and in the adult (FIG. 4 A). In addition to the patchy cardiac myocyte expression in the adult, *184Ncx1*-β-*gal* reporter activity was detected in the sinus venosus, and the pulmonary and caval veins, which was never observed in any of the *1831Ncx1* transgenic mouse lines.[17] Koban *et al.* showed for the rat *Ncx1* promotors, that expression of the transgene could be detected in the conduction system.[21] Interestingly, even though reporter gene expression is extremely patchy in the adult the *184Ncx1*-β-*gal* transgene expression is consistently found within the cardiac myocytes that appeared to be part of the adult conduction system (FIG. 4 B). Positive *lacZ* staining could be detected in cells surrounding papillary muscles and in the wall of the left ventricle (FIG. 4 B, arrowheads). These data indicate that the *184Ncx1*-β-*gal* transgene contains the enhancer elements sufficient to

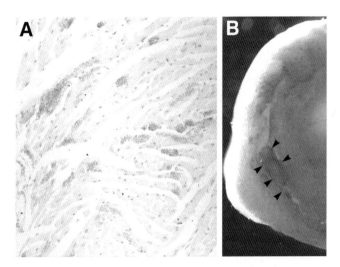

FIGURE 4. Expression of the *184Ncx1-β-gal* transgene in young adult hearts. (**A**) Transverse histological section of the left ventricular free wall showing mosaic *LacZ* staining in cardiomyocytes. (**B**) Whole mount *LacZ* staining of a central region of a heart showing *LacZ*-positive cells in and around a papillary muscle as well as the right ventricular free wall (*arrowheads*).

drive the correct cardiac myocyte-specific expression early in development but that it is missing some of the elements that contribute to the persistence of ubiquitous cardiomyocyte expression and extracardiac expression in the neonate and adult. What appears to be affected is the probability of the expression of the reporter gene in any particular cardiocyte not the level of expression. This probability of expression may be highest in specialized cardiomyocytes that make up the conduction system, though we have not yet confirmed this using a specific marker against specialized conduction system myocytes. Additionally, most of the cells that continue to express the transgene in late development and in the adult appear to be subendocardial rather than subepicardial.

Although the minimal promotor is sufficient to direct the cardiomyocyte-restricted expression at the proper time in development, it is not sufficient to mediate upregulation in response to pressure overload.[37] Clearly, these data indicate that although elements within the proximal 184 bases may be required for its expression there are additional distal elements necessary for *Ncx1* regulation. We are currently analyzing the role of distal elements to determine which are required for upregulation and to maintain levels of expression in all cardiocytes. In spite of the loss of regulation by pressure overload and persistence of ubiquitous cardiocyte expression in the adult, it is remarkable that such a small region of the *Ncx1* promotor is capable of driving the correct temporal- and tissue-specific expression.

REFERENCES

1. KENT, R.L., J.D. ROZICH, P.L. McCOLLAM, *et al*. 1993. Am. J. Physiol. **265:** H1024–H1029.
2. HASENFUSS, G., H. REINECKE, R. STUDER, *et al*. 1994. Circ. Res. **75:** 434–442.
3. POGWIZD, S.M., M. QI, W. YUAN, *et al*. 1999. Circ. Res. **85:** 1009–1019.
4. HOBAI, I.A. & B. O'ROURKE. 2001. Circulation **103:** 1577–1584.
5. MENICK, D.R., K.V. BARNES, U.F. THACKER, *et al*. 1996. Ann. N. Y. Acad. Sci. **779:** 489–501.
6. CHENG, G., T.P. HAGEN, M.L. DAWSON, *et al*. 1999. J. Biol. Chem. **274:** 12819–12826.
7. HASENFUSS, G., M. MEYER, W. SCHILLINGER, *et al*. 1997. Basic. Res. Cardiol. **92:** 87–93.
8. HASENFUSS, G., W. SCHILLINGER, S.E. LEHNART, *et al*. 1999. Circulation **99:** 641–648.
9. SHANNON, T.R., S.M. POGWIZD & D.M. BERS. 2003. Circ. Res. **93:** 592–594.
10. MARX, S.O., S. REIKEN, Y. HISAMATSU, *et al*. 2000. Cell **101:** 365–376.
11. POGWIZD, S.M., K. SCHLOTTHAUER, L. LI, *et al*. 2001. Circ. Res. **88:** 1159–1167.
12. WEBER, C.R., V. PIACENTINO, III, S.R. HOUSER & D.M. BERS. 2003. Circulation **108:** 2224–2229.
13. SIEGMUND, B., A. KOOP, T. KLIETZ, *et al*. 1990. Am. J. Physiol. **258:** H285–H291.
14. EIGEL, B.N. & R.W. HADLEY. 2001. Am. J. Physiol. Heart Circ. Physiol. **281:** H2184–H2190.
15. SCHAFER, C., Y. LADILOV, J. INSERTE, *et al*. 2001. Cardiovasc. Res. **51:** 241–250.
16. OHTSUKA, M., H. TAKANO, M. SUZUKI, *et al*. 2004. Biochem. Biophys. Res. Commun. **314:** 849–853.
17. MULLER, J.G., Y. ISOMATSU, S.V. KOUSHIK, *et al*. 2002. Circ. Res. **90:** 158–164.
18. QUEDNAU, B.D., D.A. NICOLL & K.D. PHILIPSON. 1997. Am. J. Physiol. **272:** C1250–C1261.
19. BARNES, K.V., G. CHENG, M.M. DAWSON & D.R. MENICK. 1997. J. Biol. Chem. **272:** 11510–11517.
20. KOUSHIK, S.V., J. BUNDY & S.J. CONWAY. 1999. Mech. Dev. **88:** 119–122.
21. KOBAN, M.U., S.A. BRUGH, D.R. RIORDON, *et al*. 2001. Mech. Dev. **109:** 267–279.
22. ZECHNER, D., D.J. THUERAUF, D.S. HANFORD, *et al*. 1997. J. Cell Biol. **139:** 115–127.
23. WANG, Y., S. HUANG, V.P. SAH, *et al*. 1998. J. Biol. Chem. **273:** 2161–2168.
24. ZECHNER, D., R. CRAIG, D.S. HANFORD, *et al*. 1998. J. Biol. Chem. **273:** 8232–8239.
25. NEMOTO, S., Z. SHENG & A. LIN. 1998. Mol. Cell Biol. **18:** 3518–3526.
26. LIANG, F. & D.G. GARDNER. 1999. J. Clin. Invest. **104:** 1603–1612.
27. LIANG, F., S. LU & D.G. GARDNER. 2000. Hypertension **35:** 188–192.
28. LIAO, P., D. GEORGAKOPOULOS, A. KOVACS, *et al*. 2001. Proc. Natl. Acad. Sci. USA **98:** 12283–12288.
29. ANDREWS, C., P.D. HO, W.H. DILLMANN, *et al*. 2003. Cardiovasc. Res. **59:** 46–56.
30. MENICK, D.R., L. XU, C. KAPPLER, *et al*. 2002. Ann. N. Y. Acad. Sci. **976:** 237–247.
31. XU, L., J.G. MULLER, P.R. WITHERS, *et al*. 2002. Ann. N. Y. Acad. Sci. **976:** 285–287.
32. XU, L., C.S. KAPPLER & D.R. MENICK. 2005. J. Mol. Cell Cardiol. **38:** 735–743.
33. LEMKE, L.E., L.J. BLOEM, R. FOUTS, *et al*. 2001. J. Mol. Cell Cardiol. **33:** 1527–1540.

34. RAKHIT, R.D., A.N. KABIR, J.W. MOCKRIDGE, *et al.* 2001. Biochem. Biophys. Res. Commun. **286:** 995–1002.
35. KRAEV, A., I. CHUMAKOV & E. CARAFOLI. 1996. Genomics **37:** 105–112.
36. SCHELLER, T., A. KRAEV, S. SKINNER & E. CARAFOLI. 1998. J. Biol. Chem. **273:** 7643–7649.
37. XU, L., L. RENAUD, J.G. MULLER, *et al.* 2006. J. Biol. Chem. **281:** 34430–34440.

Cyclosporin A-Dependent Downregulation of the Na^+/Ca^{2+} Exchanger Expression

H. RAHAMIMOFF,[a,b] B. ELBAZ,[a,b] A. ALPEROVICH,[a]
C. KIMCHI-SARFATY,[b,c] M. M. GOTTESMAN,[b] Y. LICHTENSTEIN,[a]
M. ESKIN-SHWARTZ,[a] AND J. KASIR[a]

[a]Department of Biochemistry, Hebrew University-Hadassah Medical School, Jerusalem, 91120 Israel

[b]Laboratory of Cell Biology, NCI, NIH, Bethesda, Maryland 20892, USA

[c]Division of Hematology, Food and Drug Administration, Bethesda, Maryland 20892, USA

ABSTRACT: Cyclosporin A (CsA) is an immunosuppressive drug commonly given to transplant patients. Its application is accompanied by severe side effects related to calcium, among them hypertension and nephrotoxicity. The Na^+/Ca^{2+} exchanger (NCX) is a major calcium regulator expressed in the surface membrane of all excitable and many nonexcitable tissues. Three genes, *NCX1*, *NCX2*, and *NCX3* code for Na^+/Ca^{2+} exchange activity. *NCX1* gene products are the most abundant. We have shown previously that exposure of NCX1-transfected HEK 293 cells to CsA, leads to concentration-dependent reduction of Na^+/Ca^{2+} exchange activity and surface expression, without a reduction in total cell-expressed NCX1 protein. We show now that the effect of CsA on NCX1 protein expression is not restricted to transfected cells overexpressing the NCX1 protein but exhibited also in cells expressing endogenously the NCX1 protein (L6, H9c2, and primary smooth muscle cells). Exposure of NCX2- and NCX3-transfected cells to CsA results also in reduction of Na^+/Ca^{2+} exchange activity and surface expression, though the sensitivity to the drug was lower than in NCX1-transfected cells. Studying the molecular mechanism of CsA–NCX interaction suggests that cyclophilin (Cyp) is involved in NCX1 protein expression and its modulation by CsA. Deletion of 426 amino acids from the large cytoplasmic loop of the protein retains the CsA-dependent downregulation of the truncated NCX1 suggesting that CsA–Cyp–NCX interaction involves the remaining protein domains.

KEYWORDS: Na^+/Ca^{2+} exchanger expression; cyclosporin A; cyclophilin A; proline mutagenesis

Address for correspondence: Hannah Rahamimoff, Department of Biochemistry, Hebrew University-Hadassah Medical School, Jerusalem, 91120 Israel. Voice: 972-2-675-8511; fax: 972-2-675-7379.
Hannah.Rahamimoff@huji.ac.il

Ann. N.Y. Acad. Sci. 1099: 204–214 (2007). © 2007 New York Academy of Sciences.
doi: 10.1196/annals.1387.046

INTRODUCTION

Immunophilins are a family of ubiquitously expressed cellular receptors that bind immunosuppressive drugs.[1] Cyclophilins (Cyp) bind cyclosporin A (CsA). Interaction of the complex Cyp–CsA with calcineurin, is the basis of the immunosuppressive process, involving inactivation of the phosphatase activity of the latter, which prevents the translocation of the phosphorylated NF-AT into the nucleus and subsequent T cell activation.[2]

In addition to its immunosuppressive action, CsA interaction with Cyp results also in inhibition of the PPIase (peptidyl–prolyl isomerase) and chaperons activities of Cyps. These two activities are of primary importance to protein maturation and folding into a function competent form. ER-based quality control mechanisms ensure that only mature and correctly folded proteins reach their target destination.[3] Impaired protein folding resulting from CsA treatment results in intracellular retention of the synthesized protein and a decrease in surface expression of several receptors, channels, and transporter.[4–7]

We have previously shown that exposure of NCX1.5[8]-transfected HEK 293 cells to CsA resulted in downregulation of surface expression and Na^+/Ca^{2+} exchange activity, without a change in the amount of total cell immunoreactive protein.[9] Intracellular retention of incorrectly processed and folded NCX1 could explain the downregulation of its surface expression and transport activity based on CsA-related interaction with Cyp. Based on this, we have decided to examine the interaction between CsA, Cyp, and Na^+/Ca^{2+} exchanger (NCX) expression.

RESULTS

The Effect of CsA on Expression of the NCX in H9c2, L6, and Primary Cultured Smooth Muscle Cells

H9c2 (ATCC CRL-1446[TM] Rattus Norvegicus; ATCC, Manassas, VA) cells derived from heart myocardium; L6 (ATCC CRL-1458[TM] Rattus Norvegicus) derived from skeletal muscle myoblasts; and primary smooth muscle cells isolated from rat aorta were chosen to study the effect of CsA on nontransfected cells, expressing NCX1 endogenously. RT-PCR was used to generate NCX transcripts using sequence-specific oligonucleotides derived from the variable region of NCX1, NCX2, and NCX3.

FIGURE 1 A–C shows specific NCX1-RT-PCR products from all three cell types tested. No other transcripts except those generated by using NCX1-specific oligonucleotides were detected. Positive controls were generated by using plasmids encoding NCX1, NCX2, and NCX3 cDNA (not shown).

FIGURE 2 A–C shows that treatment of H9c2, L6, and primary cultured smooth muscle cells with CsA, results in a concentration-dependent downregulation of transport activity. The downregulation of transport activity

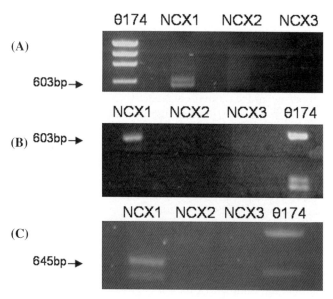

FIGURE 1. Identification of NCX-mRNA in primary cultured aortic cells, L6 and H9c2 cells. Total RNA was extracted from primary cultured aortic cells (**A**), L6 (**B**), and H9c2 (**C**) using Tri Reagent (Sigma). cDNA synthesis and amplification was carried out with RobusT II RT-PCR kit (Finnzymes, Espoo, Finland). Sequence-specific oligonucleotides were derived from the variable region in the large intracellular loop of *NCX1*, *NCX2*, and *NCX3*. The following sense (shown 5' →3') and corresponding antisense oligonucleotides were used to generate NCX-specific transcripts: NCX1 TCTTCAGAAGTCTCGGAAGAT; NCX2 GCGTGTGGGCGATGCTCA; NCX3 CTGGAAGAGGGGATGACCC.

correlated with the downregulation of surface expression (shown for L6 cells in Fig. 3).

These results suggest, that the CsA-dependent downregulation of both NCX1 transport activity and surface expression is not a phenomenon restricted to protein overexpression in transfected cells, but is manifested also in cells expressing the protein without overexpression.

Is CsA-Dependent Downregulation of NCX Transport Activity and Surface Expression Characteristic of NCX1 or, Shared Also by NCX2 and NCX3 Gene Products?

NCX2 and NCX3 are brain-specific isoforms of the Na^+/Ca^{2+} exchanger.[10,11] They share about 65% homology with NCX1.[12] Since CsA was shown to cross the blood–brain barrier,[13] it is of particular interest to examine whether the drug has a similar effect on the expression of NCX2 and NCX3 proteins as that observed for NCX1. We have inserted the Flag epitope instead of N34 in NCX2 and N45 in NCX3. HEK 293 cells were transfected

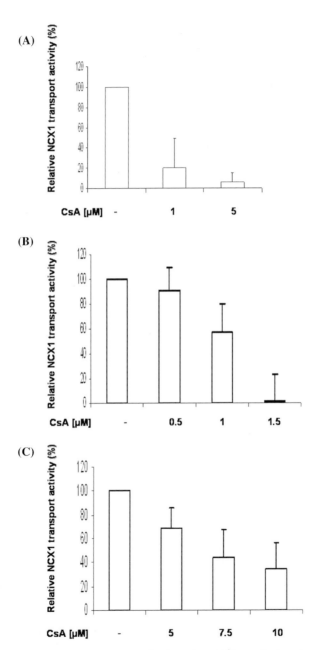

FIGURE 2. The effect of CsA on Na⁺- dependent Ca²⁺ uptake in primary cultured aortic cells (**A**), L6 (**B**), and H9c2 cells (**C**). Cells were grown in DMEM containing 10% (**A, C**) or 2% (**B**) FCS. Appropriate concentrations of CsA in DMSO or equal volume of DMSO were added to the culture media for 7 days (**A, B**) or 3 days (**C**) after which Na⁺-dependent Ca²⁺ uptake was determined essentially as described in Reference 9.

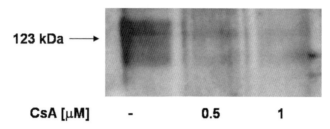

FIGURE 3. Surface expression of NCX1 protein in L6 cells. L6 cells were treated with CsA as described in FIGURE 2 B. Surface biotinylation was carried out as described in Reference 17.

with plasmids encoding N-Flag-tagged NCX2 and N-Flag-tagged NCX3. The transfected cells were exposed to 10–30 μM CsA. Transport activity was determined 24 h post transfection. CsA treatment resulted in a dose-dependent decrease in Na^+/Ca^{2+} exchange activity and surface expression. The sensitivity of the brain isoforms to CsA was lower than that of cells transfected with NCX1. The decrease in transport activity in cells treated with 10 μM CsA is shown FIGURE 4. FACS analysis, using M2 monoclonal antibody (Sigma, Rehovoth, Israel) was done to determine surface expression as described in

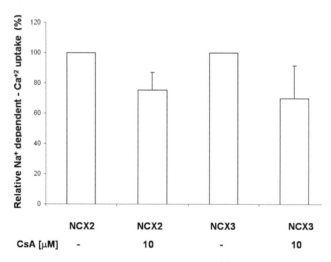

FIGURE 4. The effect of CsA on Na^+-dependent Ca^{2+} uptake in N-Flag-tagged NCX2 and N-Flag-tagged NCX3-transfected HEK 293 cells. HEK 293 cells were transfected with the mammalian expression vector pcDNA3.1 encoding N-Flag-tagged NCX2 or N-Flag-tagged NCX3 and the cells were exposed to CsA as described in Reference 9. Na^+-dependent Ca^{2+} uptake has been determined 24 h post transfection. The effect of 10 μM CsA in DMSO or equal volume of DMSO is shown. Na^+-dependent Ca^{2+} uptake without CsA treatment is taken in each experiment as 100% and all other data are calculated in relative values.

Reference 9. Median values were determined using CellQuest software (Becton Dickinson, Mountain View, CA) histogram analysis for each curve. Control median values (no CsA) were taken as 100% and those obtained from cells treated with CsA are given in relative values. NCX2 surface expression following treatment of the transfected cells with 10 μM CsA decreased to 74.69% (SD = 14.1) and that of NCX3 to 64.46% (SD = 7.6).

Is PPIase Involved in Maturation and Folding of NCX1

Peptide bonds between all 19 of the 20 amino acids are planar, rigid, and with the carbonyl group of the amino acid in position n being trans to the amino group of the amino acid in position n + 1 forming this bond. Proline is an exception, since about 15% of the peptide bonds X-Pro are in cis configuration. Cis proline peptide bonds are found mostly in loops and β turns. Isomerization of X-Pro peptide bonds between cis and trans configurations is a rate-limiting step in acquisition of functional conformation of proteins.[14]

NCX1 protein has 29 conserved proline residues (FIG. 5). Truncation of a segment of 428 amino acids from the large cytoplasmic loop of NCX1 between amino acid 240 and 668 results in a truncated exchanger that retains about 70% of the transport activity relative to its parent NCX1.5. FIGURE 6 A shows that exposure of cells expressing NCX1.5 or NCX1.5Δ428 to 10 μM CsA results in reduction of the relative transport activities to 50% and 33%,

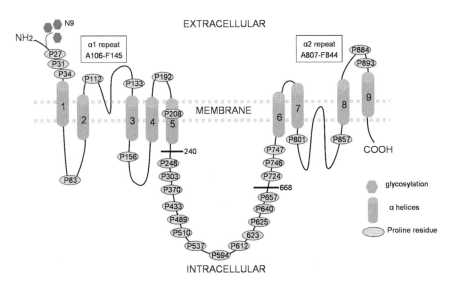

FIGURE 5. Topological model of NCX1. The model is based on References 15 and 18. The positions of the proline residues along the polypeptide chain are shown. Numbering corresponds to NCX1.5 isoform.[8,19]

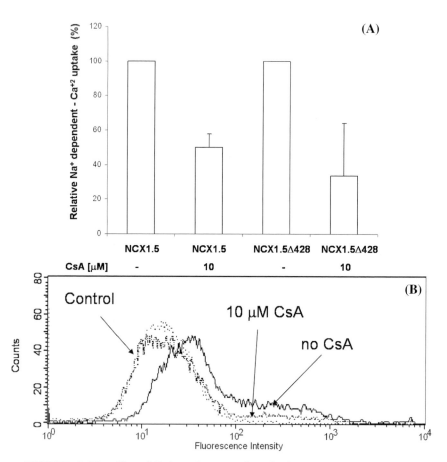

FIGURE 6. The effect of CsA on the expression of NCX1.5Δ428 in HEK 293 cells. HEK 293 cells were transfected with the N-Flag-tagged NCX1.5 and its mutant in which the large cytoplasmic loop was shortened by truncation between I241 and I669. The transfected cells were treated with CsA in DMSO or an equal volume of DMSO. Na$^+$-dependent Ca^{2+} uptake (**A**) and surface expression (**B**) using M2 the anti-Flag monoclonal antibody (Sigma) were measured by FACS analysis as described in Reference 9.

respectively. FACS analysis of surface-expressed NCX1.5 and NCX1.5Δ426 reveals parallel reduction in surface expression. FIGURE 6 B shows the reduction of the surface expression of cells expressing NCX1.5Δ426 in cells treated with 10 μM CsA.

The effect of CsA on the expression of the truncated NCX1 suggests, that the complex CsA–Cyp can interact not only with the WT exchanger, but also with the remaining polypeptide chain after truncation of 428 amino acids from the large cytoplasmic loop.

Based on this, we have exchanged one by one each of the 16 remaining proline residues of NCX1.5Δ428 to Ala or Ser and examined the transport activity, surface expression, and the total immunoreactive protein expression (not shown) of the mutant protein.

Transfection of HEK 293 cells with NCX1.5/P112A, NCX1.5/P208A, and NCX1.5/P746A resulted in the expression of significantly lower transport

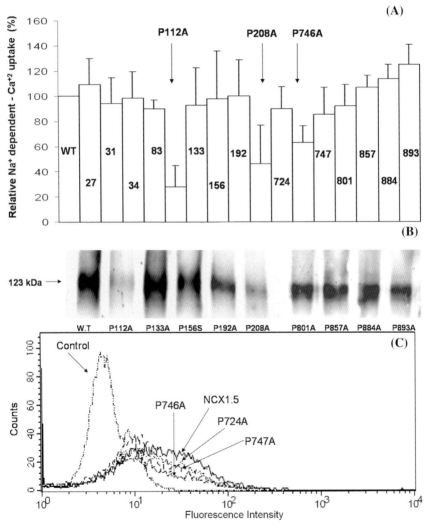

FIGURE 7. Expression of single proline mutants in HEK 293 cells. HEK 293 cells were transfected with the N-Flag-tagged NCX1.5 or with mutants in which a single proline was exchanged to Alanine (except P156 that was exchanged to Serine). Relative Na$^+$-dependent Ca^{2+} uptake (**A**), surface expression by surface biotinylation[17] (**B**), or by FACS analysis[9] (**C**) were determined.

activities and corresponding surface expression than that of the WT parent exchanger. FIGURE 7 shows the relative transport activities (A) and some of the corresponding surface expression by surface biotinylation (B) and by FACS analysis (C) of these mutants.

P112 is within the highly conserved α repeat region[15] of NCX1 protein. To differentiate between a specific requirement for proline in position 112 and the possibility that mutation of P112 to A112 was unacceptable for expression of a function competent NCX1 protein we have also exchanged P112 to G112 and L112. Surprisingly, expression of both mutants in HEK 293 cells resulted in very low transport activities, even lower than NCX1.5/P112A. But whereas the relative surface expression of NCX1.5/P112A and NCX1.5/P112G correlated with the transport activity, the surface expression of the nonfunctional P112L was WT-like.

Cyp A Is Involved in Expression of NCX1

HEK 293 cells express a large repertoire of immunophilins.[16] Yet it is possible, that following transfection, the amount of NCX1 protein expressed exceeds

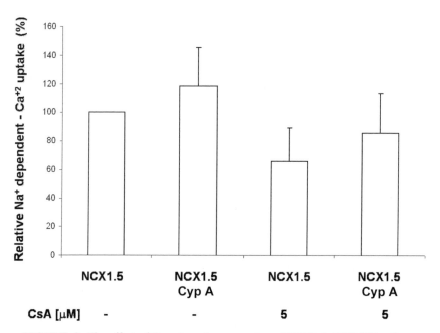

FIGURE 8. The effect of Cyp A on the expression of NCX1.5. HEK 293 cells were transfected with pcDNA3.1 encoding the N-Flag-tagged NCX1.5 or with N-Flag-tagged NCX1.5 and Cyp A. Twenty-four hours post transfection Na^+-dependent Ca^{2+} uptake was determined as described in Reference 9.

the capacity of endogenous Cyp to mediate proper maturation and folding. To study the involvement of Cyp A in NCX1 expression, we have cotransfected NCX1.5 together with different amounts of cloned Cyp A. Preliminary experiments suggest, that cotransfection of Cyp A with NCX1 increases consistently the transport activity (though not dramatically) of NCX1.5. The major effect, however, which Cyp A coexpression with NCX1.5 exerts, is a reduction of the downregulation of the Na^+/Ca^{2+} exchange activity by CsA (FIG. 8).

ACKNOWLEDGMENTS

Some of the results presented in this work were supported by research grants from the Israel Science Foundation and from the Israel Ministry of Health to H.R.; B.E. was supported by a travel fellowship from the Authority of Research Students of the Hebrew University. Cloned Cyp A was a generous gift of Drs. Patrick and Helekar, Baylor College of Medicine, Houston, TX and cloned NCX2 and NCX3 were the generous gift of Drs. Philipson and Nicoll, Cardiovascular Research Laboratory, UCLA, CA.

REFERENCES

1. IVERY, M.T. 2000. Immunophilins: switched on protein binding domains? Med. Res. Rev. **20:** 452–484.
2. KLEE, C.B., H. REN & X. WANG. 1998. Regulation of the calmodulin-stimulated protein phosphatase, calcineurin. J. Biol. Chem. **273:** 13367–13370.
3. KOPITO, R.R. 1997. ER quality control: the cytoplasmic connection. Cell **88:** 427–430.
4. CHEN, H. *et al.* 1998. Cyclosporin A selectively reduces the functional expression of Kir2.1 potassium channels in *Xenopus oocytes*. FEBS Lett. **422:** 307–310.
5. HELEKAR, S.A. *et al.* 1994. Prolyl isomerase requirement for the expression of functional homo-oligomeric ligand-gated ion channels. Neuron **12:** 179–189.
6. HELEKAR, S.A. & J. PATRICK. 1997. Peptidyl prolyl cis-trans isomerase activity of cyclophilin A in functional homo-oligomeric receptor expression. Proc. Natl. Acad. Sci. USA **94:** 5432–5437.
7. TRAN, T.T., W. DAI & H.K. SARKAR. 2000. Cyclosporin A inhibits creatine uptake by altering surface expression of the creatine transporter. J. Biol. Chem. **275:** 35708–35714.
8. FURMAN, I. *et al.* 1993. Cloning of two isoforms of the rat brain Na^+-Ca^{2+} exchanger gene and their functional expression in HeLa cells. FEBS Lett. **319:** 105–109.
9. KIMCHI-SARFATY, C. *et al.* 2002. Transport activity and surface expression of the Na^+-Ca^{2+} exchanger NCX1 is inhibited by the immunosuppressive agent cyclosporin A and the non-immunosupressive agent PSC833. J. Biol. Chem. **277:** 2505–2510.
10. LI, Z. *et al.* 1994. Cloning of the NCX2 isoform of the plasma membrane Na^+-Ca^{2+} exchanger. J. Biol. Chem. **269:** 17434–17439.

11. NICOLL, D.A. *et al.* 1996. Cloning of a third mammalian Na^+-Ca^{2+} exchanger, NCX3. J. Biol. Chem. **271:** 24914–24921.
12. ANNUNZIATO, L., G. PIGNATARO & G.F. DI RENZO. 2004. Pharmacology of brain Na+/Ca2+ exchanger: from molecular biology to therapeutic perspectives. Pharmacol. Rev. **56:** 633–654.
13. UCHINO, H. *et al.* 1998. Amelioration by cyclosporin A of brain damage in transient forebrain ischemia in the rat. Brain Res. **812:** 216–226.
14. FISCHER, G. *et al.* 1989. Cyclophilin and peptidyl-prolyl cis-trans isomerase are probably identical proteins [see comments]. Nature **337:** 476–478.
15. NICOLL, D.A. *et al.* 1999. A new topological model of the cardiac sarcolemmal Na^+-Ca^{2+} exchanger. J. Biol. Chem. **274:** 910–917.
16. RAHAMIMOFF, H. *et al.* 2002. NCX1 surface expression: a tool to identify structural elements of functional importance. Ann. N. Y. Acad. Sci. **976:** 176–186.
17. KASIR, J. *et al.* 1999. Truncation of the C-terminal of the rat brain Na^+-Ca^{2+} exchanger RBE-1 (NCX1.4) impairs surface expression of the protein. J. Biol. Chem. **274:** 24873–24880.
18. IWAMOTO, T. *et al.* 1999. Unique topology of the internal repeats in the cardiac Na^+/Ca^{2+} exchanger. FEBS Lett. **446:** 264–268.
19. FURMAN, I. *et al.* 1995. The putative amino- terminal signal peptide of the cloned rat brain Na^+-Ca^{2+} exchanger gene (RBE 1) is not mandatory for functional expression. J. Biol. Chem. **270:** 19120–19127.

Functional Significance of Na^+/Ca^{2+} Exchangers Co-localization with Ryanodine Receptors

ANNA A. SHER,[a] ROBERT HINCH,[b] PENELOPE J. NOBLE,[b]
DAVID J. GAVAGHAN[a] AND DENIS NOBLE[b]

[a]Computational Biology Group and [b]Cardiac Electrophysiology Group,
University of Oxford, Oxford OX1 3QD, UK

ABSTRACT: Co-localization of Na^+/Ca^{2+} exchangers (NCX) with ryanodine receptors (RyRs) is debated. We incorporate local NCX current in a biophysically detailed model of L-type Ca^{2+} channels (LCCs) and RyRs and study the effect of NCX on the regulation of Ca^{2+}-induced Ca^{2+} release and the shape of the action potential. In canine ventricular cells, under pathological conditions, e.g., impaired LCCs, local NCXs become an enhancer of sarcoplasmic reticulum release. Under such conditions incorporation of local NCXs is critical to accurately capture mechanisms of excitation–contraction coupling.

KEYWORDS: Na^+/Ca^{2+} exchanger; Ca^{2+}-induced Ca^{2+} release; co-localization with ryanodine receptors

LOCAL CONTROL THEORY AND Na^+/Ca^{2+} EXCHANGER

In ventricular myocytes, sarcoplasmic reticulum (SR) release within dyads is known to be triggered by elevated Ca^{2+} entry via L-type Ca^{2+} channels (LCC) located in the T-tubule membrane. Tight regulation of the interaction between the LCC and the Ca^{2+} release channels (the ryanodine receptors, RyR) is well studied both experimentally and via modeling. However, there is a debate on the role of Na^+/Ca^{2+} exchanger (NCX) as a trigger or an enhancer of SR release in the local control of Ca^{2+}-induced Ca^{2+} release (CICR). In particular, it is well accepted that NCX in its forward mode regulates diastolic Ca^{2+} levels by extruding Ca^{2+} out of the cell, whereas the role of NCX during the initial phases of the action potential (AP) is controversial.[1–3]

Experimental data show that NCX are primarily (60–70%) located within T-tubules,[4] however, co-localization with RyR is not well established (e.g.,

Address for correspondence: David J. Gavaghan, Computational Biology Group, University of Oxford Computing Laboratory, Wolfson Building, Parks Road, Oxford OX1 3QD, Oxford, UK. Voice: 01865 281899; fax: 01865 273839.

david.gavaghan@comlab.ox.ac.uk

Ann. N.Y. Acad. Sci. 1099: 215–220 (2007). © 2007 New York Academy of Sciences.
doi: 10.1196/annals.1387.047

Moore *et al.*[5] find only ~6% of NCX to be co-localized with RyR in normal rat hearts).

Our objective was to introduce NCX current within the dyad and to study the effect of local NCX on the CICR and on the shape of the AP via mathematical modeling. In order to gain insight into the functional significance of potential NCX co-localization with RyR, we modeled the extreme case scenario when all NCX located in T-tubules are co-localized with RyR.

MODELING NCX IN THE DYAD

Our model is derived from the canine Greenstein *et al.*, 2006 model.[6] The time evolution of the dyadic Ca^{2+} is modified to include NCX current

$$V_{ds}dCa_{ds}/dt = J_{LCC} + J_{NCX} + J_{RyR} - J_D,$$

where V_{ds} is the dyadic volume, J_{LCC}, $J_{RyR,}$ and J_{NCX} are fluxes within the dyad via LCC, RyR, and NCX, respectively, and J_D denotes Ca^{2+} flux from the dyadic cleft to the bulk myoplasm (FIG. 1). Our model incorporates

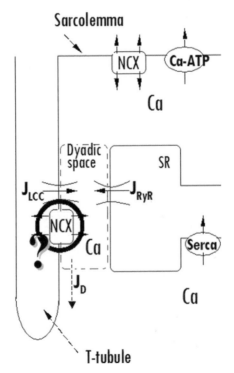

FIGURE 1. Schematic diagram of Ca^{2+} homeostasis. LCC and RyR contribute to local J_{LCC} and J_{RyR} fluxes within the dyadic cleft, SERCA reuptakes Ca^{2+} back into the SR, Ca^{2+}-ATPases and NCX pump Ca^{2+} out of the cell. NCX may also be present in the dyadic space but that is controversial. *Arrows* represent the direction in which Ca^{2+} flows.

local NCX current in both uniform and nonuniform fashion, which is the key difference from previous models,[6,7] and captures the key properties of CICR including graded release and voltage dependence of excitation–contraction (EC) coupling gain.

SIMULATION RESULTS AND DISCUSSION

Mathematical modeling allowed us to examine the effect of the NCX presence within dyads on the shape of an AP and on the I_{NCX} and I_{CaL} dynamics under physiological and pathological conditions.

Our simulations (FIG. 2) showed that co-localization of NCX with RyR would yield a transient inward NCX current during the notch phase of an AP in the canine ventricular cells. A significant transient inward current is observed even when only 10% of local NCX are present within the dyads. In 2003 Armoundas *et al.*[1] modeled local NCX, however, they assumed all NCX within proximity to the release sites, and based on their observations they concluded that NCX is unlikely to sense Ca^{2+} in the dyads. Nevertheless,

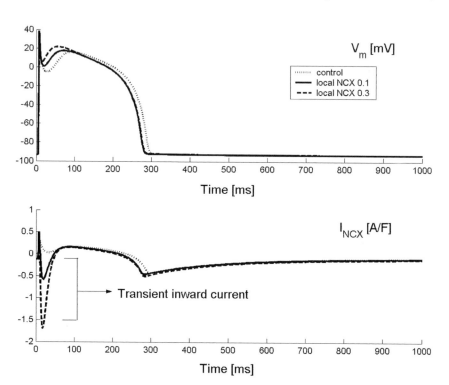

FIGURE 2. The upper panel illustrates the AP and the lower panel shows total I_{NCX} current, which is the summation of local and global NCX currents. The presence of local NCX yields transient inward I_{NCX} current.

in their discussion Armoundas *et al.* state that they cannot rule out the possibility that early in the AP NCX may transiently shift to forward mode due to the subsarcolemmal Ca^{2+} gradients. Moreover, they note that such findings would be more in agreement with Weber *et al.*,[3] who in 2002, suggested that the driving force for NCX during an AP may be much more inward than previously determined from the experiments. Our simulations support the idea that NCX sensing much higher than bulk $[Ca^{2+}]$ produce a transient depolarization current.

FIGURE 3 illustrates that the effect of local NCX on I_{CaL} is phase-dependent. We did not find a pronounced decrease in the amplitude of I_{CaL} peak in the presence of local NCX in the canine model (FIG. 3), whereas experimental data for mouse myocytes[8] suggest a substantial increase in the peak of I_{CaL} in the NCX knockout mice. The lack of a significant decrease in the peak amplitude in our model suggests that the role of local NCX is species-dependent. Importantly, the qualitative trend of a decrease in I_{CaL} amplitude is preserved. Our results also suggest that in the canine model, under physiological conditions, the relative contribution to SR release by local NCX is minor compared to LCC. Local NCX for the first few milliseconds pump Ca^{2+} in and then quickly

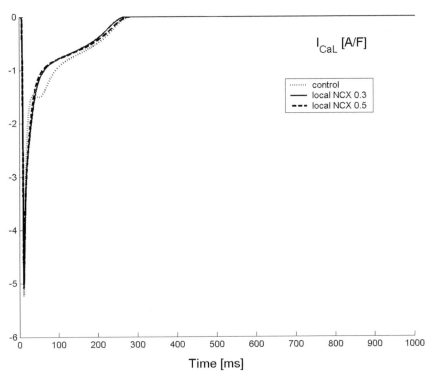

FIGURE 3. I_{CaL} dynamics in the presence/absence of local NCX.

reverse upon sensing high dyadic Ca^{2+} and contribute to Ca^{2+} efflux, decrease Ca^{2+}-dependent inactivation (CDI) of LCC, hence indirectly increasing SR release during the initial state of the AP by enhancing I_{CaL} activity.

FIGURE 4 illustrates the AP in the presence and absence of local NCX under normal and reduced activity of LCC. The impaired LCC were modeled by reduced mean LCC open time (40% decrease in the transition rate to the open-state constant) and increased CDI (30% increase in the transition rate to CDI-state constant). The EC gain is defined as the ratio of peak SR Ca^{2+} release flux to peak LCC Ca^{2+} influx. In the absence of local NCX, simulating impaired LCC yielded a 42.8% decrease in EC gain and in the presence (even when only a small proportion of NCX (10%) was co-localized with RyR), our model predicted a significantly smaller decrease of only 10.3% in EC gain. Our simulations demonstrated that, when the activity of LCC is reduced, the decrease in the EC coupling efficiency is significantly less pronounced in the presence of local NCX. This implies, if LCC were impaired, the presence of local NCX improves the EC coupling efficiency.

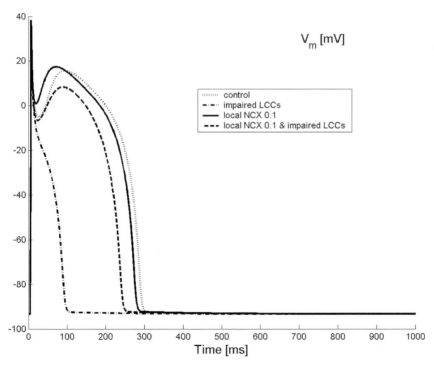

FIGURE 4. The AP in the presence/absence of local NCX under control conditions and when LCC are impaired, that is, decreased mean LCC open time (40% decrease in the transition rate to the open-state constant) and increased CDI (30% increase in the transition rate to CDI-state constant).

The effect of the NCX co-localization with RyR discussed above was studied in the canine ventricular cells. We would like to emphasize that the role of NCX is species-dependent and we are currently examining the significance of NCX–RyR co-localization within other species, for example, guinea pigs and rats, where I_{NCX} is predominantly inward during the plateau phase of an AP.

Our results demonstrated the importance of CICR regulation by NCX when LCC are impaired, hence suggesting that under pathological conditions the incorporation of local NCX into the models is critical to accurately capture mechanisms of EC coupling.

REFERENCES

1. ARMOUNDAS, A.A. *et al.* 2003. Role of sodium-calcium exchanger in modulating the action potential of ventricular myocytes from normal and failing hearts. Circ. Res. **93:** 46–53.
2. LITWIN, S.E., J. LI & J.H.B. BRIDGE. 1998. Na-Ca exchanger and the trigger for sarcoplasmic reticulum Ca release: studies in adult rabbit ventricular myocytes. Biophys. J. **75:** 359–371.
3. WEBER, C.R. *et al.* 2002. Na^+-Ca^{2+} exchange current and submembrane $[Ca^{2+}]$ during the cardiac action potential. Circ. Res. **90:** 182–189.
4. DESPA, S. *et al.* 2003. Na/Ca exchange and Na/K-ATPase function are equally concentrated in transverse tubules of rat ventricular myocytes. Biophys. J. **85:** 3388–3396.
5. SCRIVEN, D.R.L., P. DAN & E.D.W. MOORE. 2000. Distribution of proteins implicated in excitation-contraction coupling in rat ventricular myocytes. Biophys. J. **79:** 2682–2691.
6. GREENSTEIN, J.L., R. HINCH & R.L. WINSLOW. 2006. Mechanisms of excitation-contraction coupling in an integrative model of the cardiac ventricular myocyte. Biophys. J. **90:** 77–91.
7. HINCH, R. *et al.* 2004. A simplified local control model of calcium-induced calcium release in cardiac ventricular myocytes. Biophys. J. **87:** 3723–3736.
8. POTT, C. *et al.* 2006. Mechanism of shortened action potential duration in Na^+-Ca^{2+} exchanger knockout mice. Am. J. Physiol. Cell Physiol.: In press.

Three Types of Muscles Express Three Sodium–Calcium Exchanger Isoforms

DMITRI O. LEVITSKY

Université de Nantes, CNRS UMR 6204, Biotechnologie, Biocatalyse et Biorégulation, Faculté des Sciences et des Techniques, 44322 Nantes Cedex 3, France

ABSTRACT: The sodium–calcium exchanger (NCX) of plasma membrane is expressed in any animal cell. The specific role of its three isoforms (NCX1–3) is not yet established. Their levels vary considerably during murine postnatal development. In particular, in skeletal muscle, NCX1 expression decreases gradually upon aging while reciprocal changes take place for NCX3. NCX2 expression is restricted to brain and smooth muscles. The data on SDS-gel mobility shifts indicate that all three isoforms undergo Ca^{2+}-dependent conformational changes, and that an exchanger regulatory Ca^{2+}-binding domain interacts directly with mutually exclusive exons A and B inducing two different NCX1 conformations.

KEYWORDS: sodium–calcium exchanger isoforms; skeletal muscle; smooth muscle; NCX1; NCX2; NCX3; development; calcium

INTRODUCTION

The plasmalemmal electrogenic sodium–calcium exchange (NCX) plays an important role in the regulation of intracellular free calcium, providing Ca^{2+} extrusion from the cytoplasm (normal mode) and its entry into the cell (reverse mode). Three Na^+/Ca^{2+} exchanger isoforms have been described.[1] Functional characteristics and molecular organization of NCX1 have been studied in great detail.[2] The transcription of *Ncx1* may, potentially, result in expression of multiple splicing variants containing mutually exclusive exons A or B.[1,3] Though NCX1 ubiquitously expressed in mammalian tissues,[1,4] it is traditionally considered as "cardiac isoform." Much less information is available on the physiological significance of two other isoforms, NCX2 and NCX3. Indeed, *Ncx2* and *Ncx3* knockouts are not lethal for mice,[5,6] which may indicate that NCX1 efficiently compensates the loss of the second and

Address for correspondence: Dmitri O. Levitsky, CNRS UMR 6204, Biotechnologie, Biocatalyse et Biorégulation, Faculté des Sciences et des Techniques, 2, rue de la Houssinière–B.P. 92208, 44322 Nantes Cedex 3, France. Voice: 33-2-40-29-89-59; fax: 33-2-51-12-56-11.

Dmitri.Levitsky@univ-nantes.fr

Ann. N.Y. Acad. Sci. 1099: 221–225 (2007). © 2007 New York Academy of Sciences.
doi: 10.1196/annals.1387.063

the third exchanger isoforms in mammals. Nevertheless, one cannot exclude that in particular cases and in some tissues and cells, NCX2 and NCX3 play quite specific roles. Indeed, NCX2 knockout mice exhibit enhanced memory and learning potentials, indicating an implication of this isoform in neuronal plasticity.[5] Furthermore, the loss of NCX3 leads to necrosis of the muscle fibers and to marked alterations at the level of neuromuscular transmission, similar to those characteristic of the myasthenic syndrome.[6] Interestingly, the lack of NCX3 does not result in overexpression of the housekeeping isoform NCX1.[6] Here I present data on expression of the three isoforms in mammalian tissues with the emphasis on their similarities in respect to Ca^{2+}-dependent conformational changes.

MATERIALS AND METHODS

Tissue homogenates from Wistar-Kyoto rats and SC57BL/10J mice were obtained after an Ultra-Thurrax homogenization in a medium containing 20 mM HEPES (pH 7.0), 1 mM sodium azide, 1 mM PMSF, 1 μg of aprotinin, and 1 mM dithiothreitol. Nuclei-free fractions were isolated after a 20-min centrifugation at 700 g. The electrophoresis samples applied to SDS-PAAG were normalized with respect to homogenate protein concentration. Electrotransfer of the proteins from the gels to nitrocellulose membranes and incubation of the blots with anti-NCX antibodies were done as previously described.[7,8] The blots were developed with an enhanced chemiluminescence detection kit (Amersham, Les Ulis, France). The anti-NCX1 monoclonal antibody R3F1, polyclonal anti-NCX2 and anti-NCX3 antibodies were kindly provided by Ken Philipson (UCLA).

Total RNA was isolated by Superscript II (Life Technologies SARL, Gergy-Pontoise, France). The contaminating DNA was eliminated using a "DNA-free" kit (Ambion Europe, Huntington, UK). The RNA was transcribed with Superscript II using random hexamers. The cDNA was amplified using Hot-Start DNA polymerase (Eurogentec, Anders, France). Amplification of variable region of mouse transcript was carried out for 35 cycles. The forward primer was 5′-caacactgccaccataacc-3′ and the reverse primer was 5′-gactcttcgatgatcacctc-3′.

RESULTS AND DISCUSSION

In contracting heart there are two sources of Ca^{2+}: an internal store of the sarcoplasmic reticulum and external one provided after the activation of sarcolemmal L-type Ca^{2+} channels; most of the calcium ions that enter the myocyte are extruded from the myoplasm via the Na^+/Ca^{2+} exchange pathway.[9] A crucial role of NCX1 in the working heart has been proven in experiments on *Ncx1* null mice embryos.[10]

In smooth muscles, multiple transport systems control the level of tonic contraction. The relative contribution of the Na^+/Ca^{2+} exchanger and the plasmalemma Ca^{2+}-ATPase for Ca^{2+} extrusion from smooth muscle cells is not yet established (for review, see Ref. 2).

It is well accepted that the functioning of mammalian adult skeletal muscle depends exclusively on intracellular calcium (for reviews, see Refs. 11,12). Logically, for their contractile activity skeletal muscles would not require such a system as a sarcolemmal Na^+/Ca^{2+} exchange. We were quite surprised to find out that in rat skeletal muscles, at any stage of postnatal development, the levels of NCX transcripts, either of NCX1 or of NCX3, were reasonably high.[7] In addition, our immunofluorescence data indicated that all types of muscle fibers expressed the exchanger, either NCX1 or NCX3.

Here the data are presented on the expression of the three exchanger isoforms at protein level. Fast and slow skeletal muscles at different developmental stages were compared with other rat and mouse tissues. In newborn muscle, two major NCX1 splicing variants were detected, of high and low electrophoretic mobility (FIG. 1 A). Upon skeletal muscle maturation, the NCX1 level decreases dramatically, and in adult muscles it is reduced to very low levels. In fast twitch muscles (EDL and tibialis), predominates a cardiac splicing variant ACDEF, while slow skeletal muscle soleus expresses shorter variants containing exon B (FIG. 1 A, and RT-PCR at right). It is to indicate that this developmental evolution of NCX1 reminds that found in mammalian heart[13] but not in the brain in which, from birth to adult age, the expression of NCX1 (and its "slow-mobility" and "fast-mobility" bands) remains constant (FIG. 1 A).

Two NCX1 variants could be detected in the blots only when Ca^{2+} was present in the loading buffer prior to electrophoresis. Upon Ca^{2+} chelation with EGTA, a single intermediate band was observed. It has been concluded[7,8,14] that gel mobility shifts of NCX1 polypeptides are due to desaturation of a high-affinity Ca^{2+}-binding domain of the protein. The latter has been shown to be present in the middle of a large cytoplasmic loop of the protein.[14] The opposite mobility shifts (FIG. 1 A) indicate different conformational states of Ca^{2+}-saturated NCX1 variants containing mutually exclusive exons (A, cardiac or exon B, slow muscle).

The gel shifts have been demonstrated as well in NCX1 peptides corresponding to the high-affinity Ca^{2+}-binding domain lacking the splicing variable region.[8,14] An inevitable conclusion is that the exons A and B interact with the Ca^{2+}-binding site(s), and that it is the mutually exclusive exons that determine the conformation of the cytoplasmic loop.

Two other NCX isoforms as well seem to have a different conformational state in Ca^{2+}-bound and Ca^{2+}-free forms (FIG. 1 B, C). Our extensive screening of rat tissues has shown that the expression of NCX3 is restricted to brain and skeletal muscles. This isoform progressively compensates NCX1 decrease during postnatal skeletal muscle development. In adult skeletal fast, slow, and

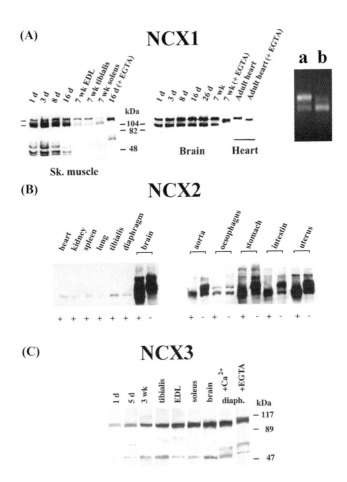

FIGURE 1. Western blot analyses of murine homogenates. Before electrophoresis, 20 to 30 μg of homogenate proteins were placed in a sample buffer supplemented with 2 mM CaCl$_2$ or 2 mM EGTA. (**A**) NCX1 expression in developing rat skeletal muscle and brain. *Right panel:* RT-PCR data on samples obtained from fast EDL (a) and slow soleus (b) muscles are shown. The predominant expression of exon A (a) and B (b) transcripts had been verified in RT-PCR experiments using corresponding A and B forward primers. (**B**) Tissue distribution of NCX2 in adult mouse. (**C**) Postnatal evolution of NCX3 expression in rat skeletal muscles.

mixed muscles, the levels of NCX3 expression are as high as in the brain (FIG. 1 C). It thus seems that NCX3 may play a key role in the regulation of free Ca^{2+} concentration in adult muscle fibers. This conclusion is in line with the findings of Sokolow *et al.*[6] showing the absence of the reverse Na$^+$/Ca^{2+} exchange activity in mouse skeletal muscles lacking NCX3.

The tissue distribution of NCX2 in adult mouse was further studied. As seen from FIGURE 1 B, this isoform is expressed exclusively in brain and smooth muscles. Interestingly, in esophagus, faint immunoreactive bands were also observed. However, these bands do not correspond to NCX2 since no differences were found between the samples pretreated with Ca^{2+} and EGTA. The absence of NCX2 in this tissue can be reasonably explained by a well-known phenomenon of transdifferentiation of esophagus musculature from smooth muscle phenotype in newborn animals to skeletal muscle in adults.

In conclusion, brain is the only murine tissue in which all three NCX isoforms are expressed. Otherwise, we can consider NCX1 as a cardiac isoform, NCX2 as a smooth muscle isoform, and NCX3 as a skeletal muscle isoform.

REFERENCES

1. QUEDNAU, B.D., D.A. NICOLL & K.D. PHILIPSON. 1997. Tissue specificity and alternative splicing of the Na^+/Ca^{2+} exchanger isoforms NCX1, NCX2, and NCX3 in rat. Am. J. Physiol. Cell Physiol. **272:** C1250–C1261.
2. BLAUSTEIN, M.P. & W.J. LEDERER. 1999. Sodium-calcium exchange: its physiological implications. Physiol. Rev. **79:** 763–854.
3. KOFUJI, P., W.J. LEDERER & D.H. SCHULZE. 1994. Mutually exclusive and cassette exons underlie alternatively spliced isoforms of the Na/Ca exchanger. J. Biol. Chem. **269:** 5145–5149.
4. LEE, S.-L., A.S.L. YU & J. LYTTON. 1994. Tissue-specific expression of Na^+-Ca^{2+} exchanger isoforms. J. Biol. Chem. **269:** 14849–14852.
5. JEON, D., Y.M. YANG, M.J. JEONG, et al. 2003. Enhanced learning and memory in mice lacking Na^+/Ca^{2+} exchanger 2. Neuron **38:** 965–976.
6. SOKOLOW, S., M. MANTO, P. GAILLY, et al. 2004. Impaired neuromuscular transmission and skeletal muscle fiber necrosis in mice lacking Na/Ca exchanger 3. J. Clin. Invest. **113:** 265–273.
7. FRAYSSE, B., T. ROUAUD, M. MILLOUR, et al. 2001. Expression of the Na^+/Ca^{2+} exchanger in skeletal muscle. Am. J. Physiol. Cell Physiol. **280:** C146–C154.
8. LEVITSKY, D.O., B. FRAYSSE, C. LÉOTY, et al. 1996. Cooperative interaction between Ca^{2+} binding sites in the hydrophilic loop of the Na^+-Ca^{2+} exchanger. Mol. Cell. Biochem. **160/161:** 27–32.
9. REEVES, J.P., M. CONDRESCU, G. CHERNAYA & J.P. GARDNER. 1994. Na^+/Ca^{2+} antiport in the mammalian heart. J. Exp. Biol. **196:** 375–388.
10. KOUSHIK, S.V., J. WANG, R. ROGERS, et al. 2001. Targeted inactivation of the sodium-calcium exchanger (*Ncx1*) results in the lack of a heartbeat and abnormal myofibrillar organization. FASEB J. **15:** 1209–1211.
11. HASSELBACH, W. 1964. Relaxing factor and the relaxation of muscle. Prog. Biophys. Mol. Biol. **14:** 167–222.
12. MELZER, W., A. HERRMANN-FRANK & H.C. LÜTTGAU. 1995. The role of Ca^{2+} ions in excitation-contraction coupling of skeletal muscle fibres. Biochim. Biophys. Acta **1241:** 59–116.
13. VETTER, R., R. STUDER, H. REINECKÉ, et al. 1995. Reciprocal changes in the postnatal expression of the sarcolemmal Na^+-Ca^{2+}-exchanger and SERCA2 in rat heart. J. Mol. Cell. Cardiol. **27:** 1689–1701.
14. LEVITSKY, D.O., D.A. NICOLL & K.D. PHILIPSON. 1994. Identification of the high affinity Ca^{2+}-binding domain of the cardiac Na^+-Ca^{2+} exchanger. J. Biol. Chem. **269:** 22847–22852.

Plasma Membrane Ca^{2+} ATPases as Dynamic Regulators of Cellular Calcium Handling

EMANUEL E. STREHLER,[a] ARIEL J. CARIDE,[a] ADELAIDA G. FILOTEO,[a] YUNING XIONG,[a] JOHN T. PENNISTON,[a,b] AND AGNES ENYEDI[a,c]

[a]Department of Biochemistry and Molecular Biology, Mayo Clinic College of Medicine, Rochester, Minnesota 55905, USA

[b]Neuroscience Center, Massachusetts General Hospital, Boston, Massachusetts 02114, USA

[c]National Medical Center, Budapest 1113, Hungary

ABSTRACT: Plasma membrane Ca^{2+} ATPases (PMCAs) are essential components of the cellular toolkit to regulate and fine-tune cytosolic Ca^{2+} concentrations. Historically, the PMCAs have been assigned a housekeeping role in the maintenance of intracellular Ca^{2+} homeostasis. More recent work has revealed a perplexing multitude of PMCA isoforms and alternative splice variants, raising questions about their specific role in Ca^{2+} handling under conditions of varying Ca^{2+} loads. Studies on the kinetics of individual isoforms, combined with expression and localization studies suggest that PMCAs are optimized to function in Ca^{2+} regulation according to tissue- and cell-specific demands. Different PMCA isoforms help control slow, tonic Ca^{2+} signals in some cells and rapid, efficient Ca^{2+} extrusion in others. Localized Ca^{2+} handling requires targeting of the pumps to specialized cellular locales, such as the apical membrane of cochlear hair cells or the basolateral membrane of kidney epithelial cells. Recent studies suggest that alternatively spliced regions in the PMCAs are responsible for their unique targeting, membrane localization, and signaling cross-talk. The regulated deployment and retrieval of PMCAs from specific membranes provide a dynamic system for a cell to respond to changing needs of Ca^{2+} regulation.

KEYWORDS: calcium homeostasis; calcium pump; calcium signaling; PMCA; splice variant

Address for correspondence: Emanuel E. Strehler, Department of Biochemistry and Molecular Biology, Mayo Clinic College of Medicine, 200 First Street S.W., Rochester, MN 55905. Voice: 507-284-9372; fax: 507-284-2384.
strehler.emanuel@mayo.edu

Ann. N.Y. Acad. Sci. 1099: 226–236 (2007). © 2007 New York Academy of Sciences.
doi: 10.1196/annals.1387.023

INTRODUCTION

Ionized calcium (Ca^{2+}) acts as universal messenger controlling cellular processes ranging from fertilization to programmed cell death.[1,2] The signaling function of Ca^{2+} requires an elaborate "toolkit" of proteins to allow Ca^{2+} influx, efflux, and buffering in and between different cellular compartments and among different cells.[3] To achieve spatial and temporal signal sensitivity appropriate types and amounts of the calcium handling proteins must be precisely localized in cellular locales to form tightly connected "calcium signalosomes." High signal sensitivity is achieved by keeping cytosolic $[Ca^{2+}]$ at very low levels during the resting state. This is accomplished by Ca^{2+}-buffering proteins as well as by membrane-intrinsic Ca^{2+} transport systems capable of removing Ca^{2+} from the cytosol even against a large concentration gradient. ATP-driven calcium pumps and ion gradient-dependent Na^+/Ca^{2+} exchangers are the major systems responsible for such "uphill" transport of Ca^{2+} across biological membranes. Plasma membrane Ca^{2+} ATPases (PMCAs) are high-affinity Ca^{2+} pumps dedicated to the expulsion of Ca^{2+} from the cytosol into the extracellular space. Because of their ubiquitous expression and low capacity, they have traditionally been thought to act as major housekeeping system responsible for setting and maintaining the normally low cytosolic $[Ca^{2+}]$. However, the discovery of a multitude of PMCA isoforms and alternative splice variants, as well as recent results on PMCA "knockout" mice and PMCA mutants show that at least some PMCAs have a bigger role in local Ca^{2+} handling. The identification of a growing number of specific PMCA-interacting proteins with regulatory, targeting, and signaling functions further supports the new paradigm that PMCAs are not only responsible for global Ca^{2+} homeostasis but are dynamic participants in spatially defined Ca^{2+} signaling.

MULTIPLE PMCA ISOFORMS AND ALTERNATIVE SPLICE VARIANTS: SPECIFIC ROLES IN CELLULAR CALCIUM REGULATION

PMCAs belong to the type IIB subfamily within the large superfamily of P-type ATPases.[4] In mammals, four separate genes code for the major PMCA isoforms 1–4. In humans, these genes (genome database nomenclature ATP2B1-ATP2B4) are found on chromosomes 12q21.3, 3p25.3, Xq28, and 1q32.1, whereas the mouse genes (database nomenclature Atp2b1-Atp2b4) are located on chromosomes 10C3, 6E3, XA7.3, and 1E4, respectively. Alternative RNA splicing further augments the diversity of PMCA isoforms: splicing at two "hotspots" named site A and site C results in the generation of over 20 PMCA variants.[5] The splice variants differ in the length of the first intracellular loop (site A splicing) and in the C-terminal tail (site C splicing). A scheme of the overall PMCA topology and the location of splice hotspots

A and C are shown in FIGURE 1. TABLE 1 lists data on the major PMCA isoforms and splice variants in humans. The various PMCA isoforms and splice
variants show developmental-, tissue-, and cell-specific patterns of expression.
The ubiquitous PMCA1x/b is detected at the earliest time point in developing
mouse embryos and is expressed in most tissues throughout life.[6] In contrast,
the splice variants PMCA1x/a and PMCA1x/c are only expressed in specific
tissues and cell types and are found mainly in differentiated neurons and striated muscle, respectively.[5,7] Similarly, PMCA isoforms 2 and 3 are almost
exclusively expressed in excitable tissues, with some splice forms, such as
PMCA2w/a being specific for select cell types (e.g., auditory and vestibular hair cells).[8] Multiple splice forms of PMCA isoforms 2 and 3 are also
expressed in insulin-secreting β cells of the endocrine pancreas.[9] Notably,
these cells are electrically excitable and display characteristic Ca^{2+} oscillations. The difference in expression pattern and abundance suggests that the
different PMCA isoforms and splice variants fulfill different roles in cellular Ca^{2+} regulation. Recent work on mice with targeted deletions or spontaneous mutations of specific PMCA genes supports this notion: homozygous

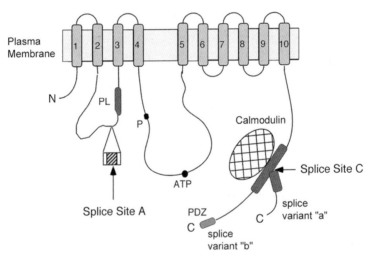

FIGURE 1. Scheme of the PMCA. The membrane-spanning regions are numbered
and shown as shaded boxes. The amino- (N) and carboxyl-terminal (C) ends are labeled,
P denotes the obligatory aspartyl-phosphate formed during the reaction cycle, and ATP
indicates the region involved in nucleotide (ATP) binding. The phospholipid-sensitive region
(PL), the calmodulin-binding region, and the PDZ-binding motif at the C terminus of all
PMCA b-splice variants are shown as gray boxes. The PMCA is represented in its activated
state with Ca^{2+}-calmodulin (Ca^{2+}-CaM) bound to the C-tail. Arrows labeled "splice site
A" and "splice site C" denote the regions affected by alternative splicing. A hatched box
indicates the peptide segment encoded by alternatively spliced exon(s) at site A; at site C
the two major splice variants ("a" and "b") are shown as separate tails to indicate their
divergent reading frames.

TABLE 1. Human PMCA isoforms and major alternative splice variants

Isoform	Major alternative splice variants	Length in amino acids	GenBank accession numbers	Tissue distribution
PMCA1	PMCA1x/a	1176	NM_001001323	Brain, nervous tissue
	PMCA1x/b	1220	NM_001682	Ubiquitous
	PMCA1x/c	1249		Skeletal muscle, heart
PMCA2	PMCA2w/a	1199	U15688*	Inner ear hair cells
	PMCA2x/a	1168		Brain (relatively rare)
	PMCA2z/a	1154		Brain (generally more abundant than 2x/a)
	PMCA2w/b	1243	NM_001001331	Brain, breast (lactating mammary gland), pancreatic β-cells
	PMCA2x/b	1212		Brain/excitable tissue
	PMCA2z/b	1198	NM_001683	Brain/excitable tissue
PMCA3	PMCA3x/a	1173	NM_021949	Brain
	PMCA3z/a	1159		Brain (cortex, thalamus, substantia nigra), pancreatic β-cells
	PMCA3x/b	1220	NM_001001344	Brain
	PMCA3z/b	1206		Brain (cortex, thalamus, substantia nigra)
	PMCA3x/f	1129		(Fast) skeletal muscle, brain (rare)
PMCA4	PMCA4x/a	1170	NM_001001396	Brain, heart, stomach
	PMCA4z/a	1158		Heart, pancreas (Islet of Langerhans)
	PMCA4x/b	1205	NM_001684	Ubiquitous
	PMCA4z/b	1193		Heart

*Splice site C alternative exon sequence only.

Atp2b1-/- mice are embryonic lethal, Atp2b2-/- mice are ataxic and profoundly deaf, and Atp2b4-/- mice show male infertility.[10–12]

DIFFERENT KINETICS OF REGULATION: PMCA ISOFORMS ARE ADAPTED TO SPECIFIC CALCIUM HANDLING NEEDS

The main regulator of PMCA function is Ca^{2+}-calmodulin ($Ca^{2+}-CaM$).[13] In the absence of CaM, the pumps are autoinhibited by a mechanism that involves binding of their C-terminal tail to the two major intracellular loops. Activation requires binding of Ca^{2+}-CaM to the C-terminal tail and a conformational change that displaces the autoinhibitory tail from the major catalytic domain. Release of autoinhibition may be facilitated by means other than CaM binding, including by acidic phospholipids, protein kinase A- or C-mediated phosphorylation of specific (Ser/Thr) residues in the C-terminal tail, partial proteolytic cleavage of the tail (e.g., by calpain or caspases), or dimerization via the C-terminal tail.[14,15] Different PMCA isoforms show significant differences in their regulation by kinases and CaM. For example, some isoforms are activated by PKC (PMCA4b) whereas others are unaffected (PMCA4a) or even slightly inhibited (PMCA2a, 3a).[16–18] The CaM affinity and the extent of activation by CaM also vary greatly among PMCA isoforms and splice variants, with the a-splice forms generally showing lower CaM affinity but higher basal (CaM-independent) activity than the b-splice forms.[19,20]

The functional differences among PMCA isoforms are relevant to how the pumps handle temporary changes in $[Ca^{2+}]_i$. Recent work on the rates of activation of individual PMCAs has revealed significant differences in their ability to respond to a sudden increase in Ca^{2+} and to decode the frequency of repetitive Ca^{2+} spikes.[21] The rate for CaM activation of PMCA4b is slow ($t_{1/2} \sim 1$ min at 0.5 μM Ca^{2+}) compared to PMCA4a ($t_{1/2} \sim 20$ s). The inactivation rate (off-rate) for PMCA4b is even slower ($t_{1/2} \sim 20$ min), indicating that this isoform may be geared toward handling slow, tonic Ca^{2+} changes in nonexcitable cells.[22] Because a change in subplasmalemmal Ca^{2+} is the crucial physiological event sensed by the PMCAs, their "reaction time" to such a stimulus should be optimal for the desired outcome of Ca^{2+} signaling. Indeed, detailed kinetic studies showed that PMCAs expressed in cells with a need for fast Ca^{2+} responses (e.g., striated muscles, neurons, sensory hair cells) are generally "fast," that is, they are activated quickly by a rise in $[Ca^{2+}]_i$.[23] Accordingly, PMCAs 2a and 3f, which are expressed mostly in excitable tissues, are very fast pumps while PMCA4b is a slow pump. The very different off-rates for CaM release (and hence, inactivation) also result in pronounced differences in the "memory" of PMCA isoforms for their previous activation. For example, PMCA2b has a very slow off-rate for CaM dissociation and retains a long memory of its recent activation.[21] In cells with repetitive Ca^{2+} spikes, PMCA2b will thus remain "preactivated" for an extended time and respond almost immediately to a new Ca^{2+} signal. This facilitation of Ca^{2+} extrusion may be crucial in neurons to maintain sensitivity to signal frequency.

PMCAs ARE INTEGRATED INTO SIGNALING PATHWAYS VIA MULTIPLE PROTEIN–PROTEIN INTERACTIONS

Besides the well-known regulation of PMCAs by CaM, the pumps are affected in less well-studied ways by multiple kinases (both Ser/Thr-specific and Tyr-specific); these probably relay important information about the cell's metabolic state to the calcium pump extrusion system.[15,24] In addition, recent work based on yeast or bacterial two-hybrid screens has identified numerous PMCA-interacting proteins that are likely to connect specific PMCA isoforms to particular signaling pathways in the cell. FIGURE 2 provides a schematic overview of the major currently known interacting proteins and the site of their interaction with the pump. Many of these protein interactions are specific for a subset of PMCA isoforms and/or splice variants. For example, all b-splice variants, but none of the a-splice variants contain a C-terminal

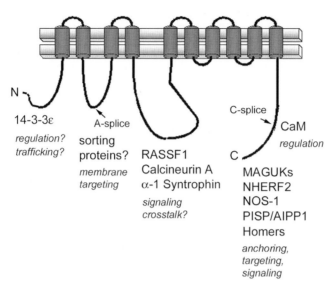

FIGURE 2. PMCA-interacting proteins and their possible roles in pump regulation and function. PMCA-interacting proteins are listed near the domain of the PMCA to which they bind. Suspected or known functional roles of these different protein interactions are indicated in italics. Note that no binding partners (sorting proteins?) have yet been identified for the fist cytosolic loop of the PMCA carrying the A-splice. Also note that except for CaM, all proteins listed as binding to the C-terminal tail of the PMCA are specific for the b-splice variants and do not interact with the a-splice variants. For details, see the text.

consensus motif for binding PDZ domains. It is perhaps not surprising that specific interacting partners are found for the most divergent regions of the PMCAs, including the alternatively spliced C-tail and the N-terminal tail. On the other hand, some of the interacting proteins are promiscuous and bind to several PMCA isoforms with similar affinity. This includes members of the membrane-associated guanylate kinase (MAGUK) family of PDZ proteins that recognize a short peptide at the C-terminus of PMCA b-splice variants,[25] and α-1 syntrophin, which was recently found to interact with a conserved domain in the major cytosolic loop of PMCA4 and PMCA1.[26] In the following sections, we will briefly summarize the main classes of recently discovered PMCA-interacting proteins and their possible roles in PMCA regulation and signaling cross-talk.

Proteins Interacting with the N-Terminal Tail

Using the N-terminal cytosolic tail of human PMCA4 as bait in a yeast two-hybrid screen, Rimessi and co-workers[27] recently identified the protein 14-3-3ε as a novel interaction partner of PMCA4. This interaction appears to be specific for the N-terminal tail of PMCA4, as the corresponding region of

PMCA2 was unable to interact with 14-3-3ε. When overexpressed in HeLa cells, 14-3-3ε had an inhibitory effect on PMCA4 function as measured by the ability of the cells to clear an agonist-induced Ca^{2+} load. While direct regulation (inhibition) of PMCA4 activity is one possibility for the functional role of the 14-3-3ε/PMCA4 interaction, 14-3-3ε is also involved in a variety of other tasks, including protein trafficking and cellular signaling. Given the large number of 14-3-3ε interacting proteins and the diverse tasks already ascribed to 14-3-3ε, a role for this protein in PMCA4 chaperoning and/or trafficking appears plausible.

Proteins Interacting with the Large Cytosolic Loops

No interactions with other proteins have yet been described for the first cytosolic loop of the PMCA. However, this region is known to be involved in the acidic phospholipid sensitivity of the PMCA and to make intramolecular contacts with the autoinhibitory C-tail of the pump.[5] This loop also contains the alternative splice site A, where the insertion of extra amino acids affects membrane targeting (FIG. 2)[28]: in PMCA2, the w-splice insert directs the pump to the apical membrane of cochlear hair cells.[8,29] This differential membrane targeting may require specific protein interactions of the loop (e.g., with apical/basolateral sorting proteins) to discriminate between the w- and the x/z-splice variants.

At least three different proteins have recently been found to physically interact with the large catalytic loop of PMCA4b (see FIG. 2): the Ras effector protein RASSF1,[30] the catalytic A-subunit of the Ca^{2+}/CaM-dependent phosphatase calcineurin,[31] and α-1 syntrophin.[26] Because these proteins interact with regions in the catalytic domain that are highly conserved among all PMCAs, they are likely promiscuous and affect all PMCA isoforms. Indeed, α-1 syntrophin was shown to bind to both PMCA1 and PMCA4.[26] Functionally, RASSF1 interaction with PMCA4b appears to downregulate Ras-mediated signaling from the EGF receptor to the Erk phosphorylation and transcriptional activation pathway. However, the physiological relevance of this observation remains to be determined as the published data were all obtained in cells overexpressing recombinant proteins.[30] Similarly, the interaction of (overexpressed) PMCA4b with calcineurin A seems to downregulate calcineurin-mediated signaling events (such as NFAT transcriptional activity), perhaps by recruiting calcineurin to the plasma membrane into microdomains of low $[Ca^{2+}]_i$.[31] α-1 Syntrophin is thought to form a ternary signaling complex with the PMCA and nitric oxide synthase-1 (NOS-1) (which binds PMCA4b independently through a PDZ domain; see below). This complex may tether the PMCA to the dystrophin complex in skeletal and heart muscle.[26] The above examples suggest that PMCAs may be directly involved in the regulation of downstream signaling events, supporting their integration in bidirectional signaling crosstalk from and to the membrane. It must be noted, however, that the impact of any of the above interactions on PMCA function has not yet been studied. Because RASSF1, calcineurin A, and α-1 syntrophin all bind to the large cytosolic loop of the PMCA that is also involved in intramolecular interactions with the C-tail, the question arises whether these external proteins compete with the C-tail for binding to the catalytic loop, and if and how their binding affects pump activation.

Proteins Interacting with the C-Terminal Tail

CaM is the "prototype" of all PMCA-interacting proteins, having been identified as major regulator of PMCA activity almost 30 years ago.[32] Besides CaM, many additional proteins have now been found to bind to the C-tail of the PMCA, and the list of new interacting proteins is still growing (FIG. 2). As already mentioned, the b-splice variants of all PMCA isoforms contain the consensus sequence E-T/S-X-L/V for binding type-I PDZ domains at their carboxyl terminus. PDZ proteins shown to interact with the PMCAs include members of the MAGUK family, such as the synapse-associated proteins PSD95/SAP90, SAP97/hDlg, SAP102, and PSD93/chapsyn-110,[25] as well as the Ca^{2+}/CaM-dependent serine protein kinase CASK.[33] These protein interactions serve to cluster specific PMCA isoforms into multiprotein signaling complexes at cellular sites, such as presynaptic nerve terminals or postsynaptic spines. By recruiting PMCAs into close proximity of Ca^{2+} entry receptors/channels (e.g., NMDA receptors), cells are able to generate microdomains of spatially confined Ca^{2+} signaling. Similarly, the interaction with Homer-1/Ania-3 may allow specific PMCAs to be functionally coupled to TrpC and IP_3 receptor Ca^{2+} channels.[34] Consistent with the theme of PMCA-mediated local Ca^{2+} control, the interaction of PMCA4b with NOS-1 downregulates NO production presumably due to a PMCA-mediated decrease of local $[Ca^{2+}]$ in the immediate vicinity of NOS-1.[35] In contrast, the functional role of the interaction between the single-PDZ protein PISP/AIPP1[36,37] and the PMCA b-splice variants is less clear. The predominant localization of PISP/AIPP1 in punctuate, presumably vesicular structures suggests that it may be involved in the trafficking of the pumps to or from the plasma membrane. Finally, the interaction of NHERF2 with PMCA2b[38] is of interest. The apically localized NHERF2 may link apically targeted PMCA2b to the underlying actin cytoskeleton (via its ezrin/radixin/moesin-interacting domain) and thereby promote local retention and enhanced apical localization of the pump.

CONCLUSIONS: PMCAs ARE A DYNAMIC SYSTEM TO RESPOND TO CHANGING Ca^{2+} HANDLING NEEDS

Work on gene-targeted mice and PMCA mutants, combined with detailed studies of the subcellular localization, regulation and kinetics, and novel protein interactions of PMCAs has shown that specific pump isoforms and splice variants fulfill precisely orchestrated roles in handling local Ca^{2+} changes. This requires their integration in multiple intracellular signaling pathways, and mandates that they act as both receiver and sender of signals. Examples include the involvement of specific PMCAs in calcineurin signaling,[31] local NO signaling,[35] and controlling the duration of action potential after-hyperpolarization in sensory neurons.[39] To respond dynamically to the changing needs of Ca^{2+} signaling, cells must be able to precisely control the type, amount, localization, and activation state of each PMCA. This control is largely exerted via proteins that specifically interact with the PMCAs, as discussed throughout this review. As dynamic participants in cellular Ca^{2+} handling, the PMCAs must be regulated at all levels. Long-term regulation involves changes of gene transcription, mRNA stability, alternative splicing, and protein translation. This type of regulation is controlled by factors, such as cell differentiation and changes in $[Ca^{2+}]_i$ itself. In the mid term, local PMCA availability is regulated by differential membrane targeting, internalization, and possibly, recycling. PDZ proteins, such as NHERF2, PISP/AIPP1, and MAGUKs, are probably involved in this control and provide isoform/splice variant specificity to the regulation. Finally, short-term regulation is provided by CaM, differential phosphorylation, interactions with other signaling molecules, and partial proteolysis (e.g., by calpain and caspases). This short-term regulation is highly isoform-specific (as illustrated by the very different CaM activation kinetics of PMCA2b and PMCA4b) and further contributes to the diversity of local Ca^{2+} handling. Multiple levels of control allow the dynamic regulation of PMCA function on time scales ranging from seconds to days, and enable the PMCAs to participate in varying tasks from fast Ca^{2+} signaling in neurons to managing slow Ca^{2+} transients and bulk Ca^{2+} movement in transepithelial Ca^{2+} flux.

ACKNOWLEDGMENTS

This work was supported in part by NIH Grant GM28835 (to E.E.S. and J.T.P.), AHA Grant 01-30531Z (to A.J.C.), and OTKA grant T049476 (to A.E.).

REFERENCES

1. BERRIDGE, M.J., M.D. BOOTMAN & P. LIPP. 1998. Calcium—a life and death signal. Nature **395:** 645–648.
2. CARAFOLI, E. *et al.* 2001. Generation, control, and processing of cellular calcium signals. Crit. Rev. Biochem. Molec. Biol. **36:** 107–260.
3. BERRIDGE, M.J., M.D. BOOTMAN & H.L. RODERICK. 2003. Calcium signalling: dynamics, homeostasis and remodelling. Nat. Rev. Mol. Cell Biol. **4:** 517–529.
4. AXELSEN, K.B. & M.G. PALMGREN. 1998. Evolution of substrate specificities in the P-type ATPase superfamily. J. Mol. Evol. **46:** 84–101.

5. STREHLER, E.E. & D.A. ZACHARIAS. 2001. Role of alternative splicing in generating isoform diversity among plasma membrane calcium pumps. Physiol. Rev. **81:** 21–50.

6. ZACHARIAS, D.A. & C. KAPPEN. 1999. Developmental expression of the four plasma membrane calcium ATPase (PMCA) genes in the mouse. Biochim. Biophys. Acta **1428:** 397–405.

7. KIP, S.N. *et al.* 2006. Changes in the expression of plasma membrane calcium extrusion systems during the maturation of hippocampal neurons. Hippocampus **16:** 20–34.

8. HILL, J.K. *et al.* 2006. Splice-site A choice targets plasma-membrane Ca^{2+}-ATPase isoform 2 to hair bundles. J. Neurosci. **26:** 6172–6180.

9. KAMAGATE, A. *et al.* 2000. Expression of multiple plasma membrane Ca^{2+}-ATPases in rat pancreatic islet cells. Cell Calcium **27:** 231–246.

10. KOZEL, P.J. *et al.* 1998. Balance and hearing deficits in mice with a null mutation in the gene encoding plasma membrane Ca^{2+}-ATPase isoform 1. J. Biol. Chem. **273:** 18693–18696.

11. OKUNADE, G.W. *et al.* 2004. Targeted ablation of plasma membrane Ca^{2+}-ATPase (PMCA) 1 and 4 indicates a major housekeeping function for PMCA1 and a critical role in hyperactivated sperm motility and male fertility for PMCA4. J. Biol. Chem. **279:** 33742–33750.

12. SCHUH, K. *et al.* 2004. Plasma membrane Ca^{2+} ATPase 4 is required for sperm motility and male fertility. J. Biol. Chem. **279:** 28220–28226.

13. PENNISTON, J.T. *et al.* 1988. Purification, reconstitution, and regulation of plasma membrane Ca^{2+}-pumps. Methods Enzymol. **157:** 340–351.

14. CARAFOLI, E. 1994. Biogenesis: plasma membrane calcium ATPase: 15 years of work on the purified enzyme. FASEB J. **8:** 993–1002.

15. PENNISTON, J.T. & A. ENYEDI. 1998. Modulation of the plasma membrane Ca^{2+} pump. J. Membr. Biol. **165:** 101–109.

16. VERMA, A.K. *et al.* 1999. Protein kinase C phosphorylates plasma membrane Ca^{2+} pump isoform 4a at its calmodulin binding domain. J. Biol. Chem. **274:** 527–531.

17. ENYEDI, A. *et al.* 1996. Protein kinase C activates the plasma membrane Ca^{2+} pump isoform 4b by phosphorylation of an inhibitory region downstream of the calmodulin-binding domain. J. Biol. Chem. **271:** 32461–32467.

18. ENYEDI, A. *et al.* 1997. Protein kinase C phosphorylates the "a" forms of plasma membrane Ca^{2+} pump isoforms 2 and 3 and prevents binding of calmodulin. J. Biol. Chem. **272:** 27525–27528.

19. ENYEDI, A. *et al.* 1994. The Ca^{2+} affinity of the plasma membrane Ca^{2+} pump is controlled by alternative splicing. J. Biol. Chem. **269:** 41–43.

20. ELWESS, N.L. *et al.* 1997. Plasma membrane Ca^{2+} pump isoforms 2a and 2b are unusually responsive to calmodulin and Ca^{2+}. J. Biol. Chem. **272:** 17981–17986.

21. CARIDE, A.J. *et al.* 2001. The plasma membrane calcium pump displays memory of past calcium spikes. Differences between isoforms 2b and 4b. J. Biol. Chem. **276:** 39797–39804.

22. CARIDE, A.J. *et al.* 1999. The rate of activation by calmodulin of isoform 4 of the plasma membrane Ca^{2+} pump is slow and is changed by alternative splicing. J. Biol. Chem. **274:** 35227–35232.

23. CARIDE, A.J. *et al.* 2001. Delayed activation of the plasma membrane calcium pump by a sudden increase in Ca^{2+}: fast pumps reside in fast cells. Cell Calcium **30:** 49–57.

24. DEAN, W.L. *et al.* 1997. Regulation of platelet plasma membrane Ca^{2+}-ATPase by cAMP-dependent and tyrosine phosphorylation. J. Biol. Chem. **272:** 15113–15119.

25. DEMARCO, S.J. & E.E. STREHLER. 2001. Plasma membrane Ca^{2+}-ATPase isoforms 2b and 4b interact promiscuously and selectively with members of the membrane-associated guanylate kinase family of PDZ (PSD-95/Dlg/ZO-1) domain-containing proteins. J. Biol. Chem. **276:** 21594–21600.

26. WILLIAMS, J.C. *et al.* 2006. The sarcolemmal calcium pump, a-1 syntrophin, and neuronal nitric-oxide synthase are parts of a macromolecular protein complex. J. Biol. Chem. **281:** 23341–23348.

27. RIMESSI, A. *et al.* 2005. Inhibitory interaction of the 14-3-3ε protein with isoform 4 of the plasma membrane Ca^{2+}-ATPase pump. J. Biol. Chem. **280:** 37195–37203.

28. CHICKA, M.C. & E.E. STREHLER. 2003. Alternative splicing of the first intracellular loop of plasma membrane Ca^{2+}-ATPase isoform 2 alters its membrane targeting. J. Biol. Chem. **278:** 18464–18470.

29. GRATI, M. *et al.* 2006. Molecular determinants for differential membrane trafficking of PMCA1 and PMCA2 in mammalian hair cells. J. Cell Sci. **119:** 2995–3007.

30. ARMESILLA, A.L. *et al.* 2004. Novel functional interaction between the plasma membrane Ca^{2+} pump 4b and the proapoptotic tumor suppressor Ras-associated factor 1 (RASSF1). J. Biol. Chem. **279:** 31318–31328.

31. BUCH, M.H. *et al.* 2005. The sarcolemmal calcium pump inhibits the calcineurin/nuclear factor of activated T-cell pathway via interaction with the calcineurin A catalytic subunit. J. Biol. Chem. **280:** 29479–29487.

32. JARRETT, H.W. & J.T. PENNISTON. 1978. Purification of the Ca^{2+}-stimulated ATPase activator from human erythrocytes: its membership in the class of Ca^{2+}-binding modulator proteins. J. Biol. Chem. **253:** 4676–4682.

33. SCHUH, K. *et al.* 2003. Interaction of the plasma membrane Ca^{2+} pump 4b/CI with the Ca^{2+}/calmodulin-dependent membrane-associated kinase CASK. J. Biol. Chem. **278:** 9778–9883.

34. SGAMBATO-FAURE, V. *et al.* 2006. The homer-1 protein Ania-3 interacts with the plasma membrane calcium pump. Biochem. Biophys. Res. Commun. **343:** 630–637.

35. SCHUH, K. *et al.* 2001. The plasmamembrane calmodulin-dependent calcium pump: a major regulator of nitric oxide synthase I. J. Cell Biol. **155:** 201–205.

36. GOELLNER, G.M., S.J. DEMARCO & E.E. STREHLER. 2003. Characterization of PISP, a novel single-PDZ protein that binds to all plasma membrane Ca^{2+}-ATPase b-splice variants. Ann. N. Y. Acad. Sci. **986:** 461–471.

37. STEPHENSON, S.E.M. *et al.* 2005. A single PDZ domain protein interacts with the Menkes copper ATPase, ATP7A. J. Biol. Chem. **280:** 33270–33279.

38. DEMARCO, S.J., M.C. CHICKA & E.E. STREHLER. 2002. Plasma membrane Ca^{2+} ATPase isoform 2b interacts preferentially with Na^+/H^+ exchanger regulatory factor 2 in apical plasma membranes. J. Biol. Chem. **277:** 10506–10511.

39. USACHEV, Y.M. *et al.* 2002. Bradykinin and ATP accelerate Ca^{2+} efflux from rat sensory neurons via protein kinase C and the plasma membrane Ca^{2+} pump isoform 4. Neuron. **33:** 113–122.

Functional Specificity of PMCA Isoforms?

TEUTA DOMI,[a] FRANCESCA DI LEVA,[a] LAURA FEDRIZZI,[a]
ALESSANDRO RIMESSI,[b] AND MARISA BRINI[a]

[a]*Departments of Biochemistry and Experimental Veterinary Sciences,
University of Padova, 35121 Padova, Italy*

[b]*Department of Experimental and Diagnostic Medicine, Interdisciplinary
Center for the Study of Inflammation, University of Ferrara,
44100 Ferrara, Italy*

ABSTRACT: In mammals, four different genes encode four PMCA isoforms. PMCA1 and PMCA4 are expressed ubiquitously. PMCA2 and
PMCA3 are expressed prevalently in the central nervous systems. More
than 30 variants are generated by mechanisms of alternative splicing. The
physiological meaning of the existence of such elevated number of isoforms is not clear, but it would be plausible to relate it to the cell-specific
demands of Ca^{2+} homeostasis. To characterize functional specificity of
PMCA variants we have investigated two aspects: the effects of the overexpression of the different PMCA variants on cellular Ca^{2+} handling
and the existence of possible isoform-specific interactions with partner
proteins using a yeast two-hybrid technique. The four basic PMCA isoforms were coexpressed in CHO cells together with the Ca^{2+}-sensitive
recombinant photoprotein aequorin. The effects of their overexpression
on Ca^{2+} homeostasis were monitored in the living cells. They had revealed that the ubiquitous isoforms 1 and 4 are less effective in reducing
the Ca^{2+} peaks generated by cell stimulation as compared to the neuron-
specific isoforms 2 and 3. To establish whether these differences were
related to different and new physiological regulators of the pump, the
90 N-terminal residues of PMCA2 and PMCA4 have been used as baits
for the search of molecular partners. Screening of a human brain cDNA
library with the PMCA4 bait specified the ε-isoform of protein 14-3-3,
whereas no 14-3-3 ε clone was obtained with the PMCA2 bait. Overexpression of PMCA4/14-3-3 ε (but not of PMCA2/14-3-3 ε) in HeLa cells
together with targeted aequorins showed that the ability of the cells to
export Ca^{2+} was impaired. Thus, the interaction with 14-3-3 ε inhibited
PMCA4 but not PMCA2. The role of PMCA2 has been further characterized by Ca^{2+} measurements in cells overexpressing different splicing variants. The results indicated that the combination of alternative

Address for correspondence: Marisa Brini, Department of Biochemistry, University of Padova, Viale
G. Colombo 3, 35121 Padova, Italy. Voice: + 39-049-8276167; fax: + 39-049-8276125.
marisa.brini@unipd.it

Ann. N.Y. Acad. Sci. 1099: 237–246 (2007). © 2007 New York Academy of Sciences.
doi: 10.1196/annals.1387.043

splicing at two different sites in the pump structure was responsible for different functional characteristics of the pumps.

KEYWORDS: calcium homeostasis; plasma membrane Ca^{2+} pumps; isoforms; aequorin

INTRODUCTION

The transport of Ca^{2+} out of the cytosol of eukaryotic cells, which is essential to the maintenance of cellular Ca^{2+} homeostasis, is accomplished by two systems: a low-affinity, high-capacity Na^+/Ca^{2+} exchanger, which is particularly active in excitable tissues, and a high-affinity, low-capacity Ca^{2+}-ATPase (the plasma membrane Ca^{2+} pump, PMCA), which is active in all eukaryotic cells. The high affinity of the ATPase enables it to interact with Ca^{2+} with adequate efficiency even when its concentration is at the very low level prevailing in the cytosol of cells at rest (100–200 nM). Thus, the PMCA pump is the fine tuner of cell Ca^{2+}: it counteracts the action of the plasma membrane channels, across which a limited and carefully controlled amount of Ca^{2+} penetrates into the cytosol.[1] The ATPase was discovered in erythrocytes in 1966[2] and was later characterized as a P-type pump.[3] In the 40 years that have elapsed since its discovery, the work has developed as in the case of other transport ATPases, gradually evolving from an initial phase focused on the properties of the transport process and on the reaction mechanism to a later phase in which the enzyme was dissected molecularly and characterized genetically. Knowledge on the PMCA pump has progressed rapidly, particularly in recent years, establishing the enzyme as a central actor in the precise control of Ca^{2+} homeostasis in the cells, and thus in their proper functioning.

The PMCA pump is the product of a multigenic family, and several variants (more than 30) are generated through mechanisms of alternative splicing of its RNA.[4,5] In mammals, four different genes encode four PMCA isoforms. PMCA1 and PMCA4 are expressed ubiquitously; PMCA2 and PMCA3 are expressed prevalently in the central nervous systems. RNA alternative splicing generates several isoforms. The pump variants generated by alternative splicing at the C-terminal region of the protein (site C) are of particular interest, since this site contains the calmodulin-binding site. This domain, in the absence of calmodulin, interacts with the protein and keeps the pump in an inactive state. Calmodulin displaces the binding with the intramolecular receptors, removing the inhibition. This type of splicing occurs in all isoforms and causes the inclusion of one (or two) additional exons, or portion of exons, leading to a truncated version of the pump, which displays a shorter regulatory domain as compared to the full-length variant. The truncated version of the pump is named *a* or CII, the full-length variant is named *b* or CI variant.

The second site of alternative splicing is located closer to the 5' end of the gene (site A) in the cytosolic portion that connects the second to the third transmembrane domain of the pump. This portion is considered to be the "transducer domain" of the pump, since it seems to be involved in the conformational changes occurring during the transport cycle. Depending on the isoform, splicing at site A introduces one or more exons.

The situation of PMCA2 is particularly complicated: three different exons, respectively of 33, 60, and 42 base pairs (bp) can be alternatively introduced. Up to date, only four types of proteins have been individuated: the *w* or AIII variant (all three exons included), the *z* or AI variant (all three exons excluded), the *x* or AII variant (where only exon of 42 bp is included), and the *y* or AIV variant (in which the exons of 33 and 60 bp are included).

The physiological rationale for the existence of such an elevated number of isoforms is not clear, but it must be related to the specific demands of Ca^{2+} homeostasis during the different phases of cell life. The common opinion, not yet supported by experimental data, is that there is no redundancy, but that each isoform could play a specific role. Furthermore, the cell could tune its repertoire according to specific Ca^{2+} homeostasis demands that, in turn, can change with time. Mutations in the gene encoding PMCA2 isoform, which is particularly abundant in neuronal cells, have reinforced the opinion. The *w/a* variant of PMCA2 is expressed at high levels in the hair cells of Corti's organ of the inner ear. This variant is located exclusively at the apical membrane and represents the only Ca^{2+} extrusion mechanism in this portion of the cells.[6-8] PMCA2 also contributes to the creation of a high-extracellular Ca^{2+} concentration around the hair bundle, that is, the mechanotransduction apparatus for auditory signals, and maintains the appropriate Ca^{2+} concentration in the endolymph. The finding that mice harboring spontaneous mutations in the gene of PMCA2 display a phenotype associated with hearing loss and defects in the coordination of the movements is of great interest.[9-11] A mutation next to the active center of the pump (V586M) was later found to depress activity of the pump and to increase loss of hearing in heterozygous human patients that also carried a homozygous mutation in cadherin 23 (CDH23).[12]

RESULTS AND DISCUSSION

To clarify the physiological meaning of the existence of such an elevated number of pump isoforms, we had studied their activity in living cells, where the physiological ambient is intact. We then decided to search for isoform-specific molecular interactors, which could modify pump activity or distribution on specific portion of the plasma membrane.

Initially, CHO cells were cotransfected with the expression plasmids for the four basic isoforms together with the plasmid encoding the Ca^{2+}-sensing probe aequorin targeted to the cytoplasm (cytAEQ). At variance with fluorescent dyes, aequorin has very low Ca^{2+}-buffering capacity, and does not significantly perturb the resting cytosolic Ca^{2+} level, which could influence the activity of the Ca^{2+} transporters.[13] CHO cells were stimulated with ATP, an inositol 1,4,5-trisphosphate ($InsP_3$) generating agonist that acts on P2Y purinergic receptors. FIGURE 1 shows cytosolic Ca^{2+} ($[Ca^{2+}]_c$) measurements, which indicate that the two ubiquitous isoforms, PMCA1 and PMCA4, were less effective than the neuronal isoforms, PMCA2 and PMCA3 in restoring the basal $[Ca^{2+}]$ level after the transient induced by ATP. While the faster clearance of the signal in PMCA-overexpressing cells could be reasonably attributed to the increased exporting activity, the decreased height of the Ca^{2+} peak could be due either to the decreased release of Ca^{2+} from the incompletely filled intracellular stores

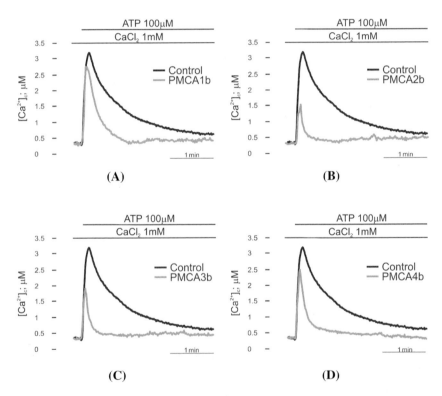

FIGURE 1. Monitoring of cytosolic $[Ca^{2+}]$ in CHO cells transfected with cytAEQ and cotransfected with cytAEQ and the full-length cDNA of PMCA isoforms. Cells were transfected with mammalian expression plasmids for the PMCA isoforms and for the Ca^{2+}-sensitive photoprotein aequorin (cytAEQ), which monitor the Ca^{2+} concentration in the cytosol, $[Ca^{2+}]_c$. The prosthetic group coelenterazine was added to the incubation medium to reconstitute active aequorin before carrying out Ca^{2+} measurements. Where indicated ATP was added as an agonist linked to generating of $InsP_3$, liberating Ca^{2+} from the ER and causing a cytosolic Ca^{2+} transient. (**A**) PMCA1b-expressing cells. (**B**) PMCA2b-expressing cells. (**C**) PMCA3b-expressing cells. (**D**) PMCA4b-expressing cells.

(i.e., the endoplasmic reticulum [ER]) or to the more efficient dissipation of the [Ca^{2+}] transient by the overexpressed PMCA. In fact, the two neuronal isoforms reduced [Ca^{2+}] in the ER by about 30%, whereas PMCA4 only decreased it by about 15%. Isoform 1 was even less effective, reducing ER [Ca^{2+}] only marginally.[14]

The matter of endogenous regulators of the activity of PMCA pumps is now attracting increasing attention. A number of studies on protein partners of the pump have appeared and recent contributions have extended the information to their effects on the activity of the pump. With one or two exception, these studies have not considered the possibility of isoform-specific effects. One interesting contribution, however, has shown that PMCA2 interacts with the Na^+/H^+ exchanger regulatory factor 2, whereas PMCA4 does not.[15] Most studies searching for PMCA pump interactors have focused on the C-terminal domain of the pump, showing its ability to bind the PDZ domain of partner proteins,[16] leading to the identification of partners like the membrane-associated guanylate kinases (MAGUK),[17] the calcium calmodulin-dependent serine protein kinase (CASK),[18] and the neuronal NO synthetase.[19] Other studies have shown interactions with the proapoptotic tumor suppressor Ras-associated factor-1 via a domain in the large cytosolic loop that contains the active site[20] and with the catalytic domain of the calcineurin A subunit.[21] We have extended the search for interactors to the pump isoforms by a yeast two-hybrid screening using the N-terminal domain of the pump, which differs most significantly among the four basic isoforms, as the bait. Two isoforms were initially chosen for the study, one ubiquitously expressed (PMCA4) and one tissue-restricted (PMCA2). The results have identified one protein partner for PMCA4 (the 14-3-3 ε protein, a member of a family of small acidic proteins found in eukaryotes that influence a very large number of cellular processes)[22] but have shown that 14-3-3 ε does not interact with PMCA2. To investigate whether the interaction with 14-3-3 ε has any effect on the activity of the pump, recombinant aequorins were cotransfected with 14-3-3 ε and PMCA4 or PMCA2 expression plasmids in HeLa cells. Because the three recombinant proteins were expressed by the same subset of cells, the risk associated with the use of stable clones or single cells was eliminated. As analyzed in detail in a previous study,[13] this approach was preferred to the isolation and analysis of stable clones coexpressing the photoprotein and the proteins under investigation. FIGURE 2 shows [Ca^{2+}]$_c$ monitored with cytAEQ in HeLa cells transfected with PMCA4/14-3-3 ε or PMCA2/14-3-3 ε or PMCA4 and PMCA2 alone. Histamine stimulation in a medium containing physiological $CaCl_2$ concentration (1 mM) generated higher cytosolic Ca^{2+} transients, higher posttransient cytosolic Ca^{2+} plateaus in the PMCA4/14-3-3 ε -expressing cells with respect to PMCA4-expressing cells, but not in cells coexpressing isoform 2 of the pump together with 14-3-3 ε . These findings suggest that the interaction of 14-3-3 ε with isoform 4 of the PMCA inhibits its activity (FIG. 2). Higher

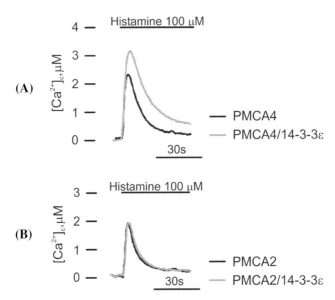

FIGURE 2. Effect of 14-3-3 ε overexpression on the activities of PMCA4 and PMCA2 pump. HeLa cells were transfected with PMCA4 (**A**) or PMCA2 (**B**) or cotransfected with 14-3-3 ε. The [Ca^{2+}]$_c$ measurements were performed with cytAEQ, and the cells were challenged with histamine where indicated.

Ca^{2+} levels were also measured in the ER lumen and in the domain beneath the plasma membrane in HeLa cells (which express endogenous PMCA4)[5] transfected with 14-3-3 ε .[23]

As mentioned in the "Introduction," PMCA2 is a peculiar isoform because of the abundance of its splicing variants and because, so far, it is the only pump for which single-point mutations have been reported to generate a pathological phenotype, that is, hearing loss. For these reasons we decided to focus on the studies on the functional characterization of PMCA isoforms on the different splicing variants of PMCA2. FIGURE 3 summarizes all the possible combinations of splicing products occurring at the two main sites: site A and site C.

To investigate and compare the activity of splice isoforms *z/b*, *z/a*, *w/b*, and *w/a*, their expression plasmids were cotransfected together with cytAEQ in CHO cells. Appropriate controls (Western blots and quantitative immunocytochemistry) established that all variants of the pump were expressed at approximately equivalent levels, and were correctly delivered to the plasma membrane (not shown). As illustrated in FIGURE 1 and in previously published work[14,24] PMCA pumps overexpression affected both the amplitude and the kinetic of the cytosolic Ca^{2+} transient generated following cells stimulation with ATP. The kinetic of the declining phase of the cytosolic Ca^{2+} transient

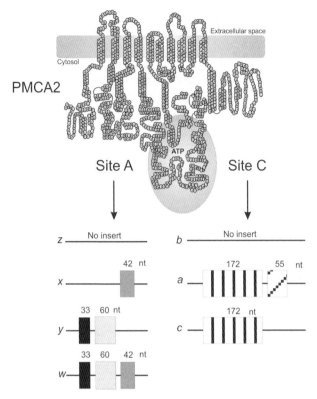

FIGURE 3. Sequence, membrane topography, and scheme summarizing splice variants at site A and site C of isoform 2 of the PMCA pump. The pump is organized in the membrane with 10 transmembrane domains connected on the external side by short loops. The cytosolic portion of the pump contains the catalytic center and other functionally important domains. The catalytic aspartic acid is enlarged. The ATP-binding domain is indicated by a gray shadowed ellipse. The scheme of the splice variants refers to the genomic structure of the pump. The sizes of alternatively spliced exons (represented by boxes) are indicated in nucleotides (nt). Introns are shown as solid lines connecting the alternatively spliced exons. The different splice options are indicated and the resulting splice variants are labeled by their lowercase symbol.

depends on the contribution of several components of the Ca^{2+} signaling machinery: the activity of the PMCA and the sarco/endoplasmic reticulum Ca^{2+} (SERCA) pump, which counteracts the Ca^{2+} influx due to the opening of the plasma membrane Ca^{2+} channels following the emptying of intracellular stores upon agonist application. To better evaluate the contribution of the PMCA pump to dissipate the Ca^{2+} transient we decided to perform cytosolic Ca^{2+} measurements in the presence of the specific SERCA pump inhibitor 2,5-di-tert-butyl-1,4 benzohydroquinone (tBuBHQ). FIGURE 4 A shows that the overexpression of all four splicing variants of the pump had clear effects on the ability of the cells to handle Ca^{2+}. The lowering of Ca^{2+} peak with respect to untransfected cells is likely to reflect the ability of pump variants to respond immediately with a burst of activation to the sudden situation of a Ca^{2+} pulse. The faster clearance of the cytosolic Ca^{2+} signal is easily explained by the increased pump activity. Two of the splice variants (*z/a* and *w/b*)

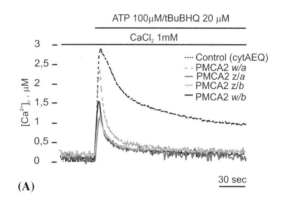

(A)

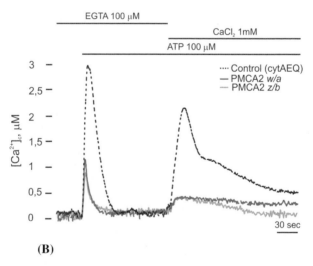

(B)

FIGURE 4. Monitoring of $[Ca^{2+}]_c$ in CHO cells transfected with cytAEQ and cotransfected with cytAEQ and the different splice variants of PMCA2 isoform. (**A**) Effects of PMCA2 splice variants expression on $[Ca^{2+}]_c$. The cells were challenged with ATP in the presence of the specific SERCA pump inhibitor 2,5-di-tert-butyl-1,4 benzohydroquinone (tBuBHQ). (**B**) Effects of PMCA *w/a* and *z/b* expression on ATP-induced internal Ca^{2+} mobilization and on Ca^{2+} influx in CHO cells transfected with cytAEQ. Where indicated the medium was supplemented with 100 μM EGTA, 100 μM ATP, and 1 mM $CaCl_2$. The first and the second peak reflect the contribution of $InsP_3$-induced Ca^{2+} mobilization and of capacitative Ca^{2+} influx from the external medium, respectively.

behaved essentially as the full-length, noninserted z/b pump, which is a very active PMCA isoform (see FIG. 1).[14] By contrast, the doubly inserted variant w/a was less able to control the height of the Ca^{2+} peak, that is, was less able to react rapidly to the incoming of a Ca^{2+} pulse.

The Ca^{2+} transient generated by ATP was due both to the Ca^{2+} release from the intracellular stores through the opening of InsP$_3$ receptors and to the Ca^{2+} influx from the extracellular medium through the store-operated calcium channels (SOCC) that open following the emptying of the intracellular Ca^{2+} stores. To dissect the two components, advantage was taken of the fact that when ATP is applied in the absence of extracellular Ca^{2+}, that is, in a medium containing 100 μM EGTA, the transient generated was exclusively shaped by the release of Ca^{2+} from the intracellular stores. The subsequent addition of 1 mM CaCl$_2$ to the medium induced its entry from the extracellular ambient, thus generating a second Ca^{2+} transient. FIGURE 4 B reports a typical experiment carried out under this experimental protocol. Isoforms w/a and z/b were coexpressed in CHO cells together with cytAEQ. The effects of their overexpression on Ca^{2+} transients were analyzed in respect with control cells. Both the PMCA variants reduced the height of the first Ca^{2+} transient, suggesting that they are able to counteract the opening of InsP$_3$ receptor, however, the w/a variant is slightly less efficient. Concerning the ability of the pumps to counteract the Ca^{2+} influx from the extracellular ambient, the w/a and the z/b appear equivalent in the initial phase, but then w/a overexpressing cells maintain higher Ca^{2+} values than the z/b expressing cells, suggesting the w/a variant is less able (or it is slower) than the z/b variant in reestablishing the basal Ca^{2+} concentrations.

ACKNOWLEDGMENTS

The work was supported by the Telethon Foundation (Project GGP04169 to M.B), the Italian Ministry of University and Research (PRIN 2003 and 2005 to M.B). The authors are indebted to Dr. E.E. Strehler (Rochester, MN) for the donation of PMCA2 and PMCA3 clones.

REFERENCES

1. CARAFOLI, E. 1987. Intracellular calcium homeostasis. Annu. Rev. Biochem. **56:** 395–433.
2. SCHATZMANN, H.J. 1966. ATP-dependent Ca^{++}-extrusion from human red cells. Experientia **22:** 364–365.
3. PEDERSEN, P. & E. CARAFOLI. 1987. Trends. Biochem. Sci. **12:** 146–150.
4. CARAFOLI, E. & M. BRINI. 2000. Calcium pumps: structural basis for and mechanism of calcium transmembrane transport. Curr. Opin. Chem. Biol. **4:** 152–161.
5. STREHLER, E.E. & D.A. ZACHARIAS. 2001. Role of alternative splicing in generating isoform diversity among plasma membrane calcium pumps. Physiol. Rev. **81:** 21–50.

6. DUMONT, R.A. *et al.* 2001. Plasma membrane Ca^{2+}-ATPase isoform 2a is the PMCA of hair bundles. J. Neurosci. **21:** 5066–5078.
7. GRATI, M. *et al.* 2006. Molecular determinants for differential membrane trafficking of PMCA1 and PMCA2 in mammalian hair cells. J. Cell. Sci. **119:** 2995–3007.
8. HILL, J.K. *et al.* 2006. Splice-site A choice targets plasma-membrane Ca^{2+}-ATPase isoform 2 to hair bundles. J. Neurosci. **26:** 6172–6180.
9. STREET, V.A. *et al.* 1998. Mutations in a plasma membrane Ca^{2+}-ATPase gene cause deafness in deaf waddler mice. Nat. Genet. **19:** 390–394.
10. KOZEL, P.J. *et al.* 1998. Balance and hearing deficits in mice with a null mutation in the gene encoding plasma membrane Ca^{2+}-ATPase isoform 2. J. Biol. Chem. **273:** 18693–18696.
11. TAKAHASHI, K. & K. KITAMURA. 1999. A point mutation in a plasma membrane Ca(2+)-ATPase gene causes deafness in Wriggle Mouse Sagami. Biochem. Biophys. Res. Commun. **261:** 773–778.
12. SCHULTZ, J.M. *et al.* 2005. Modification of human hearing loss by plasma-membrane calcium pump PMCA2. N. Engl. J. Med. **352:** 1557–1564.
13. BRINI, M. *et al.* 1995. Transfected aequorin in the measurement of cytosolic Ca^{2+} concentration ($[Ca^{2+}]c$). A critical evaluation. J. Biol. Chem. **270:** 9896–9903.
14. BRINI, M. *et al.* 2003. A comparative functional analysis of plasma membrane Ca^{2+} pump isoforms in intact cells. J. Biol. Chem. **278:** 24500–24508.
15. DEMARCO, S.J., M.C. CHICKA & E.E. STREHLER. 2002. Plasma membrane Ca^{2+} ATPase isoform 2b interacts preferentially with Na^+/H^+ exchanger regulatory factor 2 in apical plasma membranes. J. Biol. Chem. **277:** 10506–10511.
16. DEMARCO, S.J. & E.E. STREHLER. 2001. Plasma membrane Ca^{2+}-ATPase isoforms 2b and 4b interact promiscuously and selectively with members of the membrane-associated guanylate kinase family of PDZ (PSD95/Dlg/ZO-1) domain-containing proteins. J. Biol. Chem. **276:** 21594–21600.
17. KIM, E. *et al.* 1998. Plasma membrane Ca^{2+} ATPase isoform 4b binds to membrane-associated guanylate kinase (MAGUK) proteins via their PDZ (PSD-95/Dlg/ZO-1) domains. J. Biol. Chem. **273:** 1591–1595.
18. SCHUH, K. *et al.* 2003. Interaction of the plasma membrane Ca^{2+} pump 4b/CI with the Ca^{2+}/calmodulin-dependent membrane-associated kinase CASK. J. Biol. Chem. **278:** 9778–9783.
19. SCHUH, K. *et al.* 2001. The plasma membrane calmodulin-dependent calcium pump: a major regulator of nitric oxide synthase I. J. Cell. Biol. **155:** 201–205.
20. ARMESILLA, A.L. *et al.* 2004. Novel functional interaction between the plasma membrane Ca^{2+} pump 4b and the proapoptotic tumor suppressor Ras-associated factor 1 (RASSF1). J. Biol. Chem. **279:** 31318–31328.
21. BUCH, M.H. *et al.* 2005. The sarcolemmal calcium pump inhibits the calcineurin/nuclear factor of activated T-cell pathway via interaction with the calcineurin A catalytic subunit. J. Biol. Chem. **280:** 29479–29487.
22. MACKINTOSH, C. 2004. Dynamic interactions between 14-3-3 proteins and phosphoproteins regulate diverse cellular processes. Biochem. J. **381:** 329–342.
23. RIMESSI, A. *et al.* 2005. Inhibitory interaction of the 14-3-3{epsilon} protein with isoform 4 of the plasma membrane Ca(2+)-ATPase pump. J. Biol. Chem. **280:** 37195–37203.
24. BRINI, M. *et al.* 2000. Effects of PMCA and SERCA pump overexpression on the kinetics of cell Ca(2+) signalling. EMBO J. **19:** 4926–4935.

Plasma Membrane Calcium ATPase and Its Relationship to Nitric Oxide Signaling in the Heart

ELIZABETH J. CARTWRIGHT, DELVAC OCEANDY, AND LUDWIG NEYSES

Division of Cardiovascular and Endocrine Sciences. University of Manchester, M13 9FT Manchester, United Kingdom

ABSTRACT: The plasma membrane calcium/calmodulin-dependent ATPase (PMCA) is a ubiquitously expressed calcium-extruding enzymatic pump. In the majority of cells the main function of PMCA is as the only system to extrude calcium from the cytosol, however, in the excitable cells of the heart it has only a minor role in the bulk removal of calcium compared to the sodium–calcium exchanger. There is increasing evidence to suggest that PMCA has an additional role as a potential modulator of a number of signal transduction pathways. Of key interest in the heart is the functional interaction between the calcium/calmodulin-dependent enzyme neuronal nitric oxide synthase (nNOS) and isoform 4 of PMCA. Nitric oxide production from nNOS is known to be important in the regulation of excitation–contraction (EC) coupling and subsequently contractility. This article will focus on recent evidence suggesting that PMCA4 has a regulatory role in the nitric oxide signaling pathway in the heart.

KEYWORDS: plasma membrane calcium/calmodulin-dependent ATPase; neuronal nitric oxide synthase; signal transduction

INTRODUCTION

Calcium transport within the cardiomyocyte and across the sarcolemma is vital in regulating two essential processes in the heart; intracellular Ca^{2+} is known to be the central regulator of cardiac contractility and it is becoming increasingly clear that calcium is a key second messenger in signal transduction pathways in the heart.

Calcium regulates excitation–contraction (EC) coupling, where it is essential for cardiac electrical activity and myofilament activation.[1] Following depolarization of the cell membrane Ca^{2+} enters the cell mainly through the L-type

Address for correspondence: Ludwig Neyses, Room 1.302 Stopford Building, University of Manchester, Oxford Road, Manchester, UK, M13 9PT. Voice: +44-161-275-1628; fax: +44-161-275-5669. ludwig.neyses@manchester.ac.uk

Ann. N.Y. Acad. Sci. 1099: 247–253 (2007). © 2007 New York Academy of Sciences.
doi: 10.1196/annals.1387.007

calcium channels, leading to the release of Ca^{2+} from the sarcoplasmic reticulum (SR) via the ryanodine receptors and a rise in free intracellular calcium ($[Ca^{2+}]_i$). The contractile machinery is then activated by Ca^{2+} binding to the myofilament protein troponin C. Relaxation of the cell is mediated by a decline in $[Ca^{2+}]_i$, with Ca^{2+} being pumped back into the SR by the SR Ca^{2+} ATPase (SERCA) and extruded from the cell by sodium–calcium exchanger (NCX) and to a minor extent by the plasma membrane calcium/calmodulin-dependent ATPase (PMCA).

Ca^{2+} has a diverse role in cellular signaling by acting as an essential second messenger following liberation from intracellular stores, as well as a first messenger acting on plasma membrane receptors, and even as a third messenger following Ca^{2+}-induced release of Ca^{2+} from intracellular stores.[2,3] This versatility enables calcium to control a number of fundamental cellular processes including transcription,[2,4] fertilization,[5] apoptosis,[6] exocytosis,[7] and synaptic plasticity.[8] In the heart, as well as controlling contraction through EC coupling,[1] Ca^{2+} regulates cell growth and hypertrophy.[9,10] It is now known that calcium/calmodulin-dependent enzymes, such as calcineurin, neuronal and endothelial nitric oxide synthases (nNOS and eNOS), and Ca^{2+}/calmodulin-dependent protein kinases (CaM kinases) are capable of carrying localized intracellular signals thereby regulating hypertrophy and redox equilibrium in the heart.[10,11]

PMCA AS A SIGNALING PROTEIN

Given that in the cardiomyocyte $[Ca^{2+}]_i$ levels fluctuate between 10 nM and 1 μM with each contraction it is as yet unclear how the myocardial cell distinguishes between the contractile and signaling roles of calcium; it is perhaps transporters not involved in the contractile cycle that are able to carry noncontractile signals. PMCA is a ubiquitously expressed calcium-extruding enzymatic pump,[12,13] which we increasingly suspect has an additional role in signal transduction.

In the majority of cell types, including cardiomyocytes, PMCA is found to be localized in caveolae,[14] which are small invaginations of the plasma membrane. These membrane regions are believed to be organizing centers for signaling molecules as they contain a concentration of receptors, signal transducers, and effectors.[15,16] The subcellular localization of PMCA and its known ability to interact with a number of signaling molecules suggests that it may be integrated into important cellular signaling networks.

The "b" splice variants of human PMCA have been described to contain a conserved consensus motif for PDZ domain interaction.[17,18] This modular protein domain was named after the three proteins in which it was first described: the synaptic protein PSD 95, the Drosophila Discs large protein, and the epithelial tight junction protein Zona Occludens-1.[19,20] Our group and others have identified a number of molecular interactions between the COOH-terminal region of PMCA isoforms and PDZ-domain containing proteins. These proteins include members of the membrane-associated guanylate kinase (MAGUK) family,[17,18] cytoskeletal proteins,[21] Na^+/H^+ exchanger regulatory factor 2,[22] nNOS,[23] calcium/calmodulin-dependent serine protein kinase (CASK),[24] and the PMCA-interacting single PDZ protein (PISP).[25] Our group has also demonstrated a number of novel protein–protein interactions involving other domains of PMCA including interactions with Ras-associated factor 1 (Rassf1),[26] calcineurin,[27] and syntrophin,[28] which suggests a role for PMCA as an organizer of macromolecular protein complexes, and a regulator of intracellular signaling pathways.

In the majority of nonexcitable cells PMCA is the only system for calcium extrusion and its primary function is to expel calcium from the cytosol. In the heart, however, PMCA clearly plays a minor role in the bulk removal of intracellular calcium compared to NCX,[29–31] which transports 10–15 times more calcium (μmol/L cytosol/s), depending on species, than PMCA.[32] Although, in the heart the role of PMCA remains to be fully elucidated there have been several reports of its involvement in the regulation of cell growth and differentiation[33–36] and a suggestion that it may be involved in calcineurin-mediated hypertrophy.[27]

ROLE OF PMCA IN SIGNALING: REGULATION OF nNOS

nNOS and its regulation is of major importance in cardiomyocyte biology and our group was the first to demonstrate that isoform 4b of human PMCA forms a PDZ-mediated interaction with nNOS.[23] This interaction has been demonstrated in neuronal cells, smooth muscle cells, as well as in cardiomyocytes.[23,28,37]

The functionality of the PMCA4–nNOS interaction has been shown in a cellular system: using HEK293 cells it has been demonstrated that overexpression of PMCA4 dramatically downregulates nNOS activity, whereas expression of a mutant PMCA4, which is unable to bind nNOS due to the deletion of the PDZ binding domain, fails to reduce nNOS activity.[23] Since nNOS is a calcium/calmodulin-dependent enzyme, it is possible therefore that calcium extrusion by PMCA is the mechanism responsible for nNOS downregulation. Indeed, overexpression of the mutant PMCA4 Asp672Glu, which reduces the calcium transport activity, failed to downregulate nNOS activity.[23]

Our group has recently found that the other isoform of PMCA expressed in the heart (PMCA1) also interacts with nNOS.[28] We demonstrated that a strong interaction occurred between PMCA1 and nNOS when overexpressed in HEK293 cells and that this interaction also occurs in mouse heart tissue;

however, this interaction appears to be weaker than that of PMCA4–nNOS. This difference may be explained by a difference between PMCA1b and PMCA4b in amino acid sequence (1b: ETSL, 4b: ETSV) at the C-terminal region, which is thought to be responsible for the PDZ ligand binding. Furthermore, we demonstrated that PMCA1 has less nNOS inhibitory capacity than PMCA4.

The involvement of nitric oxide in the regulation of cardiovascular function has been very well documented in the past decade.[38] However, it has only recently been determined that nNOS-derived NO has a vital role in cardiac physiology. Two independent groups have shown that nNOS regulates EC coupling both in isolated cardiomyocytes and *in vivo* in the heart.[39,40] Furthermore, nNOS has been described to be involved in the regulation of the β-adrenergic inotropic response,[40] redox equilibrium,[11] and development of heart failure.[41,42] This emerging evidence, together with the finding that both PMCA and nNOS are localized to caveolae,[14,43] raises the question whether PMCA plays a role in the nNOS signaling axis in the heart.

The first experimental evidence to address whether the PMCA–nNOS interaction is functional *in vivo* comes from studies using transgenic mice that overexpress PMCA4 in arterial smooth muscle under the control of the SM22α promtor. Our group and others have found that SM22α-PMCA4b transgenic mice displayed higher blood pressure than control animals.[37,44] This was an unexpected result since the conventional hypothesis would be that overexpression of PMCA4 would reduce intracellular calcium levels and subsequently lower blood pressure due to relaxation of vascular muscle cells. Essentially, PMCA4-overexpressing myocytes exhibited normal calcium transient and calcium sensitivity; but the levels of cGMP, which is a marker of nNOS activity, was significantly reduced.[44] This suggests that the transgenic PMCA4 may regulate nNOS activity in the vascular cells by controlling local calcium levels and is not directly involved in the regulation of cytosolic calcium concentration.

In keeping with vascular data, recent findings from our PMCA4 gene knockout[45] and cardiac-specific transgenic mice support the functional role of the PMCA4–nNOS interaction. PMCA4 null mutant mice displayed increased basal cardiac contractility whereas transgenic mice overexpressing PMCA4 in the heart showed an attenuated β-adrenergic contractile response. Experiments using a transgenic mouse line overexpressing a mutant of PMCA4, which lacks the C-terminal 120 amino acids, and therefore the capacity to bind to nNOS, as well as experiments in which specific NOS isoforms were pharmacologically inhibited suggested that modulation of nNOS function by PMCA4 was likely the mechanism involved in the regulation of cardiac contractility.

It is evident therefore that the plasma membrane calcium pump should no longer be regarded as merely a minor player in the regulation of diastolic calcium but as a key signal transduction molecule acting within the cardiomyocyte.

REFERENCES

1. BERS, D.M. 2002. Cardiac excitation-contraction coupling. Nature **415:** 198–205.
2. CARAFOLI, E. 2002. Calcium signaling: a tale for all seasons. Proc. Natl. Acad. Sci. USA **99:** 1115–1122.
3. BERRIDGE, M.J., M.D. BOOTMAN & H.L. RODERICK. 2003. Calcium signalling: dynamics, homeostasis and remodelling. Nat. Rev. Mol. Cell Biol. **4:** 517–529.
4. CARAFOLI, E. *et al.* 2001. Generation, control, and processing of cellular calcium signals. Crit. Rev. Biochem. Mol. Biol. **36:** 107–260.
5. STRICKER, S.A. 1999. Comparative biology of calcium signaling during fertilization and egg activation in animals. Dev. Biol. **211:** 157–176.
6. ORRENIUS, S., B. ZHIVOTOVSKY & P. NICOTERA. 2003. Regulation of cell death: the calcium-apoptosis link. Nat. Rev. Mol. Cell Biol. **4:** 552–565.
7. RETTIG, J. & E. NEHER. 2002. Emerging roles of presynaptic proteins in Ca++-triggered exocytosis. Science **298:** 781–785.
8. ZUCKER, R.S. 1999. Calcium- and activity-dependent synaptic plasticity. Curr. Opin. Neurobiol. **9:** 305–313.
9. FREY, N. *et al.* 2004. Hypertrophy of the heart: a new therapeutic target? Circulation **109:** 1580–1589.
10. MOLKENTIN, J.D. & I.G. DORN II. 2001. Cytoplasmic signaling pathways that regulate cardiac hypertrophy. Annu. Rev. Physiol. **63:** 391–426.
11. HARE, J.M. 2004. Nitroso-redox balance in the cardiovascular system. N. Engl. J. Med. **351:** 2112–2114.
12. STREHLER, E.E. & D.A. ZACHARIAS. 2001. Role of alternative splicing in generating isoform diversity among plasma membrane calcium pumps. Physiol. Rev. **81:** 21–50.
13. CARAFOLI, E. 1991. Calcium pump of the plasma membrane. Physiol. Rev. **71:** 129–153.
14. FUJIMOTO, T. 1993. Calcium pump of the plasma membrane is localized in caveolae. J. Cell. Biol. **120:** 1147–1157.
15. KURZCHALIA, T.V. & R.G. PARTON. 1999. Membrane microdomains and caveolae. Curr. Opin. Cell. Biol. **11:** 424–431.
16. MAXFIELD, F.R. 2002. Plasma membrane microdomains. Curr. Opin. Cell. Biol. **14:** 483–487.
17. KIM, E. *et al.* 1998. Plasma membrane Ca2+ ATPase isoform 4b binds to membrane-associated guanylate kinase (MAGUK) proteins via their PDZ (PSD-95/Dlg/ZO-1) domains. J. Biol. Chem. **273:** 1591–1595.
18. DEMARCO, S.J. & E.E. STREHLER. 2001. Plasma membrane Ca2+-ATPase isoforms 2b and 4b interact promiscuously and selectively with members of the membrane-associated guanylate kinase family of PDZ (PSD95/Dlg/ZO-1) domain-containing proteins. J. Biol. Chem. **276:** 21594–21600.
19. HUNG, A.Y. & M. SHENG. 2002. PDZ domains: structural modules for protein complex assembly. J. Biol. Chem. **277:** 5699–5702.
20. SHENG, M. & C. SALA. 2001. PDZ domains and the organization of supramolecular complexes. Annu. Rev. Neurosci. **24:** 1–29.
21. ZABE, M. & W.L. DEAN. 2001. Plasma membrane Ca(2+)-ATPase associates with the cytoskeleton in activated platelets through a PDZ-binding domain. J. Biol. Chem. **276:** 14704–14709.

22. DeMarco, S.J., M.C. Chicka & E.E. Strehler. 2002. Plasma membrane Ca2+ ATPase isoform 2b interacts preferentially with Na+/H+ exchanger regulatory factor 2 in apical plasma membranes. J. Biol. Chem. **277:** 10506–10511.
23. Schuh, K. *et al.* 2001. The plasma membrane calmodulin-dependent calcium pump: a major regulator of nitric oxide synthase I. J. Cell. Biol. **155:** 201–205.
24. Schuh, K. *et al.* 2003. Interaction of the plasma membrane Ca2+ pump 4b/CI with the Ca2+/calmodulin-dependent membrane-associated kinase CASK. J. Biol. Chem. **278:** 9778–9783.
25. Goellner, G.M., S.J. DeMarco & E.E. Strehler. 2003. Characterization of PISP, a novel single-PDZ protein that binds to all plasma membrane Ca2+-ATPase b-splice variants. Ann. N Y Acad. Sci. **986:** 461–471.
26. Armesilla, A.L. *et al.* 2004. Novel functional interaction between the plasma membrane Ca2+ pump 4b and the proapoptotic tumor suppressor Ras-associated factor 1 (RASSF1). J. Biol. Chem. **279:** 31318–31328.
27. Buch, M.H. *et al.* 2005. The sarcolemmal calcium pump inhibits the calcineurin/nuclear factor of activated T-cell pathway via interaction with the calcineurin A catalytic subunit. J. Biol. Chem. **280:** 29479–29487.
28. Williams, J.C. *et al.* 2006. The sarcolemmal calcium pump, alpha-1 syntrophin and neuronal nitric oxide synthase are part of a macromolecular protein complex. J. Biol. Chem. **281:** 23341–23348.
29. Choi, H.S. & D.A. Eisner. 1999. The role of sarcolemmal Ca2+-ATPase in the regulation of resting calcium concentration in rat ventricular myocytes. J. Physiol. **515:** 109–118.
30. Bers, D.M., J.W. Bassani & R.A. Bassani. 1993. Competition and redistribution among calcium transport systems in rabbit cardiac myocytes. Cardiovasc. Res. **27:** 1772–1777.
31. Philipson, K.D. & D.A. Nicoll. 2000. Sodium-calcium exchange: a molecular perspective. Annu. Rev. Physiol. **62:** 111–133.
32. Bers, D.M. 2000. Calcium fluxes involved in control of cardiac myocyte contraction. Circ. Res. **87:** 275–281.
33. Hammes, A. *et al.* 1996. Expression of the plasma membrane Ca2+-ATPase in myogenic cells. J. Biol. Chem. **271:** 30816–30822.
34. Husain, M. *et al.* 1997. Regulation of vascular smooth muscle cell proliferation by plasma membrane Ca(2+)-ATPase. Am. J. Physiol. **272:** C1947–C1959.
35. Piuhola, J. *et al.* 2001. Overexpression of sarcolemmal calcium pump attenuates induction of cardiac gene expression in response to ET-1. Am. J. Physiol. Regul. Integr. Comp. Physiol. **281:** R699–R705.
36. Hammes, A. *et al.* 1998. Overexpression of the sarcolemmal calcium pump in the myocardium of transgenic rats. Circ. Res. **83:** 877–888.
37. Schuh, K. *et al.* 2003. Regulation of vascular tone in animals overexpressing the sarcolemmal calcium pump. J. Biol. Chem. **278:** 41246–41252.
38. Stamler, J.S., S. Lamas & F.C. Fang. 2001. Nitrosylation. The prototypic redox-based signaling mechanism. Cell **106:** 675–683.
39. Sears, C.E. *et al.* 2003. Cardiac neuronal nitric oxide synthase isoform regulates myocardial contraction and calcium handling. Circ. Res. **92:** e52–e59.
40. Barouch, L.A. *et al.* 2002. Nitric oxide regulates the heart by spatial confinement of nitric oxide synthase isoforms. Nature **416:** 337–339.
41. Damy, T. *et al.* 2004. Increased neuronal nitric oxide synthase-derived NO production in the failing human heart. Lancet **363:** 1365–1367.

42. BENDALL, J.K. *et al.* 2004. Role of myocardial neuronal nitric oxide synthase-derived nitric oxide in beta-adrenergic hyporesponsiveness after myocardial infarction-induced heart failure in rat. Circulation **110:** 2368–2375.

43. VENEMA, V.J. *et al.* 1997. Interaction of neuronal nitric-oxide synthase with caveolin-3 in skeletal muscle. Identification of a novel caveolin scaffolding/inhibitory domain. J. Biol. Chem. **272:** 28187–28190.

44. GROS, R. *et al.* 2003. Plasma membrane calcium ATPase overexpression in arterial smooth muscle increases vasomotor responsiveness and blood pressure. Circ. Res. **93:** 614–621.

45. SCHUH, K. *et al.* 2004. Plasma membrane Ca2+ ATPase 4 is required for sperm motility and male fertility. J. Biol. Chem. **279:** 28220–28226.

Functional Importance of PMCA Isoforms in Growth and Development of PC12 Cells

LUDMILA ZYLINSKA,[a] ANNA KOZACZUK,[a] JANUSZ SZEMRAJ,[b] CHRISTOS KARGAS,[a] AND IWONA KOWALSKA[a]

[a]Department of Molecular Neurochemistry, Medical University, 92-215 Lodz, Poland

[b]Department of Medical Biochemistry, Medical University, 92-215 Lodz, Poland

ABSTRACT: Intracellular Ca^{2+} in neuronal cells is an essential regulatory ion responsible for excitability, synaptic plasticity, and neurite outgrowth. Plasma membrane calcium ATPase (PMCA) is the most sensitive enzyme in decreasing of the Ca^{2+} concentration. The diverse PMCA isoforms composition in the membranes suggests their specific function in the cell, and whereas PMCA1 and 4 appear to be ubiquitous, PMCA2 and 3 are characteristic isoforms for excitable cells. The aim of our study was to elucidate if and how the elimination of neuron-specific isoforms affects the pattern of cell growth and development. We have obtained stable-transfected PC12 cell lines with a suppressed expression of PMCA2, PMCA3, or both neuron-specific isoforms. The modified profile of PMCA generated considerable changes in morphology of examined PC12 lines, suggesting the activation of a differentiation process to pseudoneuronal phenotype. Experiments with Fura-2/AM-loaded cells revealed an increased cytosolic Ca^{2+} concentration in the cell lines with blocked PMCA2 isoform. The suppression of PMCA2 concomitantly altered expression of sarco/endoplasmic Ca^{2+}-ATPase 2 isoform (SERCA2) at the protein level. Comparative flow cytometry analysis, using Annexin V/PI conjugate, showed the difference in the mean percentage of apoptotic cells in modified PC12 lines. Our data suggest that specific PMCA isoforms presence can regulate the intact cell development; however, it may involve multiple unidentified yet signaling pathways.

KEYWORDS: Ca^{2+}-ATPase; isoforms; PC 12 cells; neuritogenesis; antisense oligonucleotides

Address for correspondence: Ludmila Zylinska, Department of Molecular Neurochemistry, Medical University, 6/8 Mazowiecka Street, 92-215 Lodz, Poland. Voice: +48-42- 678-06-20; fax: +48-42-678-24-65.

luska@csk.umed.lodz.pl

Ann. N.Y. Acad. Sci. 1099: 254–269 (2007). © 2007 New York Academy of Sciences.
doi: 10.1196/annals.1387.008

INTRODUCTION

The plasma membrane Ca^{2+}-ATPase (PMCA) constitutes a high affinity system extruding Ca^{2+} outside cell, and in the resting state maintains intracellular free calcium in the nanomolar range. The enzyme is coded by four separate genes (PMCA1-4), and alternative splicing can create more than 20 variants.[1] Their expression is cell and tissue specific and developmentally regulated.[2,3] Therefore, it could be modified directly in response to a calcium-mediated second messenger pathway.[4] It is thought that different PMCA isoforms and variant composition in excitable and nonexcitable cells influence the final cell response on physiological stimuli.[5,6] Distinct properties of the isoforms, their characteristic distribution in the cells and tissues, diverse density in cell membrane, and the dynamic regulation of PMCA expression suggest their unique function in the maintenance of calcium homeostasis throughout the cell life.[7,8] PMCA1 and 4 isoforms are present at almost all tissues, whereas PMCA2 and 3 are found in more specialized cell types, such as cerebellar Purkinje cells, inner ear cells, liver, uterus, or kidney.[1] Differences in a structure and localization of PMCA isoforms are assumed to correlate with specific regulatory properties, and may have consequences for a proper Ca^{2+} signaling. To establish the specific role of individual isoform, its overexpression or downregulation in the cell culture is examined, mutated enzymes with truncated domains important for proper functioning are incorporated to the cells, and recently, transgenic animals are created.[9] Using different models, several studies have shown that PMCA2 absence caused deafness and balance disorders in mice,[10] the blocking of PMCA1 expression in vascular endothelial cells increased the resting calcium concentration,[11] and prolonged treatment with antisense oligonucleotides caused apoptosis in vascular smooth muscle cells.[12] PMCA1 knockout mice embryos died, and PMCA4 knockout mice males were sterile because of sperm immobility.[13] In the PC12 line with overexpressed PMCA4 isoform the cells were less vulnerable for calcium-induced cell death[14]; however, suppression of endogenous PMCA4 isoform by an antisense method afforded better protection against Ca^{2+}-mediated cell death in the presence of nerve growth factor (NGF).[15] Inhibition of PMCA1 with antisense RNA in PC6 cells (a subclone of PC12 cells), decreased by about 37% of total PMCA protein and impaired the ability of the cells to extend neurites in response to NGF.[16] Up to now, little is known about the functional importance of neuron-specific isoforms PMCA2 and 3 during growth and development of PC12 cells.

Rat pheochromocytoma PC12 cell line is one of the most frequently used models of a neuronal cell, because upon the neurotrophin exposure it differentiates into sympathetic-like neurons, becomes electrically excitable, expresses neuronal markers, and extends neurites.[17] Undifferentiated PC12 cells possess all four main PMCA isoforms with several splice variants (1b, 2b, 3a, 3b, 3c, 4b),[18] and PMCA4b has been shown to constitute a major calcium pump isoform in PC12.[14] Differentiation process induces the specific expression of some other variants (1c, 2a, 2c, 4a), but the physiological significance of these forms remains still unclear.[18]

To evaluate the potential role of neuron-specific PMCA isoforms in differentiated PC12 cells, in our previous study we transiently blocked the PMCA 2 and 3 isoforms, using the phosphothioate antisense oligonucleotides.[19] The first observation was that the transfection brought about the morphological changes and diminished survival of the cells. Transfected cells exhibited altered kinetic behavior of the calcium pump, when compared to the control line. The enzyme affinity to Ca^{2+} was 3.5 times lower than that of control cells, and in transfected PC12 membranes the enzyme was insensitive for stimulation by naturally existing regulator—calmodulin. Moreover, from comparison with the synaptosomal and erythrocyte membranes (possessing all four PMCA isoforms and PMCA 1 and 4, respectively), we concluded that the enzyme activity could be related not only to the PMCA isoforms presence, but also to the membrane types. These results gave us the background to analyze some aspects of Ca^{2+} homeostasis regulation by PMCA isoforms composition in the cells with potency to transform to pseudoneuronal ones. To address this question, we obtained stable-transfected undifferentiated PC12 cell lines with a suppressed expression of PMCA2, PMCA3, and both neuron-specific isoforms.

EXPERIMENTAL PROCEDURES

Cell Culture

PC12 rat pheochromocytoma cell line (ATCC, Teddington, UK) was routinely grown on a collagen (type I from rat tail)-coated dishes in 85% RPMI 1640 medium supplemented with 25 mM HEPES buffer, 10% heat-inactivated horse serum, 5% heat-inactivated calf serum, 2 mM L-glutamine, 1mM sodium pyruvate, 2 g/L D-(+) glucose, 25 U/mL penicillin, and 25 μg/mL streptomycin, at 37°C in 5% CO_2 in a humidified incubator. The medium was exchanged every 48 h.

RNA Purification, cDNA Synthesis, and Polymerase Chain Reaction (PCR) Amplification

Total cellular RNAs were extracted from PC12 cells using the Trizol reagent method as described previously.[19] The concentration and purity of eluted mRNA were determined by spectrophotometric readings at 260 and 280 nm. One microgram of total RNA was then used for first-strand cDNA synthesis with SuperScript II RNase H Transcriptase (Invitrogen Life Technologies, Paisley, UK), using oligo (dT) 12–18 primers. cDNA was amplified with primers specific for PMCA isoforms 1–4.[20] The oligonucleotide primers were as follows:

PMCA 1-3: 5′-CGG CTC TGA ATC TTC TAT CC-3′
PMCA 1-5: 5′-TAG GCA CCT TTG TGG TGC AG-3′
PMCA 2-3: 5′-GCT CGA GTT CTG CTT GAG CGC-3′
PMCA 2-5: 5′-AAG ATC CAC GGC GAG CGT AAC-3′
PMCA 3-3: 5′-CGT TGT TGT TCT GGT TAG GG-3′
PMCA 3-5: 5′-GTC CAA TTT GGA GGG AAG CC-3′
PMCA 4-3: 5′-CAG CAT CCG ACA GGC GCT TG-3′
PMCA 4-5: 5′-ATG CCG AGA TGG AGC TTC GC-3

and β–actin forward 5′-ATCCGTAAAGACCTCTATGC-3′, reverse 5′-AACGCAGCTCAGTAACAGTC-3′ as a control of mRNA quantity PCR amplification. The final amplification products were separated by electrophoresis in 6% polyacrylamide gels in TAE buffer and bands were visualized by UV light. The results were recorded photographically and analyzed densitometrically using an LKB Ultrascan XL Enhanced Laser Densitometer (Pharmacia Biotech, Piscataway, NJ).

Plasmids Construction

The cDNA encoding regions of the 5′end of rat plasma membrane Ca^{2+}-ATPase isoform 2 and 3 were obtained by using reverse transcriptase-coupled PCR (RT-PCR). The oligonucleotide primers were as follows: 5′GTGGATCCGGTGAGCTTGCCCTGAAGC3′ and 5′GTGGAATTCG AGCAATGAAGGATGTG3′ (for PMCA2), and 5′GTGGATCCACCCG GCCATCCACAACG3′ and 5′GTGGAATTCGACTTGGAGAAACGCAG3′ (for PMCA3).[21] PCR conditions were as follows: 94°C, 30 s; 54°C, 30 s; 72°C, 30 s for 35 cycles. The PCR products (669 bp for PMCA2 and 949 bp for PMCA3) were ligated using the PCR T7 TOPO TA cloning system (Invitrogen Life Technologies). For expression, the PMCA2 and PMCA3 inserts were subcloned using *Bam*HI/ EcoRI site in opposite orientation into pcDNA3.1(+) vector (Invitrogen Life Technologies) following the human cytomegalovirus immediate-early (CMV) promotor, which provides high-level of the plasmid. The resulting constructs, anti-PMCA2 and anti-PMCA3, were next sequenced to verify the integrity and orientation in the vector. The plasmid pcDNA3.1(+) (without the cDNA insert) was used as a control. One microgram of each anti-PMCA constructs or vector alone were transfected into PC12 cells by electroporation method. The neomycin-resistant cells were selected with 800 μg/mL G418 (Invitrogen Life Technologies).

Trypan Blue Staining

Cell viability was examined using the trypan blue test. In brief, PC12 cells were removed from the culture dishes and centrifuged (1000 g for 5 min). Cell suspension was diluted by phosphate-buffered saline (PBS) (2×10^5–4×10^5 cells/mL) and then, mixed in the 1:1 ratio with trypan blue solution (Sigma, Schnelldorf, Germany). Viable cells were counted under the microscope using a hemacytometer.

Immunodetection of Ca^{2+}-ATPases

Membranes from PC12 cells were obtained as earlier.[19] For Western blotting, 20–40 μg of the membrane proteins were separated on 7.5% SDS-PAGE, transferred to immobilon polyvinylidene fluoride (PVDF) membrane, and next incubated for 12 h at 4°C with the appropriate antibodies: general monoclonal 5F10 (Sigma) that detects all four isoforms of PMCA, polyclonal anti-PMCA1, monoclonal anti-PMCA4 (AffinityBioreagents, Golden, CO), polyclonal anti-PMCA2, polyclonal anti-PMCA3 (Upstate Biotechnology, Lake Placid, NY), monoclonal anti-SERCA2, and polyclonal anti-SERCA3 (Abcam, Cambridge, UK). For staining, the secondary antibodies were coupled with alkaline phosphatase or peroxidase. The reaction was developed with NBT and BCIP (Sigma) or using ECL system (Pierce, Cramlington, UK), according to the manufacturer's procedures. Quantification of the proteins content was performed after densitometric scanning of immunoblots or autoradiograms using GelDoc^{TM} EQ system with Quantity One software (BioRad, Hercules, CA).

Measurement of Intracellular Ca^{2+} Concentration

Intracellular-free Ca^{2+} concentration was measured in Fura-2/AM-loaded PC12 cells using dual-wavelength spectrofluorometry.[22] In brief, the cells were incubated with 0.5 μM Fura-2/AM buffer for 45 min at 37°C, washed twice, and resuspended in buffer with $CaCl_2$ (1.2 mM). Fluorescence measurements were made on 600 μL aliquots of cells maintained at 37°C and constantly stirred. After 100 s 1 μM thapsigargin (TG) was added. Changes in intracellular-free Ca^{2+} concentration were monitored as variations in the fluorescence ratio of the 340/380-nm excitation wavelength in a Perkin–Elmer LS-50 spectrofluorimeter. Calcium concentrations were calculated using The Intracellular Biochemistry Application (Perkin–Elmer, Beaconsfield, UK).

Flow Cytometry Analysis

For measuring apoptosis and necrosis in PC12 lines, the cells were prepared with Annexin–V–FLUOS Staining Kit (Roche Diagnostics GmbH, Mannheim,

Germany), according to the manufacturer's instruction. Both signals from Annexin V and PI were simultaneously recorded from 10^4 cells in each sample
using Becton Dickinson FACSCalibur (Warsaw, Poland). The data were analyzed with Becton Dickinson CELLQuest software.

Statistics

Data presented in all the figures are from 3 to 6 independent cell cultures.
Statistical analyses were done using Statgraphics Plus 5.1 (Statistical Graph-
ics Corp., Princeton, NJ) computer program. Student's *t*-test was used where
applicable, and *P*-values less than 0.05 were considered to indicate statistically
significant differences.

RESULTS AND DISCUSSION

To examine the role of PMCA2 and 3 in the growth and development of
PC12 cells we have suppressed one or both neuron-specific isoforms using
anti-PMCA2 and/or anti-PMCA3 constructs. Vectors were introduced to the
cells by electroporation method, and then, neomycin-resistant cell clones were
selected in the presence of G418. Stable-transfected cell lines under the control
of a CMV promotor were obtained, and PCR analyses confirmed decreased
amount of mRNA of PMCA2, PMCA3, or both isoforms in examined cell
lines (FIG. 1). The main products were 430 bp for *PMCA*1, 560 bp for *PMCA*2,

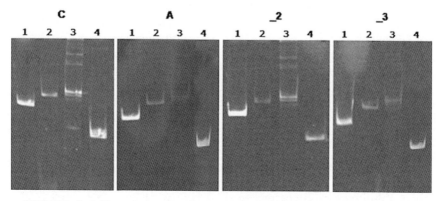

FIGURE 1. PCR analysis of PMCA1-4 isoforms. PC12 cells were transfected with
eukaryotic vectors containing antisense sequences designed to PMCA2 and PMCA3 or
both isoforms. The main products were 430 bp for *PMCA*1, 560 bp for *PMCA*2, 579 bp
and 646 bp for *PMCA*3, and 260 bp for *PMCA*4. C-control PC12 line, A-cell line with
suppressed PMCA2 and 3 isoforms, _2-cell line with suppressed PMCA2 isoform, _3-cell
line with suppressed PMCA3 isoform.

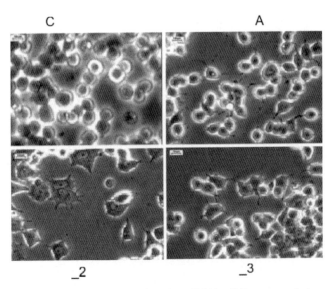

FIGURE 2. PC12 cell lines morphology. Four PC12 cell lines were photographed after 72 h using an inverted microscopy (Olympus, Warsaw, Poland). The scale bars, 10 μm. C-control PC12 line, A-cell line with suppressed PMCA2 and 3 isoforms, _2-cell line with suppressed PMCA2 isoform, _3-cell line with suppressed PMCA3 isoform.

579 bp and 646 bp for *PMCA*3, and 260 bp for *PMCA*4, which respond PMCA 1b, 2b, 3a, 3c, and 4b variants, respectively.[18]

Morphological changes in four PC12 cell lines were monitored and photographed after 72 h using inverted microscopy (FIG. 2). Suppression of neuron-specific isoforms altered the cells structure, inducing a process similar to the beginning of neuritogenesis. We observed a significant transformation, including neurite outgrowth and creation of synaptic connections between cells in the transfected PC12 lines. Reconstructed 3D pictures of PC12 cell line with suppressed both isoforms, obtained by using ChemiGenius software (Syngene, Cambridge, UK), showed the appearance of growth cone-like areas after 48 h (FIG. 3). Viability of all PC12 cell lines was assessed by the trypan blue dye exclusion test. In contrast to our previous study on transiently blocked differentiated PC12 cells, we did not observe the decreased ability of modified PC12 cell lines to survive, when compared with control cells; however, the lines transfected with anti-PMCA2 and with both constructs exhibited the lowered ability for proliferation.

The amount of PMCA isoforms at the protein level in modified and control lines was determined using general (recognizing all isoforms) and isoform-specific antibodies (FIG. 4 A–E). The total amount of Ca^{2+}-ATPase in the plasma membranes was significantly lowered (up to 50%) in all transfected PC12 lines. The PMCA1 protein level was unchanged in the membranes of

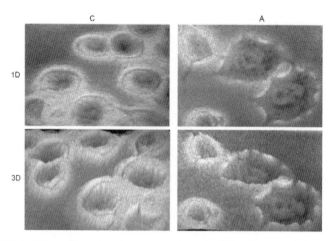

FIGURE 3. Three-dimensional reconstruction of PC12 cells. Three-dimensional picture of the control cells (C), and cells transfected with anti-PMCA2 and anti-PMCA3 vectors (A) was taken after 48 h, using ChemiGenius image system. Magnification × 400.

all examined lines, which confirms the housekeeping role of this isoform. The PMCA2 protein was lowered in the lines with downregulated PMCA2 isoform; however, the blocking of PMCA3 resulted in diminished immunoreactivity not only in PMCA3-suppressed lines but, unexpectedly, the lowered PMCA3 presence was detected in the cells transfected exclusively with anti-PMCA2 vector. Since in the PMCA2-suppressed line we did not observe differences in PMCA3 mRNA level, one can speculate that either the half-life of mRNA shortened or the proteolytic degradation of PMCA3 protein occurred. Even though we could not notice higher amounts of the proteolytic products by Western blotting, because of possible degradation of the epitope recognized by the antibody used, this possibility cannot be ruled out. Nonetheless, both hypotheses need further study. Surprisingly, the PC12 line with blocked expression of PMCA2 and 3 isoforms exhibited the increase of PMCA4 level by 25–30%. It could suggest that suppression of both neuron-specific PMCA isoforms induces the compensatory increase in the main calcium pump in PC12 cells, as a putative adaptive mechanism.

To evaluate the functional effect of PMCA suppression on calcium homeostasis, we determined cytosolic Ca^{2+} in the transfected PC12 lines. The resting intracellular Ca^{2+} concentration and depletion of endoplasmic reticulum (ER) stores by an application of TG were examined using Fura-2/AM dye (FIG. 5). The amount of Ca^{2+} determined upon TG addition reflected the Ca^{2+} concentration in ER. Both, the resting and TG-stimulated levels of intracellular Ca^{2+} were increased in the lines transfected only with anti-PMCA2 and with both vectors. A little, but statistically significant, increase in cytosolic Ca^{2+} by 30–50 nM correlated with a higher level of TG-sensitive pool of Ca^{2+}.

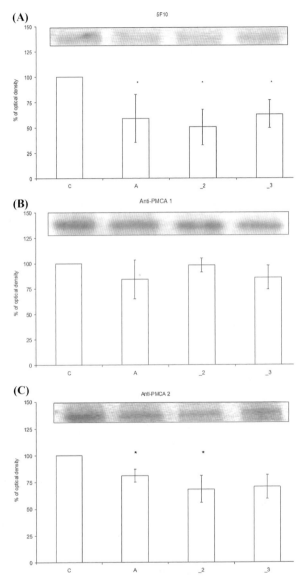

FIGURE 4. (A–E) Immunocharacteristics of plasma membrane calcium pump in PC12 lines. Membrane fractions (20–40 μg) from each PC12 lines were separated on 7.5% SDS-PAGE, electroblotted on Millipore PVDF membrane (Vienna, Austria), and incubated with antibodies, which recognized all PMCA isoforms (5F10) or particular isoforms, as described in the "Experimental Procedures" section. For quantitative analysis, density of each protein band was measured by densitometry scanning using GelDoc[TM] EQ system with Quantity One software (BioRad). The amount of PMCA isoforms in the control membranes was taken as 100%. The results are from 4–6 independent cultures. C-control PC12 line, A-cell line with suppressed PMCA2 and 3 isoforms, _2-cell line with suppressed PMCA2 isoform, _3-cell line with suppressed PMCA3 isoform.

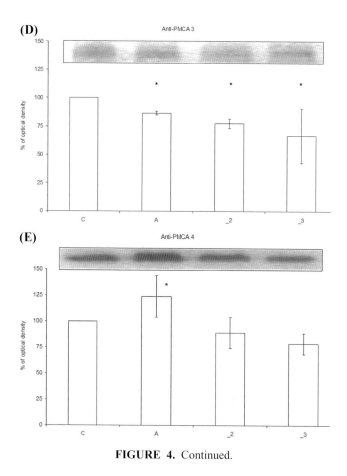

FIGURE 4. Continued.

Based on these results, the functional relationship between plasma membrane and sarco/endoplasmic ATPases in control and transfected PC12 cell lines was explored. The family of sarco/endoplasmic ATPases (SERCA) comprises products of three genes, and due to alternative splicing a number of enzyme variants can exist.[23–25] SERCA1 variants are expressed exclusively in skeletal muscle, while SERCA2b is expressed ubiquitously. The SERCA3 isoform is expressed at a significant amount only in a selected number of tissues, and has a much lower apparent affinity for Ca^{2+} than the other isoforms.[26] Increasing cytosolic Ca^{2+} was shown to regulate the SERCA3 expression, suggesting that Ca^{2+} might be a factor that contributes to the selective expression of SERCA3 in environments where it is suited to function.[27,28] SERCA2b, SERCA3a, and 3b/c have been identified in PC12 cells, and these two main isoforms appeared to be co-localized.[29] Using the antibodies that recognized SERCA2 and 3 isoforms we found higher SERCA2 immunoreactivity in PC12 lines with lowered PMCA2 expression, whereas SERCA3 protein was at a similar level in all examined cell lines (FIG. 6). Thus, downregulation of PMCA2 and

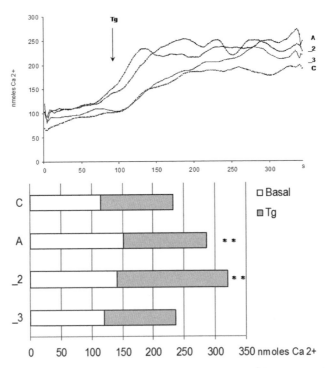

FIGURE 5. Fluorometric measurement of intracellular Ca^{2+} concentration. *Upper panel.* Control or transfected PC12 cells were loaded with Fura-2/AM as described in the "Experimental Procedures" section. Cytosolic Ca^{2+} concentration was measured at the resting state, and 1 μM TG was added at 100 s, to deplete intracellular Ca^{2+} stores. *Lower panel.* Quantitative analysis of basal and TG-induced cytosolic Ca^{2+} concentration. The Ca^{2+} changes were measured from the base line to when the elevation of Ca^{2+} reached its peak. The results are means from three independent cultures. C-control PC12 line, A-cell line with suppressed PMCA2 and 3 isoforms, _2-cell line with suppressed PMCA2 isoform, _3-cell line with suppressed PMCA3 isoform.

higher amount of SERCA2 resulted in the increase of TG-sensitive pool of Ca^{2+} in ER. The above suggests that in the cells with diminished level of PMCA2 (the isoform that exhibits the highest affinity for calcium[30,31]), Ca^{2+} may upregulate the SERCA2 expression. Recently, a link between ER calcium concentration and neurite outgrowth was established by showing that overexpression of SERCA2b in PC12 cells reduced neurite outgrowth in the Bcl-2-overexpressing cells and that inhibition of SERCA increased neurite outgrowth.[22] This was confirmed by measuring calcium concentration in the Bcl-2-overexpressing PC12 cells, which showed increased cytosolic calcium concentration in response to various stimuli. Increased Ca^{2+} activated CREB and ERK, and supported both, survival and axon/neurite growth. In our models, the sustained increase in cytosolic Ca^{2+} was reached due to suppression of

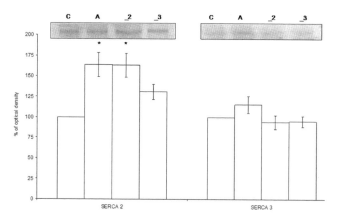

FIGURE 6. Immunocharacteristics of sarco/endoplasmic calcium pump in PC12 lines. Membrane fractions (20–40 μg) from each PC12 lines were separated on 7.5% SDS-PAGE, electroblotted on Millipore PVDF membrane, and incubated with antibodies, which recognized SERCA2 or SERCA3 isoforms. For quantitative analysis, density of each protein band was measured by densitometry scanning using GelDoc™ EQ system with Quantity One software (BioRad). The amount of SERCA isoforms in the control membranes was taken as 100%. The results are from three independent cultures. C-control PC12 line, A-cell line with suppressed PMCA2 and 3 isoforms, _2-cell line with suppressed PMCA2 isoform, _3-cell line with suppressed PMCA3 isoform.

PMCA2 isoform, and upregulation of SERCA2 appeared insufficient in the decrease of Ca^{2+} to the control level. Moreover, it could suggest that the observed Ca^{2+} increase might promote neuritogenesis by activation of CREB and ERK pathways. However, this assumption should be verified in a more detailed study.

To assess further the characteristics of transfected PC12 lines, the flow cytometry analyses were performed using double Annexin V/PI staining (FIG. 7). The viable cells were negative for both PI and Annexin V. The cells at early stage of apoptosis were positive for Annexin V and negative for PI, whereas at the late stage of apoptosis displayed both high Annexin V and PI labeling. Nonviable, necrotic cells were positive for PI and negative for Annexin V. The first observation was the similar percentage of necrotic cells in all examined lines. In comparison to the control, no difference in the amounts of viable cells was detected in the line transfected only with anti-PMCA3 vector. Neither anti-PMCA2 nor anti-PMCA3 transfection alone did induce apoptosis, but suppression of both PMCA isoforms increased the apoptosis, and nearly 15% of the cells were at late apoptotic stage. It was shown that both, the transcript and protein of PMCA3 are present in the cells at low levels, and the specific role of PMCA3 in less known.[32] It is plausible that PMCA3 could play a supportive role in the maintenance of calcium homeostasis, or its function may

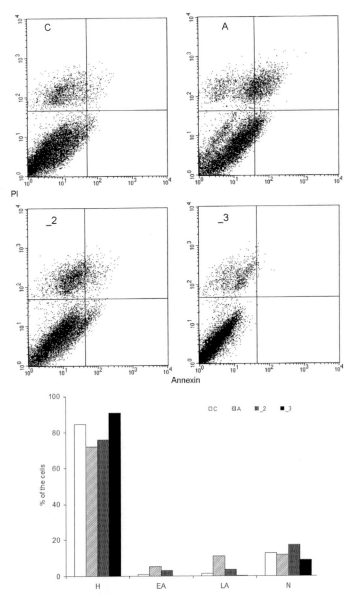

FIGURE 7. Flow cytometry analysis of PC12 cell lines. *Upper panel*. Using the Annexin V/PI double staining viable, early and late apoptotic/dead cells were visualized on the flow. The representative dot plot pattern is shown. *Lower panel*. The histograms show the percentage of viable, apoptotic, and necrotic cell populations in control and transfected PC12 lines quantitated using Cellquest software (Becton Dickinson). Data were collected from 10,000 cells per sample, and the results are means from four independent cultures. C-control PC12 line, A-cell line with suppressed PMCA2 and 3 isoforms, _2-cell line with suppressed PMCA2 isoform, _3-cell line with suppressed PMCA3 isoform. H-viable cells, EA-early apoptosis, LA-late apoptosis, N-necrosis.

depend on the membrane localization. The distribution of PMCA3 remains unknown yet, but membrane PMCA localization may be critical in control Ca^{2+} transient during the cell development and differentiation processes. Despite the fact that PMCA isoforms differ in sensitivity to Ca^{2+} and susceptibility to specific mode of regulation, the presence of both, PMCA2 and PMCA3 in the cells involving fast recovery of the resting Ca^{2+} may suggest their cooperation in the control of calcium signaling inside the cell. Our results demonstrated the induction of neuritogenesis in PC12 cells with suppressed PMCA2; however, the concomitant suppression of PMCA3 isoform resulted also in the increase of apoptotic cells. PMCA was shown to be degraded by caspases in neuronal and non-neuronal cells undergoing apoptosis[33,34]; however, this process involved higher Ca^{2+} concentration than observed in our study. In PC12 cells with suppressed isoforms the weaker Ca^{2+} extrusion and the prolonged calcium signal may influence cellular processes, and while in the differentiated cells it could be a "death signal," in undifferentiated ones it may trigger the gene expression program necessary for neuritogenesis.

CONCLUSIONS

All examined PC12 lines were maintained and analyzed under similar conditions and the differences observed in functional and developmental patterns are likely due to the altered presence of PMCA2 and/or PMCA3 isoforms. Recently, the targeting of PMCA2b and 4b isoforms to subcellular membrane compartments, known as caveolae or signalosomes, was demonstrated and these functionally specific membrane microdomains may be responsible for the local organization of Ca^{2+} signaling.[35,36] The tentative, similar cooperation of PMCA2 and 3 isoforms in the membranes remains an open question. However, to elucidate the molecular mechanisms responsible for the observed changes, characteristics of putative Ca^{2+}-dependent signaling pathways, triggering of new genetic programs, or modification of existing ones, further studies are needed.

ACKNOWLEDGMENTS

We are grateful to Bozena Ferenc, MSc. for her excellent technical assistance. Supported by the grants 2 P05A 03529, PBZ-MIN-012/P04/2004 from the Ministry of Education and Science, Poland, 503-6806-2 and 502-16-197 from the Medical University of Lodz.

REFERENCES

1. STREHLER, E.E. & D.A. ZACHARIAS. 2001. Role of alternative splicing in generating isoform diversity among plasma membrane calcium pumps. Physiol. Rev. **81:** 21–50.

2. STAUFFER, T.P., D. GUERINI & E. CARAFOLI. 1995. Tissue distribution of the four gene products of the plasma membrane Ca pump. J. Biol. Chem. **270:** 12184–12190.

3. BRANDT, P. & R.L. NEVE. 1992. Expression of plasma membrane calcium-pumping ATPase mRNAs in developing rat brain and adult brain subregions: evidence for stage-specific expression. J. Neurochem. **59:** 1566–1569.

4. GUERINI, D., E. GARCIA-MARTIN, *et al.* 1999. The expression of plasma membrane Ca^{2+} pump isoforms in cerebellar granule neurons is modulated by Ca^{2+}. J. Biol. Chem. **274:** 1667–1676.

5. STREHLER, E.E. & M. TREIMAN. 2004. Calcium pumps of plasma membrane and cell interior. Curr. Mol. Med. **4:** 323–335.

6. LEHOTSKY, J., P. KAPLAN, *et al.* 2002. The role of plasma membrane Ca^{2+} pumps (PMCA) in pathologies of mammalian cells. Front. Biosci. **1:** d53–d84.

7. BRINI, M., L. COLETTO, *et al.* 2003. A comparative functional analysis of plasma membrane Ca^{2+} pump isoforms in intact cells. J. Biol. Chem. **278:** 24500–24508.

8. BRINI, M. & E. CARAFOLI. 2000. Calcium signalling: a historical account, recent developments and future perspectives. Cell. Mol. Life Sci. **57:** 354–370.

9. SHULL, G.E. 2000. Gene knockout studies of Ca^{2+}-transporting ATPases. Eur. J. Biochem. **267:** 5284–5290.

10. KOZEL, P.J., R.A. FRIEDMAN, *et al.* 1998. Balance and hearing deficits in mice with a null mutation in the gene encoding plasma membrane Ca^{2+}-ATPase isoform 2. J. Biol. Chem. **273:** 18693–18696.

11. NAKAO, M., K. FURUKAWA, *et al.* 2000. Inhibition by antisense oligonucleotides of plasma membrane Ca^{2+}-ATPase in vascular endothelial cells. Eur. J. Pharmacol. **387:** 273–277.

12. SASAMURA, S., K. FURUKAWA, *et al.* 2002. Antisense-inhibition of plasma membrane Ca^{2+} pump induces apoptosis in vascular smooth muscle cells. Jpn. J. Pharmacol. **90:** 164–172.

13. OKUNADE, G.W., M.L. MILLER, *et al.* 2004. Targeted ablation of plasma membrane Ca^{2+}-ATPase isoforms 1 and 4 indicates a major housekeeping function for PMCA1 and a critical role in hyperactivated sperm motility and male fertility for PMCA4. J. Biol. Chem. **279:** 33742–33750.

14. GARCIA, M.L., Y.M. USACHEV, *et al.* 2001. Plasma membrane calcium ATPase plays a role in reducing Ca^{2+}-mediated cytotoxicity in PC12 cells. J. Neurosci. Res. **64:** 661–669.

15. USACHEV, Y.M., S.J. DEMARCO, *et al.* 2002. Bradykinin and ATP accelerate Ca^{2+} efflux from rat sensory neurons via protein kinase C and the plasma membrane Ca^{2+} pump isoform 4. Neuron **33:** 113–122.

16. BRANDT, P.C., J.E. SISKEN, *et al.* 1996. Blockade of plasma membrane calcium pumping ATPase isoform I impairs nerve growth factor-induced neurite extension in pheochromocytoma cells. Proc. Natl. Acad. Sci. USA **93:** 13843–13848.

17. GREENE, L.A. & D.R. KAPLAN. 1995. Early events in neurotrophin signalling via Trk and p75 receptors. Curr. Opin. Neurobiol. **5:** 579–587.

18. HAMMES, A., S. OBERDORF, *et al.* 1994. Differentiation-specific isoform mRNA expression of the calmodulin-dependent plasma membrane Ca^{2+}-ATPase. FASEB J. **8:** 428–435.

19. SZEMRAJ, J., I. KAWECKA, *et al.* 2004. The effect of antisense oligonucleotide treatment of plasma membrane Ca^{2+}-ATPase in PC12 cells. Cell. Mol. Biol. Lett. **9:** 451–464.

20. REINHARDT, T.A. & R.L. HORST. 1999. Ca^{2+}-ATPases and their expression in the mammary gland of pregnant and lactating rats. Am. J. Physiol. (Cell Physiol. 45) **276:** C796–C802.

21. STAHL, W.L., T.J. EAKIN, *et al.* 1992. Plasma membrane Ca^{2+}-ATPase isoforms: distribution of mRNAs in rat brain by in situ hybridization. Mol. Brain Res. **16:** 223–231.

22. JIAO, J., X. HUANG, *et al.* 2005. Bcl-2 enhances Ca^{2+} signaling to support the intrinsic regenerative capacity of CNS axons. EMBO J. **24:** 1068–1078.

23. BABA-AISSA, F., L. RAEYMAEKERS, *et al.* 1998. Distribution and isoform diversity of the organellar Ca^{2+} pumps in the brain. Mol. Chem. Neuropathol. **33:** 199–208.

24. VANGHELUWE, P., L. RAEYMAEKERS, *et al.* 2005. Modulating sarco(endo)plasmic reticulum Ca^{2+}ATPase 2 (SERCA2) activity: cell biological implications. Cell Calcium **38:** 291–302.

25. CASPERSEN, C., P.S. PEDERSEN & M. TREIMAN. 2000. The sarco/endoplasmic reticulum calcium-ATPase 2b is an endoplasmic reticulum stress-inducible protein. J. Biol. Chem. **275:** 22363–22372.

26. WUYTACK, F., L. DODE, *et al.* 1995. The SERCA3-type of organellar Ca^{2+} pumps. Biosci. Rep. **15:** 299–306.

27. KUO, T.H., B.F. LIU, *et al.* 1997. Co-ordinated regulation of the plasma membrane calcium pump and the sarco(endo)plasmic reticular calcium pump gene expression by Ca^{2+}. Cell Calcium **21:** 399–408.

28. KELLER, D. & A.K. GROVER. 2000. Nerve growth factor treatment alters Ca^{2+} pump levels in PC12 cells. Neuroreport **11:** 65–68.

29. ROONEY, E. & J. MELDOLESI. 1996. The endoplasmic reticulum in PC12 cells. Evidence for a mosaic of domains differently specialized in Ca^{2+} handling. J. Biol. Chem. **271:** 29304–29311.

30. ELWESS, N.L., A.G. FILOTEO, *et al.* 1997. Plasma membrane Ca^{2+} pump isoforms 2a and 2b are unusually responsive to calmodulin and Ca^{2+}. J. Biol. Chem. **272:** 17981–17986.

31. CARIDE, A.J., A.G. FILOTEO, *et al.* 2001. Delayed activation of the plasma membrane calcium pump by a sudden increase in Ca^{2+}: fast pumps reside in fast cells. Cell Calcium **30:** 49–57.

32. FILOTEO, A.G., A. ENYEDI, *et al.* 2000. Plasma membrane Ca^{2+} pump isoform 3f is weakly stimulated by calmodulin. J. Biol. Chem. **275:** 4323–4328.

33. PASZTY, K., A.K. VERMA, *et al.* 2002. Plasma membrane Ca^{2+} ATPase isoform 4b is cleaved and activated by caspase-3 during the early phase of apoptosis. J. Biol. Chem. **277:** 6822–6829.

34. SCHWAB, B.L., D. GUERINI, *et al.* 2002. Cleavage of plasma membrane calcium pumps by caspases: a link between apoptosis and necrosis. Cell Death Differ. **9:** 818–831.

35. DEMARCO, S.J. & E.E. STREHLER. 2001. Plasma membrane Ca^{2+}-ATPase isoforms 2b and 4b interact promiscuously and selectively with members of the membrane-associated guanylate kinase family of PDZ (PSD-95/Dlg/ZO-1) domain-containing proteins. J. Biol. Chem. **276:** 21594–21600.

36. RAZANI, B., S.E. WOODMAN & M.P. LISANTI. 2002. Caveolae: from cell biology to animal physiology. Pharmacol. Rev. **54:** 431–467.

Na$^+$/Ca^{2+} Exchanger Knockout Mice

Plasticity of Cardiac Excitation–Contraction Coupling

CHRISTIAN POTT,[a,b] SCOTT A. HENDERSON,[a]
JOSHUA I. GOLDHABER,[a] AND KENNETH D. PHILIPSON[a]

[a]Departments of Physiology and Medicine and the Cardiovascular Research
Laboratories, David Geffen School of Medicine at UCLA, Los Angeles,
California 90095-1760, USA

[b]Clinic C for Internal Medicine, Department of Cardiology and Angiology,
University Hospital Muenster, 48149 Münster, Germany

ABSTRACT: The Na$^+$/Ca^{2+} exchanger (NCX) is the main Ca^{2+} extrusion
mechanism of the cardiac myocyte. Nevertheless, cardiac-specific NCX
knockout (KO) mice are viable to adulthood. We have identified two
adaptations of excitation–contraction coupling (ECC) to the absence of
NCX in these animals: (a) a reduction of the L-type Ca^{2+} current (I$_{Ca}$)
with an increase in ECC gain and (b) a shortening of the action potential
(AP) to further limit Ca^{2+} influx. Both mechanisms contribute to Ca^{2+}
homeostasis by reducing Ca^{2+} influx while maintaining contractility.
These adaptations may comprise important feedback mechanisms by
which cardiomyocytes may be able to limit Ca^{2+} influx in situations of
compromised Ca^{2+} extrusion capacity.

KEYWORDS: Na$^+$/Ca^{2+} exchanger; L-type Ca^{2+} channel; excitation–
contraction coupling; genetically altered mice; cardiac myocytes

INTRODUCTION

The cardiac contractile cycle is initiated by an influx of Ca^{2+} via voltage-
dependent L-type Ca^{2+} channels (I$_{Ca}$). This Ca^{2+} activates ryanodine recep-
tors to induce release of Ca^{2+} from the sarcoplasmic reticulum (SR).[1] The
Na$^+$/Ca^{2+} exchanger (NCX) has an essential role in this process as it removes
Ca^{2+} that enters the cardiomyocyte and thereby maintains a balance between
Ca^{2+} influx and efflux.[2,3] Not surprisingly, global knockout (KO) of NCX

Address for correspondence: Kenneth D. Philipson, Cardiovascular Research Laboratories, MRL
3645, David Geffen School of Medicine at UCLA, 675 Charles E. Young Dr. South, Los Angeles, CA
90095. Voice: 310-825-7679; fax: 310-206-5777.

kphilipson@mednet.ucla.edu

Ann. N.Y. Acad. Sci. 1099: 270–275 (2007). © 2007 New York Academy of Sciences.
doi: 10.1196/annals.1387.015

is lethal in a murine model.[4-6] Recently, however, we introduced a cardiac-specific murine NCX KO model that was generated using cre/lox technology. Strikingly, these animals are viable to adulthood with only modest cardiac dysfunction. We did not detect changes in the expression levels of other proteins that are involved in cardiac excitation–contraction coupling (ECC), such as the sarcoendoplasmatic Ca^{2+} ATPase (SERCA), the plasma membrane Ca^{2+} ATPase (PMCA), the dihydropyridine receptor (DHPR), or the ryanodine receptor (RyR).[7] Nevertheless, consistent with the KO of NCX, we demonstrated that Ca^{2+} extrusion is drastically reduced (to about 20%) in caffeine release experiments.[8]

We discuss below how functional ECC can be maintained under these extraordinary conditions. We describe a set of molecular mechanisms by which NCX KO mice adapt to the absence of NCX. These adaptations may comprise important regulatory mechanisms that could also enable the cardiomyocyte to reduce Ca^{2+} influx in other situations of impaired Ca^{2+} extrusion.

REDUCTION OF L-TYPE Ca^{2+} CURRENT

In ventricular myocytes isolated from NCX KO mice, the amplitude of I_{Ca} is reduced to about 50% and inactivation kinetics are accelerated despite normal expression levels of the DHPR.[7] This is an important adaptive mechanism as the reduced Ca^{2+} influx via I_{Ca} can compensate for reduced Ca^{2+} extrusion capacity. These alterations in I_{Ca} are eliminated when SR Ca^{2+} stores are depleted with thapsigargin (Tg) and ryanodine (Ry) and Ba^{2+} is used instead of Ca^{2+} as the charge carrier. Ba^{2+} is readily conducted by the DHPR but, unlike Ca^{2+}, is ineffective in inducing I_{Ca} inactivation. Similarly, the changes in I_{Ca} amplitude and decay kinetics are eliminated when cytosolic Ca^{2+} is heavily buffered by dialyzing the myocytes with the fast Ca^{2+} chelator BAPTA.[9]

These data demonstrate that in NCX KO myocytes, Ca^{2+}-dependent mechanisms are responsible for the reduction of I_{Ca} amplitude and for the accelerated inactivation rate of I_{Ca}. It therefore follows that in the absence of NCX, subsarcolemmal Ca^{2+} is elevated, leading to inactivation of some L-type Ca^{2+} channels.

ACCELERATION OF ACTION POTENTIAL REPOLARIZATION

Action potentials (APs) recorded from KO myocytes repolarize more rapidly than those in wild-type (WT) and thus are of shorter duration.[7] When we applied KO- and AP-like waveforms as voltage commands to patch clamped WT myocytes, we observed a reduction of the Ca^{2+} influx via I_{Ca} due to the shorter KO AP waveform (FIG. 1 A, B). When this experiment was repeated in KO myocytes, a further reduction of Ca^{2+} influx was observed (FIG. 1 C, D).[8]

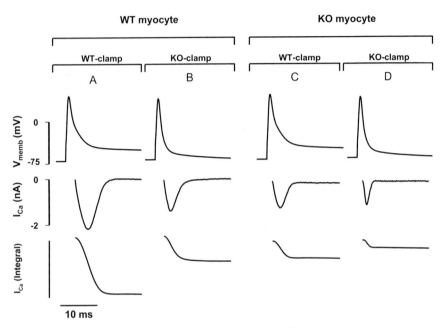

FIGURE 1. The shortened AP drastically reduces Ca^{2+} influx via I_{Ca}. *Top panel*: Voltage clamp commands simulating a WT AP (WT-clamp) or a KO AP (KO-clamp). *Middle panel*: I_{Ca} induced by the respective clamp command measured as the Cd^{2+}-sensitive current. *Lower panel*: integral of I_{Ca} ($\int I_{Ca}$) as a measure of Ca^{2+} entry in WT and KO myocytes. (**A**) I_{Ca} and $\int I_{Ca}$ in a WT myocyte induced by a WT clamp. (**B**) Stimulation of the same WT myocyte with a KO-clamp. (**C** and **D**) The experiment was repeated using a KO myocyte. Integral is given in arbitrary units. Modified from Pott *et al.*, 2005.[8]

Quantification of these data revealed that the combination of decreased I_{Ca} activity and the abbreviated AP reduced Ca^{2+} influx in KO myocytes to about 20% of that in WT myocytes.

What is the mechanism that underlies the abbreviated AP duration in KO? In principle, both the elimination of NCX inward current (I_{NCX}) and the reduction of I_{Ca} could be responsible for the AP abbreviation, as they are both repolarizing currents.[10,11] However, these currents become active during the plateau phase of the AP whereas we observe an accelerated repolarization in the very initial phase of the KO AP. In this early repolarization phase the transient outward current (I_{to}) is especially dominant.[12] In NCX KO myocytes, there is indeed an increase of the expression level of the I_{to} generating K^+ channel subunit Kv4.2 and the K^+ channel-interacting protein (KChip), which is accompanied by an increase in I_{to} to about 180%. Using a computer model of the murine AP[13] we found that this upregulation of I_{to} is the major cause of the shortened AP. The reduced I_{Ca} and elimination of I_{NCX} had little effect on the AP duration. These observations are consistent with a number of studies that have recently demonstrated a close interaction between cardiac Ca^{2+} handling and the expression and the activity level of repolarizing currents in cardiac tissue.[14-16] Future experimental work is required to identify the exact causal chain that leads from KO of NCX to the upregulation of I_{to}.

MAINTENANCE OF CONTRACTILITY IN THE ABSENCE OF NCX

The combination of reduced I_{Ca} and the abbreviation of the AP severely limit the amount of trigger Ca^{2+} entering the cell under conditions of an unchanged SR load.[8] Nevertheless, KO myocytes have similar Ca^{2+} transients as their WT littermates when paced via external field stimulation.[7] By simultaneously measuring I_{Ca} and the Ca^{2+} transient using a standard voltage clamp protocol, we demonstrated that the reduced I_{Ca} observed in KO was able to trigger a similar amount of Ca^{2+} release from the SR as in WT. This indicates an increase in the gain of ECC in KO cells.[8] The mechanism that underlies the increased gain is unclear. A reduced RyR threshold due to elevated subsarcolemmal Ca^{2+} or spatial rearrangements of the diadic cleft may be responsible, though this remains for future studies to resolve.

The abbreviation of the KO AP further limits the amount of trigger Ca^{2+} but also does not change the magnitude of the Ca^{2+} transient.[8] This is consistent with other studies demonstrating that a faster repolarization of the AP can improve the effectiveness of cardiac ECC.[17,18]

SUMMARY

Our findings indicate that NCX KO myocytes limit Ca^{2+} influx to about 20% of that in WT by reducing I_{Ca} and by abbreviating the AP. This provides an explanation for how Ca^{2+} homeostasis can be maintained in the absence of NCX as transsarcolemmal Ca^{2+} fluxes are now reduced to a degree where the PMCA[19,20] can handle all Ca^{2+} extrusion from the cell. Contractility is maintained by an increase in the gain of ECC. The increased gain is due to an as yet unidentified mechanism, though a more rapid repolarization of the AP certainly is a major contributor. These adaptations not only reveal an amazing plasticity of cardiac ECC but may also represent regulatory mechanisms that enable the cardiomyocyte to balance a limited Ca^{2+} extrusion capacity with a reduction of Ca^{2+} influx under pathological conditions.

ACKNOWLEDGMENTS

The authors appreciate helpful discussions with Dr. R. Olcese and Dr. L. Xie. This research was supported by Köln Fortune and the German Research

Foundation (DFG PO 1004/1-1 and PO 1004/1-2) (C.P.), National Institutes of Health grants HL70828 (J.I.G.) and HL48509 (K.D.P.), and the Laubisch Foundation.

REFERENCES

1. BERS, D.M. 2002. Cardiac excitation-contraction coupling. Nature **415**: 198–205.
2. BRIDGE, J.H., J.R. SMOLLEY & K.W. SPITZER. 1990. The relationship between charge movements associated with I_{Ca} and I_{Na-Ca} in cardiac myocytes. Science **248**: 376–378.
3. PHILIPSON, K.D. & D.A. NICOLL. 2000. Sodium-calcium exchange: a molecular perspective. Annu. Rev. Physiol. **62**: 111–133.
4. KOUSHIK, S.V., J. WANG, R. ROGERS, *et al.* 2001. Targeted inactivation of the sodium-calcium exchanger (Ncx1) results in the lack of a heartbeat and abnormal myofibrillar organization. FASEB J. **15**: 1209–1211.
5. REUTER, H., T. HAN, C. MOTTER, *et al.* 2003. Cardiac excitation-contraction coupling in the absence of Na^+ - Ca^{2+} exchange. Cell Calcium **34**: 19–26.
6. POTT, C., J.I. GOLDHABER & K.D. PHILIPSON. 2004. Genetic manipulation of cardiac Na^+/Ca^{2+} exchange expression. Biochem. Biophys. Res. Commun. **322**: 1336–1340.
7. HENDERSON, S.A., J.I. GOLDHABER, J.M. SO, *et al.* 2004. Functional adult myocardium in the absence of Na^+-Ca^{2+} exchange: cardiac-specific knockout of NCX1. Circ. Res. **95**: 604–611.
8. POTT, C., K.D. PHILIPSON & J.I. GOLDHABER. 2005. Excitation-contraction coupling in Na^+-Ca^{2+} exchanger knockout mice: reduced transsarcolemmal Ca^{2+} flux. Circ. Res. **97**: 1288–1295.
9. POTT, C., M. YIP, J.I. GOLDHABER & K.D. PHILIPSON. 2006. Ca^{2+} current in Na^+-Ca^{2+} exchanger knockout mice functional coupling of the Ca^{2+} channel and the Na^+-Ca^{2+} exchanger. Biophys. J. In press. Biophys. J., Abstract Issue. Abstract.
10. SPENCER, C.I. & J.S. SHAM. 2003. Effects of Na^+/Ca^{2+} exchange induced by SR Ca^{2+} release on action potentials and afterdepolarizations in guinea pig ventricular myocytes. Am. J. Physiol. Heart Circ. Physiol. **285**: 2552–2562.
11. TAKAMATSU, H., T. NAGAO, H. ICHIJO, *et al.* 2003. L-type Ca^{2+} channels serve as a sensor of the SR Ca^{2+} for tuning the efficacy of Ca^{2+}-induced Ca^{2+} release in rat ventricular myocytes. J. Physiol. **552**: 415–424.
12. NERBONNE, J.M. 2000. Molecular basis of functional voltage-gated K^+ channel diversity in the mammalian myocardium. J. Physiol. **525**: 285–298.
13. BONDARENKO, V.E., G.P. SZIGETI, G.C. BETT, *et al.* 2004. Computer model of action potential of mouse ventricular myocytes. Am. J. Physiol. Heart Circ. Physiol. **287**: 1378–1403.
14. XU, Y., P.H. DONG, Z. ZHANG, *et al.* 2002. Presence of a calcium-activated chloride current in mouse ventricular myocytes. Am. J. Physiol. Heart Circ. Physiol. **283**: 302–314.
15. XU, Y., Z. ZHANG, V. TIMOFEYEV, *et al.* 2005. The effects of intracellular Ca^{2+} on cardiac K^+ channel expression and activity: novel insights from genetically altered mice. J. Physiol. **562**: 745–758.

16. PERRIER, E., R. PERRIER, S. RICHARD, *et al.* 2004. Ca^{2+} controls functional expression of the cardiac K+ transient outward current via the calcineurin pathway. J. Biol. Chem. **279:** 40634–40639.

17. ZAHRADNIKOVA, A., Z. KUBALOVA & J. PAVELKOVA. 2004. Activation of calcium release assessed by calcium release-induced inactivation of calcium current in rat cardiac myocytes. Am. J. Physiol. Cell. Physiol. **286:** 330–341.

18. SAH, R., R.J. RAMIREZ, G.Y. OUDIT, D. GIDREWICZ, *et al.* 2003. Regulation of cardiac excitation-contraction coupling by action potential repolarization: role of the transient outward potassium current I_{to}. J. Physiol. **546:** 5–18.

19. BASSANI, J.W., R.A. BASSANI & D.M. BERS. 1994. Relaxation in rabbit and rat cardiac cells: species-dependent differences in cellular mechanisms. J. Physiol. **476:** 279–293.

20. BASSANI, R, J.W. BASSANI & D.M. BERS. 1995. Relaxation in ferret ventricular myocytes: role of the sarcolemmal Ca ATPase. Pflugers Arch. **430:** 573–578.

Distinct Phenotypes among Plasma Membrane Ca²⁺-ATPase Knockout Mice

VIKRAM PRASAD,[a] GBOLAHAN OKUNADE,[a] LI LIU,[b]
RICHARD J. PAUL,[b] AND GARY E. SHULL[a]

[a]Departments of Molecular Genetics, Biochemistry, and Microbiology,
University of Cincinnati College of Medicine, Cincinnati, Ohio 45267, USA

[b]Department of Molecular and Cellular Physiology, University of Cincinnati
College of Medicine, Cincinnati, Ohio 45267, USA

ABSTRACT: Ca^{2+} gradients across the plasma membrane, required for Ca^{2+} homeostasis and signaling, are maintained in part by plasma membrane Ca^{2+}-ATPase (PMCA) isoforms 1–4. Gene targeting has been used to analyze the functions of PMCA1, PMCA2, and PMCA4 in mice. PMCA1 null mutant embryos die during the preimplantation stage, and loss of a single copy of the *PMCA1* gene contributes to apoptosis in vascular smooth muscle. PMCA2 deficiency in sensory hair cells of the inner ear causes deafness and balance defects, most likely by affecting both intracellular Ca^{2+} and extracellular Ca^{2+} in the endolymph. PMCA2 is required for viability of certain neurons, consistent with a major role in maintenance of intracellular Ca^{2+}. Surprisingly, loss of PMCA2 in lactating mammary glands causes a sharp reduction in milk Ca^{2+}, consistent with a macrocalcium secretory function. Although PMCA4 is widely expressed and is the most abundant isoform in some tissues, null mutants appear healthy. However, male PMCA4 null mutants are infertile due to a failure of hyperactivated sperm motility resulting from the absence of PMCA4 in the sperm tail, and Ca^{2+} signaling in B lymphocytes, involving interactions between PMCA4, CD22, and the tyrosine phosphatase SHP-1, is defective. Studies of bladder smooth muscle from PMCA4 null mutants and PMCA1 heterozygous mice suggest that PMCA1 and PMCA4 play different roles in smooth muscle contractility, with PMCA1 contributing to overall Ca^{2+} clearance and PMCA4 being required for carbachol-stimulated contraction. These phenotypes indicate that PMCA1 serves essential housekeeping functions, whereas PMCA4 and particularly PMCA2 serve more specialized physiological functions.

KEYWORDS: embryonic stem cells; knockout; *ATP2B1*; *ATP2B2*; *ATP2B3*; *ATP2B4*

Address for correspondence: Gary E. Shull, Department of Molecular Genetics, Biochemistry, and Microbiology, University of Cincinnati College of Medicine, 231 Bethesda Avenue, ML 524, Cincinnati, OH 45267-0524. Voice: 513-558-0056; fax: 513-558-1885.
shullge@ucmail.uc.edu

Ann. N.Y. Acad. Sci. 1099: 276–286 (2007). © 2007 New York Academy of Sciences.
doi: 10.1196/annals.1387.029

INTRODUCTION

The ability of Ca^{2+} to function as a major intracellular signaling molecule in mammalian tissues requires that its concentration in the cytoplasm be maintained far below that occurring in extracellular compartments. Two groups of transporters, the Na^+/Ca^{2+} exchangers[1] and the calmodulin-sensitive plasma membrane Ca^{2+}-ATPases (PMCA)[2–4] mediate extrusion of Ca^{2+} from the cell, thereby maintaining a high Ca^{2+} concentration gradient across the plasma membrane. Na^+/Ca^{2+} (and $Na^+/K^+/Ca^{2+}$) exchangers are often viewed as mediating bulk extrusion of Ca^{2+} from cells in which high Ca^{2+} fluxes occur across the plasma membrane, whereas PMCA are generally regarded as the enzymes that provide fine tuning of intracellular Ca^{2+} concentrations both within the cell and in the subplasma membrane compartments that might affect various Ca^{2+}-sensitive receptors, channels, and signaling molecules.

The mammalian PMCA are encoded by four different genes (*ATP2B1-4* in humans), and alternative splicing of the primary transcripts leads to at least several dozen proteins, with major differences in biochemical characteristics and membrane locations.[2–14] The tissue distributions of the four isoforms differ substantially, with PMCA1 and PMCA4 being broadly expressed[6,10,13] and PMCA2 and PMCA3 being expressed in only a limited number of tissues and cell types.[6,13] Effective isoform-specific PMCA inhibitors are not available, so the functions of individual isoforms and splice variants have been difficult to determine. To provide this information, we and others have been developing and analyzing knockout mouse models in which individual isoforms have been ablated. Here we describe the phenotypes of mice in which the *PMCA1*, *PMCA2*, and *PMCA4* genes have been disrupted.

PMCA1

PMCA1 is expressed in all tissues and cell types analyzed,[6,11,13] leading to the suggestion that it serves the major "housekeeping function" of maintaining low resting levels of intracellular Ca^{2+} in most if not all mammalian cells. It should be noted that a housekeeping function would not necessarily mean that PMCA1 is essential for viability of all cells since it is frequently coexpressed with PMCA4 and other isoforms, nor would it rule out the possibility that it serves important tissue- and cell-type-specific functions, such as modulation of signaling events. In fact, there is evidence that PMCA1, as well as PMCA4, interacts with α-1 syntrophin and regulates signaling by neuronal nitric oxide synthase.[15]

A PMCA1 null mutant mouse model was prepared by targeted deletion of sequences encoding the catalytic phosphorylation site.[16] Null mutants were present during the preimplantation stages of embryogenesis, but only heterozygous mutants were seen at 8.5 days of gestation and later.[17] The presence of live null mutant embryos during early embryogenesis should not be taken as an indication that PMCA1 is unnecessary during this period because the early embryo is likely to contain some PMCA1 that was produced in the heterozygous primary oocyte. Thus, these data show that PMCA1 is essential during embryogenesis and are consistent with a major housekeeping function. Further support for a major housekeeping role comes from the observation that loss of one copy of the *PMCA1* gene in a PMCA4 null background can lead to apoptosis of vascular smooth muscle cells,[16] presumably due to Ca^{2+} overload, and studies of PMCA1 heterozygous mutant mice indicate that PMCA1 contributes to Ca^{2+} extrusion in bladder smooth muscle following stimulation by a receptor-mediated pathway with carbachol or by depolarization with KCl.[18]

PMCA2

PMCA2 has a restricted tissue distribution with high levels of expression in sensory hair cells of the inner ear,[19,20] certain neurons,[21,22] and lactating mammary glands,[23] and lower levels of expression in heart.[6] The physiological functions of PMCA2 have been studied extensively using a knockout mouse model[24] and a number of naturally occurring mouse mutants. Much of the discussion below comes from analyses of the gene-targeted PMCA2 mutant, however, similar studies have also been performed with several deaf waddler mutants[25] and the Wriggle Mouse Sagami mutant,[26] all of which contain mutations in the *PMCA2* gene. As indicated by the names of the naturally occurring mutants, the major apparent phenotypes are deafness and ataxia.

In the gene-targeted PMCA2 knockout, measurements of auditory brain stem responses showed that null mutants were profoundly deaf, even at a very young age, and that heterozygous mice developed a significant hearing deficit as they were aged.[24] The latter finding was apparently due to interactions with other recessive gene loci, as backcrossing for one generation onto a CAST/Ei background, which lacks a recessive age-related hearing loss locus, eliminated the enhanced age-related hearing loss.[27] However, PMCA2 heterozygous mice on the CAST/Ei background were susceptible to noise-induced hearing loss.[27] Thus, depending on the genetic background, PMCA2 haploinsufficiency can predispose mice to either age-induced hearing loss or noise-induced hearing loss. A correlate of this finding in humans is the recent report of a family in which hearing loss caused by homozygous mutations in cadherin 23 was exacerbated by heterozygosity for a mutation in the *PMCA2* gene.[28] Thus, mutations in a single copy of the *PMCA2* gene, which by itself does not cause a hearing deficit, can enhance the severity of hearing loss caused by other conditions, including genetic background and noise.

The occurrence of balance problems in PMCA2 null mice[24] was consistent with defects of the vestibular system; however, PMCA2 is also expressed at high levels in cerebellar Purkinje cells,[21] which can also affect balance.

Histological analyses revealed some alterations in the cerebellum, including an increase in the number of Purkinje cells, a reduction in the number of granule cells, and thinning of the molecular layer.[24] These changes, however, were not sufficient to account for the severe balance defect. In the vestibular system, there were no apparent changes in the semicircular canels, cristae ampulares, nerves, or sensory epithelium.[24] However, a consistent finding in the knockout was the absence of otoconia,[24] the calcium carbonate crystals normally embedded in the gelatinous otolithic membrane overlying the sensory epithelium of the saccule and utricle, which sense linear acceleration and gravity. The sections of inner ear analyzed in this study were subjected to a decalcification protocol, so the extent of the deficit in otoconia formation is uncertain and should be further investigated. Nevertheless, the absence of any remnants of the otoconia in wild-type inner ears indicates that at least part of the balance defect is due to a loss or reduction in the mass of the otoconia. A possible mechanism is that much of the Ca^{2+} present in the endolymph, which is very low relative to most extracellular fluids, is normally provided by PMCA2 and that, in the absence of this pump, Ca^{2+} concentrations in the endolymph are insufficient to support normal formation of the otoconia. Consistent with this hypothesis, a recent study showed that Ca^{2+} concentrations in the endolymph of the cochlear duct were reduced from 20–25 μM in wild-type mice to 6–7 μM in PMCA2 mutants.[29]

A range of histopathology was observed in the auditory system of PMCA2 null mice,[24,25] including loss of support cells, pillar cells, hair cells, and nerve cells, although in some regions of the cochlear duct these structures were relatively normal, suggesting that the profound hearing loss was not due to these histological abnormalities. A number of observations suggest that the primary defect in hearing, and perhaps in balance as well, is due to the loss of PMCA2 in hair bundles of the sensory hair cells, with subsequent loss of hair cell function. An earlier study of amphibian vestibular hair cells showed that hair bundles contained exceptionally high levels of PMCA protein that were capable of generating a significant outwardly directed Ca^{2+} current that might, in fact, create a high-localized Ca^{2+} concentration in the endolymph surrounding the hair bundles.[30] In mammals, PMCA2 mRNA is expressed at high levels in outer hair cells of the organ of Corti[20] and PMCA2 protein has been detected in both inner and outer hair cells.[19,31] Studies of autoacoustic emissions in homozygous and heterozygous deaf waddler mutants have confirmed that outer hair cell function is defective.[32]

The observations discussed above have led to the suggestion that PMCA2 may serve some unusual functions for a plasma membrane Ca^{2+}-ATPase. Although PMCA are normally considered to be the premiere regulators of intracellular Ca^{2+}, PMCA2 may serve both this function and others, including regulation of the overall Ca^{2+} concentration of the endolymph. Also, because of its very high density and electrogenic activity, it might also contribute to the membrane potential of the hair bundle and to a high-localized Ca^{2+}

concentration surrounding the hair bundle,[17,30] both of which would affect Ca^{2+} influx during mechanoelectrical transduction.

One of the clearest examples of an unusual function for a plasma membrane Ca^{2+}-ATPase is the observation that PMCA2 is required for high levels of Ca^{2+} in milk. It was long thought that Ca^{2+} in milk was provided via the secretory pathway,[33] however, PMCA2 was upregulated ~100-fold in lactating mammary glands.[23] Studies of PMCA2 null mice showed that loss of PMCA2 reduced the concentration of Ca^{2+} in milk by $\sim60\%$, whereas milk proteins were unaffected.[34] Sarco(endo)plasmic reticulum Ca^{2+}-ATPase isoform 2 (SERCA2), an endoplasmic reticulum Ca^{2+} pump, and secretory pathway Ca^{2+}-ATPase isoform 1 (SPCA1), a Golgi Ca^{2+} pump, were upregulated in mammary glands of the PMCA2 knockout, indicating that much of the remaining Ca^{2+} in the milk of PMCA2 null mice was provided via the secretory pathway. These data showed that PMCA2 serves a macrocalcium secretory function.[34]

Despite the unusual functions of PMCA2 in the inner ear and mammary glands, there is evidence that it also serves the more conventional function of regulating intracellular Ca^{2+} levels. Studies of an experimental autoimmune encephalomyelitis model showed that PMCA2 was sharply downregulated in spinal neurons.[22] Subsequent experiments using both the gene-targeted PMCA2 null mouse and a deaf waddler mutant showed a significant reduction in the number of spinal cord motor neurons.[35] These and other data suggested that PMCA2 plays a major role in Ca^{2+} clearance in these neurons, and that its absence leads to neuronal damage and death.

PMCA4

PMCA4 mRNA is expressed at relatively high levels in most adult tissues[10,13] and in some tissues, including the aorta, portal vein, urinary bladder, diaphragm, seminal vesicles, and testes, it is the most abundant PMCA mRNA.[16] The nearly ubiquitous tissue distribution and robust expression levels for PMCA4 indicated that it might play an important housekeeping role, similar to that proposed for PMCA1. However, unlike in the case of PMCA1, PMCA4 null mutants survived, appeared outwardly normal, and exhibited no apparent histopathological evidence of *in vivo* cell death or structural abnormalities when tissues were isolated from freshly euthanized mice and processed immediately for routine histology.[16] These observations were consistent with the possibility that PMCA4 serves more specialized functions and was not essential for the general housekeeping function of preventing intracellular Ca^{2+} overload, although, as discussed below, it may contribute to this process.

In studies of smooth muscle contractility, we observed an absence of phasic contractions in portal veins isolated from PMCA4$^{-/-}$ mice of the original mixed background strain in which the null mutant was established.[16] Initially, this seemed to indicate a major role for PMCA4 in phasic contractions, however, this was not the case as histological analyses revealed a high incidence of apoptosis among the smooth muscle cells after it was removed from the animal; phasic contractions were lost because the tissue was dying. After backcrossing onto a Black Swiss background, the apoptosis phenotype disappeared and phasic contractions were normal in PMCA4$^{-/-}$ portal veins, indicating that the phenotype was specific for the original 129SvJ/Black Swiss mixed background. The same *in vitro* phenotype was, however, observed in PMCA4 null mice of the Black Swiss background that were also heterozygous for the PMCA1 null allele.[16] These observations suggest that in portal vein smooth muscle, both PMCA4 and PMCA1 may contribute to the housekeeping function of preventing Ca^{2+} overload. It remains to be determined whether PMCA1 activity is essential for preventing Ca^{2+} overload in smooth muscle cells of the portal vein and other tissues; a tissue-specific PMCA1 null mouse will be required to test this possibility.

Another important effect of the loss of PMCA4 in mice was the inability of null mutant sperm to achieve hyperactivated motility, an impairment that resulted in the sperm being unable to traverse the female genital tract to fertilize the egg.[16,36] Loss of PMCA4 therefore caused male infertility, even though spermatogenesis and mating behavior were normal in PMCA4$^{-/-}$ males, and the mutant sperm were alive and were competent for both nonhyperactivated motility[16] and *in vitro* fertilization.[36] Immunoblot analysis of testes and sperm proteins revealed that PMCA4 accounted for >90% of the total PMCA protein in the sperm.[16] PMCA4 was predominantly expressed in the principal piece of the sperm tail,[16,36] and localized to the same region as CatSper, the sperm cation channel.[37] Genetic ablation of CatSper has been shown to cause male infertility, resulting from impaired sperm motility and an inability of CatSper null sperm to fertilize intact eggs.[37] Ultrastructural analyses of the PMCA4$^{-/-}$ sperm tail indicated that the motility apparatus was normal; however, there was an increase in mitochondrial condensation, suggesting that the loss of PMCA4 leads to Ca^{2+} overload and mitochondrial dysfunction.[16] Whether signaling processes involved in sperm motility are also impaired has not been analyzed, although it is possible that the impairment of mitochondrial function is responsible for the sperm motility defect.

In excitable cells, PMCA has been presumed to play a secondary role in the maintenance of low diastolic Ca^{2+} levels, overshadowed by more robust extrusion systems, such as the sodium–calcium exchanger (NCX) and intracellular Ca^{2+}-sequestering activity of sarco(endo)plasmic reticulum Ca^{2+}-ATPases (SERCA), such as SERCA2. However, recent reports suggest that PMCA4 is intimately involved in the regulation of excitation–contraction coupling and force generation in bladder smooth muscle.[18,38] Using PMCA4 null and double gene-targeted mice that were heterozygous for the PMCA1 null mutation and homozygous for the PMCA4 null mutation (PMCA1$^{+/-}$/PMCA4$^{-/-}$ mice), the data showed that while NCX activity was predominant, PMCA activity accounted for 25–30% of relaxation of bladder smooth muscle.[38] Furthermore, these studies revealed an apparent cooperativity between NCX and PMCA that might also contribute to the interesting observation that the half-time for force development in response to KCl was significantly prolonged in *PMCA* gene-targeted bladders. The increase in contraction half-time in the mutant bladder could be the result of increased subplasmalemmal Ca^{2+} levels in these myocytes, which stimulates NCX activity, thereby countering the initial Ca^{2+} influx during contraction. This idea was supported by the finding that inhibition of NCX significantly shortened the contraction half-time in *PMCA* gene-targeted bladders to levels comparable to wild-type controls.[38]

An intriguing divergence was revealed during investigations into the roles of the PMCA isoforms in Ca^{2+} homeostasis and smooth muscle contractility in bladder.[18] While responses to carbachol were greater in $PMCA1^{+/-}$ bladders, $PMCA4^{-/-}$ and $PMCA1^{+/-}/PMCA4^{-/-}$ bladders were found to have significantly smaller responses to carbachol when compared to wild-type controls. These observations could not be explained simply as resulting from increased intracellular Ca^{2+} caused by reduced PMCA-mediated Ca^{2+} clearance in the gene-targeted bladders; KCl-elicited contractility was increased in $PMCA1^{+/-}/PMCA4^{-/-}$ bladders, resembling the $PMCA1^{+/-}$ bladders. Therefore, only the responses to carbachol were attenuated upon loss of the PMCA4 isoform, suggesting that while PMCA1 is involved in overall Ca^{2+} clearance, PMCA4 plays a more intricate role, modulating acetylcholine receptor signaling. The exact mechanism remains to be determined.

There is increasing evidence that PMCA4, through its Ca^{2+}-efflux activity, might play a more direct role in signaling as a modulator of signaling pathways. Members of the membrane-associated guanylate kinase family have been found to be associated with PMCA4b via a PDZ domain-mediated interaction.[39] Furthermore, PMCA4b has been shown to be a negative regulator of nitric oxide synthase 1 (NOS-1), an activity that required both the Ca^{2+}-transporting activity of PMCA4 and a PDZ domain-mediated interaction between PMCA4b and NOS-1.[40] More recently, PMCA4b has been localized to the syntrophin–dystrophin complex in the heart, a macromolecular complex that includes NOS-1[15] and the disruption of which has been associated with degeneration of muscle and acquired dilated cardiomyopathy.[41,42] Coexpression of PMCA4 along with α-1 syntrophin in HEK293 cells was shown to synergistically enhance the negative regulation of NOS-1 activity. PMCA4 has also been implicated in the negative regulation of Ca^{2+} signaling in B lymphocytes, induced by binding of antigen to B cell receptors (BCR). The increase in intracellular Ca^{2+} induced by BCR cross-linking is essential for B cell activation and is attenuated by CD22, a transmembrane glycoprotein. CD22 was shown to enhance PMCA4-mediated Ca^{2+} efflux from activated B cells.[43] While coimmunoprecipitation studies showed that CD22 associated with PMCA4, there were indications that other modulators and adaptors might be included in this functional complex, such as SHP-1 (Src homology 2 domain-containing tyrosine phosphatase 1), which was found to be necessary for CD22-mediated regulation of Ca^{2+} efflux via PMCA4.[43]

CONCLUSIONS

Gene-targeting studies have now been conducted for three of the four PMCA isoforms. The death of PMCA1 null embryos supports the view that PMCA1, which appears to be expressed in all cell types, serves critical housekeeping functions. This by no means excludes its involvement in more specialized tissue-specific functions or the possibility that compensation for the loss of its housekeeping functions could occur in certain cell types, particularly those in which PMCA4 has been shown to be more abundant. A gene-targeted mouse model that allows tissue-specific ablation of PMCA1 will be required to rigorously address these issues. Perhaps the most unique isoform studied so far is PMCA2. In certain neurons of the central nervous system, its absence leads to loss of cell viability, consistent with the common role of PMCA in the regulation of intracellular Ca^{2+} homeostasis and prevention of Ca^{2+} overload. It serves unusual functions in mammary glands, where it is involved in secretion of Ca^{2+} into the milk, and in sensory hair cells of the inner ear, where it is critical for mechanoelectrical transduction, a process that appears to involve significant contributions to the maintenance of intracellular Ca^{2+} in the stereocilia, the membrane potential of the stereocilia, and Ca^{2+} concentrations of the extracellular endolymphatic fluid. Regarding PMCA4, the data indicate that it contributes, at least in some cell types, to the housekeeping function of preventing Ca^{2+} overload, and that it also serves tissue-specific functions, including sperm motility, modulation of smooth muscle contractility, and regulation of Ca^{2+} signaling in B lymphocytes and other tissues via receptor-mediated pathways.

ACKNOWLEDGMENTS

Portions of this work were supported by NIH grants HL61974 and HL66044.

REFERENCES

1. PHILIPSON, K.D. & D.A. NICOLL. 2000. Sodium-calcium exchange: a molecular perspective. Annu. Rev. Physiol. **62:** 111–133.
2. CARAFOLI, E. 1994. Biogenesis: plasma membrane calcium ATPase: 15 years of work on the purified enzyme. FASEB J. **13:** 993–1002.

3. GUERINI, D. *et al.* 1998. The calcium pump of the plasma membrane: membrane targeting, calcium binding sites, tissue-specific isoform expression. Acta Physiol. Scand. Suppl. **643:** 265–273.

4. STREHLER, E.E. & D.A. ZACHARIAS. 2001. Role of alternative splicing in generating isoform diversity among plasma membrane calcium pumps. Physiol. Rev. **81:** 21–50.

5. SHULL, G.E. & J. GREEB. 1988. Molecular cloning of two isoforms of the plasma membrane Ca^{2+}-transporting ATPase from rat brain. J. Biol. Chem. **263:** 8646–8657.

6. GREEB, J. & G.E. SHULL. 1989. Molecular cloning of a third isoform of the calmodulin-sensitive plasma membrane Ca^{2+}-transporting ATPase that is expressed predominantly in brain and skeletal muscle. J. Biol. Chem. **264:** 18569–18576.

7. VERMA, A.K. *et al.* 1988. Complete primary structure of a human plasma membrane Ca^{2+} pump. J. Biol. Chem. **263:** 14152–14159.

8. STREHLER, E.E. *et al.* 1990. Peptide sequence analysis and molecular cloning reveal two calcium pump isoforms in the human erythrocyte membrane. J. Biol. Chem. **265:** 2835–2842.

9. HEIM, R. *et al.* 1992. Microdiversity of human-plasma-membrane calcium-pump isoform 2 generated by alternative RNA splicing in the N-terminal coding region. Eur. J. Biochem. **205:** 333–340.

10. KEETON, T.P. & G.E. SHULL. 1995. Primary structure of rat plasma membrane Ca^{2+}-ATPase isoform 4 and analysis of alternative splicing patterns at splice site A. Biochem. J. **306:** 779–785.

11. KEETON, T.P., S.E. BURK & G.E. SHULL. 1993. Alternative splicing of exons encoding the calmodulin-binding domains and C termini of plasma membrane Ca^{2+}-ATPase isoforms 1, 2, 3, and 4. J. Biol. Chem. **268:** 2740–2748.

12. CHICKA, M.C. & E.E. STREHLER. 2003. Alternative splicing of the first intracellular loop of plasma membrane Ca^{2+}-ATPase isoform 2 alters its membrane targeting. J. Biol. Chem. **278:** 18464–18470.

13. STAUFFER, T.P. *et al.* 1993. Quantitative analysis of alternative splicing options of human plasma membrane calcium pump genes. J. Biol. Chem. **268:** 25993–26003.

14. HILL, J.K. *et al.* 2006. Splice-site A choice targets plasma-membrane Ca^{2+}-ATPase isoform 2 to hair bundles. J. Neurosci. **26:** 6172–6180.

15. WILLIAMS, J.C. *et al.* 2006. The sarcolemmal calcium pump, alpha-1 syntrophin, and neuronal nitric-oxide synthase are parts of a macromolecular protein complex. J. Biol. Chem. **281:** 23341–23348.

16. OKUNADE, G.W. *et al.* 2004. Targeted ablation of plasma membrane Ca^{2+}-ATPase isoforms 1 and 4 indicates a major housekeeping function for PMCA1 and a critical role in hyperactivated sperm motility and male fertility for PMCA4. J. Biol. Chem. **279:** 33742–33750.

17. PRASAD, V. *et al.* 2004. Phenotypes of SERCA and PMCA knockout mice. Biochem. Biophys. Res. Commun. **322:** 1192–1203.

18. LIU, L. *et al.* 2006. Distinct roles of PMCA isoforms in Ca^{2+}-homeostasis of bladder smooth muscle: evidence from PMCA gene-ablated mice. Am. J. Physiol. Cell Physiol. **292:** C423–C431.

19. DUMONT, R.A. *et al.* 2001. Plasma membrane Ca^{2+}-ATPase isoform 2a is the PMCA of hair bundles. J. Neurosci. **21:** 5066–5078.

20. FURUTA, H. *et al.* 1998. Evidence for differential regulation of calcium by outer versus inner hair cells: plasma membrane Ca-ATPase gene expression. Hearing Res. **123:** 10–26.
21. STAHL, W.L. *et al.* 1992. Plasma membrane Ca^{2+}-ATPase isoforms: distribution of mRNAs in rat brain by in situ hybridization. Mol. Brain Res. **16:** 223–231.
22. NICOT, A. *et al.* 2003. Regulation of gene expression in experimental autoimmune encephalomyelitis indicates early neuronal dysfunction. Brain **126:** 398–412.
23. REINHARDT, T.A. *et al.* 2000. Ca^{2+}-ATPase protein expression in mammary tissue. Am. J. Physiol. Cell Physiol. **279:** C1595–C1602.
24. KOZEL, P.J. *et al.* 1998. Balance and hearing deficits in mice with a null mutation in the gene encoding plasma membrane Ca^{2+}-ATPase isoform 2. J. Biol. Chem. **273:** 18693–18696.
25. STREET, V.A. *et al.* 1998. Mutations in a plasma membrane Ca^{2+}-ATPase gene cause deafness in deaf waddler mice. Nat. Genet. **19:** 390–394.
26. TAKAHASHI, K. & K. KITAMURA. 1999. A point mutation in a plasma membrane Ca2+ ATPase gene causes deafness in Wriggle Mouse Sagami. Biochem. Biophys. Res. Comm. **261:** 773–778.
27. KOZEL, P.J. *et al.* 2002. Deficiency in plasma membrane calcium ATPase isoform 2 increases susceptibility to noise-induced hearing loss in mice. Hearing Res. **164:** 231–239.
28. SCHULTZ, J.M. *et al.* 2004. Modification of human hearing loss by plasma-membrane calcium pump PMCA2. N. Engl. J. Med. **352:** 1557–1564.
29. WOOD, J.D. *et al.* 2005. Low endolymph calcium concentrations in deaf waddler2J mice suggest that PMCA2 contributes to endolymph calcium maintenance PMCA2. J. Assoc. Res. Otolaryngol. **5:** 99–110.
30. YAMOAH, E.N. *et al.* 1998. Plasma membrane Ca^{2+}-ATPase extrudes Ca^{2+} from hair cell stereocilia. J. Neurosci. **18:** 610–624.
31. APICELLA, S. *et al.* 1997. Plasmalemmal ATPase calcium pump localizes to inner and outer hair bundles. Neuroscience **79:** 1145–1151.
32. KONRAD-MARTIN, D. *et al.* 2001. Effects of PMCA2 mutation on DPOAE amplitudes and latencies in deaf waddler mice. Hearing Res. **151:** 205–220.
33. SHENNAN, D.B. & M. PEAKER. 2000. Transport of milk constituents by the mammary gland. Physiol. Rev. **80:** 925–951.
34. REINHARDT, T.A. *et al.* 2004. Null mutation in the gene encoding plasma membrane Ca^{2+}-ATPase isoform 2 impairs calcium transport into milk. J. Biol. Chem. **279:** 42369–42373.
35. KURNELLAS, M.P. *et al.* 2005. Plasma membrane calcium ATPase deficiency causes neuronal pathology in the spinal cord: a potential mechanism for neurodegeneration in multiple sclerosis and spinal cord injury. FASEB J. **19:** 298–300.
36. SCHUH, K. *et al.* 2004. Plasma membrane Ca^{2+} ATPase 4 is required for sperm motility and male fertility. J. Biol. Chem. **279:** 28220–28226.
37. REN, D. *et al.* 2001. A sperm ion channel required for sperm motility and male fertility. Nature **413:** 603–609.
38. LIU, L. *et al.* 2005. Role of plasma membrane Ca^{2+}-ATPase in contraction-relaxation processes of the bladder: evidence from PMCA gene-ablated mice. Am. J. Physiol. Cell Physiol. **290:** C1239–C1247.
39. KIM, E. *et al.* 1998. Plasma membrane Ca^{2+} ATPase isoform 4b binds to membrane-associated guanylate kinase (MAGUK) proteins via their PDZ (PSD-95/Dlg/ZO-1) domains. J. Biol. Chem. **273:** 1591–1595.

40. SCHUH, K. *et al.* 2001. The plasma membrane calmodulin-dependent calcium pump: a major regulator of nitric oxide synthase I. J. Cell Biol. **155:** 201–205.
41. CAMPBELL, K.P. 1995. Three muscular dystrophies: loss of cytoskeleton-extracellular matrix linkage. Cell **80:** 675–679.
42. HEIN, S. *et al.* 2000. The role of the cytoskeleton in heart failure. Cardiovasc. Res. **45:** 273–278.
43. CHEN, J. *et al.* 2004. CD22 attenuates calcium signaling by potentiating plasma membrane calcium-ATPase activity. Nat. Immunol. **5:** 651–657.

Role of Plasma Membrane Calcium ATPase Isoform 2 in Neuronal Function in the Cerebellum and Spinal Cord

MICHAEL P. KURNELLAS, AMANDA K. LEE,
KAROLYNN SZCZEPANOWSKI, AND STELLA ELKABES

*Department of Neurology and Neuroscience, New Jersey Medical School,
University of Medicine and Dentistry of New Jersey, Newark,
New Jersey 07103, USA*

Neurology Service, Veterans Affairs, East Orange, New Jersey 07018, USA

ABSTRACT: The distinct role of plasma membrane calcium ATPase 2 (PMCA2) in the function of different neuronal subpopulations in the central nervous system is not well understood. We found that lack of PMCA2 leads to a reduction in the number of motor neurons in the spinal cord of PMCA2-null mice and to abnormal changes in molecular pathways in Purkinje cells. Thus, PMCA2 may have unique, nonredundant functions in spinal cord and cerebellar neurons. Our results suggest that anomalous alterations in PMCA2 activity or expression may induce pathology in some neuronal populations, a possibility that will be the focus of future investigations.

KEYWORDS: ATP2b2; glutamate receptors; axonal/neuronal pathology

Plasma membrane calcium ATPase (PMCA2) is a calcium pump, which is expressed primarily in neurons, including motor neurons of the spinal cord and Purkinje cells of the cerebellum.[1–3] Although a number of studies have implicated dysregulation of PMCA2 in pathological conditions,[4] the importance and the unique contribution of PMCA2 to the function or dysfunction of distinct neuronal subpopulations in the central nervous system is not well understood.

Earlier studies in our laboratory indicated a significant decrease in the levels of PMCA2 in the inflamed spinal cord of rats and mice affected by experimental autoimmune encephalomyelitis (EAE), an animal model of multiple sclerosis.[5,6] Subsequently, we showed that a reduction in PMCA activity causes

Address for correspondence: Stella Elkabes, Ph.D., Department of Neurology and Neuroscience, MSB, H-506, New Jersey Medical School-UMDNJ, 185 South Orange Ave, Newark, NJ 07103. Voice: 973-676-1000; ext.: 1156 (office) or 3589 (laboratory); fax: 973-395-7233.
elkabest@umdnj.edu

Ann. N.Y. Acad. Sci. 1099: 287–291 (2007). © 2007 New York Academy of Sciences.
doi: 10.1196/annals.1387.025

pathology of spinal cord neurons, *in vitro*.[7] Inhibition of PMCA activity delayed the clearance of depolarization-induced calcium transients. This was followed by beading of dendrites and axons, cytoskeletal abnormalities, and, finally, death of cultured spinal cord neurons. Motor neurons were most vulnerable to inhibition of PMCA activity, *in vitro*, as they were the first cells to die. In accordance with these findings, quantification of motor neurons in the lumbar spinal cord of the adult PMCA2-null mouse indicated a significant reduction in their number as compared with that in wild-type littermates.[7] Further studies are necessary to determine whether the decrease in the number of motor neurons is due to death and/or developmental abnormalities in the proliferation or differentiation of these cells in the knockout mice. Nevertheless, these findings pinpoint the importance of PMCA2 in motor neuron function.

In addition to hindlimb weakness, which may be partially attributed to motor neuron loss, the phenotype of the PMCA2-null mouse includes ataxia, movement dyscoordination, and balance deficits. Whereas some of these anomalies are due to alterations in vestibular function, as reported previously,[8] they also raise the possibility of cerebellar dysfunction. In fact, the morphometric studies of Kozel *et al.*[8] indicated an increase in the density of Purkinje neurons and a decrease in the density of granule cells. Reductions in the thickness of the molecular layer were also reported, which may be due to loss of dendritic spines or projecting axons, such as parallel and climbing fibers, which innervate

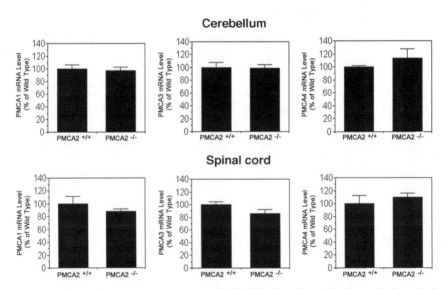

FIGURE 1. Semiquantitative analysis of PMCA1, 3, and 4 mRNA levels in the spinal cord and cerebellum of the PMCA2[+/+] and PMCA2[−/−] mice by reverse transcription-polymerase chain reaction. No significant differences were found. Graphs represent the mean ± SEM. The experiments were repeated twice and yielded similar results; n = 6.

Purkinje cells or interneurons. PMCA2 levels are severalfold higher in Purkinje cells as compared to other neuronal populations in the brain[1] and the lack of PMCA2 does not appear to induce compensatory increases in the mRNA expression of other PMCA isoforms either in the spinal cord or in the cerebellum of the PMCA2-null mouse (FIG. 1). Although it is not yet known whether the protein levels or the activity of other PMCAs are modified in the absence of PMCA2, it is possible to speculate that the lack of a major calcium pump in these neurons together with the absence of compensatory changes in other PMCAs may lead to calcium dyshomeostasis, a trigger that can affect Purkinje cell signaling. Therefore, we initiated studies to investigate the role of PMCA2 in Purkinje neurons. Proteomic analysis using either the isobaric tags for relative and absolute quantitation (iTRAQ) or two-dimensional gel electrophoresis followed by mass spectroscopy identified a number of differentially expressed proteins in the cerebellum of the PMCA2-null mouse, including inositol-3-phosphate receptor 1 (IP3R1)[9] and Homer 3, a scaffold protein that links IP3Rs to metabotropic glutamate receptors (mGluRs). IP3R1 is a downstream effector of mGluR1, which has been implicated in plasticity at the Purkinje cell-parallel fiber synapse, cerebellum-dependent associative learning, such as classical eyeblink conditioning and movement coordination.[10–12] As

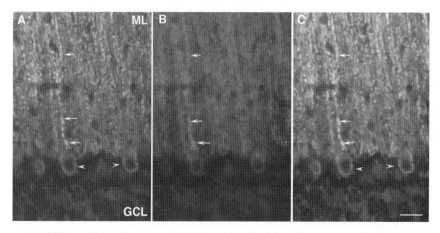

FIGURE 2. Colocalization of PMCA2 with mGluR1 in the mouse cerebellum by immunocytochemistry. (**A**) A coronal section through the cerebellum showing distribution of PMCA2 in somata (arrowheads) and dendrites (arrows) of Purkinje neurons. A FITC-tagged secondary antibody was used for visualization (green). (**B**) The same section was labeled with an antibody against mGluR1, which stained dendrites (arrows) but not cell bodies of Purkinje neurons. A Texas Red tagged secondary antibody was used for visualization (red). (**C**) Merged picture showing colocalization of mGluR1 with PMCA2 in dendrites (yellow; arrows). The arrowhead points to a Purkinje somata, which is labeled only with the PMCA2 antibody (green). ML: molecular layer, GCL: granule cell layer. Bar represents 75 μM. Color picture available online.

mGluR1-Homer-IP3R form a complex in the cerebellum,[13,14] we hypothe-sized that PMCA2 may play a role in mGluR1 signaling by associating with the receptor and its signaling complex. This would implicate colocalization of mGluR1 with PMCA2. To investigate this possibility, we performed im-munocytochemical studies on mouse cerebellar sections. PMCA2 immunore-activity was localized to dendrites and plasma membranes of Purkinje somata (FIG. 2 A). mGluR1 staining was at background level in cell bodies but strong in dendrites of Purkinje cells (FIG. 2 B). The distribution of PMCA2 and mGluR1 immunostaining was similar, and the merged picture indicated colo-calization in dendrites of Purkinje neurons (FIG. 2 C). In agreement with these results, preliminary studies have shown that PMCA2 coimmunoprecipitates with mGluR1, Homer 3, and IP3R1[15] suggesting that PMCA2 is a component of mGluR1 signaling complex.

In conclusion, PMCA2 appears to play an important function in the integrity of spinal cord neurons and might contribute to mGluR signaling in the cere-bellum. Our results, taken together, suggest that PMCA2 may have unique and nonredundant functions in spinal cord neurons and Purkinje cells. Future studies will define the exact role of PMCA2 in mGluR1 signaling and will identify the mechanisms and pathways that are affected in the spinal cord and cerebellum when PMCA2 activity is dysregulated.

ACKNOWLEDGMENTS

This work was supported by grants 01-3008-SCR-S-0 from NJCSCR and NS 046363 from NIH/NINDS to SE.

REFERENCES

1. STAHL, W.L., T.J. EAKIN, J.W. OWENS, JR., *et al.* 1992. Plasma membrane Ca(2+)-ATPase isoforms: distribution of mRNAs in rat brain by in situ hybridization. Brain Res. Mol. Brain Res. **16:** 223–231.
2. STAUFFER, T.P., D. GUERINI, M.R. CELIO & E. CARAFOLI. 1997. Immunolocalization of the plasma membrane Ca^{2+} pump isoforms in the rat brain. Brain Res. **748:** 21–29.
3. TACHIBANA, T., H. OGURA, A. TOKUNAGA, *et al.* 2004. Plasma membrane calcium ATPase expression in the rat spinal cord. Brain Res. Mol. Brain Res. **131:** 26–32.
4. LEHOTSKY, J., P. KAPLAN, R. MURIN & L. RAEYMAEKERS. 2002. The role of plasma membrane Ca^{2+} pumps (PMCAs) in pathologies of mammalian cells [Review]. Front. Biosci. **7:** D53–D84.
5. NICOT, A., M. KURNELLAS & S. ELKABES. 2005. Temporal pattern of plasma mem-brane calcium ATPase 2 expression in the spinal cord correlates with the course of clinical symptoms in two rodent models of autoimmune encephalomyelitis. Eur. J. Neurosci. **21:** 2660–2670.

6. NICOT, A., P.V. RATNAKAR, Y. RON, *et al.* 2003. Regulation of gene expression in experimental autoimmune encephalomyelitis indicates early neuronal dysfunction. Brain **126:** 398–412.
7. KURNELLAS, M.P., A. NICOT, G.E. SHULL & S. ELKABES. 2005. Plasma membrane calcium ATPase deficiency causes neuronal pathology in the spinal cord: a potential mechanism for neurodegeneration in multiple sclerosis and spinal cord injury. FASEB J. **19:** 298–300. Epub 2004 Dec 2.
8. KOZEL, P.J., R.A. FRIEDMAN, L.C. ERWAY, *et al.* 1998. Balance and hearing deficits in mice with a null mutation in the gene encoding plasma membrane Ca^{2+}-ATPase isoform 2. J. Biol. Chem. **273:** 18693–18696.
9. HU, J., J. QIAN, O. BORISOV, *et al.* 2006. Optimized proteomic analysis of a mouse model of cerebellar dysfunction using amine-specific isobaric tags. Proteomics **6:** 4321–4334.
10. ICHISE, T., M. KANO, K. HASHIMOTO, *et al.* 2000. mGluR1 in cerebellar Purkinje cells essential for long-term depression, synapse elimination, and motor coordination. Science **288:** 1832–1835.
11. KISHIMOTO, Y., R. FUJIMICHI, K. ARAISHI, *et al.* 2002. mGluR1 in cerebellar Purkinje cells is required for normal association of temporally contiguous stimuli in classical conditioning. Eur. J. Neurosci. **16:** 2416–2424.
12. KNÖPFEL, T. & P. GRANDES. 2002. Metabotropic glutamate receptors in the cerebellum with a focus on their function in Purkinje cells [Review]. Cerebellum **1:**19–26.
13. BRAKEMAN, P.R., A.A. LANAHAN, R. O'BRIEN, *et al.* 1997. Homer: a protein that selectively binds metabotropic glutamate receptors. Nature **386:** 284–288.
14. TU, J.C., B. XIAO, J.P. YUAN, *et al.* 1998. Homer binds a novel proline-rich motif and links group 1 metabotropic glutamate receptors with IP3 receptors. Neuron **21:** 717–726.
15. KURNELLAS, M.P., A.K. LEE, H. LI, L. DENG, D. EHRLICH & S. ELKABES. 2007. Molecular alterations in the cerebellum of the plasma membrane calcium ATPase 2 (PMCA2)-null mouse indicate abnormalities in Purkinje neurons. Mol. Cell Neurosci. **34:** 178–188. Epub 2006 Dec. 5.

Increased Tolerance to Ischemic Neuronal Damage by Knockdown of Na$^+$–Ca^{2+} Exchanger Isoform 1

JING LUO,[a,b] YANPING WANG,[a] XINZHI CHEN,[a,c] HAI CHEN,[a,c]
DOUGLAS B. KINTNER,[a] GARY E. SHULL,[d] KENNETH D. PHILIPSON,[e]
AND DANDAN SUN[a,b]

[a]Department of Neurosurgery, University of Wisconsin Medical School,
Madison, Wisconsin 53792, USA

[b]Department of Physiology, University of Wisconsin Medical School, Madison,
Wisconsin 53792, USA

[c]Department of Neuroscience Training Program, University of Wisconsin
Medical School, Madison, Wisconsin 53792, USA

[d]Department of Molecular Genetics, Biochemistry and Microbiology, University
of Cincinnati, Cincinnati, Ohio 45267, USA

[e]Department of Physiology and Medicine, University of California, Los Angeles,
California 90095, USA

ABSTRACT: We hypothesize that stimulation of Na$^+$–K$^+$–Cl$^+$ cotransporter (NKCC1) causes Na$^+$ overload that may lead to reversal of
Na$^+$–Ca^{2+} exchanger isoform 1 (NCX1) and ischemic neuronal damage. NCX1 protein expression and Ca^{2+} influx via reversal of NCX were
decreased by ∼70% in NCX1$^{+/-}$ neurons. Compared to NCX1$^{+/+}$ neurons, NCX1$^{+/-}$ neurons exhibited significantly less cell death (∼30%)
after 3 h oxygen and glucose deprivation (OGD) and 21 h reoxygenation.
Additional neuroprotection was found in NCX1$^{+/-}$ neurons treated with
NCX inhibitor KB-R7943. Moreover, expression of NCX1 protein was
∼40% lower in NCX1$^{+/-}$ brains than in NCX1$^{+/+}$ brains. However,
there was no significant reduction in cerebral infarction in NCX1$^{+/-}$
mice following middle cerebral artery occlusion (MCAO). These data
suggest that moderate reduction of NCX1 protein may be not enough
to exert protection. We used small RNA-interference (siRNA) approach
to further elucidate the role of NCX1 in ischemic cell damage. Efficacy of anti-NCX1 siRNA was tested in astrocytes and ∼50% knockdown of NCX1 protein expression was achieved after 24–72 h transfection. Reduction in NCX1 protein expression was also found in brains
of NCX1$^{+/-}$ mice after the siRNA injection. NCX1$^{+/-}$ mice treated

Address for correspondence: Dandan Sun, M.D., Ph.D., Department of Neurological Surgery,
University of Wisconsin Medical School, H4/332 Clinical Sciences Center, 600 Highland Ave.,
Madison, WI 53792. Voice: 608-263-4060; fax: 608-263-1409.
sun@neurosurg.wisc.edu

Ann. N.Y. Acad. Sci. 1099: 292–305 (2007). © 2007 New York Academy of Sciences.
doi: 10.1196/annals.1387.016

with siRNA showed $\sim$20% less MCAO-induced infarction, compared to NCX1$^{+/-}$ mice. Approximately 50% neuroprotection was detected in NKCC1$^{+/-}$/NCX1$^{+/-}$ mice following MCAO. In conclusion, these data suggest that NCX1 plays an important role in ischemia/reperfusion-induced neuronal injury.

KEYWORDS: Ca^{2+} overload; reversal of NCX; neuronal death; focal ischemia; infarction

INTRODUCTION

Disruption of ionic homeostasis plays a central role in ischemic cerebral damage. Numerous recent studies suggest that non-NMDA receptor-mediated mechanisms are important in perturbation of intracellular Na$^+$ and Ca^{2+} homeostasis and ischemic cell damage. These include glutamate-independent, Ca^{2+}-permeable acid-sensing ion channels,[1] Na$^+$–K$^+$–Cl$^-$ cotransporter isoform 1 (NKCC1),[2] Na$^+$–Ca^{2+} exchangers (NCXs),[3–5] and Na$^+$–H$^+$ exchanger.[6–8] In particular, we recently reported that NKCC1-mediated Na$^+$ overload triggers reversal of NCX in astrocytes following *in vitro* ischemia.[9] These data imply that ion transporters may be functionally connected in cellular microdomains and that they act in concert to contribute to ischemic cell damage. In the present study, we demonstrate that either knockdown of Na$^+$/Ca^{2+} exchanger isoform 1 (NCX1) expression or reduction of both NKCC1 and NCX1 expression provides neuronprotection against ischemic damage.

MATERIALS AND METHODS

Materials

Eagle's modified essential medium (EMEM) and Hanks balanced salt solution (HBSS) were from Mediatech Cellgro (Herndon, VA). Fetal bovine sera (FBS) were obtained from Valley Biomedical (Knoxville, TN). 1-[6-Amino-2-(5-carboxy-2-oxazolyl)-5-benzofuranyloxy]-2-(2-amino-5-methylphenoxy)ethane-N,N,N′,N′-tetraacetic acid (fura-2 AM), Lipofectamine and Opti MEM were obtained from Invitrogen (Eugene, OR). Antibody for β tubulin type III was from Promega (Madison, WI). NKCC1 monoclonal antibody (T4) was from Developmental Studies Hybridoma Bank (Iowa City, IA). NCX1 monocolonal antibody was from Swant (Bellinzona, Switzerland). KB-R7943 was obtained from Tocris Cookson Inc. (Ellisville, MI).

Animal Preparation

The NCX1 transgenic mouse (SV129/Black Swiss) and NKCC1 transgenic mouse (SV129/Black Swiss) were established previously.[10] The genotype of each mouse was determined by a polymerase chain reaction (PCR) of DNA from tail biopsies as described earlier.[11] NKCC1$^{+/-}$ and NCX$^{+/-}$ breeders were bred to obtain a double NKCC1$^{+/-}$/NCX1$^{+/-}$ heterogyzous mouse.

Primary Cultures of Mouse Cortical Neurons or Astrocytes

Cortical Neuron Cultures

E14–16 pregnant mice were anesthetized with 5% halothane and euthanized as described in our recent study.[6] Fetuses were removed and rinsed in cold HBSS. Each mouse fetus was genotyped using fetus tail biopsies with the PCR method. The cortices were removed and minced. The tissues were treated with trypsin at 37°C for 25 min. After centrifugation, the cell suspension was diluted in EMEM containing 5% FBS and 5% HS. The cells from individual fetal cortices were seeded separately in poly-D-lysine–coated plates or coverslips incubated at 37°C in an incubator with 5% CO_2 and atmospheric air. After 96 h in culture, 1 mL of fresh media containing 8 μM cytosine 1-b-D arabinofuranoside was added. The media were replaced as described earlier.[2] DIV 10–14 cultures (days in culture) were used in the study.

Cortical Astrocyte Cultures

Dissociated cortical astrocyte cultures were established as described earlier.[12] Briefly, cerebral cortices were removed from 1-day-old mice and incubated in a trypsin solution for 25 min at 37°C. The dissociated cells were rinsed and resuspended in EMEM containing 10% FBS. Viable cells were plated in collagen type 1–coated plates or coverslips. Cultures were maintained in a 5% CO_2 atmosphere at 37°C for 7 days.

Oxygen and Glucose Deprivation (OGD) Treatment

DIV 10–14 neuronal cultures were rinsed with an isotonic OGD solution (pH 7.4), as described earlier.[6] Cells were incubated in 0.5 mL OGD solution in a hypoxic incubator (model 3130 from Thermo Forma, Marietta, OH) containing 94% N_2, 1% O_2, and 5% CO_2 for 3 h. The oxygen level in the medium of cultured cells in 24-well plates was monitored with an oxygen probe (Model M1-730, Microelectrodes, Bedford, NH, UK) and decreased to ∼2–3% after 60 min in the hypoxic incubator. For reoxygenation (REOX), the cells were incubated for 21 h in 0.5 mL of EMEM containing 5.5 mM glucose at 37°C in the incubator with 5% CO_2 and atmospheric air. Normoxic control cells for cell viability assay were performed in sister cultures.[6]

Measurement of Cell Death

Cell viability was assessed by propidium iodide (PI) uptake and retention of calcein using a Nikon TE 300 inverted epifluorescence microscope (Tokyo, Japan). Cultured neurons were rinsed with HEPES-MEM and incubated with 1 $\mu g/mL$ calcein-AM and 10 $\mu g/mL$ PI in the same buffer at 37°C for 30 min. For cell counting, cells were rinsed with the isotonic control buffer and visualized using a Nikon 20X objective lens. Calcein and PI fluorescences were visualized using FITC filters and Texas Red filters as described before.[13] Images were collected using a Princeton Instruments MicroMax CCD camera (Trenton, NJ). In a blind manner, a total of 1000 cells/condition were counted using MetaMorph image-processing software (Universal Imaging Corp., Downingtown, PA). Cell mortality was expressed as the ratio of PI-positive cells to the sum of calcein-positive and PI-positive cells.

Gel Electrophoresis and Western Blotting

Cells were scraped from the plates and lysed in PBS (pH 7.4) containing 2 mM EDTA and protease inhibitors by 30 s sonication at 4°C.[14] Protein content was determined by BCA.[6] Protein samples (15 $\mu g/lane$) and pre-stained molecular mass markers (Bio-Rad, Hercules, CA) were denatured in SDS 2 × sample buffer. The samples were then electrophoretically separated on 6% SDS gels and the resolved proteins were electrophoretically transferred to a PVDF membrane.[14] The blots were incubated in 7.5% nonfat dry milk in Tris-buffered saline (TBS) overnight at 4°C, and then incubated for 1 h with a primary antibody. The blots were rinsed with TBS and incubated with horseradish peroxidase-conjugated secondary IgG for 1 h. Bound antibody was visualized using the enhanced chemiluminescence assay (Amersham Corp, Piscataway, NJ). Monocolonal T4 antibody against NKCC1 (1:4000) and anti-NCX1 monoclonal antibody (1:1000) were used for detection of NKCC1 and NCX1, respectively.

Intracellular Ca^{2+} Measurement

Neurons grown on coverslips were incubated with 5 μM fura-2 AM for 45 min.[6,11] The cells were washed and the coverslips placed in the open-bath imaging chamber containing HEPES-MEM at ambient temperature. Using a Nikon TE 300 inverted epifluorescence microscope and a 40X Super Fluor oil immersion objective lens, neurons were excited every 10 s at 345 and 385 nm and the emission fluorescence at 510 nm recorded. Images were collected and analyzed with the MetaFluor image-processing software. At the end of each experiment, the cells were exposed to 1 mM $MnCl_2$ in Ca^{2+}-free HEPES-MEM.

The Ca^{2+}-insensitive fluorescence was subtracted from each wavelength before calculations.[6] The $MnCl_2$-corrected 345/385 emission ratios were converted to concentration using the Grynkiewicz equation,[15] as described previously.[11]

Reverse mode NCX was induced in neurons as described by Hoyt et al.[16] Neurons were exposed to Ca^{2+}-free HEPES buffer for 1 min. Reverse mode NCX was initiated by exposing cells to a Na^+-free buffer (1.2 mM Ca^{2+}) for 30–40 s, which triggered a rise in $[Ca^{2+}]_i$. Cells quickly regulated Ca^{2+}_i to baseline values when they were returned to control HEPES-MEM.

Small RNA-Interference-Mediated Knockdown of NCX1 In Vitro and In Vivo

Astrocytes cultures (30–40% confluent) were reefed with antibiotic-free medium 1 day before transfection. Small RNA-interference (siRNA) corresponding to mouse NCX1 (NM˙011406) was purchased from Invitrogen. SiRNA for NCX1 (forward: 5′-GGACCAAGAUGAUGAGGAA-3′ and reverse: 5′-UUCCUCAUCAUCUUGGUCC-3′) targeted 1041–1060 nucleotides downstream of the start codon of open reading frame. A scrambled control contained the same nucleotides with an irregular sequence (forward: 5′-GGAAGAAUAGUGAGCCGAA-3′ and reverse: 5′-UUCGGCUCACUAUUCUUCC-3′). Lipofectamine served as a vehicle control. Transfection mixture contained Opti MEM I–reduced serum medium (0.5 mL), 0.1 nM siRNA, and 5 μL Lipofectamine. Transfection mixture (500 μL) was added to each well (1.5 mL complete EMEM medium without antibiotics). After 24, 48, or 72 h incubation, the effect of siRNA on expression of NCX1 was determined by Western blotting.

For in vivo injection of siRNA, male mice (25–30 g) were anesthetized as described before.[2] All subsequent procedures were performed with sterile technique. After a midline incision over the bregma (1–2 cm) was made, a hole was drilled through the skull over the middle cerebral artery (MCA). A micromanipulator attached to a stereotaxic frame was used to locate injection position (0.02 mm anterior-posterior; 2.5 mm lateral to bregma, according to an atlas).[17] A needle (22–26 gauge) attached to a 10 μL syringe glass was inserted into the brain to a depth of 2.5 mm. Lipofectamine (6 μL), scrambled control siRNA (10 μM, 3 μL) or siRNA (100 μM, 3 μL) in Lipofectamine (3 μL) was injected at 0.25 μL/min using a stoelting syringe pump. After injection, sterile bone wax was used to seal the hole on the skull and the skin incision was sutured. After 10-h recovery, animals were returned to the animal care facility and checked daily. NCX1 expression was tested 1–3 days after siRNA injection.

Focal Ischemic Model

Focal cerebral ischemia in mice was induced by occlusion of the left MCA, as described previously.[2] Briefly, the left common carotid artery was exposed and the occipital artery branches of the external carotid artery were isolated and coagulated. The internal carotid artery was isolated and the extracranial branch was dissected and ligated. A polyamide resin glue-coated suture (6–0 monofilament nylon) was used to block the MCA blood flow for 30 or 60 min. For reperfusion, the suture was withdrawn after the middle cerebral artery occlusion (MCAO). The incision was closed and the mice recovered under a heating lamp to ensure that the core temperature (36.0–37.0 °C) was maintained during recovery. After recovery (0.5 h), animals were returned to their cages with free access to food and water. At 24 or 72 h of reperfusion, the animals were sacrificed and infarction volume measured.

A total of 14 adult mice were used in this study. All animal procedures used in this study were conducted in strict compliance with the National Institutes of Health Guide for the Care and Use of Laboratory Animals and approved by the University of Wisconsin Center for Health Sciences Research Animal Care Committee.

Infarction Size Measurement

After 24 and 72 h reperfusion, mice were anesthetized with 5% halothane vaporized in N_2O and O_2 (3:2) and then decapitated. Brains were removed and frozen at $-80°C$ for 5 min. Two-millimeter coronal slices were made with a rodent brain matrix (Ted Pella Inc., Redding, CA). The sections were stained for 20 min at $37°C$ with 2% 2,3,5-triphenyltetrazolium chloride monohydrate. Infarction volume was calculated as described previously[2] to compensate for brain swelling in the ischemic hemisphere.

RESULTS AND DISCUSSION

Reduced Expression of Full-Length NCX1 and Activity in $NCX1^{+/-}$ Neurons

Expression of NCX1 full-length protein was examined in $NCX1^{+/+}$ and $NCX1^{+/-}$ neuronal cultures. As shown in FIGURE 1 A, a NCX1 band ($\sim$ 116 kDa) was detected in $NCX1^{+/+}$ neuronal cultures. A faint 116 kDa band was shown in $NCX1^{+/-}$ neuronal cultures. Densitometry analysis revealed that NCX1 protein was decreased by $\sim$70% in $NCX1^{+/-}$ neuronal cultures ($P < 0.05$, FIG. 1 A, B). β III tubulin was expressed at the same level in the two genotype samples. No significant difference in NKCC1 protein expression was found in $NCX1^{+/+}$ and $NCX1^{+/-}$ neuronal cultures.

We then determined whether NCX1 activity was decreased in NCX1$^{+/-}$ neuronal cultures by measuring reversed mode operation of NCX1. As seen in FIGURE 1 C, following a brief exposure to Ca^{2+}-free buffer (a), reversal of NCX in NCX1$^{+/+}$ neurons was triggered by returning the cells to the

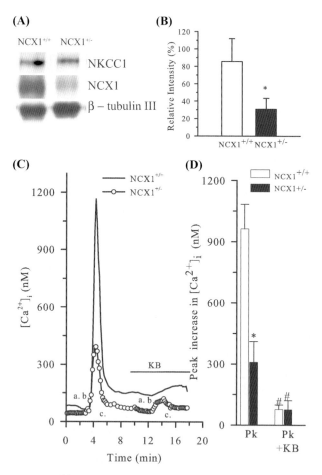

FIGURE 1. NCX1$^{+/-}$ neurons exhibit reduced NCX1 expression and activity. **(A)** Cellular lysates from NCX1$^{+/+}$ and NCX1$^{+/-}$ neuronal cultures were separated electrophoretically as described in Methods. Expression of NCX1, NKCC1, and β III tubulin proteins in DIV 11 cultures is shown on the same blot. **(B)** Summary of NCX1 and β III tubulin expression. Data were presented as a ratio of NCX1/β III tubulin intensities (means ± SD, n = 3–5). **(C)** Reverse mode operation of NCX was induced by exposing NCX1$^{+/+}$, or NCX1$^{+/-}$ neurons to Ca^{2+}-free HEPES buffer for 1 min **(a)**, and subsequently to a Na$^+$-free buffer (1.2 mM Ca^{2+}) for 30–40 s **(b)**. Cells were then returned to normal HEPES-MEM. The same protocol was repeated in the presence of 10 μM KB-R9743. **(D)** Summary of peak [Ca^{2+}]$_i$ in **C**. Data are mean ± SE, n = 3–4. *P < 0.05 versus NCX1$^{+/+}$. #P < 0.05 versus corresponding controls without KB-R.

Na$^+$- free buffer (1.2 mM Ca^{2+}, Fig. 1 C, B) resulting in a large increase in [Ca^{2+}]$_i$ (Fig. 1 C). 98% of the rise in [Ca^{2+}]$_i$ was blocked by 10 μM KB-R7943, a preferred inhibitor of NCX1 reverse exchange. In contrast, NCX1$^{+/-}$ neurons exhibited significantly reduced Ca^{2+} influx when NCX reverse exchange was activated. The peak values of [Ca^{2+}]$_i$ was reduced to 308 ± 102 nM (compared to 963 ± 121 nM in NCX1$^{+/+}$ neurons, Fig. 1 C, D). The increase in [Ca^{2+}]$_i$ in NCX1$^{+/-}$ neurons was also sensitive to KB-R7943 (Fig. 1 C, D).

Knockdown of NCX1 Expression Is Neuroprotective in Neuronal Cultures

We investigated whether reduction of NCX1 expression would provide neuronal protection following OGD/REOX. As shown in Figure 2, the cell death basal level in NCX1$^{+/+}$ neurons was ~15 ± 5%. Three hour OGD and 21 h REOX led to 78 ± 4% cell death in NCX1$^{+/+}$ neurons ($P < 0.05$, Fig. 2). Treatment of NCX1$^{+/+}$ neurons with 1 μM KB-R7943 either during the first 3 h REOX or the entire 21 h REOX reduced cell death to 65 ± 18% or 64 ± 20%, respectively ($P > 0.05$, Fig. 2). Moreover, NCX1$^{+/-}$ neurons exhibited significantly less cell death (60 ± 19%, $P < 0.05$, Fig. 2). Additional cell protection was found in NCX1$^{+/-}$ neurons when they were treated with 1 μM

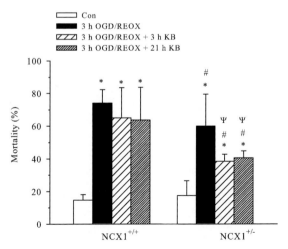

FIGURE 2. Reduction of NCX1 in neurons increases tolerance to ischemic damage. Cell mortality was assessed in neuronal cultures (DIV 12–15) after 3 h OGD and 21 h REOX. Control cultures were incubated for 24 h in normoxic control buffers for 24 h (CON). Either during the first 3 h REOX or during the entire 21 h REOX, 1 μM KB-R7943 was added. At the end of the experiment, cells were stained with PI and calcein-AM. Data are means ± SD (n = 3–8 cultures). *$P < 0.05$ versus CON; #$P < 0.05$ versus NCX1$^{+/+}$ OGD/REOX; Ψ $P < 0.002$ versus NCX1$^{+/-}$ OGD/REOX.

KB-R7943 for 3 h REOX or for 21 h REOX ($38 \pm 5\%$ or $40 \pm 4\%$, respectively, $P < 0.002$, FIG. 2).

No Significant Reduction in Brain Damage of NCX1$^{+/-}$ Mice following Focal Ischemia

We further investigated whether knockdown of NCX1 expression in NCX1$^{+/-}$ mice could lead to less ischemic damage following focal ischemia. Expression of NCX1 protein was reduced by $\sim$42% in NCX1$^{+/-}$ brains, compared to NCX1$^{+/+}$ brains ($P < 0.05$, FIG. 3 A, B). However, no significant protection in infarct volume was found in NCX1$^{+/-}$ mice, measured at 24 reperfusion following 60 min MCAO (86 ± 19 mm^3 in NCX1$^{+/+}$ mice versus 84 ± 10 mm^3 in NCX1$^{+/-}$ mice, FIG. 3 C). The data imply that if reversal of NCX1 has detrimental effects during focal ischemia, reducing NCX exchange by 50% in NCX1$^{+/-}$ brains is insufficient to block it.

Neuroprotection in siRNA-Treated NCX1$^{+/-}$ Mice or NKCC1$^{+/-}$/NCX1$^{+/-}$ Mice

Specificity of anti-NCX1 siRNA was established in NCX1$^{+/+}$ astrocytes. Lipofectamine alone or scrambled siRNA did not affect NCX1 expression in NCX1$^{+/+}$ astrocytes after 48 h transfection (FIG. 4 A). In contrast, expression of

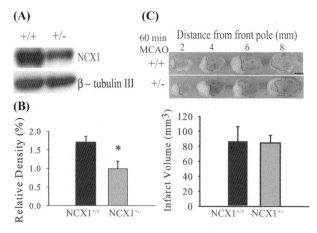

FIGURE 3. Reduced NCX1 expression in NCX1$^{+/-}$ brains. (**A**) Protein homogenates from NCX1$^{+/+}$ and NCX1$^{+/-}$ brains were used in immunoblotting. Blots were probed with anti-NCX1 or anti-β III tubulin antibodies. (**B**) Summary data were shown as a ratio of NCX1/β-tubulin III band intensity. Data are means $\pm$ SD (n = 3). *$P < 0.05$ versus NCX1$^{+/+}$. (**C**) Infarct volume in NCX1$^{+/+}$ and NCX1$^{+/-}$ mice was measured following 60 min MCAO and 24 h reperfusion (TTC staining). Data are mean $\pm$ SD (n = 3).

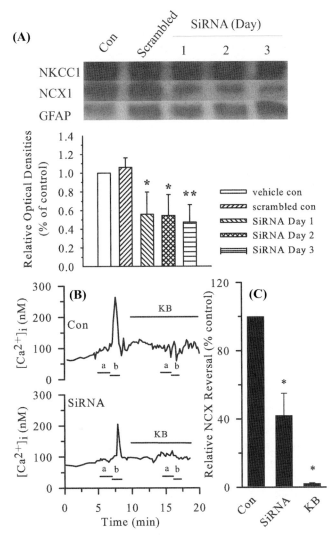

FIGURE 4. SiRNA-mediated knockdown of NCX1 expression and activity in astrocytes. Astrocytes were transfected with either scrambled siRNA or anti-NCX1 siRNA in Lipofectamine (50 nM). (**A**) Expression of NCX1, NKCC1, and GFAP proteins was detected by immunoblotting after 24–72 h transfection. Summary data were shown in a bar graph. Data are means $\pm$ SD. n = 3. $^*P < 0.05$ versus Lipofectamine (vehicle) control; $^{**}P < 0.01$. (**B**) Ca^{2+} influx was induced by reverse mode operation of NCX. NCX1$^{+/+}$ or siRNA-treated NCX1$^{+/+}$ astrocytes (24–48 h) were exposed to Ca^{2+}-free HEPES buffer for 1 min (**a**), and subsequently to a Na^+-free buffer (1.2 mM Ca^{2+}) for 30–40 s (**b**). Cells were returned to normal HEPES-MEM. The same protocol was repeated in the presence of 10 μM KB-R9743. (**C**) Summary of peak $Ca^{2+}{}_i$. Data are mean $\pm$ SE, n = 3–4. $^*P < 0.05$ versus NCX1$^{+/+}$ control.

NCX1 was reduced by $\sim$ 50% after 24, 48, or 72 h transfection with anti-NCX1 siRNA ($P < 0.05$, FIG. 4 A, B). No significant changes in expression of GFAP or NKCC1 were detected in anti-NCX1 siRNA-treated cells (FIG. 4 A, B). Moreover, knockdown of NCX1 in NCX1$^{+/+}$ astrocytes for 48 h with siRNA attenuated reversal activity of NCX by 60% (FIG. 4 B, C, $P < 0.05$). The remaining NCX-mediated Ca^{2+} influx was further inhibited by 10 μM KB-R7943.

SiRNA-mediated knockdown of NCX1 protein expression in cerebral cortex was examined in NCX1$^{+/+}$ mice. As shown in FIGURE 5 A, NCX1 protein was reduced by $\sim$40% in cerebral cortex of NCX1$^{+/+}$ mice after 24 h or 48 h injection. No significant decrease in NCX1 expression was found in NCX1$^{+/+}$ mice after 72 h injection.

To test whether knockdown of NCX1 in NCX1$^{+/-}$ mice was neuroprotective, we examined infarct volume in scrambled siRNA-control and siRNA-injected NCX1$^{+/-}$ mice following 24 h injection. There was $\sim$20% reduction in infarct volume in siRNA-treated NCX1$^{+/-}$ mice at 24 h and 72 h reperfusion, compared to NCX1$^{+/-}$ mice (FIG. 5 B, C).

We then investigated whether reduction of NKCC1 protein, a protein that causes intracellular Na^+ overload and subsequently triggers reversal of NCX1,[9] could have synergistic protective effects in NCX1$^{+/-}$ mice. Significant decrease in infarct volume was found in NKCC1$^{+/-}$/NCX1$^{+/-}$ mice at 72 h reperfusion following 30 min MCAO (FIG. 5 D). Taken together, these data suggest that NKCC1 and NCX1 play a synergistic role in ischemic damage.

DISCUSSION

In the present study, NCX1 protein expression was reduced by $\sim$70% in NCX1$^{+/-}$ neurons and by $\sim$40% in NCX1$^{+/-}$ brains. A similar level of decrease in NCX1 protein expression in cerebral cortex was achieved by the siRNA approach. The decrease in NCX1 protein expression in neurons and astrocytes was accompanied by a reduced ion exchange function as measured by Ca^{2+} influx through reversal of NCX. No significant changes in expression of other proteins such as NKCC1 and ß III tubulin were detected in NCX1$^{+/-}$ neurons and NCX1$^{+/-}$ brains. In *in vitro* and *in vivo* ischemic models, 20–30% neuroprotection was observed in NCX1$^{+/-}$ neurons or in NCX1$^{+/-}$ brains treated with anti-NCX1 siRNA.

This moderate neuroprotective effect following knockdown of NCX1 suggests that the role for NCX in cerebral ischemic damage is complex. Several studies have shown that forward mode operation of NCX and Ca^{2+} extrusion is crucial for neuronal survival following excitotoxicity or ischemic cell damage.[3,18] Knockdown of NCX3 in cereballar granule cells with siRNA or reduction of NCX1 and NCX3 expression in cerebral cortex with antisense oligodeoxynucleotides exacerbates excitotoxicity or cell damage

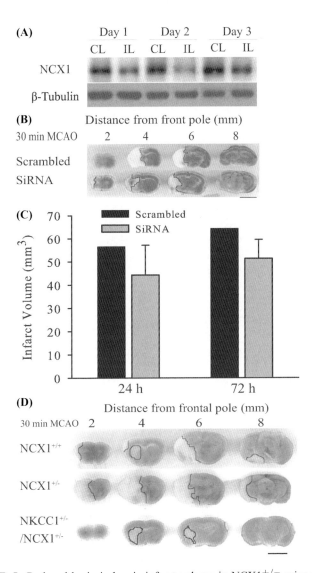

FIGURE 5. Reduced brain ischemic infarct volume in NCX1$^{+/-}$ mice treated with siRNA anti-NCX1 siRNA or scrambled control (100 μM in Lipofectamine) was injected into left hemispheres of NCX1$^{+/+}$ mice as described in Methods. Brain homogenates from contralateral (CL) or ipsilateral (IL) hemispheres were used for immunoblotting. (**A**) Expression of NCX1 and β III tubulin proteins was measured after 24–72 h siRNA injection. (**B**) Representative images of infarct volume in NCX1$^{+/-}$ mice with or without siRNA treatment after 30 min MCAO and 72 h reperfusion. (**C**) After 30 min MCAO followed by 24 or 72 h reperfusion, infarct volumes in NCX1$^{+/-}$ mice with or without siRNA injection were determined by TTC staining. Data are means ± SD. n = 1–2. Scale bar: 5 mm. (**D**) Representative images of infarct volume in NCX1$^{+/+}$, NCX1$^{+/-}$, and NKCC1$^{+/-}$/NCX1$^{+/-}$ mice after 30 min MCAO and 72 h reperfusion. Scale bar: 5 mm.

following permanent focal cerebral ischemia.[19] Therefore, one possible explanation for the moderate protection via knockdown of NCX1 is that it may affect both forward mode and reverse mode operation of NCX1 and the former effect is detrimental to neurons. However, we found that in $NCX1^{+/-}$ neurons, inhibition of NCX with KB-R7943 caused additional neuroprotection. Moreover, when we decreased NKCC1 expression in $NKCC1^{+/-}/NCX1^{+/-}$ neurons or $NKCC1^{+/-}/NCX1^{+/-}$ brains, a synergistic neuroprotective effect was found. These data favor for the explanation that NCX1 tends to operate in a reverse mode under ischemia/reperfusion conditions. Reducing NKCC1-mediated Na^+ overload can decrease Na^+-mediated toxicity and provide protection in $NKCC1^{+/-}/NCX1^{+/-}$ neurons or in $NKCC1^{+/-}/NCX1^{+/-}$ brains. But decrease of Na^+ overload could also facilitate forward mode operation of NCX1. In deed, $\sim86\%$ of NCX reversal function was reduced in $NKCC1^{+/-}/NCX1^{+/-}$ neurons. Thus, inhibition of synergistic effect between $NKCC1^{+/-}$ and $NCX1^{+/-}$ may underlie the significant neuroprotection in $NKCC1^{+/-}/NCX1^{+/-}$ neurons or in $NKCC1^{+/-}/NCX1^{+/-}$ brains.

ACKNOWLEDGMENTS

This work was supported in part by an NIH grant R01NS38118 and AHA EIA 0540154 (D. Sun).

REFERENCES

1. XIONG, Z.G. *et al.* 2004. Neuroprotection in ischemia: blocking calcium-permeable acid-sensing ion channels. Cell **118:** 687–698.
2. CHEN, H. *et al.* 2005. Na^+-dependent chloride transporter (NKCC1)-null mice exhibit less gray and white matter damage after focal cerebral ischemia. J. Cereb. Blood Flow Metab. **25:** 54–66.
3. BANO, D. *et al.* 2005. Cleavage of the plasma membrane Na^+/Ca^{2+} exchanger in excitotoxicity. Cell **120:** 275–285.
4. CZYZ, A. & L. KIEDROWSKI. 2002. In depolarized and glucose-deprived neurons, Na^+ influx reverses plasmalemmal K^+-dependent and K^+-independent Na^+/Ca^{2+} exchangers and contributes to NMDA excitotoxicity. J. Neurochem. **83:** 1321–1328.
5. ANNUNZIATO, L. *et al.* 2004. Pharmacology of brain Na^+/Ca^{2+} exchanger: from molecular biology to therapeutic perspectives. Pharmacol. Rev. **56:** 633–654.
6. LUO, J. *et al.* 2005. Decreased neuronal death in Na^+/H^+ exchanger isoform 1-null mice after *in vitro* and *in vivo* ischemia. J. Neurosci. **25:** 11256–11268.
7. SHELDON, C. *et al.* 2004. Sodium influx pathways during and after anoxia in rat hippocampal neurons. J. Neurosci. **24:** 11057–11069.
8. CHESLER, M. 2005. Failure and function of intracellular pH regulation in acute hypoxic-ischemic injury of astrocytes. Glia **50:** 398–406.

9. KINTNER, D.B. *et al.* 2006. Na^+-K^+-Cl^- cotransport and Na^+/Ca^{2+} exchange in astrocyte mitochondrial dysfunction in response to *in vitro* ischemia. Am. J. Physiol Cell Physiol. doi: 10:1152/ajpcell.00412.2006C-00412-2006R1.

10. FLAGELLA, M. *et al.* 1999. Mice lacking the basolateral Na-K-2Cl cotransporter have impaired epithelial chloride secretion and are profoundly deaf. J. Biol. Chem. **274:** 26946–26955.

11. LENART, B. *et al.* 2004. Na-K-Cl cotransporter-mediated intracellular Na^+ accumulation affects Ca^{2+} signaling in astrocytes in an *in vitro* ischemic model. J. Neurosci. **24:** 9585–9597.

12. SU, G. *et al.* 2002. Astrocytes from Na^+-K^+-Cl^- cotransporter-null mice exhibit absence of swelling and decrease in EAA release. Am. J. Physiol. Cell Physiol. **282:** C1147–C1160.

13. BECK, J. *et al.* 2003. Na-K-Cl cotransporter contributes to glutamate-mediated excitotoxicity. J. Neurosci. **23:** 5061–5068.

14. KINTNER, D.B. *et al.* 2004. Increased tolerance to oxygen and glucose deprivation in astrocytes from Na^+/H^+ exchanger isoform 1 null mice. Am. J. Physiol. Cell Physiol. **287:** C12–C21.

15. GRYNKIEWICZ, G. *et al.* 1985. A new generation of Ca^{2+} indicators with greatly improved fluorescence properties. J. Biol. Chem. **260:** 3440–3450.

16. HOYT, K.R. *et al.* 1998. Reverse Na^+/Ca^{2+} exchange contributes to glutamate-induced intracellular Ca^{2+} concentration increases in cultured rat forebrain neurons. Mol. Pharmacol. **53:** 742–749.

17. FRANKLIN, K.B.J. & G. PAXIONS. 1997. The Mouse Brain in Stereotaxic Coordinates. Acadamic Press. San Diego.

18. BOSCIA, F. *et al.* 2005. Permanent focal brain ischemia induces isoform-dependent changes in the pattern of Na(+)/Ca(2+) exchanger gene expression in the ischemic core, periinfarct area, and intact brain regions. J. Cereb. Blood Flow Metab. **26:** 502–517.

19. PIGNATARO, G. *et al.* 2004. Two sodium/calcium exchanger gene products, NCX1 and NCX3, play a major role in the development of permanent focal cerebral ischemia. Stroke **35:** 2566–2570.

Resistance of Cardiac Cells to NCX Knockout

A Model Study

DENIS NOBLE,[a] NOBUAKI SARAI,[b] PENELOPE J. NOBLE,[a]
TSUTOMU KOBAYASHI,[c] SATOSHI MATSUOKA,[b]
AND AKINORI NOMA[b]

[a]*Department of Physiology, Anatomy and Genetics, University of Oxford,
Parks Road, Oxford, OX1 3PT, United Kingdom*

[b]*Department of Physiology and Biophysics, Kyoto University, Kyoto 606-8501,
Japan*

[c]*Pharmacological Research Laboratory, Tanabe Seiyaku Co Ltd,
Osaka 335-8505, Japan*

ABSTRACT: The effects of NCX knockout were determined in a variety
of cardiac cell models. Those of the mouse and rat ventricles, and of
atrial cells in other species behave similarly to the experiments on mouse
ventricle showing only small effects, and considerable tolerance of NCX
knockout. Models of ventricular cells with high action potential plateaus,
however, are more sensitive and require compensatory mechanisms to
adjust other conductance parameters to enable the cells to resist NCX
knockout. The effects can therefore be expected to be species-specific,
and it is not possible to extrapolate the mouse results to those that may
occur in the Guinea pig or human.

KEYWORDS: modeling sodium–calcium exchange; sodium–calcium
exchange knockout; cardiac cell models

Cardiac-specific knockout of sodium–calcium exchange in mice does not result
in large changes either in action potential or intracellular calcium transient[1] (see
also papers in this volume). This is surprising because NCX is normally responsible for about 90% of calcium efflux. We have incorporated the CellML version (www.cellml.org) of the model of a mouse ventricular cell[2] into the software package, COR—*Cellular Open Resource* (http://cor.physiol.ox.ac.uk) to
show that virtually the same result is obtained in that model (FIG. 1). Compensatory mechanisms are therefore represented in this model, although it was
evident in discussion at the meeting (Christian Pott, personal communication)

Address for correspondence: Denis Noble, Department of Physiology, Anatomy and Genetics,
University of Oxford, Parks Road, Oxford, OX1 3PT, UK. Voice: 01865 272528; fax: 01865 272554.
Denis.noble@physiol.ox.ac.uk

Ann. N.Y. Acad. Sci. 1099: 306–309 (2007). © 2007 New York Academy of Sciences.
doi: 10.1196/annals.1387.018

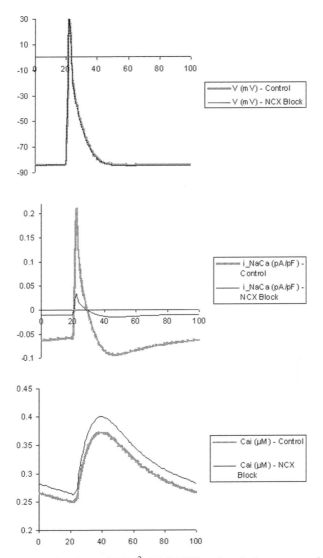

FIGURE 1. Bondarenko *et al.* 2004[2]: 85% NCX knockout in the mouse model showing negligible change in action potential and similar calcium transients.

that the model will need further development both to represent correctly the relative roles of NCX and PMCA, and because the dyadic space modeling is inadequate.

It is interesting to determine the extent to which models of ventricular and atrial cells in other species are resistant to NCX knockout. Studies using the Kyoto Guinea pig ventricular cell model have already shown that the model is less resistant than the mouse model, but that it can be "rescued" (i.e., be almost normally functional) by upregulation of PMCA and downregulation of I_{CaL}.[3]

These studies were extended to include the Oxford Guinea pig cell models,[4,5] rabbit and human atrial cell models,[6,7] rat ventricular cells,[8] and a human ventricular cell model.[9] Except for the rat ventricular cell model, which behaves like the mouse model, the results confirm the work of Sarai et al.[3] in showing that models of species showing high "square" plateau action potential require varying degrees of compensation by up- or downregulation of other conductance mechanisms to protect the model from NCX knockout (FIG. 2). The atrial cell models are intermediate. The late low-level plateau in this case is maintained by sodium–calcium exchange current so it is not surprising that the late plateau is abolished by NCX knockout. This was also shown to be the case in this meeting for experiments on mouse ventricular action potentials when they show a late plateau (see paper by Philipson's group at this meeting).

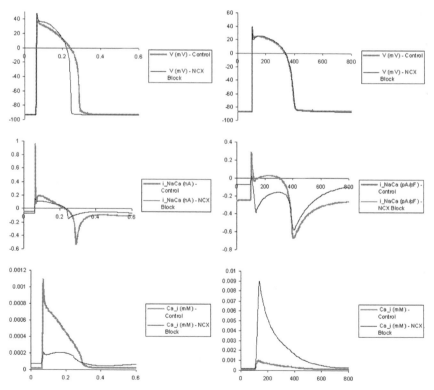

FIGURE 2. Noble et al. 1998[5] Guinea pig (left) and TenTusscher and Panfilov, 2006[9] human (right) ventricular cell models: 85% NCX knockout in the Guinea pig model causes a reduction in APD and a fall in the calcium transient. The same knockout in the human model results in little change in action potential but a large increase in the calcium transient.

Our conclusion therefore is that the results of NCX knockout cannot be extended to other species. If the models are correct, then the effect of NCX knockout will be species-specific. From a comparison of models and species with different action potential shapes, this difference is largely correlated with the high "square" plateau. Cells with short action potentials and very low-late plateaus, like the mouse and rat, and atrial action potentials in other species can be expected to behave like the mouse ventricle. It is therefore important also to investigate NCX knockout in species other than mouse.

ACKNOWLEDGMENTS

This work was supported by the EU BioSim Consortium, The Wellcome Trust, the Leading Project for Biosimulation, and a Grant-in-Aid for Scientific Research from the Ministery of Education, Culture, Sports, Science, and Technology of Japan.

REFERENCES

1. HENDERSON, S.A., J.I. GOLDHABER, J.M. SO, *et al.* 2004. Functional adult myocardium in the absence of Na^+-Ca^{2+} exchange: cardiac-specific knockout of NCX1. Circ. Res. **95:** 604–611.
2. BONDARENKO, V.E., G.P. SZIGETI, G.C. BETT, *et al.* 2004. A computer model for the action potential of mouse ventricular myocytes. Am. J. Physiol. **287:** H1378–H1403.
3. SARAI, N., T. KOBAYASHI, S. MATSUOKA, *et al.* 2006. A simulation study to rescue the Na^+-Ca^{2+} exchanger knockout mice. J. Physiol. Sci. **56:** 211–217.
4. NOBLE, D., S.J. NOBLE, G.C.L. BETT, *et al.* 1991. The role of sodium-calcium exchange during the cardiac action potential. Ann. N. Y. Acad. Sci. **639:** 334–353.
5. NOBLE, D., A. VARGHESE, P. KOHL, *et al.* 1998. Improved guinea-pig ventricular cell model incorporating a diadic space, iKr & iKs, and length- & tension-dependent processes. Can. J. Cardiol. **14:** 123–134.
6. EARM, Y.E. & D. NOBLE. 1990. A model of the single atrial cell: relation between calcium current and calcium release. Proc. R. Soc. B. **240:** 83–96.
7. NYGREN, A., L.J. LEON & W.R. GILES. 2001. Simulations of the human atrial action potential. Phil. Trans. R. Soc. Lond. A. **359:** 1111–1125.
8. PANDIT, S.V., R.B. CLARK, W.R. GILES, *et al.* 2001. A mathematical model of action potential heterogeneity in adult rat left ventricular myocytes. Biophys. J. **81:** 3029–3051.
9. TENTUSSCHER, K.H.W.J. & A.V. PANFILOV. 2006. Alternans and spiral breakup in a human ventricular tissue model. Am. J. Physiol. **291:** H1088–H1100.

Homozygous Overexpression of the Na^+-Ca^{2+} Exchanger in Mice

Evidence for Increased Transsarcolemmal Ca^{2+} Fluxes

CHRISTIAN POTT,[a,b] JOSHUA I. GOLDHABER,[a]
AND KENNETH D. PHILIPSON[a]

[a]Departments of Physiology and Medicine and the Cardiovascular Research Laboratories, David Geffen School of Medicine at UCLA, Los Angeles, California 90095-1760, USA

[b]Clinic C for Internal Medicine, Department of Cardiology and Angiology, University Hospital Muenster, 48149 Münster, Germany

ABSTRACT: Mice with homozygous overexpression of the Na^+-Ca^{2+} exchanger (NCX) exhibit threefold levels of NCX expression and an increased Ca^{2+} extrusion rate. To investigate how Ca^{2+} homeostasis is maintained in this model, we have characterized Ca^{2+} influx under these conditions. We find that L-type Ca^{2+} currents (I_{Ca}) inactivate slower due to a reduction of Ca^{2+}-dependent inactivation. Additionally, NCX-overexpressing animals exhibit a prolongation of the action potential (AP). We conclude that transsarcolemmal Ca^{2+} fluxes in NCX-overexpressing myocytes are balanced by an increase in Ca^{2+} influx via (a) slowed inactivation of I_{Ca} and (b) a prolongation of the AP to compensate for increased Ca^{2+} efflux.

KEYWORDS: Na^+-Ca^{2+} exchanger; L-type Ca^{2+} channel; excitation–contraction coupling; transgenic mice; cardiac myocytes

INTRODUCTION

The Na^+-Ca^{2+} exchanger (NCX) is the main Ca^{2+} efflux mechanism in cardiac myocytes.[1,2] Homozygous overexpression of NCX in transgenic mice (HOM) leads to a threefold increase in protein expression of NCX resulting in a substantial acceleration of Ca^{2+} extrusion capacity.[3] Possibly, accelerated Ca^{2+} extrusion is balanced by an increase in Ca^{2+} influx. Here, we follow-up on

Address for correspondence: Kenneth D. Philipson, Cardiovascular Research Laboratories, MRL 3645, David Geffen School of Medicine at UCLA, 675 Charles E. Young Dr. South, Los Angeles, CA 90095. Voice: 310-825-7679; fax: 310-206-5777.
kphilipson@mednet.ucla.edu

Ann. N.Y. Acad. Sci. 1099: 310–314 (2007). © 2007 New York Academy of Sciences.
doi: 10.1196/annals.1387.019

this hypothesis and investigate the underlying mechanisms by characterizing Ca^{2+} influx under these conditions.

METHODS

We used the whole cell patch-clamp technique in cardiac myocytes isolated from wild-type (WT) and HOM mice as described previously.[4,5] The pipette solution contained (in mmol/L): 120 CsCl, 10 TEA-Cl, 10 NaCl, 20 HEPES, 5 MgATP, 0.05 cAMP, pH 7.2 with CsOH. The bath solution contained (in mmol/L): 136 NaCl, 5.4 KCl, 10 HEPES, 1.0 $MgCl_2$, 0.33 NaH_2PO_4, 1.0 $CaCl_2$, 10 glucose, pH 7.4 with NaOH with modifications described below. To record L-type Ca^{2+} currents, cells were held at –75 mV and then briefly depolarized to –40 mV to inactivate Na^+ currents. I_{Ca} was then measured during a 200 ms depolarization to 0 mV. Prepulses were given to ensure steady-state Ca^{2+} load of the sarcoplasmic reticulum (SR). For action potential (AP) recordings, the pipette solution contained (in mmol/L): KCl 110, MgCl 5, NaCl 10 and HEPES 20, pH 7.2 with KOH.

RESULTS AND DISCUSSION

Ca^{2+}-Dependent Inactivation of I_{Ca} Is Decreased in HOM

To investigate Ca^{2+} influx, we measured the L-type Ca^{2+} current (I_{Ca}) in HOM and WT myocytes. Inactivation of I_{Ca} was slowed in HOM ($\tau = 44.5 \pm 2.3$ ms; $n = 18$) when compared to WT ($\tau = 33.5 \pm 1.7$ ms; $n = 14$; $P < 0.01$; FIG. 1 A, B), which is consistent with a decrease of Ca^{2+}-dependent inactivation of I_{Ca}. To deplete cytosolic Ca^{2+}, we replaced Ca^{2+} in the bath with Ba^{2+} (3 mM), which then served as the charge carrier, and included ryanodine (0.2 μM, 10 min) and thapsigargin (10 μM, 10 min) to deplete SR Ca^{2+}. Under these conditions, the differences in I_{Ca} inactivation between WT ($\tau = 85.3 \pm 8.6$ ms; $n = 9$) and HOM ($\tau = 95.8 \pm 5.3$ ms; $n = 13$; $P > 0.05$; FIG. 1 A, B) were eliminated. Similarly, when myocytes were dialyzed with the fast Ca^{2+} chelator BAPTA (10 mM) via the patch pipette, no difference in inactivation rate was observed between WT ($\tau = 54.4 \pm 1.8$ ms; $n = 5$) and HOM ($\tau = 49.3 \pm 5.7$ ms; $n = 3$; $P > 0.05$; FIG. 2). These observations imply that Ca^{2+}-dependent inactivation of I_{Ca} is responsible for the differences in I_{Ca} between WT and HOM. We conclude that in NCX overexpressor mice, I_{Ca} inactivation kinetics are slowed because of reduced Ca^{2+}-dependent inactivation, which is most likely the result of decreased diadic cleft Ca^{2+} due to accelerated removal by increased NCX activity.

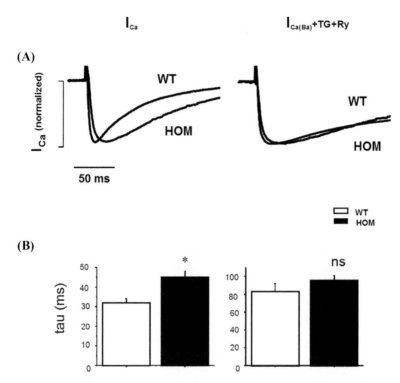

FIGURE 1. Ca^{2+}-dependent inactivation of I_{Ca} is decreased in HOM myocytes. (**A**) A slowed inactivation of I_{Ca} is observed in HOM (*left*). This difference is eliminated when SR Ca^{2+} is depleted by thapsigargin and ryanodine and Ba^{2+} is used as a charge carrier (*right*). Traces shown are normalized in peak I_{Ca}. (**B**) Quantification of experiments demonstrated in A.

Prolonged Action Potential Duration in HOM

APs were recorded in HOM ($n = 20$) and WT ($n = 19$) myocytes during steady-state pacing (FIG. 3). The resting potential (WT: 68.4 ± 2.6 mV; HOM: 69.8 ± 0.7 mV; $P > 0.05$) and amplitude (WT: 111.4 ± 2.6 mV; HOM: 110.5 ± 2.2; mV; $P > 0.05$) were similar but the late repolarization phase was significantly delayed in HOM (time to 90% repolarization of the AP (APD$_{90}$): WT: 88.2 ± 10.4 mV; HOM: 159.5 ± 18.7 ms; $P < 0.01$), which may result in prolonged Ca^{2+} influx during the AP.

CONCLUSIONS

Our observations suggest that Ca^{2+} influx is increased in HOM mice via two mechanisms: (*a*) slowed inactivation of I_{Ca} and (*b*) prolonged activity of I_{Ca}

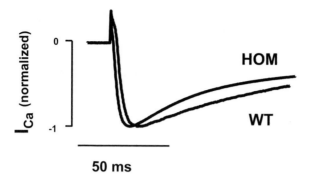

$$I_{Ca} + BAPTA$$

FIGURE 2. Effects of Ca^{2+} buffering on I_{Ca}. When myocytes are dialyzed with the fast Ca^{2+} chelator BAPTA, inactivation rates are similar in HOM and WT. Traces shown are normalized to peak I_{Ca}.

expected during the longer AP in HOM. These two mechanisms independently increase Ca^{2+} influx to compensate for the additional Ca^{2+} efflux caused by the increased NCX activity. This may represent an important regulatory mechanism allowing the cardiomyocyte to increase Ca^{2+} influx in situations characterized by increased Ca^{2+} efflux.

ACKNOWLEDGMENTS

The authors appreciate helpful discussions with Dr. R. Olcese and Dr. L. Xie. This research was supported by the German Research Foundation (DFG

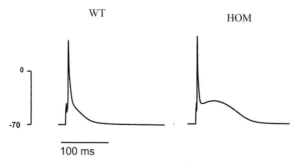

FIGURE 3. AP measurements. Resting potential and amplitude are similar in HOM and WT but the plateau phase is more pronounced in HOM. This results in a significant prolongation of the APD_{90}.

PO 1004/1-1 and PO 1004/1-2) (C.P.), National Institutes of Health grants HL70828 (J.I.G.) and HL48509 (K.D.P.), and the Laubisch Foundation.

REFERENCES

1. BERS, D.M. 2002. Cardiac excitation-contraction coupling. Nature **415:** 198–205.
2. PHILIPSON, K.D. & D.A. NICOLL. 2000. Sodium-calcium exchange: a molecular perspective. Annu. Rev. Physiol. **62:** 111–133.
3. REUTER, H., T. Han, C. MOTTER, *et al.* 2004. Mice overexpressing the cardiac sodium-calcium exchanger: defects in excitation-contraction coupling. J. Physiol. **554:** 779–789.
4. POTT, C., K.D. PHILIPSON & J.I. GOLDHABER. 2005. Excitation-contraction coupling in Na+-Ca2+ exchanger knockout mice: reduced transsarcolemmal Ca^{2+} flux. Circ. Res. **97:** 1288–1295.
5. HENDERSON, S.A., J.I. GOLDHABER, J.M. SO, *et al.* 2004. Functional adult myocardium in the absence of Na^{+}-Ca^{2+} exchange: cardiac-specific knockout of NCX1. Circ. Res. **95:** 604–611.

Na/Ca Exchange

Regulator of Intracellular Calcium and Source of Arrhythmias in the Heart

L. A. VENETUCCI, A. W. TRAFFORD, S. C. O'NEILL, AND D. A. EISNER

Unit of Cardiac Physiology, University of Manchester, 3.18 Core Technology Facility, Manchester M13 9NT, United Kingdom

ABSTRACT: The major effect of Na/Ca exchange (NCX) on the systolic Ca transient is secondary to its effect on the Ca content of the sarcoplasmic reticulum (SR). SR Ca content is controlled by a mechanism in which an increase of SR Ca produces an increase in the amplitude of the systolic Ca transient. This, in turn, increases Ca efflux on NCX as well as decreasing entry on the L-type current resulting in a decrease of both cell and SR Ca content. This control mechanism also changes the response to other maneuvers that affect excitation–contraction coupling. For example, potentiating the opening of the SR Ca release channel (ryanodine receptor, RyR) with caffeine produces an immediate increase in the amplitude of the systolic Ca transient. However, this increases efflux of Ca from the cell on NCX and then decreases SR Ca content until a new steady state is reached. Changing the activity of NCX (by decreasing external Na) changes the level of SR Ca reached by this mechanism. If the cell and SR are overloaded with Ca then Ca waves appear during diastole. These waves activate the electrogenic NCX and thereby produce arrhythmogenic-delayed afterdepolarizations. A major challenge is how to remove this arrhythmogenic Ca release *without* compromising the normal systolic release. We have found that application of tetracaine to decrease RyR opening can abolish diastolic release while simultaneously potentiating the systolic release.

KEYWORDS: NCX; cardiac muscle; sarcoplasmic reticulum

There are two sources of the Ca that activates contraction in cardiac muscle: influx of Ca from the extracellular fluid, largely via the L-type Ca current and release from the sarcoplasmic reticulum (SR) through a Ca release channel known as the ryanodine receptor (RyR); for reviews see References 1 and 2.

Address for correspondence: D. A. Eisner, Unit of Cardiac Physiology, University of Manchester, 3.18 Core Technology Facility, 46 Grafton Street, Manchester M13 9NT, UK. Voice: +44-161-275-2702; fax; +44-161-275-2703.

eisner@man.ac.uk

Ann. N.Y. Acad. Sci. 1099: 315–325 (2007). © 2007 New York Academy of Sciences.
doi: 10.1196/annals.1387.033

Under normal conditions the amount of Ca that is released from the SR is much more than that entering via the L-type current. Relaxation occurs by $[Ca^{2+}]_i$ being lowered to resting levels via a combination of extrusion from the cell largely on Na/Ca exchange (NCX) (with a small contribution from the plasma membrane Ca-ATPase, PMCA) and being taken back up into the SR via the SR Ca-ATPase (SERCA). In the steady state the amount of Ca pumped out of the cell must equal that which entered and the amount taken back into the SR by SERCA must equal that released via the RyR.[2,3] It therefore follows that the activity of SERCA must be considerably greater than that of NCX. This has the consequence that SERCA rather than NCX activity determines the kinetics of individual Ca transients.[4-6]

The major effect of NCX on systolic Ca is therefore an indirect one. The activity of NCX is one of the factors that determine the Ca content of the cell and therefore the SR. Impairing the ability of NCX to pump Ca out of the cell therefore leads to an increase of SR Ca content and the amplitude of the Ca transient.[7-9]

How Does NCX Regulate SR Ca Content?

At first sight a sarcolemmal transporter, such as NCX, might not seem to be the most obvious mechanism involved in regulating SR Ca content. That this does work depends upon the following facts. (1) An increase of SR Ca content increases the amplitude of the systolic Ca transient.[10-12] (2) An increase of $[Ca^{2+}]_i$ increases the efflux of Ca from the cell on NCX.[13] (3) This increased Ca efflux from the cell will result in a decrease of SR Ca content. Some of the evidence supporting this scheme is shown in FIGURE 1.[12] The SR had initially been emptied by application of caffeine. Caffeine was then removed just before the period shown in FIGURE 1 A. Since the SR was empty the Ca transients produced by depolarization are very small. As the SR gradually refills with Ca the amplitude of the systolic Ca transient increases. Specimen membrane current records are shown in FIGURE 1 B. The right hand record (transient b) was obtained in the steady state when the SR had refilled with Ca. The current record shows an L-type Ca current on depolarization (Ca influx) and NCX current (Ca efflux) on repolarization. (The NCX current can be seen more easily in the expanded records of FIG. 1 C). The lower panel of FIGURE 1 B shows the change of cellular Ca calculated from the L-type Ca current and NCX fluxes. It is clear that (as predicted above) the Ca entry on the L-type current is equal to the efflux on NCX. A very different picture is seen for the first pulse in the experiment (transient a). Here the influx of Ca on the L-type current is increased (largely as a result of decreased Ca-dependent inactivation of this current). The efflux on NCX is decreased as a direct result of the decrease of systolic $[Ca^{2+}]_i$ (see FIG. 1 C). The net result as shown in the bottom panel of FIGURE 1 B is that for this pulse Ca influx is larger than efflux and the cell

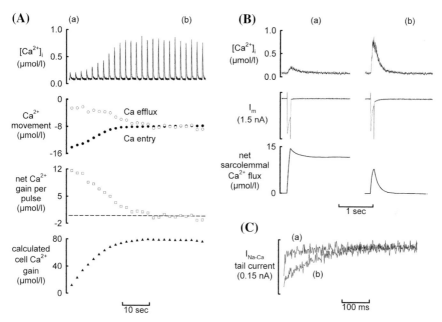

FIGURE 1. Changes of SR Ca content affect sarcolemmal Ca fluxes. (**A**) Time course of experiment. Traces show (from top to bottom): [Ca^{2+}]$_i$; calculated Ca entry (on L-type Ca current) and efflux (on NCX); net Ca gain per pulse; cumulative Ca gain. The cell had been exposed to 10 mM caffeine to empty the SR and caffeine was removed just before the period shown. (**B**) Specimen records. Traces show (from top to bottom): [Ca^{2+}]$_i$; membrane current; calculated cellular gain of Ca (obtained by integrating Ca entry and efflux). Traces a and b were obtained as the times indicated on **A**. (**C**) NCX currents on repolarization shown on an expanded scale. (Modified from Ref. 12.)

gains Ca. Calculations of influx and efflux for all the pulses in this experiment are shown in the lower panels of FIGURE 1 A. It is clear that as the amplitude of the systolic Ca transient increases there is a decrease in the entry of Ca via the L-type Ca current and an increase in the efflux (on NCX). Eventually, in the steady state, influx and efflux are equal. These changes of fluxes result in the calculated increase of SR Ca content shown in the bottom panel of FIGURE 1 A.

In the experiment of FIGURE 1 the decrease of SR Ca content results in a decreased systolic Ca transient. This produces an increase of Ca influx and a decrease of efflux until SR Ca content is restored. This mechanism for controlling SR content is important in many other situations. The data of FIGURE 2 show the effect of potentiating the opening of the RyR.[14–16] In this experiment a low concentration of caffeine was added to increase the open probability of the RyR. This produces an immediate increase of the systolic Ca transient as expected from such a maneuver. However, the amplitude of the systolic Ca transient quickly returns to control levels. The explanation of this result

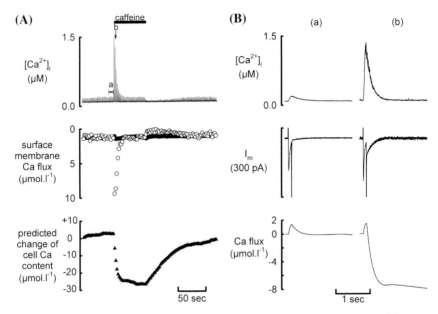

FIGURE 2. Potentiation of the RyR produces only a transient change of $[Ca^{2+}]_i$. The cell was voltage clamped (holding potential –40 mV), and 100-ms duration step depolarizations were applied to 0 mV at 0.5 Hz. (**A**) Time course. Caffeine (500 μM) was applied for the time indicated by the bar above the record. Traces show (from top to bottom): $[Ca^{2+}]_i$; pulse-by-pulse analysis of membrane Ca fluxes. Ca influx (•) and efflux (○) were obtained, respectively, by integrating the L-type Ca current during depolarization and the NCX exchange tail current on repolarization; predicted changes of SR Ca content. The differences between Ca influx and efflux on each pulse were summed to obtain a cumulative net change in cell Ca content. (**B**) Specimen records obtained before (a) and at the peak (b) of the caffeine response. Traces show (from top to bottom): $[Ca^{2+}]_i$; current; calculated Ca fluxes. (Taken from Reference 36.)

is provided by the traces of FIGURE 2 B. Trace a, obtained following steady-state stimulation before adding caffeine, shows a small Ca transient (top). This is accompanied by an inward L-type Ca current during depolarization and a NCX "tail" current on repolarization. The integrated traces (bottom) show that the calculated Ca influx during depolarization is equal to the efflux on repolarization. In contrast, trace b corresponds to the peak of the response to caffeine. The increase in the amplitude of the systolic Ca transient results in an increase of Ca efflux on NCX. This, in turn, produces a decrease of SR Ca content (FIG. 2 A). The decrease of SR content decreases the amplitude of the next Ca transient. This process carries on until the SR content has decreased to a level at which the Ca transient amplitude is the same as control. Only under these conditions will the Ca efflux on NCX be the same as in control. This experiment therefore shows that simply increasing the open probability of the RyR has no effect on the amplitude of the systolic Ca transient. The general conclusion from this result is that in the steady state the Ca efflux from the cell must equal the influx. If (as is the case in this experiment) influx is constant then the efflux must also be constant. If the properties of NCX are unaffected then the requirement of a constant efflux is satisfied by a constant amplitude of the systolic Ca transient. In other experiments we have found similar results when using 2,3-butanedione monoxime (BDM) to increase the open probability of the RyR.[17] Correspondingly, maneuvers, such as application of the local anesthetic tetracaine or intracellular acidification produce transient decreases of the amplitude of the Ca transient with no steady-state effect.[18,19]

It is useful to take the opportunity to discuss the above results in a more general context. First, the constancy of the Ca transient will only hold over a certain range. Thus it is well known that high concentrations of caffeine result in a large decrease of the amplitude of the systolic Ca transient.[20,21] Presumably the increase of RyR open probability decreases the SR Ca content to such a level that even if 100% of the SR Ca is released, the Ca transient will be smaller than control. The question then arises as to how the requirement (reviewed above) for a constant Ca efflux can be satisfied? This occurs because the Ca transient decays more slowly and therefore there is more time for Ca to be pumped out of the cell. Second, it should be noted that anything that changes the properties of NCX would be expected to change the amplitude of the systolic Ca transient. For example, if 50% of NCX is inhibited then a given-sized Ca transient will result in less efflux of Ca from the cell during the systolic Ca transient. This will result in an increase of cell and therefore SR Ca and thence an increase of the amplitude of the systolic Ca transient until the combination of increased Ca transient and depressed NCX activity results in a new steady state where influx and efflux are again in balance.

An experimental example of the effects of modulating NCX is shown in FIGURE 3. Here the external Na concentration was reduced to 60 mM.[9] Consistent with previous work[22] this reduction of the sarcolemmal Na gradient increased the amplitude of the systolic Ca transient (FIG. 3 A). The membrane current trace shows an outward shift of membrane current due to increased "reverse mode" NCX. It is this calculated Ca entry that is responsible for the increase of SR Ca content (not shown but see original paper[9]) and that of the systolic Ca transient amplitude. Another interesting question concerns how well NCX can remove Ca from the cell under these conditions. An estimate of this can be obtained by plotting NCX current as a function of $[Ca^{2+}]_i$. FIGURE 3 B (a) shows the expected linear dependence of current on $[Ca^{2+}]_i$.[13] FIGURE 3 B (b) was obtained immediately after reducing external Na and shows a flattening of the slope indicating that NCX is less effective at removing Ca from the cell. Panel c was obtained after a longer period in reduced external Na and shows that the slope has recovered toward control level. We attribute the recovery of the slope to a decrease of intracellular Na concentration.

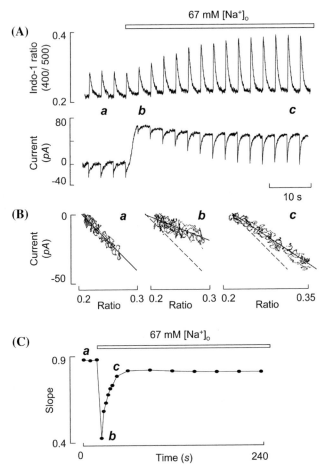

FIGURE 3. Transient effects of Na removal on NCX and systolic $[Ca^{2+}]_i$. (**A**) Time course of the experiment. Traces show: top, $[Ca^{2+}]_i$; bottom, current. For clarity, the capacity currents and the currents during depolarization have been removed so that the NCX exchange current is the only visible time-dependent current. (**B**) Plots of the NCX exchange tail current as a function of Indo 1 ratio for the three pulses identified in **A**. The continuous lines are linear regressions to the data. The dashed lines in b and c show the regression line from control (a) for comparison. (**C**) Plot of the slope of currents versus ratio as a function of duration of exposure to reduced Na solution. (Taken from Reference 9.)

Na/Ca Exchange and Abnormal Activity

It has been known for many years that Ca-overloaded conditions result in abnormal pacemaker activity.[23,24] It is now established that this arises because an increase of SR Ca content results in spontaneous waves of Ca-induced Ca release.[25] These waves activate the electrogenic NCX[26] thereby producing delayed afterdepolarizations (DAD). NCX therefore has two roles in the generation of DAD[27]: (*a*) it controls the level of cell and then SR Ca and (*b*) it produces the depolarizing current. This dual role will therefore be expected to have consequences for the effects of NCX inhibitors on arrhythmias. Such a decrease of NCX may increase SR content and thereby increase the probability of spontaneous Ca waves occurring. Set against this, however, is the fact that the depolarizing current produced by any waves will be reduced. Another important issue concerns the therapeutic management of such arrhythmias. This is a particularly difficult issue since the abnormal Ca release occurs through the same mechanism as the normal systolic release and at first sight it is not obvious how one could inhibit the arrhythmogenic Ca release without at the same time also stopping the normal cardiac contraction. One solution to this problem was suggested by experiments in which we have investigated the effects of tetracaine, a drug known to decrease the opening probability of the RyR[28,29] and thereby decrease Ca release from the SR.[18,30] FIGURE 4 shows Ca transients produced by electrical stimulation. The application of isoproterenol (middle panel) shows the well-known increase of the amplitude of the systolic Ca transient. This is also accompanied by the occurrence of Ca release during diastole. We then applied the local anesthetic tetracaine at either 25 (top) or 50 (bottom) μM. This resulted in either a decrease in frequency (top) or abolition (bottom) of the diastolic Ca release.[31] Most interestingly, there was an *increase* in the amplitude of the systolic Ca transient. We attribute this increase

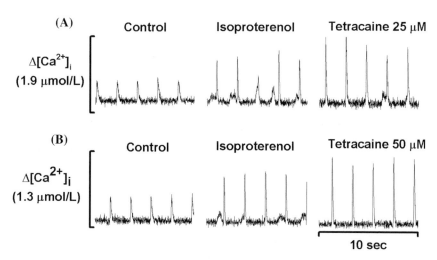

FIGURE 4. Effects of tetracaine on Ca waves. In both panels the traces show $[Ca^{2+}]_i$ in response to voltage clamp depolarizations applied from a holding potential of -40 mV to 0 mV at 0.5 Hz. From left to right the traces were obtained in the following conditions: control; isoproterenol (1 μM); isoproterenol plus the indicated tetracaine concentration. The tetracaine concentration was 25 μM in **A** and 50 μM in **B**. (Taken from Reference 31.)

to the fact that a diastolic Ca release decreases the amplitude of a subsequent systolic release.[31–33] This may result from a combination of various effects including decreased L-type Ca current, depletion of SR Ca content, and inactivation of the RyR. Irrespective of the mechanism, the important point in the current context is that it is possible to remove arrhythmogenic Ca waves *without* interfering with systolic Ca release. The outcome of these effects on NCX and other sarcolemmal Ca fluxes are shown in FIGURE 5. Recent work [34] has shown that the compound JTV-519 that prevents dissociation of accessory protein FKBP12.6 from the RyR is also antiarrhythmic[34,35] suggesting that targeting the RyR may be a productive antiarrhythmic strategy.

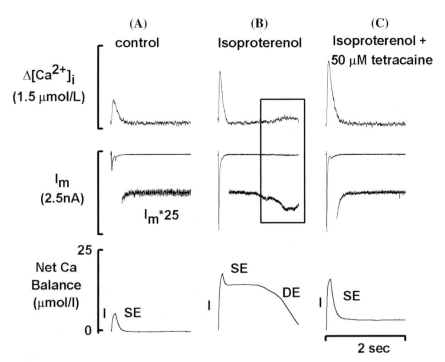

FIGURE 5. Effects of tetracaine on sarcolemmal Ca fluxes and Ca balance. Traces show (from top to bottom): $[Ca^{2+}]_i$; membrane current (NB the current on repolarization is also shown amplified $\times$ 25); calculated Ca flux (upward represents net Ca uptake into cell). The various components identified are: influx (I); Ca efflux associated with the systolic Ca transient (SE); and Ca efflux mediated by diastolic Ca release (DE). These were calculated by integration of the membrane currents. Systolic efflux was assumed to end when the level of $[Ca^{2+}]_i$ had returned to within 5% of the lowest level reached in diastole and the calculation of diastolic efflux then began. The panels show: (**A**), control; (**B**), isoproterenol (1 μM); (**C**), isoproterenol + tetracaine (50 μM). The box in **B** highlights the diastolic Ca release and accompanying transient inward current. (Taken from Reference 31.)

REFERENCES

1. BERS, D.M. 2001. Excitation-Contraction Coupling and Cardiac Contractile Force. Kluwer Academic Publishers. Dordrecht/Boston/London.
2. EISNER, D.A., H.S. CHOI, M.E. DÍAZ, *et al.* 2000. Integrative analysis of calcium cycling in cardiac muscle. Circ. Res. **87:** 1087–1094.
3. EISNER, D.A., A.W. TRAFFORD, M.E. DÍAZ, *et al.* 1998. The control of Ca release from the cardiac sarcoplasmic reticulum: regulation versus autoregulation. Cardiovasc. Res. **38:** 589–604.
4. NEGRETTI, N., S.C. O'NEILL & D.A. EISNER. 1993. The relative contributions of different intracellular and sarcolemmal systems to relaxation in rat ventricular myocytes. Cardiovasc. Res. **27:** 1826–1830.
5. CHOI, H.S. & D.A. EISNER. 1999. The role of the sarcolemmal Ca-ATPase in the regulation of resting calcium concentration in rat ventricular myocytes. J. Physiol. (Lond.) **515:** 109–118.
6. BERS, D.M., J.W.M. BASSANI & R.A. BASSANI. 1993. Competition and redistribution among calcium transport systems in rabbit cardiac myocytes. Cardiovasc. Res. **27:** 1772–1777.
7. SMITH, G.L., M. VALDEOLMILLOS, D.A. EISNER & D.G. ALLEN. 1988. Effects of rapid application of caffeine on intracellular calcium concentration in ferret papillary muscles. J. Gen. Physiol. **92:** 351–368.
8. BERS, D.M. 1987. Mechanisms contributing to the cardiac inotropic effect of Na pump inhibition and reduction of extracellular Na. J. Gen. Physiol. **90:** 479–504.
9. MEME, W., S.C. O'NEILL & D.A. EISNER. 2001. Low sodium inotropy is accompanied by diastolic Ca^{2+} gain and systolic loss in isolated guinea-pig ventricular myocytes. J. Physiol. (Lond.) **530:** 487–495.
10. JANCZEWSKI, A.M., H.A. SPURGEON, M.D. STERN & E.G. LAKATTA. 1995. Effects of sarcoplasmic reticulum Ca^{2+} load on the gain function of Ca^{2+} release by Ca^{2+} current in cardiac cells. Am. J. Physiol. **268:** H916–H920.
11. BASSANI, J.W.M., W. YUAN & D.M. BERS. 1995. Fractional SR Ca release is regulated by trigger Ca and SR Ca content in cardiac myocytes. Am. J. Physiol. **268:** C1313–C1329.
12. TRAFFORD, A.W., M.E. DÍAZ, N. NEGRETTI & D.A. EISNER. 1997. Enhanced calcium current and decreased calcium efflux restore sarcoplasmic reticulum Ca content following depletion. Circ. Res. **81:** 477–484.
13. BARCENAS-RUIZ, L., D.J. BEUCKELMANN & W.G. WIER. 1987. Sodium-calcium exchange in heart: membrane currents and changes in $[Ca^{2+}]_i$. Science **238:** 1720–1722.
14. O'NEILL, S.C. & D.A. EISNER. 1990. A mechanism for the effects of caffeine on Ca^{2+} release during diastole and systole in isolated rat ventricular myocytes. J. Physiol. (Lond.) **430:** 519–536.
15. TRAFFORD, A.W., M.E. DÍAZ & D.A. EISNER. 1998. Stimulation of Ca-induced Ca release only transiently increases the systolic Ca transient: measurements of Ca fluxes and s.r. Ca. Cardiovasc. Res. **37:** 710–717.
16. TRAFFORD, A.W., M.E. DÍAZ, G.C. SIBBRING & D.A. EISNER. 2000. Modulation of CICR has no maintained effect on systolic Ca^{2+}: simultaneous measurements of sarcoplasmic reticulum and sarcolemmal Ca^{2+} fluxes in rat ventricular myocytes. J. Physiol. (Lond.) **522:** 259–270.

17. ADAMS, W.A., A.W. TRAFFORD & D.A. EISNER. 1998. 2,3-butanedione monoxime (BDM) decreases sarcoplasmic reticulum Ca content by stimulating Ca release in isolated rat ventricular myocytes. Pflügers Arch. **436:** 776–781.
18. OVEREND, C.L., S.C. O'NEILL & D.A. EISNER. 1998. The effect of tetracaine on stimulated contractions, sarcoplasmic reticulum Ca^{2+} content and membrane current in isolated rat ventricular myocytes. J. Physiol. (Lond.) **507:** 759–769.
19. CHOI, H.S., A.W. TRAFFORD, C.H. ORCHARD & D.A. EISNER. 2000. The effect of acidosis on systolic Ca and sarcoplasmic reticulum Ca content in isolated rat ventricular myocytes. J. Physiol. (Lond.) **529:** 661–668.
20. ALLEN, D.G. & S. KURIHARA. 1980. Calcium transients in mammalian ventricular muscle. Eur. Heart J. **1:** 5–15.
21. NEGRETTI, N., S.C. O'NEILL & D.A. EISNER. 1993. The effects of inhibitors of sarcoplasmic reticulum function on the systolic Ca^{2+} transient in rat ventricular myocytes. J. Physiol. (Lond.) **468:** 35–52.
22. ALLEN, D.G., D.A. EISNER, M.J. LAB & C.H. ORCHARD. 1983. The effects of low sodium solutions on intracellular calcium concentration and tension in ferret ventricular muscle. J. Physiol. (Lond.) **345:** 391–407.
23. FERRIER, G.R., J.H. SAUNDERS & C. MENDEZ. 1973. A cellular mechanism for the generation of ventricular arrhythmias by acetylstrophanthidin. Circ. Res. **32:** 600–609.
24. ROSEN, M.R., H. GELBAND & B.F. HOFFMAN. 1973. Correlation between effects of ouabain on the canine electrocardiogram and transmembrane potentials of isolated Purkinje fibers. Circulation **47:** 65–72.
25. CHENG, H., M.R. LEDERER, W.J. LEDERER & M.B. CANNELL. 1996. Calcium sparks and $[Ca^{2+}]_i$ waves in cardiac myocytes. Am. J. Physiol. **270:** C148–C159.
26. MECHMANN, S. & L. POTT. 1986. Identification of Na-Ca exchange current in single cardiac myocytes. Nature **319:** 597–599.
27. SIPIDO, K.R., A. VARRO & D. EISNER. 2006. Sodium calcium exchange as a target for antiarrhythmic therapy. Handbook Exp. Pharmacol. **171:** 159–199.
28. XU, L., R. JONES & G. MEISSNER. 1993. Effects of local anesthetics on single channel behavior of skeletal muscle release channel. J. Gen. Physiol. **101:** 207–233.
29. GYÖRKE, S., V. LUKYANENKO & I. GYÖRKE. 1997. Dual effects of tetracaine on spontaneous calcium release in rat ventricular myocytes. J. Physiol. (Lond.) **500:** 297–309.
30. OVEREND, C.L., D.A. EISNER & S.C. O'NEILL. 1997. The effect of tetracaine on spontaneous Ca release and sarcoplasmic reticulum calcium content in rat ventricular myocytes. J. Physiol. (Lond.) **502:** 471–479.
31. VENETUCCI, L., A.W. TRAFFORD, M.E. DIAZ, et al. 2006. Reducing ryanodine receptor open probability as a means to abolish spontaneous Ca^{2+} release and increase Ca^{2+} transient amplitude in adult ventricular myocytes. Circ. Res. **98:** 1299–1305.
32. ALLEN, D.G., D.A. EISNER, G.L. SMITH & S. WRAY. 1985. The effects of an extract of the foxglove (*Digitalis purpurea*) on tension and intracellular calcium concentration in ferret papillary muscle. J. Physiol. (Lond.) **365:** 55P.
33. CAPOGROSSI, M.C., B.A. SUÁREZ-ISLA & E.G. LAKATTA. 1986. The interaction of electrically stimulated twitches and spontaneous contractile waves in single cardiac myocytes. J. Gen. Physiol. **88:** 615–633.

34. LEHNART, S.E., C. TERRENOIRE, S. REIKEN, *et al.* 2006. Stabilization of cardiac ryanodine receptor prevents intracellular calcium leak and arrhythmias. Proc. Natl. Acad. Sci. USA **103:** 7906–7910.

35. WEHRENS, X.H.T., S.E. LEHNART, S. REIKEN, *et al.* 2005. Enhancing calstabin binding to ryanodine receptors improves cardiac and skeletal muscle function in heart failure. Proc. Natl. Acad. Sci. USA **102:** 9607–9612.

36. EISNER, D.A. & A.W. TRAFFORD. 2000. No role for the ryanodine receptor in regulating cardiac contraction? News Physiol. Sci. **15:** 275–279.

Na:Ca Stoichiometry and Cytosolic Ca-Dependent Activation of NCX in Intact Cardiomyocytes

DONALD M. BERS AND KENNETH S. GINSBURG

Department of Physiology, Loyola University Chicago, Maywood, Illinois 60153, USA

ABSTRACT: We are studying both Na:Ca exchange stoichiometry and cytosolic [Ca] ($[Ca]_i$)-dependent regulation of Na–Ca exchange (NCX) in intact rabbit ventricular myocytes. Analysis of NCX fluxes in subcellular systems strongly supports a dominant 3Na:1Ca exchange, and our measurements in intact cells confirm this. However, in intact native cells, local ion gradients and other factors complicate the process of inferring stoichiometry. From a functional viewpoint, NCX stoichiometry is near 3:1 but is affected by ion accumulation/depletion as well as non-NCX fluxes. We and others have viewed $[Ca]_i$-dependent NCX regulation as a static process (dependent on instantaneous local $[Ca]_i$). However, evidence from subcellular and expression systems shows the process to be dynamic, and our observations confirm this to be the case in intact cardiac cells as well.

KEYWORDS: calcium; Na–Ca exchange; sodium; Ca-dependent regulation; Na:Ca stoichiometry

STUDY OF Na–Ca EXCHANGE (NCX) IN INTACT CARDIOMYOCYTES

As a thermodynamically dissipative process, NCX abets the adjustment of a cell's electrochemical conditions, namely membrane potential E_m, cytosolic Na ($[Na]_i$), and cytosolic Ca ($[Ca]_i$), to a new steady state or equilibrium after any disturbance.[1] Key biophysical properties of NCX have been established via study of ion concentrations and E_m at equilibrium (usually identified by current reversal and considered to represent flux balance), mainly in subcellular and/or expression systems. Complementing this is the analysis of currents and ion fluxes due to experimentally imposed disturbances in E_m and/or ion concentrations, as NCX relaxes toward equilibrium. This flux approach can be

Address for correspondence: Donald M. Bers, Ph.D., Department of Physiology, Loyola University Chicago, Stritch School of Medicine, 2160 South First Avenue, Maywood, IL 60153. Voice: 708-216-6305; fax: 708-216-6308.

dbers@lumc.edu

Ann. N.Y. Acad. Sci. 1099: 326–338 (2007). © 2007 New York Academy of Sciences.
doi: 10.1196/annals.1387.060

valuable in native preparations like intact cardiac myocytes, for two reasons. First, in cells not under voltage clamp, which are producing action potentials (APs) and contracting normally under physiological conditions (excitation–contraction coupling, ECC), NCX working together with other conductances copes quite effectively with intrinsic ion gradients, and in this context of a regulated control system, the actual activity and role of NCX are difficult to observe. Second, even under voltage clamp, establishing that a particular set of ion concentrations and E_m is constant, is known accurately, and actually represents equilibrium can be difficult.[2,3]

FLUX ANALYSIS SUPPORTS 3:1 Na:Ca EXCHANGE STOICHIOMETRY IN INTACT CARDIOMYOCYTES

We have been using flux analysis approaches to study both the exchange stoichiometry and the dynamics of $[Ca]_i$-dependent allosteric activation of NCX in rabbit ventricular myocytes. In intact cardiomyocytes, the approach must consider that substantial local Ca^4 and Na^5 gradients develop near NCX during normal ECC. Local Ca accumulation due to release from SR and influx via I_{Ca} shifts NCX current (I_{NCX}) more inward than predicted from global $[Ca]_i$, likely limiting Ca influx mediated specifically by NCX.[4] Even in the absence of $[Na]_i$ gradients, which occur during regulation by Na/K ATPase,[6] local $[Na]_i$ gradients due to I_{Na}-mediated influx[7] appear during the AP upstroke.[8]

Early work using tracer flux analysis,[1,9] electrophysiology,[10,11] or ion-selective electrodes[12] suggested a 3:1 Na:Ca exchange, or possibly 4:1 based on modeling[13] or flux reversal,[14] and this duality has continued with modern current reversal [15] or ion-selective electrode measurements,[16] with our work using electrophysiology and fluorescence supporting 3:1, while other current reversal studies support 4:1 or variable stoichiometries.[17,18]

The logic common to studies of NCX thermodynamics is that the exchange stoichiometry (hereafter n), $[Na]_i,[Na]_o,[Ca]_i,[Ca]_o$ and E_m form a fully determined set, so that any one of these quantities is determined by the other five as noted previously.[13] As described below, we have treated n as the dependent variable.

In view of the dissipative nature of NCX, and as noted previously,[3] whenever NCX current (I_{NCX}) flows, $[Na]_i$ and $[Ca]_i$ must change. We took advantage of this to study NCX stoichiometry.[19] Using acutely isolated rabbit ventricular myocytes, with SR uptake, K currents, I_{Ca}, Na/K pumping, and $I_{Cl}(Ca)$ all blocked, in ruptured patch voltage clamp (36°C), we simultaneously recorded $[Na]_i$, $[Ca]_i$, and I_{NCX} (10 mM Ni-sensitive).[20] Initially, we set $[Na]_o = 140$ mM, $[Ca]_o = 2$ mM and holding $E_m = -53$ mV (predicted E_{rev} for $n = 3$).

On changing holding E_m or $[Na]_o$, we followed changes in global $[Ca]_i$ and $[Na]_i$. To assess local $[Ca]_i$ effects we measured E_{rev} with infrequent ramps. From ion concentrations we obtained thermodynamic $E_{NCX} = (nE_{Na}\text{-}2E_{Ca})/(n\text{-}2)$, where $E_{Na} = RT/F^*\ln([Na]_o/[Na]_i)$ and $E_{Ca} = RT/2F^*\ln([Ca]_o/[Ca]_i)$. After a few to several minutes, a near steady state was attained. Thermodynamic E_{NCX} had shifted in accord with shifts in E_m and matched it quite well if E_{NCX} was calculated using $n = 3$ (approximate confidence bounds $2.8 < n < 3.5$). This experimental result also showed that even with heavy intracellular Ca buffering, NCX can change both $[Ca]_i$ and $[Na]_i$ appreciably and relatively rapidly to approach a new steady state.

In near steady state, after E_m was changed, E_{rev} also moved toward the thermodynamically expected E_{NCX}, as seen previously,[15] but did not reach E_m within our observation time. E_{rev} was positive of expectation when inward current flowed (at more negative E_m) and negative of expectation when outward current flowed (at more positive E_m). This cannot be explained by local ion accumulation/depletion during NCX current flow, which would dissipatively shift E_{rev} oppositely to what we observed. Moreover, ion accumulation effects (at least the effects of Ca during ECC) appear to decay within <1 s ($\tau < 0.2$ s was measured).[4,21]

Indeed, on short time scales (~ 1 s), overly positive E_{rev} shifts were seen with depolarizing voltage steps versus ramps, and were attributed to Ca accumulation.[22] Similarly, during our flux studies, we applied infrequent ramp probes, with descending then ascending phases (1 s each). We found oppositely to Convery and Hancox that ascending ramps reversed positive of descending ramps. We expect that in our intact cardiomyocyte experiments, ramp E_{rev} depended on ramp direction, duration, and temporal order (descending then ascending) in a way that reflected local ion accumulation and depletion, tempered in part by Ca buffering.

Each quintuple of ion concentrations and E_{rev} in our experiments yielded a point estimate of n, which can be thought of as an apparent or effective value under the particular condition at the time. E_{rev} was consistently positive of the prediction for 3:1. Averaged over all individual measurements, n was 3.099. At near steady state, n decreased with more positive holding E_m, consistent with E_{rev} being less positive than the thermodynamically predicted E_{NCX} for $n = 3$ as noted above. By regression prediction (versus E_m) n was 3.31 at -100 mV and 2.99 at $+50$.

Away from steady state we examined initial descending ramps recorded within seconds of establishing a new E_m or external [Na] by a step change. To reveal flux-dependent ion accumulation/depletion effects we studied n versus I_{NCX} size and direction. Stoichiometry n was larger for outward versus inward I_{NCX} (by regression prediction 3.27 at $+200$ pA and 2.74 at 200 pA), consistent with our expectation for short-term Ca accumulation/Na depletion during outward I_{NCX} and the converse during inward I_{NCX} (thermodynamically dissipative).

Summarizing our thermodynamic data in intact myocytes, in the aggregate they are consistent with 3:1 stoichiometry. The apparent stoichiometry was shifted initially upon step changes in E_m or $[Na]_o$ that induced NCX flux, directionally consistent with local ion accumulation or depletion near the NCX. By the time near steady state had been reached, n appeared smaller than the average value (3.099) at positive holding E_m, and larger at negative E_m. The reason for this awaits further study (but see below).

HOW CONSTANT IS NCX STOICHIOMETRY AND HOW WELL CAN IT BE DETERMINED IN INTACT MYOCYTES?

Both short- and long-term Ca flux balance in cardiac cells can be explained more easily if the NCX transport ratio is ~3 and not some other value. A value near 2 would be inconsistent with known electrogenicity. A value near 4 would mismatch known E_{rev} and would predict aggressive Ca removal at rest, unless NCX inactivated very severely at $[Ca]_i$ near the normal diastolic level.

Nonetheless n may vary under particular conditions. A propensity to non-electrogenic transport was seen under extracellular acidification ($pH_o = 5$), with cytosolic pH maintained. Inward I_{NCX} (in response to flash-photolytic Ca transients) was inhibited without loss of either external Na dependence or Ca extrusion capacity.[23] The voltage dependence of Ca extrusion remained normal, so it appears that the specifics of the transport cycle were modified by H^+ in a way not involving actual transport of H^+.

Firm support for >3:1 transport (specifically 3.14:1) rather than exactly 3:1 has been adduced from analysis of local ion fluxes through giant patches with ion-selective microelectrodes,[16] where it was proposed that cardiac NCX transport sites could infrequently bind 1Na + 1Ca instead of 3Na or 1Ca, so that Ca transport, instead of being associated with the normal charge translocation under 3:1 stoichiometry ($1e^-/Ca$), either disappeared (1Ca+1Na:1Ca) or became electroneutral (3Na:1Ca+1Na). The effective result was charge transfer of ~$1.2e^-/Ca$, or a stoichiometry of 3.14. A significant consequence of minority NCX cycles that are not 3Na:1Ca is that NCX current reversal occurs at E_m positive of the E_m for Na:Ca flux balance.[16] The extra NCX-mediated Na influx, which was proposed to occur even while NCX is keeping Ca in thermodynamic balance, might contribute to the sensitive inotropic response of cardiac tissue to even small increases in $[Na]_i$.[24] Stoichiometry n increased slightly at negative E_m,[16] at least consistent with our steady-state observations, despite radically different approaches.

Beyond these biophysical indications that NCX stoichiometry is not fully deterministic, several factors may affect the measurement and functional impact of NCX stoichiometry in intact cells. While our non-steady-state data with non-NCX fluxes nominally blocked were consistent with local ion accumulation and depletion due directly to NCX fluxes, the local inhomogeneities developed during normal cardiac cell ECC must be much larger. During the cardiac cycle, initial SR Ca release transiently increases local Ca seen by NCX severalfold, curtailing NCX-mediated influx and forcing early initiation of efflux.[4] Once SR Ca release ends and relaxation ensues, SR reuptake opposes and retards efflux via NCX. It has been shown that intact cells can support substantial and complex spatiotemporal inhomogeneities in $[Na]_i$.[8,25] Any assay of local Ca accumulation or depletion depends on extrinsic cytosolic buffering (including fluorophores), which we and others[22] have included in experiments. Our finding that Ca accumulation or depletion affected E_{rev} on a 1–2 s time scale presumably results from the specific buffering conditions we used. NCX can change $[Ca]_i$ even with heavy fast buffering (20 mM BAPTA)[15]; essentially no amount of buffering can guarantee fully to suppress flux-dependent changes in local ion concentrations.

In functioning intact cells, NCX must of course be considered in the context of all operating Ca and Na transport systems, but non-NCX transport remains an issue even where every attempt is made to isolate NCX. There is little prospect of blocking nonelectrogenic sarcolemmal Ca pumping completely. Mitochondria can remove substantial amounts of Ca and may mediate additional NCX regulation.[26] Background Ca influx, though its basis remains unclear,[27] and even possibly store-operated channels[28] are also present and a completely selective block of these versus NCX would be difficult.

We found that a mostly Ni-sensitive residual inward current flowed during inter-ramp intervals, whereas the thermodynamic expectation is that NCX net flux would dissipate to the point where $I_{NCX} = 0$. This current is difficult to interpret unambiguously. In functioning cells where resting E_m is established by E_K, not E_{NCX}, inward I_{NCX} may flow to rebalance Ca in the face of the forced non-NCX fluxes mentioned above. With most all non-NCX flux sources blocked, it may be argued that all the non-NCX fluxes are "small," and it is true that even their electrogenic components may not be discriminable in current records against parasitic effects, such as seal leakage, incompletely cancelled capacitative currents or pipette offset currents. However, the initial holding E_m is necessarily chosen using a prediction of NCX E_{rev} based on the very quantity under study. It may be asked how well any observed current reversal represents an equilibrium state. If NCX is really at equilibrium, with no net ion flux and all ion concentrations constant, then the net current is necessarily 0, but the converse while possible is not necessarily true. If as concluded in Reference 16, n is 3.14 and not 3 and this is in part due to a net Na influx transport mode, then an inward current will flow when E_m is set to E_{NCX} for $n = 3$, or Ca influx will occur if E_m is set to the (more positive) observable E_{rev}.

As pointed out previously,[2] under conditions where the available NCX conductance or NCX density is small, even a large electrochemical potential difference will drive very little current, so that the measurement of NCX current reversal becomes uncertain. Any active non-NCX flux pathways would exacerbate this uncertainty. Thus reliable association of current reversal with NCX equilibrium requires that NCX be both dominant and have a reasonably large availability.

Indirect evidence against large fluxes via NCX during normal ECC functions, at least in mouse, is provided by our own Ca removal flux distribution measurements,[29] as well as by the apparently normal survival and function of mice with a cardiac-specific NCX knockout.[30] Further affirming that NCX must work in the context of other Ca and Na handling processes, even in diastole, thermodynamics would dictate that if controlled entirely by NCX, $[Ca]_i$ at equilibrium ($E_m = E_{NCX}$) would depend on $[Na]_i$ raised to the power n.[13] Submillimolar changes in $[Na]_i$ would then be enough to destabilize ECC by changing $[Ca]_i$ radically.

None of these considerations belie the established importance of NCX in cardiac Ca regulation. It is axiomatic that Ca influx and efflux, each much larger in cardiac cells versus neuronal or other tissues, must remain in long-term balance.[13] Reflecting this, small imposed changes in NCX driving force can shift myocytes between postrest decay or potentiation of Ca transients and contraction.[31] It may be fairest to say that NCX fluxes are not normally prominent insofar as normal ECC represents a stable control system. However, the quantitatively large capacity of NCX transport can become apparent under perturbing conditions. Among diverse outcomes supporting the view that NCX has a "reserve" capacity, brief (< 1 s) Na_o deprivations increase resting Ca spark frequency[32] (although this can be attributed to NCX-mediated increased local [Ca] only to the extent that SR loading and global $[Ca]_i$ are truly constant). In heart failure, spontaneous activities and afterdischarges occur when NCX is upregulated and I_{K1} is less active,[33] while under specific destabilizing stimulus conditions (repeated small depolarizations), large inward NCX transients can occur.[34]

A final issue affecting the practical study of NCX stoichiometry is $[Ca]_i$-dependent activation. E_{rev} measurements may be affected as inward I_{NCX} (Ni-sensitive) became imperceptible at lower $[Ca]_i$, <200 nM.[10] As was done previousl,[15] we observed the growth of I_{NCX} and measured n during the removal of Ni-mediated NCX block using a fast solution switch. Logically, NCX should transiently support ion fluxes revealing its true stoichiometry, up until the preexisting ion gradients dissipate and I_{NCX} decays toward 0, and indeed $n \approx 3$ was revealed. However, the success of this protocol depends on an initially perturbed ion balance and a large available NCX conductance. Since Ca-dependent allosteric NCX activation in intact cells is key to the NCX contribution to control of ECC, we consider it further below.

We conclude that ion concentrations relevant to establishing NCX n in intact cells are established in a complex space, which is far from an isotropic global space and yet far from a highly localized compartment or specific binding domain. Strong consideration must be given to dominant globally active non-NCX flux pathways in functioning cells. We propose that NCX stoichiometry be regarded as a functional measure, which must retain consistency with established biophysical analyses, but must also consider NCX in its milieu in intact functioning cardiomyocytes.

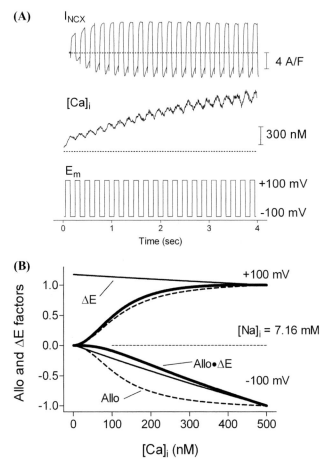

FIGURE 1. (**A**) "Envelope" protocol and representative difference current (Ni-subtracted) showing progressive activation of NCX. (**B**) [Ca]$_i$-dependence of I$_{NCX}$, shown normalized to maximal activation (*bold lines*), along with its separation into thermodynamic driving force for NCX, showing inward shift with increasing [Ca]$_i$ (*solid lines*), and allosteric increase in fractional NCX availability (*dashed lines*). Upper line of pair for each style represents +100 mV; lower line represents –100 mV.

DYNAMICS OF [Ca]$_i$-DEPENDENT NCX ACTIVATION IN INTACT MYOCYTES

It is evident that the state of [Ca]$_i$-dependent activation of NCX[35] can increase with Ca influx, by positive feedback. In a cell initially at rest then repeatedly depolarized, thermodynamics predicts that NCX should shift inward as [Ca]$_i$ increases, but more outward I$_{NCX}$ also develops, a definitive signature of increased NCX availability, as we have reported.[36]

We have followed NCX activation in isolated rabbit ventricular myocytes using recording conditions somewhat similar to those above for NCX stoichiometry, where SR function and non-NCX fluxes were blocked as well as possible but $[Ca]_i$ was not buffered and we did not measure $[Na]_i$. In initially rested cells, we activated NCX (10 mM Ni-sensitive) by repeated ± 100 mV steps (FIG. 1, panel A; typical cycle length 300 ms, with 100 ms at $+100$ and 200 ms at -100). This protocol allowed us to rapidly probe both thermodynamic shifts and NCX activation while $[Ca]_i$, increased gradually as a result of positive feedback.[36] We analyzed the envelope (borders representing maximum inward and outward current) of current traces separately for $+100$ and -100 mV using the equation:

$$I_{Na/Ca} = \frac{(V\text{max}/(1 + \{K_{mAllo}/[Ca]_i\}^2)([Na]_i^3[Ca]_o e^{\eta VF/RT} - [Na]_o^3[Ca]_i e^{(\eta-1)VF/RT})}{\left[K_{mCao}[Na]_i^3 + K_{mNao}^3[Ca]_i + K_{mNai}^3[Ca]_o(1 + [Ca]_i/K_{mCai}) + K_{mCai}[Na]_o^3(1 + \{[Na]_i/K_{mNai}\}^3) + [Na]_i^3[Ca]_o + [Na]_o^3[Ca]_i \right]} \times \left[1 + k_{sat} e^{(\eta-1)VF/RT} \right]$$

in which V_{max} sets the magnitude of I_{NCX}, V is the transmembrane voltage (E_m), the various K_m are the inside and outside transport site association constants, η is a factor setting the steepness of E_m dependence, and k_{sat} adjusts for shallower voltage dependence at more negative E_m. The $[Ca]_i$ and I_{NCX} were observed and parameters set as reported.[36] V_{max} was calculated from I_{NCX} measured in the maximally activated state, since the relation between I_{NCX} and V_{max} was linear. The extent of allosteric activation by $[Ca]_i$ is indicated by the factor $1/(1+\{K_{mAllo}/[Ca]_i\}^2)$ in the numerator. Thus we can represent I_{NCX} as the product of this activation factor (Allo) and a thermodynamic transport term (ΔE; the rest of the equation) as: $I_{NCX} = (\text{Allo}) (\Delta E)$. Note that Allo acts as an instantaneous separable scaling factor and this model has no time-dependence. Seeming to support this static view of activation, massive caffeine-induced Ca release activated NCX (at $+100$ mV) with no apparent kinetic delay, by mass action. The $[Ca]_i$ dependences of both electrochemical and activation-dependent contributions to I_{NCX} at ± 100 mV are illustrated in the FIGURE 1, panel B.

Despite the above, evidence from subcellular and expression systems shows that an instantaneous model is inadequate. Even in the original description of Ca-dependent NCX regulation, a sigmoidal growth of NCX over several seconds appears (FIG. 9 of Ref. 35), consistent with a positive feedback multistate system. Qualitatively similar sigmoidal time courses of Ca-dependent activation of NCX Ca influx have been extensively documented.[37] Dynamic, time-dependent control of NCX activation might have the benefit of maintaining long-term ion balances while minimally snubbing rapid ECC-related changes.[24]

We have more recently studied NCX activation using the same buffered $[Ca]_i$ condition as in the NCX stoichiometry studies, affording examination of activation over a longer time course.[38] Here to induce and follow activation we applied periodic step depolarizations followed after 10 ms by 0.1 s descending ramps, during which we measured local submembrane [Ca] ($[Ca]_{sm}$) via the ramp E_{rev}. It should be pointed out that the NCX electrochemical equation (above) predicts (at physiological $[Na]_i$ near 10 mM) that E_{rev} should be a highly sensitive measure of changes in local $[Ca]_i$. We represented the NCX activation state as ramp slope (V_{max}, which implicitly contains activation), normalized to its maximum value.

We expected the growth of NCX activation to follow $[Ca]_{SM}$ as measured by E_{rev}, but it systematically lagged behind. It was also shown previously[37] that NCX activation was not locked to $[Ca]_i$; once activated, NCX did not revert to an inactivated state on return of $[Ca]_i$ to rest levels; nor did we find that NCX deactivated lawfully within tens of seconds of rest.

While we do not know the mechanisms for dynamic Ca activation, several factors may contribute. Previous experiments measuring activation via Ca influx[35,37] as well as our own, inherently produced lagged responses due to their positive feedback structure. Dynamic and somewhat persistent Ca-activation of Ca efflux was also observed.[39] In our own experiments, $[Na]_i$ may gradually increase due to Na/K ATPase block, thermodynamically abetting enough $[Ca]_i$ accumulation to promote activation despite strong Ca buffering. A further possibility is the abrogation or circumvention of Ca inactivation by another NCX regulatory pathway. Recently, NCX activation was observed after Na loading (especially at ≥ 20 mM $[Na]_i$) in NCX mutants where the normal Ca-dependent regulation was severely desensitized.[40] The evident $[Na]_i$-dependent activation mode, and possibly Ca-dependent activation itself[41] may respond to PIP_2 levels; such lipid kinase regulation was described earlier.[42]

A MORE PHYSIOLOGICAL VIEW OF NCX
ACTIVATION DYNAMICS

We are also studying NCX activation in rabbit myocytes under more nearly physiological conditions, using perforated patch (amphotericin or β-escin) or ruptured patch voltage clamp, without extrinsic $[Ca]_i$ buffering, at 36°C.[43] SR function is left intact and we have set $[Na]_i = 10$ mM and $[Ca]_o = 2$ mM. NMG or Cs is substituted for K, but no other channel blockers are used.

To initialize NCX to a low activation state, cells were rested in 0.25 mM Ca (≥ 30 min) and clamped with holding $E_m = -90$ or -100 mV. To activate NCX we switched to 2 mM $[Ca]_o$ with Ni present. We applied AP clamp waveforms (APCs, derived from an average of several previously recorded APs) with preceding and following miniramps (of 10–20 ms duration, between –65 and –90 mV). To commence NCX activation we quickly removed Ni (using a fast solution switcher). These miniramps were designed to activate I_{NCX} as selectively as possible while keeping interfering currents, such as Na/K ATPase current minimized or constant. We repeated APCs at 0.1–2 Hz, and included rest pauses. We represented NCX activation as the linear slope (chord conductance, G_{chord}) of the miniramps (corrected for non-NCX conductance recorded with Ni present), and tracked changes in submembrane $[Ca]_i$ via ramp E_{rev}.

On stimulation, NCX activated and reached a saturating level that reverted at most slightly on rest (within tens of seconds). Activation, assumed zero initially, developed even at rest, consistent with the expectation of partial NCX activation at $[Ca]_i$ (~100 nM). During the time course of an experiment, activation apparently saturated, not increasing with higher frequency stimulation, as if the net Ca signal promoting activation ran down, inactivated, or otherwise lost its effectiveness over time. This activating signal may be presumed to represent a mixture of Ca influx and Ca release, which underwent adjustment as steady state was attained.[44]

As we found when assaying activation with buffered $[Ca]_i$, ramp E_{rev} did not predict activation state closely; in fact E_{rev} tended to remain stable. In rough consistency with our observations in the NCX stoichiometry studies, it was usually at a few millivolts positive of its expected value (the holding E_m between APCs, based on the maximum diastolic potential during APs, –81 mV).

CONCLUSIONS

While Na:Ca exchange stoichiometry and $[Ca]_i$-dependent activation are increasingly understood as biophysical phenomena, challenging effort remains if we are to understand their physiological consequences. The situation of NCX appears to be somewhat dichotomous. As we noted, NCX may not be highly visible, despite its important quantitative role in Ca homeostasis. For instance, during steady-state AP production and ECC, NCX is providing effective $[Ca]_i$ and $[Na]_i$ regulation, but I_{NCX} is usually small in amplitude. However, NCX must be sufficiently expressed and a sufficient reserve capacity must be available if it is to contribute effectively to maintaining the long-term $[Ca]_i$ and $[Na]_i$ economy of cardiac myocytes. This is particularly so in heart failure, where derangement of Na/K ATPase pumping, Na channels, and/or Na:H exchange lead to increased $[Na]_i$ and a rebalancing of NCX-mediated fluxes toward Ca influx occurs, with I_{NCX} either outward shifted or frankly outward during a significant fraction of the cardiac cycle.

REFERENCES

1. REEVES, J.P. & C.C HALE. 1984. The stoichiometry of the cardiac sodium-calcium exchange system. J. Biol. Chem. **259:** 7733–7739.
2. AXELSEN, P.H. & J.H BRIDGE. 1985. Electrochemical ion gradients and the Na/Ca exchange stoichiometry. Measurements of these gradients are thermodynamically consistent with a stoichiometric coefficient greater than or equal to 3. J. Gen. Physiol. **85:** 471–475.
3. HINATA, M. & J. KIMURA. 2004. Forefront of $Na^+/Ca2^+$ exchanger studies: stoichiometry of cardiac $Na^+/Ca2^+$ exchanger; 3:1 or 4:1? J. Pharmacol. Sci. **96:** 15–18.
4. WEBER, C.R., V. PIACENTINO III, K.S. GINSBURG, et al. 2002. Na/Ca exchange current and submembrane [Ca] during the cardiac action potential. Circ. Res. **90:** 182–189.
5. BARRY, W.H. 2006. Na"Fuzzy space": does it exist, and is it important in ischemic injury? J. Cardiovasc. Electrophysiol. **17**(Suppl 1): S43–S46.
6. DESPA, S. & D.M. BERS. 2003. Na/K pump current and $[Na]_i$ in rabbit ventricular myocytes: local [Na](i) depletion and Na buffering. Biophys. J. **84:** 4157–4166.
7. WEBER, C.R., K.S. GINSBURG & D.M. BERS. 2003. Cardiac submembrane $[Na^+]$ transients sensed by Na^+-Ca2^+ exchange current. Circ. Res. **92:** 950–952.
8. WENDT-GALLITELLI, M.F., T. VOIGT & G. ISENBERG. 1993. Microheterogeneity of subsarcolemmal sodium gradients. Electron probe microanalysis in guinea-pig ventricular myocytes. J. Physiol. **472:** 33–44.
9. PITTS, B.J.R. 1979. Stoichiometry of sodium-calcium exchange in cardiac sarcolemmal vesicles. J. Biol. Chem. **254:** 6232–6235.
10. EHARA, T., S. MATSUOKA & A. NOMA. 1989. Measurement of reversal potential of Na^+-Ca2^+ exchange current in single guinea-pig ventricular cells. J. Physiol. **410:** 227–249.
11. CRESPO, L.M., C.J. GRANTHAM & M.B. CANNELL. 1990. Kinetics, stoichiometry and role of the Na-Ca exchange mechanism in isolated cardiac myocytes. Nature **345:** 618–621.
12. SHEU, S.S. & H.A. FOZZARD. 1982. Transmembrane Na and Ca electrochemical gradients in cardiac muscle and their relation to force development. J. Gen. Physiol. **80:** 325–351.
13. MULLINS, L.J.. 1979. The generation of electric currents in cardiac fibers by Na/Ca exchange. Am. J. Physiol. **236:** C103–C110.
14. LEDVORA, R.F. & C. HEGYVARY. 1983. Dependence of sodium-calcium exchange and calcium-calcium exchange on monovalent cations. Biochim. Biophys. Acta **729:** 123–136.
15. HINATA, M., H. YAMAMURA, L. LI, et al. 2002. Stoichiometry of Na^+-Ca2^+ exchange is 3:1 in guinea-pig ventricular myocytes. J. Physiol. **545**(Pt 2): 453–461.
16. KANG, T.M. & D.W. HILGEMANN. 2004. Multiple transport modes of the cardiac $Na^+/Ca2^+$ exchanger. Nature **427:** 544–548.
17. FUJIOKA, Y., M. KOMEDA & S. MATSUOKA: 2000. Stoichiometry of Na^+-Ca^{2+} exchange in inside-out patches excised from guinea-pig ventricular myocytes. J. Physiol. **523:** 339–351.
18. DONG, H., J. DUNN & J. LYTTON. 2002. Stoichiometry of the cardiac $Na^+/Ca2^+$ exchanger NCX1.1 measured in transfected HEK cells. Biophys. J. **82:** 1943–1952.

19. GINSBURG, K.S., C.R. WEBER, S. DESPA & D.M. BERS. 2003. NCX stoichiometry inferred from simultaneously measured $[Na]_i$, $[Ca]_i$, and INCX in intact cardiac myocytes. Biophys. J. **84:** 259a.

20. GINSBURG, K.S., C.R. WEBER, S. DESPA & D.M. BERS. 2002. Simultaneous measurement of $[Na]_i$, $[Ca]_i$, and I_{NCX} in intact cardiac myocytes. Ann. N. Y. Acad. Sci. **976:** 157–158.

21. TRAFFORD, A.W., M.E. DIAZ, S.C. O'NEILL & D.A. EISNER. 1995. Comparison of subsarcolemmal and bulk calcium concentration during spontaneous calcium release in rat ventricular myocytes. J. Physiol. **488**(Pt 3):577–586.

22. CONVERY, M.K. & J.C. HANCOX. 1999. Comparison of Na^+-$Ca2^+$ exchange current elicited from isolated rabbit ventricular myocytes by voltage ramp and step protocols. Pflugers Arch. **437:** 944–954.

23. EGGER, M. & E. NIGGLI. 2000. Paradoxical block of the Na^+-Ca^{2+} exchanger by extracellular protons in guinea-pig ventricular myocytes. J. Physiol. **523:** 353–366.

24. HILGEMANN, D.W. 2004. New insights into the molecular and cellular workings of the cardiac Na^+/Ca^{2+} exchanger. Am. J. Physiol. Cell Physiol. **287:** C1167–C1172.

25. DESPA, S., J. KOCKSKAMPER, L.A. BLATTER & D.M. BERS. 2004. Na/K pump-induced $[Na]_i$ gradients in rat ventricular myocytes measured with two-photon microscopy. Biophys. J. **87:** 1360–1368.

26. OPUNI, K. & J.P. REEVES. 2000. Feedback inhibition of sodium/calcium exchange by mitochondrial calcium accumulation. J. Biol. Chem. **275:** 21549–21554.

27. KUPITTAYANANT, P., A.W. TRAFFORD, M.E. DIAZ & D.A. EISNER. 2006. A mechanism distinct from the L-type Ca current or Na-Ca exchange contributes to Ca entry in rat ventricular myocytes. Cell Calcium **39:** 417–423.

28. HUANG, J., C. VAN BREEMEN, K.H. KUO, et al. 2006. Store-operated Ca^{2+} entry modulates sarcoplasmic reticulum Ca^{2+} loading in neonatal rabbit cardiac ventricular myocytes. Am. J. Physiol. Cell Physiol. **290:** C1572–C1582.

29. BRITTSAN, A., K.S. GINSBURG, G. CHU, et al. 2003. Chronic SR Ca^{2+}-ATPase inhibition causes adaptive changes in cellular Ca^{2+} transport. Circ. Res. **92:** 769–776.

30. HENDERSON, S.A., J.I. GOLDHABER, J.M. SO, et al. 2004. Functional adult myocardium in the absence of Na^+-$Ca2^+$ exchange: cardiac-specific knockout of NCX1. Circ. Res. **95:** 604–611.

31. BASSANI, R.A. & D.M. BERS. 1994. Na-Ca exchange is required for rest-decay but not for rest-potentiation of twitches in rabbit and rat ventricular myocytes. J. Mol. Cell. Cardiol. **26:** 1335–1347.

32. GOLDHABER, J.I., S.T. LAMP, D.O. WALTER, et al. 1999. Local regulation of the threshold for calcium sparks in rat ventricular myocytes: role of sodium-calcium exchange. J. Physiol. **520:** 431–438.

33. POGWIZD, S.M., K. SCHLOTTHAUER, L. LI, et al. 2001. Arrhythmogenesis and contractile dysfunction in heart failure: roles of sodium-calcium exchange, inward rectifier potassium current and residual β-adrenergic responsiveness. Circ. Res. **88**(11):1159–1167.

34. DIAZ, M.E., S.C. O'NEILL & D.A. EISNER. 2004. Sarcoplasmic reticulum calcium content fluctuation is the key to cardiac alternans. Circ. Res. **94:** 650–656.

35. HILGEMANN, D.W., A. COLLINS & S. MATSUOKA. 1992. Steady-state and dynamic properties of cardiac sodium-calcium exchange. Secondary modulation by cytoplasmic calcium and ATP. J. Gen. Physiol. **100:** 933–961.

36. WEBER, C.R., K.S. GINSBURG, K.D. PHILIPSON, *et al.* 2001. Allosteric regulation of Na/Ca exchange current by cytosolic Ca in intact cardiac myocytes. J. Gen. Physiol. **117:** 119–131.
37. REEVES, J.P. & M. CONDRESCU. 2003. Allosteric activation of sodium-calcium exchange activity by calcium: persistence at low calcium concentrations. J. Gen. Physiol. **122:** 621–639.
38. WEBER, C.R., K.S. GINSBURG & D.M. BERS. 2005. Allosteric regulation of cardiac Na/Ca exchange by cytosolic Ca: dynamics in intact myocytes. Biophys. J. **88:** 136a.
39. CHERNYSH, O., M. CONDRESCU & J.P. REEVES. 2004. Calcium-dependent regulation of calcium efflux by the cardiac sodium/calcium exchanger. Am. J. Physiol. Cell Physiol. **287:** C797–C806.
40. URBANCZYK, J., O. CHERNYSH, M. CONDRESCU & J.P. REEVES. 2006. Sodium-calcium exchange does not require allosteric calcium activation at high cytosolic sodium concentrations. J. Physiol. **575:** 693–705.
41. CHERNYSH, O. & J.P. REEVES. 2006. PIP_2 and hysteresis of allosteric Ca activation of Na/Ca exchange: inhibition of persistent Ca activation by the PI-kinase inhibitor wortmannin. Biophys. J. **90:** 511a.
42. HILGEMANN, D.W., R. BALL. 1996. Regulation of cardiac Na^+,Ca^{2+} exchange and K_{ATP} potassium channels by PIP_2. Science **273:** 956–959.
43. GINSBURG, K.S., C.R. WEBER & D.M. BERS. 2005. Allosteric regulation of Na/Ca exchange by cytosolic Ca under physiological conditions in intact cardiomyocytes. Biophys. J. **88:** 136a.
44. TRAFFORD, A.W., M.E. DÍAZ, N. NEGRETTI & D.A. EISNER. 1997. Enhanced Ca^{2+} current and decreased Ca^{2+} efflux restore sarcoplasmic reticulum Ca^{2+} content after depletion. Circ. Res. **81:** 477–484.

Na/Ca Exchange and Cardiac Ventricular Arrhythmias

KARIN R. SIPIDO,[a] V. BITO,[a] G. ANTOONS,[b] PAUL G. VOLDERS,[c] AND MARC A. VOS[b]

[a]Laboratory of Experimental Cardiology, University of Leuven, B-3000 Leuven, Belgium

[b]Department of Medical Physiology, University Medical Center Utrecht, Utrecht, the Netherlands

[c]Department of Cardiology, Cardiovascular Research Institute Maastricht, Academic Hospital Maastricht, Maastricht, the Netherlands

ABSTRACT: Ventricular arrhythmias are a major cause of death in cardiovascular disease. Ca^{2+} removal from the cell by the electrogenic Na/Ca exchanger is essential for the Ca^{2+} flux balance during excitation–contraction coupling but also contributes to the electrical events. "Classic" views on the exchanger in arrhythmias include its well-recognized role as depolarizing current underlying delayed afterdepolarizations (DADs) during spontaneous Ca^{2+} release and the alterations in expression in certain forms of cardiac hypertrophy and heart failure. "Novel" views relate to more subtle roles for the exchanger in arrhythmias. Na/Ca exchange function in disease could be modulated indirectly, through phosphorylation or anchoring proteins. Ongoing studies relate Na/Ca exchange to variability in action potential duration (APD) and early afterdepolarizations (EADs) in a dog model of cardiac hypertrophy and arrhythmias. Further research on drugs that target Na/Ca exchange will have to carefully examine the effects on Ca^{2+} balance.

KEYWORDS: cardiac myocytes; arrhythmia; Na/Ca exchange; calcium; delayed after depolarizations; calcium channel; sarcoplasmic reticulum

INTRODUCTION

Ventricular tachycardias are most often "malignant," meaning they impair the proper pump function of the heart and result in reduced cardiac output, shock, and eventual death. Major underlying mechanisms are abnormal impulse formation in the ventricles (automatic and triggered arrhythmias) or a

Address for correspondence: Karin R. Sipido, M.D., Ph.D., Laboratory of Experimental Cardiology, KUL, Campus Gasthuisberg O/N1, 704, Herestraat 49, B-3000 Leuven, Belgium. Voice: 32-16-347153; fax: 32-16-345844.

Karin.Sipido@med.kuleuven.be

Ann. N.Y. Acad. Sci. 1099: 339–348 (2007). © 2007 New York Academy of Sciences.
doi: 10.1196/annals.1387.066

reentry phenomenon, which is in itself often triggered by an extra beat. Reentry can, among other factors, be promoted by regionally heterogeneous action potential prolongation, while abnormal automaticity originates in cells that are normally quiescent. Triggered arrhythmias are based on afterdepolarizations and reflect disturbances in intracellular Ca^{2+} handling. A potential role for the Na/Ca exchanger in arrhythmogenesis lies in its contribution to the action potential shape and time course and to afterdepolarizations. Early insights into the Na/Ca exchanger and triggered arrhythmias go back to the studies of digitalis intoxication as recently reviewed.[1] A rekindled interest in the role of the exchanger follows the observations of increased Na/Ca exchange activity in cardiac hypertrophy and heart failure, conditions associated with an increased arrhythmogenic risk, and the development of drugs to specifically suppress Na/Ca exchange.[2-6]

In this article we will briefly address the general background and current views on Na/Ca exchange and arrhythmias, illustrate some of our recent and ongoing studies, and identify new research directions.

THE Na/Ca EXCHANGE CURRENT IN THE CARDIAC CYCLE

A transient rise in $[Ca^{2+}]_i$ is a key event in the cardiac cycle and the cellular excitation–contraction coupling process. Ca^{2+} influx via voltage-sensitive (L-type) Ca^{2+} channels is the major source of Ca^{2+} activating the ryanodine receptor (RyR), the Ca^{2+} release channel of the sarcoplasmic reticulum, the internal Ca^{2+} store. This Ca^{2+} release is the major source of Ca^{2+} for the $[Ca^{2+}]_i$ transient and eventual contraction. Ca^{2+} extrusion by Na/Ca exchange (forward mode, inward current) across the sarcolemma is the major mechanism to balance the Ca^{2+} influx through the L-type Ca^{2+} channel. A small amount of Ca^{2+} influx during depolarization is also contributed by the Na/Ca exchanger (reverse mode, outward current). During the largest fraction of the cardiac cycle, however, the exchanger acts in the forward mode, to maintain the transsarcolemmal Ca^{2+} balance,[7] as illustrated in FIGURE 1. The amplitude of this inward current varies during the cycle, with membrane potential and $[Ca^{2+}]_i$. Both the outward and inward currents contribute to the shape of the action potential. The relative dominance of either component can vary in function of alterations in Na^+ and Ca^{2+} handling, which affect the reversal potential.[8]

Na/Ca EXCHANGE, THE ACTION POTENTIAL, AND DELAYED AFTERDEPOLARIZATIONS (DADs)

The role of the Na/Ca exchanger in arrhythmogenesis derives from its contribution to the action potential time course, from its contribution to afterdepolarizations, and from its role in setting the sarcoplasmic reticulum calcium

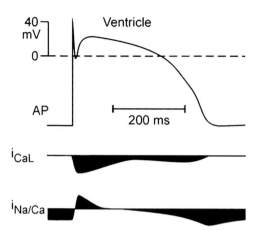

FIGURE 1. Schematic time course of the action potential in a ventricular cardiac myocyte and of the L-type Ca^{2+} current and the Na/Ca exchange current. (Modified from Carmeliet & Vereecke.[37])

content. All of these aspects were recently reviewed, in the perspective of arrhythmias in cardiac disease.[2,6,9] Earlier studies suggested that the suppression of Na/Ca exchange shortened the action potential,[10] others that it could both shorten and lengthen the action potential depending on the $[Na^+]_i$,[8] and lastly, a recent study using a Na/Ca exchange blocker found no change in action potential shape after block of Na/Ca exchange.[11] Our own data (see below) suggest that in dog ventricular myocytes the inward component during Ca^{2+} release contributes to the action potential plateau duration and its stability. If, in the presence of high $[Na^+]_i$ and low $[Ca^{2+}]_i$, as in heart failure, there is a large Ca^{2+} influx via the exchanger during the action potential,[12] the balance of fluxes requires that this influx will be removed during the diastolic period. Given that the Na/Ca exchange is the most important efflux pathway, this would lead to a sustained inward current during diastole, which would contribute to instability of the resting membrane potential. The effect of this inward current would be reenforced by a reduction of the stabilizing K^+ current (I_{K1}) as reported in animal models of heart failure.[13,14]

The best-known role of Na/Ca exchange is in DADs. These arise from spontaneous release of calcium from the sarcoplasmic reticulum. The spontaneous release occurs because of excessive calcium load and/or alterations in the RyR gating properties, as recently proposed for some of the mutations associated with catecholaminergic polymorphic ventricular tachycardia.[15] In response to the rise in $[Ca^{2+}]_i$ an inward current is generated. The amplitude of this current will depend on the membrane potential, being larger at diastolic potentials, than with a spontaneous Ca^{2+} release at the end of the action potential plateau. The effect of the resulting inward current in itself depends on the overall ion channel activity. At the resting potential, a strong inward rectifying I_{K1} will counteract depolarization, while at the depolarized plateau potential fewer channels are open and a small current may have more impact. Experimental studies have examined the amplitude of Ca^{2+} release required to reach the threshold for triggering sufficient depolarizations to trigger action potentials in a heart failure model.[13] This study, however, used caffeine pulses to induce Ca^{2+} release resulting in a more synchronized event than during spontaneous release, which is characterized by slowly traveling waves. At the wavefront, $[Ca^{2+}]_i$ is locally high with local Na/Ca exchange current activated, but the overall current is not necessarily large enough to trigger an action potential. In isolated myocytes, which exhibit spontaneous release, DADs are readily observed but triggered action potentials are more rare. Suppression of I_{K1} will facilitate DADs by lowering the Ca^{2+} release needed to reach the threshold; in the rat with myocardial infarction, the suppression of I_{K1} has been linked to the spontaneous release of Ca^{2+} itself.[13]

A lot of attention has recently been paid to the alterations in the RyR, through mutations or alterations in FKBP12.6 binding, as a cause for arrhythmogenic spontaneous Ca^{2+} release.[16] Drugs like JTV519 that target FKBP12.6 and stabilize the RyR gating were shown to prevent arrhythmias.[17] Yet this approach has not been equivocally successful.[18] Recent data show that the role of RyR gating in spontaneous Ca^{2+} release cannot be viewed independently of the changes in sarcoplasmic reticulum Ca^{2+} load, and that drugs affecting RyR gating also affect Ca^{2+} load.[19,20]

MECHANISMS OF INCREASED/DECREASED Na/Ca EXCHANGE CURRENTS IN CARDIAC DISEASE

The most often offered explanation of increased (or decreased) Na/Ca exchange current in cardiac hypertrophy or heart failure is a change of gene expression. Concordant changes have indeed been observed in a number of animal models or human studies, but this is not always the case. In the dog with chronic atrioventricular block (cAVB) we did not find a significant change in Na/Ca exchange expression despite larger outward and inward currents.[21] We can ascribe the increase in outward current to the higher levels of $[Na^+]_i$ we found.[22] The increase in inward current is more difficult to explain since a shift in reversal would actually lead to a decrease in inward current for similar levels of $[Ca^{2+}]_i$. Possibly, the larger Na/Ca exchange currents are the result of Ca-dependent regulation, as described by Weber et al.[23] The dog with cAVB has an enhanced contractility and Ca^{2+} cycling, and the protocols we used to study inward currents as function of $[Ca^{2+}]_i$ led to considerably higher levels of $[Ca^{2+}]_i$ in the cAVB myocytes (FIG. 2). This would lead to overestimation of the intrinsic levels of expression of Na/Ca exchange, though not of the functional activity in the dog with cAVB.

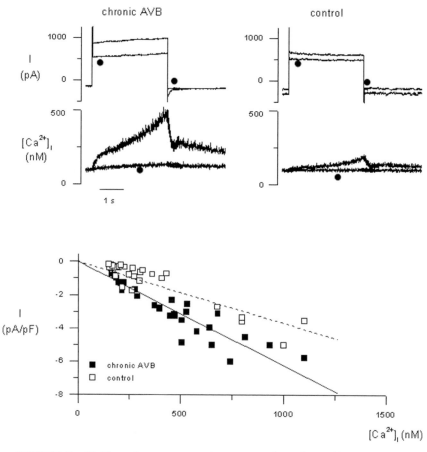

FIGURE 2. Na/Ca exchange currents in myocytes from dogs with AVB. Top panels, membrane currents and $[Ca^{2+}]_i$ recorded during steps from –45 to +60 mV with Ca^{2+} channels blocked and all K^+ replaced with Cs^+, before and after addition of 2.5 mmol/L $NiCl_2$ (marked with *solid circle*). The lower graph shows the amplitude of the inward Ni-sensitive current on repolarization as a function of the $[Ca^{2+}]_i$ at that time. The *full line* is the fit to the cAVB data points ($R^2 = 0.83$), and the *dashed line* to the data points for control cells ($R^2 = 0.86$). (Modified from Sipido *et al.*[21])

Differences in phosphorylation rather than expression levels are well-known modulators for phospholamban and Ca^{2+} channels, and may also apply to the Na/Ca exchanger. The experimental data are currently, however, still limited, though altered increased basal Na/Ca exchange current due to phosphorylation was reported in remodeling after myocardial infarction in the pig.[24]

There are few reports of mutations in the Na/Ca exchange gene associated with arrhythmic disease, but again the link may be rather through regulation than changes in the Na/Ca exchange expression itself. Mutations in ankyrin affect the Na/Ca exchange subcellular distribution pattern and have been found in patients with arrhythmogenic long QT syndrome.[25] Functional studies of the resultant Na/Ca exchange current and action potential changes are still ongoing.

Transgenic mice with cardiac-specific knockout of the Na/Ca exchange have shown (surprisingly) to be not only viable but also remarkably normal in the global contraction and Ca^{2+} handling of myocytes.[26,27] A reduced Ca^{2+} load because of reduced Ca^{2+} current and perhaps as yet unidentified alternative Ca^{2+} transporters explains this observation. The absence of Na/Ca exchange can offer advantage during ischemia/reperfusion,[28] and the knockout mouse data suggest that drugs that target the Na/Ca exchange could be tolerated and may have potential.

ACTION POTENTIAL DURATION (APD), INSTABILITY, AND EARLY AFTERDEPOLARIZATIONS (EADs)—INTERACTION BETWEEN Na/Ca EXCHANGE AND REACTIVATION OF Ca^{2+} CURRENT

The onset of arrhythmias in the cAVB dog is marked by beat-to-beat variability of the APD even before the appearance of early EADs.[29] Instability may reflect a reduction in repolarization reserve and has been proposed as a major characteristic of drugs with proarrhythmic potential.[30] Ongoing studies by Antoons et al.[31] (manuscript in preparation) have quantified beat-to-beat variability in myocytes of the cAVB dog as the short-term variability (STV) of the APD at 90% of repolarization. STV increases with action potential prolongation but the relation is not linear and STV increases more steeply at the longest APD. Suppression of sarcoplasmic reticulum Ca^{2+} release with thapsigargin reduces the APD, indicating removal of a net inward current. The major currents that are dependent on Ca^{2+} release are the transient inactivation and recovery of the Ca^{2+} current, the transient outward Cl^- current and the inward Na/Ca exchange current. These results thus suggest that the inward Na/Ca exchange current is a major factor in the APD and, at least indirectly, of instability.

EADs can result from reactivation of Ca^{2+} window current and in the case of altered gating of the Na^+ channel, also from Na^+ current.[32,33] The Na/Ca exchange current was also proposed to have a role in the initiation of EADs, in particular for EADs that are evoked during adrenergic stimulation.[34] Spontaneous Ca^{2+} release and the accompanying inward Na/Ca exchange current would then provide the "priming" depolarizing current upon which the window Ca^{2+} current could develop the EAD.

In the dog with cAVB, the lack of an increase in I_{Ks} under adrenergic stimulation enhances the incidence of EADs.[35] The upregulated Na/Ca exchange

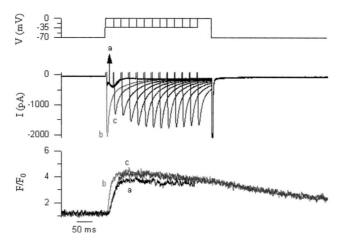

FIGURE 3. Time course of Na/Ca exchange current and dynamic modulation of the L-type Ca^{2+} current during Ca^{2+} release from the sarcoplasmic reticulum. The top panel illustrates the voltage clamp protocol, below are shown the membrane currents and $[Ca^{2+}]_i$. From a holding voltage of –70 mV, a depolarizing step is given to –35 mV (current and transient shown as "a"); during subsequent pulses a step to 0 mV is given at various time points to test for the availability of Ca^{2+} channels (current and transient shown as "b" for the step at 0 ms, as "c" for the step at 20 ms). (Modified from Antoons *et al.*[36])

current in the presence of enhanced Ca^{2+} release may facilitate the occurrence of EADs by providing an additional inward current and further prolonging the action potential, allowing sufficient time for Ca^{2+} currents to recover from inactivation. In a recent study we made a first attempt to separate the Ca^{2+} current inactivation and its subsequent recovery from the Na/Ca exchange current during Ca^{2+} release. Trace "a" of FIGURE 3 shows the net current during the Ca^{2+} transient labeled "a" evoked by a step from –70 to –35 mV (Na^+ current has been blocked). The first peak inward current is followed by a second slower inward current that peaks around the time of the peak Ca^{2+} transient. This current component is not seen in Na-free solutions (data not shown, but see Ref. 36) and represents the Na/Ca exchange current. The results of another voltage clamp protocol used to identify the time course of Ca^{2+} current inactivation and recovery are shown superimposed. The currents of "b" and "c" represent an immediate step to 0 mV and a step to 0 mV after a 25 ms delay; the peak inward current here reflects the maximal availability of Ca^{2+} current, which is initially large but sharply declines during the Ca^{2+} transient, because of Ca^{2+}-dependent inactivation. Subsequent steps to 0 mV at later time points during the step to –35 mV illustrate the subsequent recovery (and availability) of the Ca^{2+} current. This protocol suggests that during Ca^{2+} release from the sarcoplasmic reticulum, activation of inward Na/Ca exchange current and recovery of Ca^{2+} current are consecutive events and lend support to the view of the Na/Ca exchanger as a priming current for EADs.

CONCLUSIONS AND PERSPECTIVES

"Classic" views on the exchanger in arrhythmias include its well-recognized role as depolarizing current during spontaneous Ca^{2+} release and the alterations in expression in certain forms of cardiac hypertrophy and heart failure. "Novel" views relate to more complex mechanisms that increase or decrease the exchanger current, such as modulation through $[Na^+]_i$, anchoring proteins, or phosphorylation, and to more subtle roles for the exchanger in arrhythmias.

The potential of the exchanger as an antiarrhythmic target is tempered by its essential role in balancing the Ca^{2+} influx during the cardiac cycle. Data from transgenic mice suggest that suppressing the exchanger is perhaps better tolerated with a simultaneous decrease of Ca^{2+} influx. Further research on drugs that target Na/Ca exchange will have to carefully examine the effects on Ca^{2+} balance and drugs with a broader range of action, for example, including Ca^{2+} channel block, may be more useful. Preventing Na/Ca exchange activation by acting on the RyR is an attractive alternative approach but also requires that the secondary effects on sarcoplasmic reticulum Ca^{2+} load are taken into consideration.

REFERENCES

1. SIPIDO, K.R. 2006. Calcium overload, spontaneous calcium release, and ventricular arrhythmias. Heart Rhythm **3:** 977–979.
2. SIPIDO, K.R., P.G. VOLDERS, M.A. VOS & F. VERDONCK. 2002. Altered Na/Ca exchange activity in cardiac hypertrophy and heart failure: a new target for therapy? Cardiovasc. Res. **53:** 782–805.
3. POGWIZD, S.M. 2003. Clinical potential of sodium-calcium exchanger inhibitors as antiarrhythmic agents. Drugs **63:** 439–452.
4. HOBAI, I.A. & B. O'ROURKE. 2004. The potential of Na^+/Ca^{2+} exchange blockers in the treatment of cardiac disease. Expert. Opin. Investig. Drugs **13:** 653–664.
5. JANSE, M.J. 2004. Electrophysiological changes in heart failure and their relationship to arrhythmogenesis. Cardiovasc. Res. **61:** 208–217.
6. SIPIDO, K.R., A. VARRO & D. EISNER. 2006. Sodium calcium exchange as a target for antiarrhythmic therapy. Handb. Exp. Pharmacol. **171:** 159–199.
7. BRIDGE, J.H.B., J.R. SMOLLEY & K.W. SPITZER. 1990. The relationship between charge movements associated with I_{Ca} and I_{Na-Ca} in cardiac myocytes. Science **248:** 376–378.
8. ARMOUNDAS, A.A., I.A. HOBAI, G.F. TOMASELLI, *et al.* 2003. Role of sodium-calcium exchanger in modulating the action potential of ventricular myocytes from normal and failing hearts. Circ. Res. **93:** 46–53.
9. HASENFUSS, G. & B. PIESKE. 2002. Calcium cycling in congestive heart failure. J. Mol. Cell. Cardiol. **34:** 951–969.

10. JANVIER, N.C. & M.R. BOYETT. 1996. The role of Na-Ca exchange current in the cardiac action potential. Cardiovasc. Res. **32:** 69–84.

11. NAGY, Z.A., L. VIRAG, A. TOTH, *et al.* 2004. Selective inhibition of sodium-calcium exchanger by SEA-0400 decreases early and delayed afterdepolarization in canine heart. Br. J. Pharmacol. **143:** 827–831.

12. POGWIZD, S.M., K.R. SIPIDO, F. VERDONCK & D.M. BERS. 2003. Intracellular Na in animal models of hypertrophy and heart failure: contractile function and arrhythmogenesis. Cardiovasc. Res. **57:** 887–896.

13. FAUCONNIER, J., A. LACAMPAGNE, J.M. RAUZIER, *et al.* 2005. Ca^{2+}-dependent reduction of I_{K1} in rat ventricular cells: a novel paradigm for arrhythmia in heart failure? Cardiovasc. Res. **68:** 204–212.

14. POGWIZD, S.M., K. SCHLOTTHAUER, L. LI, *et al.* 2001. Arrhythmogenesis and contractile dysfunction in heart failure: roles of sodium-calcium exchange, inward rectifier potassium current, and residual ß-adrenergic responsiveness. Circ. Res. **88:** 1159–1167.

15. JIANG, D., B. XIAO, D. YANG, *et al.* 2004. RyR2 mutations linked to ventricular tachycardia and sudden death reduce the threshold for store-overload-induced Ca^{2+} release (SOICR). Proc. Natl. Acad. Sci. USA **101:** 13062–13067.

16. MARKS, A.R., S. PRIORI, M. MEMMI, *et al.* 2002. Involvement of the cardiac ryanodine receptor/calcium release channel in catecholaminergic polymorphic ventricular tachycardia. J. Cell. Physiol. **190:** 1–6.

17. WEHRENS, X.H., S.E. LEHNART, S.R. REIKEN, *et al.* 2004. Protection from cardiac arrhythmia through ryanodine receptor-stabilizing protein calstabin2. Science **304:** 292–296.

18. LIU, N., B. COLOMBI, M. MEMMI, *et al.* 2006. Arrhythmogenesis in catecholaminergic polymorphic ventricular tachycardia: insights from a RyR2 R4496C knock-in mouse model. Circ. Res. **99:** 292–298.

19. VENETUCCI, L.A., A.W. TRAFFORD, M.E. DIAZ, *et al.* 2006. Reducing ryanodine receptor open probability as a means to abolish spontaneous Ca^{2+} release and increase $^{2+}$ transient amplitude in adult ventricular myocytes. Circ. Res. **98:** 1299–1305.

20. VENETUCCI, L.A., A.W. TRAFFORD & D.A. EISNER. 2007. Increasing ryanodine receptor open probability alone does not produce arrhythmogenic calcium waves: threshold sarcoplasmic reticulum calcium content is required. Circ. Res. **100:** 105–111.

21. SIPIDO, K.R., P.G.A. VOLDERS, S.H. DE GROOT, *et al.* 2000. Enhanced Ca^{2+} release and Na/Ca exchange activity in hypertrophied canine ventricular myocytes: a potential link between contractile adaptation and arrhythmogenesis. Circulation **102:** 2137–2144.

22. VERDONCK, F., P.G.A. VOLDERS, M.A. VOS & K.R. SIPIDO. 2003. Increased Na^+ concentration and altered Na/K pump activity in hypertrophied canine ventricular cells. Cardiovasc. Res. **57:** 1035–1043.

23. WEBER, C.R., K.S. GINSBURG, K.D. PHILIPSON, *et al.* 2001. Allosteric regulation of Na/Ca exchange current by cytosolic Ca in intact cardiac myocytes. J. Gen. Physiol. **117:** 119–132.

24. WEI, S.K., A. RUKNUDIN, S.U. HANLON, *et al.* 2003. Protein kinase A hyperphosphorylation increases basal current but decreases beta-adrenergic responsiveness of the sarcolemmal Na^+-Ca^{2+} exchanger in failing pig myocytes. Circ. Res. **92:** 897–903.

25. MOHLER, P.J., J.J. SCHOTT, A.O. GRAMOLINI, *et al.* 2003. Ankyrin-B mutation causes type 4 long-QT cardiac arrhythmia and sudden cardiac death. Nature **421:** 634–639.

26. POTT, C., K.D. PHILIPSON & J.I. GOLDHABER. 2005. Excitation-contraction coupling in Na^+-Ca^{2+} exchanger knockout mice: reduced transsarcolemmal Ca^{2+} flux. Circ. Res. **97:** 1288–1295.

27. REUTER, H., S.A. HENDERSON, T. HAN, *et al.* 2003. Cardiac excitation-contraction coupling in the absence of Na^+ - Ca^{2+} exchange. Cell Calcium **34:** 19–26.

28. IMAHASHI, K., C. POTT, J.I. GOLDHABER, *et al.* 2005. Cardiac-specific ablation of the Na^+-Ca^{2+} exchanger confers protection against ischemia/reperfusion injury. Circ. Res. **97:** 916–921.

29. THOMSEN, M.B., S.C. VERDUYN, M. STENGL, *et al.* 2004. Increased short-term variability of repolarization predicts d-sotalol-induced torsades de pointes in dogs. Circulation **110:** 2453–2459.

30. HONDEGHEM, L.M., L. CARLSSON & G. DUKER. 2001. Instability and triangulation of the action potential predict serious proarrhythmia, but action potential duration prolongation is antiarrhythmic. Circulation **103:** 2004–2013.

31. ANTOONS, G., M. STENGL, M.B. THOMSEN, *et al.* 2005. Sarcoplasmic reticulum Ca^{2+} release and repolarization lability in myocytes from the dog with chronic atrioventricular block (cAVB) Biophys. J. **88:** 298A.

32. JANUARY, C.T. & S. SHOROFSKY. 1990. Early afterdepolarizations: newer insights into cellular mechanisms. J. Cardiovasc. Electrophysiol. **1:** 161–169.

33. BOUTJDIR, M., M. RESTIVO, Y. WEI, *et al.* 1994. Early afterdepolarization formation in cardiac myocytes: analysis of phase plane patterns, action potential, and membrane currents. J. Cardiovasc. Electrophysiol. **5:** 609–620.

34. VOLDERS, P.G., M.A. VOS, B. SZABO, *et al.* 2000. Progress in the understanding of cardiac early afterdepolarizations and torsades de pointes: time to revise current concepts.. Cardiovasc. Res. **46:** 376–392.

35. STENGL, M., C. RAMAKERS, D.W. DONKER, *et al.* 2006. Temporal patterns of electrical remodeling in canine ventricular hypertrophy: focus on IKs downregulation and blunted beta-adrenergic activation. Cardiovasc. Res. **72:** 90–100.

36. ANTOONS, G., P.G. VOLDERS, T. STANKOVICOVA, *et al.* 2006. Window Ca^{2+} current and its modulation by Ca^{2+} release in hypertrophied cardiac myocytes from dogs with chronic atrioventricular block. J. Physiol. **579:** 147–160.

37. CARMELIET, E. & J. VEREECKE. 2002. Cardiac Cellular Electrophysiology. Kluwer Academic Publishers. Boston, MA.

The Role of the Cardiac Na^+/Ca^{2+} Exchanger in Reverse Remodeling

Relevance for LVAD–Recovery

CESARE M. N. TERRACCIANO, MAREN U. KOBAN, GOPAL K. SOPPA,
URSZULA SIEDLECKA, JOON LEE, MARK A. STAGG,
AND MAGDI H. YACOUB

Heart Science Centre, Imperial College London, National Heart and Lung Institute, Harefield Hospital, Harefield UB9 6JH, United Kingdom

ABSTRACT: Different strategies can, at least in certain conditions, prevent or reverse myocardial remodeling due to heart failure and induce myocardial functional improvement. Na^+/Ca^{2+} exchanger (NCX) is considered a major player in the pathophysiology of heart failure but its role in reverse remodeling is unknown. A combination of mechanical unloading by left ventricular assist devices (LVADs) and pharmacological therapy has been shown to induce clinical recovery in a limited number of patients with end-stage heart failure. In myocytes isolated from these patients we found that, after LVAD treatment, NCX1/SERCA2a mRNA was 38% higher than at device implant. We studied the ability of NCX to extrude Ca^{2+} during caffeine-induced SR Ca^{2+} release in isolated ventricular myocytes from these patients. The time constant of decline was slower in heart failure. In myocytes from patients with clinical recovery following mechanical and pharmacological treatment, NCX1-mediated Ca^{2+} extrusion was faster compared with myocytes from patient who, despite identical treatment, did not recover. We propose that increased NCX function may be associated with reverse remodeling in patients and that factors that regulate NCX function (i.e., phosphorylation or intracellular $[Na^+]$) other than NCX expression levels alone, may have detrimental consequences on cardiac function.

KEYWORDS: heart failure; Na/Ca exchanger; mechanical devices; calcium; remodeling

Address for correspondence: Dr. Cesare M.N. Terracciano, Laboratory of Cell Electrophysiology, Imperial College London, Heart Science Centre, Harefield Hospital, Harefield, Middlesex UB9 6JH, UK. Voice: +44-1895-453-874; fax: +44-1895-828-900.
c.terracciano@imperial.ac.uk

Ann. N.Y. Acad. Sci. 1099: 349–360 (2007). © 2007 New York Academy of Sciences.
doi: 10.1196/annals.1387.061

INTRODUCTION

Heart failure represents a major disease burden in both Western and developing countries. The disease is progressive and, when severe, has a very poor prognosis.[1] This, combined with its adverse effects on both quality of life and survival,[2,3] demands the need for effective treatment. Currently, transplantation is the most effective form of treatment in advanced heart failure but this still remains an inadequate option due to the limited numbers of donor organs.[4] There is therefore a need to develop alternative strategies.

LVAD and Reverse Remodeling

Emerging interest focuses on the use of left ventricular assist devices (LVADs) for the management of severe chronic heart failure. LVADs are pumps that support heart function by working in parallel with the normal circulation. These mechanical devices are predominantly employed as a "bridge to transplantation," or destination therapy for patients with contraindications to cardiac transplantation.[5] Some studies have shown that mechanical unloading induced by LVADs has beneficial effects on the recipient myocardium, suggesting that these devices can induce reverse remodeling and lead to myocardial recovery.[6] In few cases it has been possible to remove the LVAD after a period of treatment without recurring to transplantation, leading to the concept of LVAD as a "bridge to recovery."[7]

Tissue from patients treated with LVADs has been extensively used to study reverse remodeling. A number of cellular, functional, and molecular changes, implicated in the development of heart failure, are reversed with LVAD therapy and are believed to contribute to this recovery.[6] Excitation–contraction (EC) coupling in particular has been studied by several groups (see FIG. 1). A return to normal values is not always observed and it is now clear that compensation pathways are activated during LVAD treatment.[8,9]

One of the major limitations of mechanical unloading in the treatment of heart failure is the consequent heart muscle atrophy[10–13] and cardiomyocyte dysfunction particularly associated with altered myofilament sensitivity to Ca^{2+}.[14,15] This may explain why the recovery rate in patients treated with LVADs alone is still low and is reported as approximately 5% only.[16] For this reason mechanical unloading alone may not be sufficient to obtain reliable reverse remodeling in heart failure, and whether the assessment of tissue from these patients in the absence of clinical recovery is a valid tool to study reverse remodeling remains debatable.

A recent study from our center has reported that the rate of clinical recovery can be improved in selected patients using LVADs in combination with specific pharmacological therapy.[17] Patients with heart failure due to nonischemic cardiomyopathy were treated with combined mechanical and pharmacological

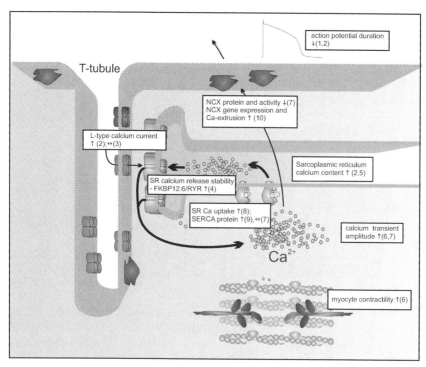

FIGURE 1. EC coupling after LVAD support. This figure shows the main findings reported for EC coupling of cardiac tissue from heart failure patients taken at the explantation of LVADs. (**1**)[31] Harding JD *et al.*, *Circulation* 2001; (**2**)[28] Terracciano CM *et al.*, *Circulation* 2004; (**3**)[32] Chen X *et al.*, *Circ. Res.* 2002; (**4**)[33] Marx SO *et al.*, *Cell* 2000; (**5**)[9] Terracciano CM *et al.*, *Eur. Heart J.* 2003; (**6**)[34] Dipla K *et al.*, *Circulation* 1998; (**7**)[8] Chaudhary KW *et al.*, *J. Am. Coll. Cardiol.* 2004; (**8**)[7] Frazier MD *et al.*, *Ann. Thor. Surg.* 1999; (**9**)[35] Heerdt PM *et al.*, *Circulation* 2000; (**10**) this article, see Results.

therapy (lisinopril, carvedilol, spironolactone, and losartan). Once regression of left ventricular enlargement had been achieved, the β_2-adrenergic-receptor agonist clenbuterol was administered to prevent myocardial atrophy. Approximately 70% of the patients had sufficient myocardial recovery to undergo explantation of the LVAD without recurrence of heart failure.[17] These data suggest the existence of alternative treatment approaches for end-stage heart failure and warrants further study for the use of LVADs as a platform to treat myocardial dysfunction.[18] Importantly, tissue from the hearts of these patients, rather than from the general population of LVAD-treated patients, represent an unique opportunity to study the pathophysiology of functional reverse remodeling; comparisons can be made between tissue before and after the treatment and between tissue from patients who present clinical improvement and those who, despite an identical treatment, do not recover.

Role of the NCX in Heart Failure

The Na^+/Ca^{2+} exchanger (NCX) has been implicated in the pathophysiology of heart failure and proposed as a target for therapy, but its role remains controversial.[19] In human heart failure the NCX has been reported to be overexpressed by some studies[20] but not others.[21] We have previously discussed the functional consequences of the overexpression of NCX[22,23] and pointed out that specific conditions, such as species, intracellular $[Na^+]$ and $[Ca^{2+}]$, phosphorylation status, concomitant status of other Ca^{2+} regulatory proteins,[24] may produce opposing results on Ca^{2+} regulation mediated by the overexpressed NCX. Some animal studies have shown that inhibiting NCX improves function in heart failure.[25] Other studies show that overexpression of NCX decreases the progression of systolic and diastolic contractile dysfunction *in vivo.*[26] Overall, the importance and the role of NCX in the development and treatment of human heart failure remain unclear.

Here, using molecular biological and electrophysiological techniques, we have tested the hypothesis that NCX expression and function are affected during recovery from heart failure in the tissue isolated from patients with myocardial recovery.

METHODS

Patients

Tissue was obtained from patients in class IV NYHA, with normal coronary arteries, during implantation of the LVAD (LVAD core). Using myocardial biopsies, tissue was also taken from patients during the LVAD explant (recovery) or from explanted hearts when the patients with LVAD support did not show clinical signs of recovery and underwent cardiac transplantation (transplanted). The clinical details of these patients are described in Birks *et al.*[17] Briefly, during LVAD support the patients were treated with a combination therapy composed of mechanical support and administration of beta-blockers, ACE-inhibitors, angiotensin II receptor antagonists, and spironolactone. This was followed by the administration of the β_2 receptor agonist, clenbuterol, which stimulates physiological hypertrophy. Biopsies obtained from donor hearts were used as control. The study was given ethical approval by the Brompton, Harefield, and NHLI Ethics Committee. Informed consent was obtained from all patients.

Cell Isolation

Ventricular tissue was placed in ice-cold cardioplegia and transported to the laboratory where cardiac myocytes were enzymatically isolated using a technique previously described.[9] Briefly, tissue was finely chopped using razor blades. It was then washed and oxygenated in a nominally zero-Ca solution containing (in mM) NaCl, 120; KCl, 5.4; $MgSO_4$, 5; pyruvate, 5; glucose, 20;

taurine, 20; HEPES, 10; nitrolotriacetic acid (NTA), 5 (bubbled with 100% O_2; pH = 6.8) for 12 min, changing the solution every 3 min. The tissue was then gently shaken in enzyme solution (in mM: NaCl, 120; KCl, 5.4; $MgSO_4$, 5; $CaCl_2$, 0.2; pyruvate, 5; glucose, 20; taurine, 20; HEPES, 10; bubbled with 100% O_2; pH =7.4) containing protease (4 U/mL, Sigma, Poole, UK) for 45 min and collagenase (1mg/mL, Sigma) for a further 45 min. A further 45-min digestion in collagenase followed. At the end of every cycle, the suspension was gently spun (500 rpm) and the pellet resuspended in enzyme solution. Cells were kept at room temperature and used within 5–6 h from isolation.

RNA Analysis

RNA was obtained from biopsy samples from donor hearts or LVAD patients (time of LVAD implantation and recovery), stored in RNAlater solution (Ambion, Huntingdon, UK), and extracted using RNAqueous kit (Ambion) as per the manufacturer's instruction. cDNA was synthesized using RETROscript (Ambion) cDNA synthesis kit. Human NCX1, SERCA2a, and phospholamban transcripts were amplified by quantitative real-time PCR using TnIc as internal control as described previously,[27] or using gene-specific oligonucleotides and probes for human SERCA2a and NCX1 in the same amplifying reaction for paired implant/recovery biopsies from LVAD patients. Fluorescently labeled primers and probes were designed using Primer Express Software (Applied Biosystems, Foster City, CA) and based on published human mRNA sequences. PCR amplifications were performed under conditions described earlier using TaqMan technology (Applied Biosystems) and an ABI-Prism 7700 sequence detector. Each sample was analyzed in duplicate. Statistical analysis was carried out on ΔCT values using a two-way *t*-test as described previously.[27]

NCX-Mediated Ca^{2+} Extrusion

The cells were superfused with normal tyrode solution containing (mM): NaCl, 140; KCl, 6; $MgCl_2$, 1; $CaCl_2$, 1; glucose, 10; HEPES, 10; pH to 7.4 with 2 M NaOH. The electrophysiological experiments were performed using an Axoclamp-2B system (Axon Instruments, Molecular Devices, Sunnyvale, CA). To avoid dialysis of the cells and to minimize the effects of changing the intracellular environment, high-resistance microelectrodes (15–25 MΩ; Clark Electromedical Instruments, Harvard Apparatus, Edenbridge, Kent, UK) were used. The microelectrode filling solution contained: KCl, 2M; EGTA; 0.1 mM; HEPES, 5 mM, pH = 7.2. Protocols were controlled with pClamp 7 software (Axon Instruments). Rapid application of caffeine followed a stimulation train of voltage clamp steps from –80 to 0 mV. A transient inward current ascribed to the NCX was recorded. The speed of decline of this current is taken as an index of NCX-mediated Ca removal and was measured by fitting a monoexponential curve using pClamp software.

Solutions

Caffeine (20 mM) was added as a solid to the final solution. The temperature of the superfusing solution was approximately 37°C. The rate of superfusion was 2–3 mL/min except during fast application of caffeine when it was 12–15 mL/min. Miniature solenoid valves (Lee, Essex, UK) were used to produce fast changes in the superfusate.

Data Acquisition and Statistical Analysis

The data obtained from the Axoclamp-2B system was recorded on a computer using pClamp7. The rate of sampling was between 0.5 and 3 kHz. To assess statistical differences between means, a one-way analysis of variance (ANOVA) with Bonferroni post test was used. $P < 0.05$ was considered significant. Unless otherwise specified, the results are expressed as mean ± standard error of the mean (SEM). The n number represents the number of patients that have undergone investigation, unless otherwise specified. Each value represents the average of the values recorded from single cells isolated from each patient. A minimum n number 4 was used.

RESULTS

The level of gene expression for NCX1 was assessed in samples from patients undergoing LVAD surgery and compared to donor tissue (FIG. 2 A). As previously reported in patients with heart failure,[20] a significant increase in NCX1 mRNA compared with donor hearts was detected (FIG. 2 A, upper graph). In the same samples, SERCA2a and phospholamban gene expression were significantly reduced (FIG. 2 A, lower graphs). In tissue taken from patients with clinical recovery after LVAD and pharmacological treatment no statistical difference was found in SERCA and NCX expression compared with tissue taken at implantation (not shown). However, the ratio between NCX1 and SERCA2a was significantly increased in patients with myocardial recovery when analyzed in the same reaction (FIG. 2 B).

We have previously reported that there is an increased SR Ca^{2+} content in myocytes from patients with clinical recovery.[28] This was not observed in myocytes from patients without recovery. Here we present a further analysis of that data to evaluate the role of NCX. We fitted monoexponential curves on the decay of the caffeine-induced transient inward current, representing the extrusion of Ca^{2+} via NCX. FIGURE 3 A, upper graph, shows that there was a slower current decay in the LVAD core group compared with the donor group. After LVAD treatment the speed of decay was increased in myocytes from patients with recovery but not from those from transplanted patients.

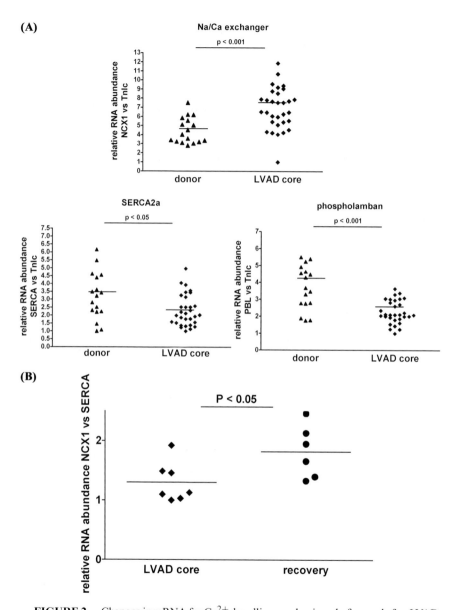

FIGURE 2. Changes in mRNA for Ca^{2+}-handling mechanisms before and after LVAD treatment. cDNA analysis was performed by quantitative real-time PCR to investigate RNA abundance for key players of the Ca^{2+} regulatory mechanisms. As previously reported in tissue from failing hearts, RNA levels of SERCA2a and phospholamban were reduced whereas NCX1 was increased (**A**) (donor $n = 17$; LVAD cores $n = 31$). In patients with recovery (paired samples), NCX1 mRNA was increased compared with SERCA2a mRNA (**B**).

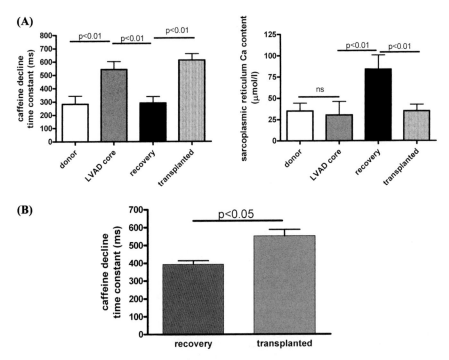

FIGURE 3. NCX-mediated Ca^{2+} extrusion is faster in myocytes from patients with recovery. Panel **A** shows values for the time constant of the monoexponential fit on the decay phase of the caffeine-induced inward current (left) and SR Ca^{2+} content calculated from the integral of the caffeine current. The values are average cellular data per patient and $n =$ patients: donor = 4, LVAD core = 7; recovery = 5; transplanted = 4. Panel **B** shows values for the time constant of the monoexponential fit only for cells with similar SR Ca^{2+} content. The data are single cellular values and $n =$ cells.

However, because the recovery group had a significant increase in SR Ca^{2+} content (FIG. 3 A, lower panel), this result may just be the consequence of larger $[Ca^{2+}]$, leading to a faster NCX-mediated Ca^{2+} extrusion, even if NCX basal activity was unchanged or reduced. For this reason we performed a further analysis of the data and studied only the cells with SR Ca^{2+} content between 30 and 50 μmol/L. Five cells from the recovery group and 11 from the transplanted group fitted these criteria and the average SR Ca^{2+} content was similar (38.4 $\pm$ 2.8 μmol/L in the recovery group and 37.7 $\pm$ 1.7 μmol/L in the transplanted group). FIGURE 3 B shows that in this subgroup of cells the time constant of decay of the caffeine-induced transient inward current in the recovery group was still smaller than the transplanted group suggesting that Ca^{2+} extrusion via the NCX is increased during functional myocardial improvement.

DISCUSSION

We studied NCX gene expression and activity in tissue from heart failure patients treated with a combined pharmacological and LVAD therapy. We showed that myocardial recovery from heart failure is associated with increased gene expression of NCX1 versus SERCA2a and with increased NCX-mediated Ca^{2+} extrusion.

Our data question the rationale of proposed strategies for heart failure aimed at reducing the expression and function of NCX. The role of NCX in heart failure remains controversial[19] and this may depend on different experimental models and conditions.[19,22] In human heart failure studies the overexpression of NCX is often reported[20] although not universally.[21] Our data show that in patients undergoing LVAD surgery the gene expression for NCX was upregulated, and this is consistent with the majority of studies. Surprisingly, patients with successful explantation of the device did not show a normalization of this parameter: when compared with the changes in the expression for SERCA2a, NCX1 gene expression was further increased. This observation supports the notion that the functional improvement observed is not due to global normalization of the parameters affected in heart failure but rather to selective changes with compensatory effects.[8,9] Unfortunately, only patients in the recovery group were studied and a comparison with transplanted patients for this parameter is not yet available.

We also studied the NCX ability to extrude Ca^{2+} in myocytes using the caffeine application technique.[29] The caffeine-induced transient inward current is carried by the NCX. While the integral of the current indicates the amount of Ca^{2+} released from the SR,[29] the time course is dependent on NCX activity. We showed that the current decay, an index of the efficacy of Ca^{2+} extrusion, was faster in the recovery compared with the transplanted group, even in the presence of similar SR Ca^{2+} release. This suggests that increased NCX function is associated with clinical recovery.

In a previous study Chaudhary *et al.* showed that NCX activity was reduced in tissue from patients at LVAD removal due to transplantation.[8] There are several possible reasons to explain this discrepancy. First, we separated the patients with recovery from those without recovery. The faster Ca^{2+} extrusion only occurred in the recovery group, which represents only around 5% of the general LVAD-treated patient population.[16] Second, our patients undergo a specific pharmacological regimen. We have recently reported that, in a rat model of heart failure, the β_2 agonist, clenbuterol, used in the combination therapy, increases NCX current after mechanical unloading (unpublished observations). Finally we used sharp microelectrodes to avoid dialysis of intracellular Na^+. It has been proposed that the increased $[Na^+]$ changes the reversal potential of the NCX in heart failure and it is possible that, during recovery, $[Na^+]$ returns to lower values.[30] This, rather than the overexpression and specific activity of the exchanger, may determine its ability to extrude Ca^{2+}.

In conclusion, we propose that reverse remodeling in human heart failure may be associated with increased NCX function and that factors that regulate NCX function (i.e., phosphorylation or intracellular $[Na^{2+}]$), other than NCX expression levels alone, may have detrimental consequences on cardiac function.

ACKNOWLEDGMENTS

We would like to thank the Wellcome Trust, the British Heart Foundation, and the Magdi Yacoub Institute for financial support.

REFERENCES

1. JESSUP, M. & S. BROZENA. 2003. Heart failure. N. Engl. J. Med. **348**: 2007–2018.
2. GIBBS, L.M., J. ADDINGTON-HALL & J.S. GIBBS. 1998. Dying from heart failure: lessons from palliative care. Many patients would benefit from palliative care at the end of their lives. BMJ **317**: 961–962.
3. STEWART, S., K. MACINTYRE, D.J. HOLE, *et al.* 2001. More 'malignant' than cancer? Five-year survival following a first admission for heart failure. Eur. J. Heart Fail. **3**: 315–322.
4. WIGHT, C., K. JAGER, G. BLOK, *et al.* 1996. Overview of the European Donor Hospital Education Program Transplant. Proc. **28**: 422–423.
5. ROSE, E.A., A.C. GELIJNS, A.J. MOSKOWITZ, *et al.* 2001. Long-term mechanical left ventricular assistance for end-stage heart failure. N. Engl. J. Med. **345**: 1435–1443.
6. WOHLSCHLAEGER, J., K.J. SCHMITZ, C. SCHMID, *et al.* 2005. Reverse remodeling following insertion of left ventricular assist devices (LVAD): a review of the morphological and molecular changes. Cardiovasc. Res. **68**: 376–386.
7. FRAZIER, O.H. & T.J. MYERS. 1999. Left ventricular assist system as a bridge to myocardial recovery. Ann. Thorac. Surg. **68**: 734–741.
8. CHAUDHARY, K.W., E.I. ROSSMAN, V. PIACENTINO, III, *et al.* 2004. Altered myocardial Ca^{2+} cycling after left ventricular assist device support in the failing human heart. J. Am. Coll. Cardiol. **44**: 837–845.
9. TERRACCIANO, C.M., S.E. HARDING, D. ADAMSON, *et al.* 2003. Changes in sarcolemmal Ca entry and sarcoplasmic reticulum Ca content in ventricular myocytes from patients with end-stage heart failure following myocardial recovery after combined pharmacological and ventricular assist device therapy. Eur. Heart J. **24**: 1329–1339.
10. KINOSHITA, M., H. TAKANO, Y. TAENAKA, *et al.* 1988. Cardiac disuse atrophy during LVAD pumping. ASAIO Trans. **34**: 208–212.
11. KOLAR, F., C. MACNAUGHTON, F. PAPOUSEK, *et al.* 1995. Changes in calcium handling in atrophic heterotopically isotransplanted rat hearts. Basic Res. Cardiol. **90**: 475–481.
12. RAKUSAN, K., M.I. HERON, F. KOLAR & B. KORECKY. 1997. Transplantation-induced atrophy of normal and hypertrophic rat hearts: effect on cardiac myocytes and capillaries. J. Mol. Cell Cardiol. **29**: 1045–1054.

13. WELSH, D.C., K. DIPLA, P.H. MCNULTY, *et al.* 2001. Preserved contractile function despite atrophic remodeling in unloaded rat hearts. Am. J. Physiol Heart Circ. Physiol. **281**: H1131–H1136.

14. RITTER, M., Z. SU, *et al.* 2000. Cardiac unloading alters contractility and calcium homeostasis in ventricular myocytes. J. Mol. Cell Cardiol. **32**: 577–584.

15. SOPPA, G.K., R.T. SMOLENSKI, N. LATIF, *et al.* 2005. Effects of chronic administration of clenbuterol on function and metabolism of adult rat cardiac muscle. Am. J. Physiol. Heart Circ. Physiol. **288**: H1468–H1476.

16. MANCINI, D.M., A. BENIAMINOVITZ, H. LEVIN, *et al.* 1998. Low incidence of myocardial recovery after left ventricular assist device implantation in patients with chronic heart failure. Circulation **98**: 2383–2389.

17. BIRKS, E.J., P.D. TANSLEY, J. HARDY, *et al.* 2006. Left ventricular assist device and drug therapy for the reversal of heart failure. N. Engl. J. Med. **355**: 1873–1884.

18. RENLUND, D.G. & A.G. KFOURY. 2006. When the failing, end-stage heart is not end-stage N. Engl. J. Med. **355**: 1922–1925.

19. SIPIDO, K.R., P.G. VOLDERS, M.A. VOS & F. VERDONCK. 2002. Altered Na/Ca exchange activity in cardiac hypertrophy and heart failure: a new target for therapy? Cardiovasc. Res. **53**: 782–805.

20. STUDER, R., H. REINECKE, J. BILGER, *et al.* 1994. Gene expression of the cardiac Na(+)-Ca^{2+} exchanger in end-stage human heart failure. Circ. Res. **75**: 443–453.

21. SCHWINGER, R.H., J. WANG, K. FRANK, *et al.* 1999. Reduced sodium pump alpha1, alpha3, and beta1-isoform protein levels and Na$^+$,K+-ATPase activity but unchanged Na$^+$-Ca^{2+} exchanger protein levels in human heart failure. Circulation **99**: 2105–2112.

22. TERRACCIANO, C. 2002. Functional consequences of Na/Ca exchanger overexpression in cardiac myocytes. Ann. N. Y. Acad. Sci. **976**: 520–527.

23. TERRACCIANO, C.M., A.I. SOUZA, K.D. PHILIPSON & K.T. MACLEOD. 1998. Na$^+$-Ca^{2+} exchange and sarcoplasmic reticular Ca^{2+} regulation in ventricular myocytes from transgenic mice overexpressing the Na$^+$-Ca^{2+} exchanger J. Physiol. **512**(Pt 3): 651–667.

24. TERRACCIANO, C.M., K.D. PHILIPSON & K.T. MACLEOD. 2001. Overexpression of the Na(+)/Ca(2+) exchanger and inhibition of the sarcoplasmic reticulum Ca(2+)-ATPase in ventricular myocytes from transgenic mice. Cardiovasc. Res. **49**: 38–47.

25. HOBAI, I.A., C. MAACK & B. O'ROURKE. 2004. Partial inhibition of sodium/calcium exchange restores cellular calcium handling in canine heart failure Circ. Res. **95**: 292–299.

26. MUNCH, G., K. ROSPORT, C. BAUMGARTNER, *et al.* 2006. Functional alterations after cardiac sodium-calcium exchanger overexpression in heart failure. Am. J. Physiol. Heart Circ. Physiol. **291**: H488–H495.

27. BARTON, P.J., E.J. BIRKS, L.E. FELKIN, *et al.* 2003. Increased expression of extracellular matrix regulators TIMP1 and MMP1 in deteriorating heart failure. J. Heart Lung Transplant. **22**: 738–744.

28. TERRACCIANO, C.M., J. HARDY, E.J. BIRKS, *et al.* 2004. Clinical recovery from end-stage heart failure using left-ventricular assist device and pharmacological therapy correlates with increased sarcoplasmic reticulum calcium content but not with regression of cellular hypertrophy. Circulation **109**: 2263–2265.

29. VARRO, A., N. NEGRETTI, S.B. HESTER & D.A. EISNER. 1993. An estimate of the calcium content of the sarcoplasmic reticulum in rat ventricular myocytes. Pflugers Arch. **423**: 158–160.

30. BERS, D.M., S. DESPA & J. BOSSUYT. 2006. Regulation of Ca^{2+} and Na^+ in normal and failing cardiac myocytes. Ann. N. Y. Acad. Sci. **1080**: 165–177.
31. HARDING, J.D., V. PIACENTINO, III, J.P. GAUGHAN, *et al.* 2001. Electrophysiological alterations after mechanical circulatory support in patients with advanced cardiac failure. Circulation **104**: 1241–1247.
32. CHEN, X., V. PIACENTINO, III, S. FURUKAWA, *et al.* 2002. L-type Ca^{2+} channel density and regulation are altered in failing human ventricular myocytes and recover after support with mechanical assist devices. Circ. Res. **91**: 517–524.
33. MARX, S.O., S. REIKEN, Y. HISAMATSU, *et al.* 2000. PKA phosphorylation dissociates FKBP12.6 from the calcium release channel (ryanodine receptor): defective regulation in failing hearts. Cell **101**: 365–376.
34. DIPLA, K., J.A. MATTIELLO, V. JEEVANANDAM, *et al.* 1998. Myocyte recovery after mechanical circulatory support in humans with end-stage heart failure. Circulation **97**: 2316–2322.
35. HEERDT, P.M., J.W. HOLMES, B. CAI, *et al.* 2000. Chronic unloading by left ventricular assist device reverses contractile dysfunction and alters gene expression in end-stage heart failure Circulation. **102**: 2713–2719.

Regulation of Na^+/Ca^{2+} Exchange Current in the Normal and Failing Heart

MICHAEL REPPEL,[a,b] BERND K. FLEISCHMANN,[c] HANNES REUTER,[d]
FRANK PILLEKAMP,[a] HERIBERT SCHUNKERT,[b]
AND JÜRGEN HESCHELER[a]

[a]Institute of Neurophysiology, University of Cologne, D-50931 Cologne,
Germany

[b]Medizinische Klinik II, University of Schleswig-Holstein, Campus Lübeck,
23538 Lübeck, Germany

[c]Institute of Physiology I, University of Bonn, Life and Brain Center,
53127 Bonn, Germany

[d]Department of Internal Medicine III, University of Cologne, D-50931 Cologne,
Germany

ABSTRACT: Cardiac NCX is modulated by diverse regulatory elements.
Although there is consensus about the regulatory function of Na^+ and
Ca^{2+} and other elements, for example, ATP, there is still a controversial debate about the functional role of cyclic nucleotides and protein
kinases. Future studies should focus on that topic since disturbances of
cAMP/cGMP concentration and kinase activity may lead to severe functional disorders in the diseased heart. S100A1 is presumably a novel
regulator of NCX.

KEYWORDS: NCX; cardiac development; electrophysiology; embryonic
heart; calcium

INTRODUCTION

Sodium–calcium exchange (NCX) depends on intracellular and extracellular $[Na^+]$ and $[Ca^{2+}]$, and also because it is electrogenic on the membrane
potential. NCX activity is also allosterically regulated by $[Ca^{2+}]_i$-dependent
activation, $[Na^+]_i$-dependent inactivation, and requires ATP.[1,2]

To date, however, the signaling pathways being involved in alterations of embryonic *NCX* function have remained elusive. Especially, the detailed mechanism of the regulation of cardiac NCX activity is complex, and depends on
many different factors.[3–6] Besides regulation by Na^+ and Ca^{2+} itself a growing
body of evidence suggests an involvement of cyclic nucleotides and protein

Address for correspondence: Michael Reppel, University of Luebeck, Medizinische Klinik II, Ratze-
burger Allee 160, D-23538 Luebeck. Voice: 0049-451-5002500; fax: 0049-451-5056130.
  akp72@uni-koeln.de

Ann. N.Y. Acad. Sci. 1099: 361–372 (2007). © 2007 New York Academy of Sciences.
doi: 10.1196/annals.1387.065

kinases. In adults, protein kinase A (PKA)-dependent phosphorylation of substrates, for example, voltage-dependent Ca^{2+} channels (VDCCs) and *NCX*, has been shown to be a critical determinant of contractility and Ca^{2+} homeostasis and to be mediated via cAMP-dependent activation of PKA.[7] S100A1, a novel regulator of cardiac contractility affecting PKA,[8] ryanodine receptor[9,10] and SR Ca^{2+} ATPase activity,[9] has also been suggested to affect NCX activity in cardiomyocytes.[11] Similar to PKA, the protein kinase C (PKC) family has been implicated in a diverse array of cellular responses.[12–14] Among others, for example, Troponin I and VDCCs,[15–19] also NCX is discussed to serve as a substrate for activated PKC.

Ca^{2+} FLUXES IN MYOCYTES FROM CONTROL AND FAILING HEARTS

In cardiac myocytes, four pathways mediate Ca^{2+} removal from the cytosol during diastole: (*a*) the SR Ca^{2+}-ATPase (SERCA), (*b*) NCX, (*c*) mitochondrial Ca^{2+} uniporter, and (*d*) sarcolemmal Ca^{2+}-ATPase.[20] In rabbit ventricle, SERCA removes 70% of Ca^{2+} and NCX 28%, whereas 2% are maintained by the mitochondrial Ca^{2+} uniporter and sarcolemmal Ca^{2+}-ATPase. In rat ventricle, the activity of SERCA is higher than in rabbits and Ca^{2+} removal through the exchanger is lower. This results in 92% of cytosolic Ca^{2+} being taken back into SR by the SERCA and only 7% being extruded via NCX. The balance of fluxes in mouse ventricle is quantitatively like rat, whereas in dog, cat, guinea pig, and human ventricle, it resembles more the rabbit.[2]

A decreased SERCA activity in heart failure combined with an increased NCX activity enables NCX to compete with the SERCA in extruding Ca^{2+} from the cytosol during relaxation. This brings the overall Ca^{2+} transported by the SERCA and NCX closer to equal in heart failure. In human end-stage heart failure, Piacentino et al.[21] found a 40% reduction of SERCA function, but unaltered NCX extrusion, resulting again in more equal Ca^{2+} extrusion via NCX and SERCA. Hasenfuss et al.[22] found a spectrum of increase in NCX and decrease in SERCA expression in individual heart failure patients, suggesting that there is individual variation as to how much the decrease in SR Ca^{2+} load in heart failure is due to increased NCX and decreased SERCA function. Taken together, reduced SERCA function and enhanced NCX function can both contribute to reduced SR Ca^{2+} content in heart failure. Therefore, regulation of these key Ca^{2+} extrusion mechanisms seems to be essential for proper cardiac function and maintenance of Ca^{2+} homeostasis.

CARDIAC MYOCYTES Ca^{2+} AND Na^+ REGULATION IN NORMAL AND FAILING HEARTS

During the course of heart failure changes in NCX function may contribute to abnormal Ca^{2+} regulation. In normal cardiomyocytes, NCX extrudes Ca^{2+}

for most of the AP. The following factors shift the direction of NCX-mediated Ca^{2+} transport in failing cardiomyocytes: (*a*) reduced systolic but increased diastolic $[Ca^{2+}]_i$, (*b*) prolonged AP duration, and (*c*) elevated $[Na]_i$.[23]

Both, diastolic $[Ca^{2+}]_i$ and $[Na^+]_i$ are elevated in heart failure.[24,25] There are two possible explanations for this: (*a*) decreased Ca^{2+} and Na^+ extrusion and (*b*) increased Ca^{2+} and Na^+ influx. Na^+ enters the cell through various pathways, including NCX, Na^+ channels, and Na^+/H^+ exchanger (NHE), whereas the sodium pump is the main route for Na^+ efflux. Na^+ and Ca^{2+} can also be extruded by NCX working in the reverse mode and vice versa. However, throughout the cardiac cycle there is net Na^+ influx and Ca^{2+} efflux via NCX.

Although structure–function studies have identified the protein regions of the exchanger subserving the Na^+- and Ca^{2+}-dependent regulatory processes, their physiological importance is unknown. A study by Maxwell and coworkers[26] examined the functional consequences of cardiospecific overexpression of the canine cardiac exchanger NCX1.1 and a deletion mutant of NCX1.1, devoid of intracellular Na^+- and Ca^{2+}-dependent regulatory properties, in transgenic mice. Normal ionic regulation was observed in patches from cardiomyocytes isolated from control and transgenic mice overexpressing NCX1.1, whereas ionic regulation was nearly abolished in mice overexpressing the deletion mutant of NCX1.1 (D680–685). When looking at postrest force development in papillary muscles from NCX1.1 and transgenic mice, postrest was found to be substantially greater in the mutants than in NCX1.1 overexpressing mice, supporting the notion that ionic regulation of Na^+/Ca^{2+} exchange is required for proper functional role of NCX during the cardiac excitation–contraction cyclus.

REGULATION OF NCX BY $[Ca^{2+}]_i$

NCX is regulated by intracellular Ca^{2+} in whole-cell recordings of the outward exchange current of intact myocytes.[27] Levitsky, Nicoll, and Philipson[28] subsequently identified and characterized a region of the cytoplasmic loop of the exchanger, which could bind $^{45}Ca^{2+}$ with high affinity. Several single-site mutations within this region markedly reduced Ca^{2+}-binding affinity and single-site mutations within two acidic clusters of the Ca^{2+}-binding domain lowered the apparent Ca^{2+} affinity at the regulatory site. Mutations had effects on both, the affinity of the exchanger loop for $^{45}Ca^{2+}$ binding and for functional Ca^{2+} regulation. All mutant exchangers with less affinities at the regulatory Ca^{2+}-binding site also have had altered kinetic properties. The outward current of the wild-type NCX declined with a significantly prolonged half time and Ca^{2+} regulation mutants responded more rapidly to Ca^{2+} application. Since the affinity of wild-type and mutant NCX for intracellular Na^+ decreased at low regulatory Ca^{2+} concentrations, Ca^{2+} regulation may modify transport properties and not only control the fraction of exchangers in an active state.

NCX is also allosterically regulated by $[Ca^{2+}]_i$ such that when $[Ca^{2+}]_i$ is low, NCX current deactivates.[29] Ca^{2+}-dependent activation of NCX was observed in ferret myocytes but not in wild-type mouse myocytes, suggesting that Ca^{2+} regulation of NCX may be species-dependent. In mice expressing canine NCX, allosteric regulation by Ca^{2+} occurs under physiological conditions. The changes in the activation state may influence SR Ca^{2+} load and cytosolic resting $[Ca^{2+}]_i$, helping to fine tune Ca^{2+} influx and efflux from cells under both normal and pathophysiological conditions.

REGULATION OF THE CARDIAC NCX BY THE ENDOGENOUS XIP REGION—ROLE OF Na^+ AND Ca^{2+}

NCX is subject to two forms of regulation designated I1 and I2.[30,31] I1 has also been termed Na^+-dependent inactivation and is manifested as a partial inactivation of outward exchange current upon application of intracellular Na^+. I2 was first described in the squid axon[32] and is also denoted as Ca^{2+}-dependent regulation. Moreover, as also mentioned above, the exchanger is regulated by intracellular Ca^{2+} at a high affinity-binding site, which is different from the Ca^{2+} transport site.[28,33]

The inactivated state of NCX forms after Na^+ binds to the intracellular transport site of the protein. After three Na^+ ions bind to the transport site a translocation of the ions is induced or a fraction of NCX inactivates.[31] The intracellular loop of the exchanger has been implicated in Na^+-dependent inactivation.[34] A portion of the intracellular loop of interest is known as the endogenous XIP region,[35] which comprises 20 amino acids at the amino terminus of the loop following the fifth transmembrane segment. It was proposed that the endogenous XIP region (FIG. 1) might have an autoregulatory function. The study by Matsuoka and co-workers[33] investigated the role of the endogenous XIP region in the exchange process by site-directed mutagenesis. The results showed that the endogenous XIP region plays indeed a central role in the Na^+-dependent inactivation and in Ca^{2+}-dependent regulation, while the major readout was found in the Na^+-dependent inactivation process. In some XIP region mutants' inactivation was markedly accelerated, whereas in others inactivation was eliminated. Regulation of the exchanger by $[Ca^{2+}]_i$ was also modified. Binding of Ca^{2+} to the intracellular loop activated the exchanger and also decreased Na^+-dependent inactivation. Regulation by Ca^{2+} was still present, whereas the apparent affinity of some mutants for regulatory Ca^{2+} was decreased. The responses of all mutants to Ca^{2+} were increased. Thus, the endogenous XIP region seems to be involved in movement of the exchanger into and out of the Na^+-induced inactivated state and in regulation by Ca^{2+}.

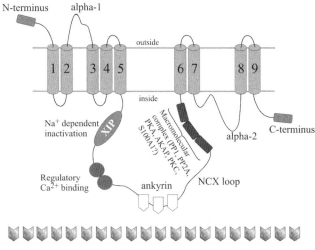

FIGURE 1. NCX signaling complex. Linear organization of the nine transmembrane spanning sequences that include possible regulatory and anchoring elements as indicated.

THE MACROMOLECULAR COMPLEX—REGULATION BY PROTEIN KINASES

Regulation of NCX by cyclic nucleotides and protein kinases, for example, PKA has remained controversial. Although earlier studies suggested that ATP-dependent regulation of NCX occurred in squid axons[32,36] and in cardiac sarcolemmal vesicles,[37] they did not distinguish between direct ATP binding and ATP-dependent phosphorylation. In contrast, Hilgemann and colleagues[30,31,38,39] investigated the question by measuring NCX currents in giant excised patches. Their studies found no functional change in cardiac NCX activity following application of catalytic subunits of PKA or PKC to the intracellular side of the giant patch. Condrescu *et al.*[40] came to similar conclusions. Other studies, however, suggested that PKA affected NCX function in heart.[41,42] The indirect way of these studies could not exclude the possibility that the measured changes were due to PKA phosphorylation of other proteins. The first proof that phosphorylation affected NCX was provided by Iwamoto and co-workers.[43] They showed that NCX from rat aorta smooth muscle cells is phosphorylated by PKC. Additionally, Iwamoto *et al.*[44] provided evidence that the intracellular loop of NCX was phosphorylated by PKC and PKA. Others showed that PKA-dependent phosphorylation of NCX increased NCX activity both in *Xenopus* oocytes expressing cardiac NCX.[45]

Local signaling complexes have been shown to regulate ion channels like the L-type Ca^{2+} channel,[46] specific K^+ channels,[47] and RyRs.[48] These complexes are composed of kinases, phosphatases, and kinase anchoring proteins (AKAPs). Accordingly, Schulze and co-workers showed that NCX is phosphorylated by PKA-dependent phosphorylation *in vitro* suggesting the existence of a local NCX "macromolecular complex"[49] (see also above). The macromolecular complex includes both the catalytic and regulatory subunits of PKA. Other regulatory enzymes are also associated with NCX, including PKC and two serine/threonine protein phosphatases, PP1 and PP2A, as well as the protein kinase A-anchoring proteins, mAKAPs (FIG. 1). These observations make local subcellular differences in response to regulatory mechanisms most likely.

REGULATION OF NCX BY β-ADRENERGIC STIMULATION

Although, as also mentioned above, some biochemical evidence of PKA phosphorylation of NCX exists,[44,45] it is not only unclear whether β-adrenergic stimulation can modulate NCX activity but also evidence of direct NCX activation is still controversial. Enhancement of NCX activity after β-adrenergic stimulation was reported in intact cells,[42,50] but not in subcellular preparations.[1] Recently, though, in contrast to all previous reports, a dramatic increase (up to fivefold) in NCX current with isoproterenol was observed in *Xenopus* oocytes expressing rat cardiac NCX[45] and in native swine myocytes.[7]

In an interesting study Ginsburg and Bers tested whether β-adrenergic stimulation with isoproterenol may regulate NCX activity in intact ventricular myocytes using different protocols.[51] The authors found that NCX function is not altered by PKA activation under any of these protocols, where intracellular conditions ranged from physiological to experimentally controlled. This does not rule out NCX modulation by PKA under all conditions, or in species other than rabbit. However, such effects are likely to be either minor (versus other PKA actions on myocyte Ca handling) or indirect, such as secondary effects dependent on altered local $[Ca^{2+}]_i$ and $[Na^+]_i$.

Linck *et al.*[52] found that Na^+-dependent Ca^{2+} uptake of baby hamster kidney cells expressing canine cardiac NCX was enhanced by forskolin, an activator of adenylate cyclase. In guinea pig ventricular myocytes, Perchenet *et al.*[42] and Zhang *et al.*[50] demonstrated that β-adrenergic stimulation enhanced NCX by 25–100%. Ruknudin *et al.*[45] (see also above) found that PKA-activating reagents phosphorylated a cardiac isoform of the NCX expressed in *Xenopus* oocytes and increased both, forward and reverse mode of NCX. Wei *et al.*[7] also reported that PKA phosphorylated the NCX protein and increased I_{NCX} by approximately 500% in control and by 100% in failing ventricular cells from pig heart, suggesting that cardiac NCX might be hyperphosphorylated in heart failure.

No stimulatory effect of β-adrenergic receptor stimulation or PKA on exchange activity was detected in giant membrane patches excised from blebs of guinea pig ventricular cells,[1] vesicle preparations of rat hearts,[53] guinea pig ventricular cells,[54] or BHK cells expressing the dog cardiac NCX1.[55] In a different study by Lin and co-workers, the effect of β-adrenergic stimulation on NCX was studied in voltage-clamped ventricular myocytes isolated from guinea pig, mouse, and rat hearts.[56] β-adrenergic stimulation did not significantly affect I_{NCX} in the ventricular cells. When I_{NCX} was recorded using a protocol similar to the previous study using Ni^{2+} the amplitude of Ni^{2+}-sensitive membrane current increased after the application of isoproterenol, suggesting an increase in I_{NCX}. This potentiation was discussed to be attributable to an increase in cystic fibrosis transmembrane conductance regulator (CFTR)-Cl current contaminating the Ni^{2+}-sensitive current. Others, however, did not find evidence of such contamination in myocytes from either the Yorkshire pig or mongrel dogs. Instead they found that the outward component of the bidirectional current was rapidly and reversibly inhibited by Ca^{2+} removal consistent with Ca^{2+}-dependence. This argues against a large contribution by CFTR-Cl.[57] Moreover, CFTR-Cl conductance has been reported to be almost absent in large species, such as humans, but strongly expressed in the guinea pig.[58] Also, other groups could not find a difference between the magnitude of the Ni^{2+}-sensitive current induced by forskolin, isoproterenol, or cAMP in failing pig myocytes, arguing against a significant contribution by CFTR-Cl in these models.[7]

REGULATION OF NCX BY S100A1

S100A1, a Ca^{2+}-binding protein of the EF hand type, is a novel regulator of cardiac contractility. S100A1 has highest expression levels in the heart[59] and is expressed early during murine embryonic heart development.[60] S100A1 is upregulated during the course of myocardial hypertrophy[61] and downregulated in the left ventricle of patients suffering from heart failure.[62] Recently, S100A1 was identified as a novel regulator of cardiac function based on the observation that overexpression of S100A1 led to a significant increase of contractility in the adult heart.[63] This was related to an increased SERCA activity and a decrease in myofilamental Ca^{2+} sensitivity suggesting involvement of PKA-dependent mechanisms.[64] Most and co-workers[11] demonstrated that extracellularly added S100A1 enhanced Ca^{2+}-transient amplitudes in neonatal ventricular cardiomyocytes through a marked decrease in diastolic bulk Ca^{2+} concentrations that was independent of SERCA activity and was presumably the result of an increased Ca^{2+} extrusion via NCX. S100A1 activated phospholipase C and PKC. We studied effects of endocytosed S100A1 (20 min extracellular application of 1 μM recombinant human S100A1) in fetal murine cardiomyocytes (bulk Ca^{2+} hold at 150 nM) and found strong stimulatory action of S100A1 on NCX (FIG. 2). At a holding potential of -120 mV (Ca^{2+} extrusion mode) NCX current density amounted to -4.01 ± 0.71 pA/pF

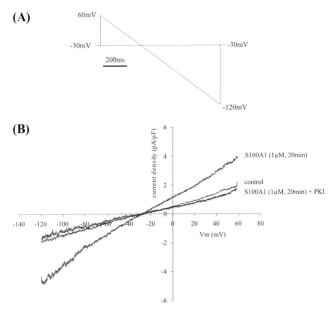

FIGURE 2. Activation of NCX current density by S100A1. I_{NCX} was measured as the bidirectional Ni^{2+} (5 mM)-sensitive current. Slow-ramp pulses (+60 mV to –120 mV; 0.09 V/s; holding potential–30 mV) at 0.1 Hz were applied as indicated schematically in (**A**). (**B**) The Ni^{2+}-sensitive current is shown in representative fetal cardiomyocyte under control conditions, after endocytotic uptake of S100A1 (1 μM) and after coapplication of PKI (via patch-pipette) in S100A1-treated cells, respectively. Note the strong increase of NCX current density after endocytotic uptake of S100A1 and the blocking effect of PKI.

($n = 9$) and at +60 mV (Ca^{2+} intrusion mode) to 4.13 ± 0.43 pA/pF, which is approximately twofold higher as reported previously for fetal embryonic cardiomyocytes.[65] As proof for PKA-mediated effects of S100A1 we applied PKI as a specific inhibitor of PKA. Coapplication of PKI and S100A1 led to a reduction of NCX exchange current density to control levels (–1.46 ± 0.33 pA/pF, –120 mV versus 1.65 ± 0.38 pA/pF, 60 mV, $n = 12$). Due to the presumed functional interaction between S100A1 and PKA and the reported activation of PKC activity by S100A1,[11] we suggest that S100A1 is an interesting candidate potentially interacting with the macromolecular complex of the intracellular loop of NCX.

REFERENCES

1. COLLINS, A., A.V. SOMLYO & D.W. HILGEMANN. 1992. The giant cardiac membrane patch method: stimulation of outward Na^+–Ca^{2+} exchange current by MgATP. J. Physiol. **454:** 27–57.

2. HILGEMANN, D.W. & R. BALL. 1996. Regulation of cardiac Na^+, Ca^{2+} exchange and KATP potassium channels by PIP_2. Science **273:** 956–959.

3. BERS, D.M. 2001. Excitation–Contraction Coupling and Cardiac Contractile Force, 2nd ed. Kluwer. Dordrecht, NL.

4. BLAUSTEIN, M.P. & W.J. LEDERER. 1999. Sodium/calcium exchange: its physiological implications. Physiol. Rev. **79:** 763–854.

5. HRYSHKO, L.V. 2002. Tissue-specific modes of Na/Ca exchanger regulation. Ann. N. Y. Acad. Sci. **976:** 166–175.

6. PHILIPSON, K.D. & D.A. NICOLL. 2000. Sodium–calcium exchange: a molecular perspective. Annu. Rev. Physiol. **62:** 111–133.

7. WEI, S.K., A. RUKNUDIN, S.U. HANLON, *et al.* 2003. Protein kinase A hyperphosphorylation increases basal current but decreases beta-adrenergic responsiveness of the sarcolemmal Na+-Ca2+ exchanger in failing pig myocytes. Circ. Res. **92:** 897–903.

8. REPPEL, M., P. SASSE, R. PIEKORZ, *et al.* 2005. S100A1 enhances the L-type Ca2+ current in embryonic mouse and neonatal rat ventricular cardiomyocytes. J. Biol. Chem. **280:** 36019–36028.

9. KETTLEWELL, S., P. MOST, S. CURRIE, *et al.* 2005. S100A1 increases the gain of excitation-contraction coupling in isolated rabbit ventricular cardiomyocytes. J. Mol. Cell Cardiol. **39:** 900–910.

10. VOLKERS, M., C.M. LOUGHREY, N. MACQUAIDE, *et al.* 2006. S100A1 decreases calcium spark frequency and alters their spatial characteristics in permeabilized adult ventricular cardiomyocytes. Cell Calcium **41:** 135–143.

11. MOST, P., M. BOERRIES, C. EICHER, *et al.* 2005. Distinct subcellular location of the Ca2+-binding protein S100A1 differentially modulates Ca2+-cycling in ventricular rat cardiomyocytes. J. Cell. Sci. **118:** 421–431.

12. TAKEISHI, Y., G. CHU, D.M. KIRKPATRICK, *et al.* 1998. *In vivo* phosphorylation of cardiac troponin I by protein kinase Cbeta2 decreases cardiomyocyte calcium responsiveness and contractility in transgenic mouse hearts. J. Clin. Invest. **102:** 72–78.

13. MALHOTRA, A., D. REICH, D. REICH, *et al.* 1997. Experimental diabetes is associated with functional activation of protein kinase C epsilon and phosphorylation of troponin I in the heart, which are prevented by angiotensin II receptor blockade. Circ. Res. **81:** 1027–1033.

14. JIDEAMA, N.M., T.A. NOLAND, JR., R.L. RAYNOR, *et al.* 1996. Phosphorylation specificities of protein kinase C isozymes for bovine cardiac troponin I and troponin T and sites within these proteins and regulation of myofilament properties. J. Biol. Chem. **271:** 23277–23283.

15. KAMP, T.J. & J.W. HELL. 2000. Regulation of cardiac L-type calcium channels by protein kinase A and protein kinase C. Circ. Res. **87:** 1095–1102.

16. PURI, T.S., B.L. GERHARDSTEIN, X.L. ZHAO, *et al.* 1997. Differential effects of subunit interactions on protein kinase A- and C-mediated phosphorylation of L-type calcium channels. Biochemistry **36:** 9605–9615.

17. BOURINET, E., F. FOURNIER, P. LORY, *et al.* 1992. Protein kinase C regulation of cardiac calcium channels expressed in *Xenopus* oocytes. Pflugers Arch. **421:** 247–255.

18. SINGER-LAHAT, D., E. GERSHON, I. LOTAN, *et al.* 1992. Modulation of cardiac Ca2+ channels in *Xenopus* oocytes by protein kinase C. FEBS Lett. **306:** 113–118.

19. BOURON, A., N.M. SOLDATOV & H. REUTER. 1995. The beta 1-subunit is essential for modulation by protein kinase C of an human and a non-human L-type Ca2+ channel. FEBS Lett. **377:** 159–162.

20. BASSANI, J.W.M., R.A. BASSANI & D.M. BERS 1994. Relaxation in rabbit and rat cardiac cells: species-dependent differences in cellular mechanisms. J. Physiol. **476:** 279–293.

21. PIACENTINO, V., III, C.R. WEBER, X. CHEN, *et al.* 2003. Cellular basis of abnormal calcium transients of failing human ventricular myocytes. Circ. Res. **92:** 651–658.

22. HASENFUSS, G., W. SCHILLINGER, S.E. LEHNART, *et al.* 1999. Relationship between Na^+-Ca^{2+}-exchanger protein levels and diastolic function of failing human myocardium. Circulation **99:** 641–648.

23. WEBER, C.R., V. PIACENTINO 3rd, S. R. HOUSER & D.M. BERS. 2003. Dynamic regulation of sodium/calcium exchange function in human heart failure. Circulation **108:** 2224–2229.

24. RUDOLPH, W. 1990. Pathophysiologic and diagnostic aspects of heart failure. Herz **15:** 147–157. Review.

25. BERS, D.M. & S. DESPA. 2006. Cardiac myocytes Ca2+ and Na+ regulation in normal and failing hearts. J. Pharmacol. Sci. **100:** 315–322. Review.

26. MAXWELL, K., SCOTT, J., A. OMELCHENKO, *et al.* 1999. Functional role of ionic regulation of Na+/Ca2+ exchange assessed in transgenic mouse hearts. Am. J. Physiol. **277:** H2212–H2221.

27. KIMURA, J., A. NOMA & H. IRISAWA. 1986. Na-Ca exchange current in mammalian heart cells. Nature **319:** 596–597.

28. LEVITSKY, D.O., D.A. NICOLL & K.D. PHILIPSON. 1994. Identification of the high affinity Ca(2+)-binding domain of the cardiac Na(+)-Ca2+ exchanger. J. Biol. Chem. **269:** 22847–22852.

29. WEBER, C.R., K.S. GINSBURG, K.D. PHILIPSON, *et al.* 2001. Allosteric regulation of Na/Ca exchange current by cytosolic Ca in intact cardiac myocytes. J. Gen. Physiol. **117:** 119–131.

30. HILGEMANN, D.W., A. COLLINS & S. MATSUOKA. 1992. Steady state and dynamic properties of cardiac sodium-calcium exchange: secondary modulation by cytoplasmic calcium and ATP. J. Gen. Physiol. **100:** 933–961

31. HILGEMANN, D.W., S. MATSUOKA, G.A. NAGEL & A. COLLINS. 1992. Steady state and dynamic properties of cardiac sodium-calcium exchange: sodium-dependent inactivation. J. Gen. Physiol. **100:** 905–932.

32. DIPOLO, R. 1979. Calcium influx in internally dialyzed squid giant axons. J. Gen. Physiol. **73:** 91–113.

33. MATSUOKA, S., D.A. NICOLL, L.V. HRYSHKO, *et al.* 1995. Regulation of the cardiac Na^+-Ca^{2+} exchanger by Ca^{2+}: mutational analysis of the Ca^{2+}-binding domain. J. Gen. Physiol. **105:** 403–420.

34. MATSUOKA, S., D.A. NICOLL, R.F. REILLY, *et al.* 1993. Initial localization of regulatory regions of the cardiac sarcolemmal Na^+-Ca^{2+} exchanger. Proc. Natl. Acad. Sci. USA **90:** 3870–3874.

35. LI, Z., D.A. NICOLL, A. COLLINS, *et al.* 1991. Identification of a peptide inhibitor of the cardiac sarcolemmal Na^+-Ca^{2+} exchanger. J. Biol. Chem. **266:** 1014–1020.

36. DIPOLO, R. & L. BEAUGE. 1994. Effects of vanadate on MgATP stimulation of Na-Ca exchange support kinase-phosphatase modulation in squid axons. Am. J. Physiol. **266:** C1382–C1391.

37. CARONI, P. & E. CARAFOLI. 1983. The regulation of the Na+ -Ca2+ exchanger of heart sarcolemma. Eur. J. Biochem. **132:** 451–460.

38. HILGEMANN, D.W., D.A. NICOLL & K.D. PHILIPSON 1991. Charge movement during Na+ translocation by native and cloned cardiac Na+/Ca2+ exchanger. Nature **352:** 715–718.

39. MATSUOKA, S. & D.W. HILGEMANN. 1994. Inactivation of outward Na(+)-Ca2+ exchange current in guinea-pig ventricular myocytes. J. Physiol. (Lond.) **476:** 443–458.

40. CONDRESCU, M., J.P. GARDNER, G. CHERNAYA, *et al.* 1995. ATP-dependent regulation of sodium-calcium exchange in Chinese hamster ovary cells transfected with the bovine cardiac sodium-calcium exchanger. J. Biol. Chem. **270:** 9137–9146.

41. HAN, X. & G.R. FERRIER. 1995. Contribution of Na(+)-Ca2+ exchange to stimulation of transient inward current by isoproterenol in rabbit cardiac Purkinje fibers. Circ. Res. **76:** 664–674.

42. PERCHENET, L., A.K. HINDE, K.C. PATEL, *et al.* 2000. Stimulation of Na/Ca exchange by the beta-adrenergic/protein kinase A pathway in guinea-pig ventricular myocytes at 37 degrees C. Pflugers Arch. Eur. J. Physiol. **439:** 822–828.

43. IWAMOTO, T., S. WAKABAYASHI & M.J. SHIGEKAWA 1995. Growth factor-induced phosphorylation and activation of aortic smooth muscle Na+/Ca2+ exchanger. Biol. Chem. **270:** 8996–9001.

44. IWAMOTO, T., Y. PAN, S. WAKABAYASHI, *et al.* 1996. Phosphorylation-dependent regulation of cardiac Na+/Ca2+ exchanger via protein kinase C. Biol. Chem. **271:** 13609–13615.

45. RUKNUDIN, A., S. HE, W.J. LEDERER & D.H. SCHULZE. 2000. Functional differences between cardiac and renal isoforms of the rat Na+-Ca2+ exchanger NCX1 expressed in *Xenopus* oocytes. J. Physiol. (Lond.) **529:** 599–610.

46. DAVARE, M.A., V. AVDONIN, D.D. HALL, *et al.* 2001. A beta2 adrenergic receptor signaling complex assembled with the Ca2+ channel Cav1.2. Science **293:** 98–101.

47. MARX, S.O., J. KUROKAWA, S. REIKEN, *et al.* 2002. Requirement of a macromolecular signaling complex for beta adrenergic receptor modulation of the KCNQ1-KCNE1 potassium channel. Science **295:** 496–499.

48. MARX, S.O., S. REIKEN, Y. HISAMATSU, *et al.* 2000. PKA phosphorylation dissociates FKBP12.6 from the calcium release channel (ryanodine receptor): defective regulation in failing hearts. Cell **101:** 365–376.

49. SCHULZE, D.H., M. MUQHAL & W.J. LEDERER. 2003. Sodium/calcium exchanger (NCX1) macromolecular complex. J. Biol. Chem. **278:** 28849–28855.

50. ZHANG, Y.H., A.K. HINDE & J.C. HANCOX. 2001. Anti-adrenergic effect of adenosine on Na+-Ca2+ exchange current recorded from guinea-pig ventricular myocytes. Cell Calcium **29:** 347–358.

51. GINSBURG, K.S. & D.M. BERS. 2005. Isoproterenol does not enhance Ca-dependent Na/Ca exchange current in intact rabbit ventricular myocytes. J. Mol. Cell. Cardiol. **39:** 972–981.

52. LINCK, B., Z. QIU, Z. HE, *et al.* 1998. Functional comparison of the three isoforms of the Na^+/Ca^{2+} exchanger (NCX1, NCX2, NCX3). Am. J. Physiol. Cell. Physiol. **274:** C415–C423.

53. BALLARD, C. & S. SCHAFFER. 1996. Stimulation of the Na^+/Ca^{2+} exchanger by phenylephrine, angiotensin II and endothelin 1. J. Mol. Cell. Cardiol. **28:** 11–17.

54. MAIN, M.J., C.J. GRANTHAM & M.B. CANNELL. 1997. Changes in subsarcolemmal sodium concentration measured by Na-Ca exchanger activity during Na-pump inhibition and β-adrenergic stimulation in guinea-pig ventricular myocytes. Pflügers Arch. **435:** 112–118.

55. HE, L.P., L. CLEEMANN, N.M. SOLDATOV & M. MORAD. 2003. Molecular determinants of cAMP-mediated regulation of the Na^+-Ca^{2+} exchanger expressed in human cell lines. J. Physiol. **548:** 677–689.

56. LIN, X., H. JO, Y. SAKAKIBARA, *et al.* 2006. Beta-adrenergic stimulation does not activate Na+/Ca2+ exchange current in guinea pig, mouse, and rat ventricular myocytes. Am. J. Physiol. Cell. Physiol. **290:** C601–C608.
57. WEI, S.K., J.M. MCCURLEY, M. SHOU, *et al.* 2005. β-Adrenergic receptor modulation of cardiac Na/Ca exchange in pig and dog myocytes (Abstract). Biophys. J. **88**(Suppl): LB110.
58. DU, X.Y., J. FINLEY & S. SOROTA. 2000. Paucity of CFTR current but modest CFTR immunoreactivity in non-diseased human ventricle. Pflügers Arch. **440:** 61–67.
59. KATO, K. & S. KIMURA. 1985. S100ao (alpha alpha) protein is mainly located in the heart and striated muscles. Biochim. Biophys. Acta **842:** 146–150.
60. KIEWITZ, R., G.E. LYONS, B.W. SCHAFER & C.W. HEIZMANN. 2000. Transcriptional regulation of S100A1 and expression during mouse heart development. Biochim. Biophys. Acta **1498:** 207–219.
61. EHLERMANN, P., A. REMPPIS, O. GUDDAT, *et al.* 2000. Right ventricular upregulation of the Ca(2+) binding protein S100A1 in chronic pulmonary hypertension. Biochim. Biophys. Acta **1500:** 249–255.
62. REMPPIS, A., T. GRETEN, B.W. SCHAFER, *et al.* 1996. Altered expression of the Ca(2+)-binding protein S100A1 in human cardiomyopathy. Biochim. Biophys. Acta **1313:** 253–257.
63. REMPPIS, A., P. MOST, E. LOFFLER, *et al.* 2002. The small EF-hand Ca^{2+} binding protein S100A1 increases contractility and Ca^{2+} cycling in rat cardiac myocytes. Basic Res. Cardiol. **97**(Suppl 1): I56–I62.
64. MOST, P., J. BERNOTAT, P. EHLERMANN, *et al.* 2001. S100A1: a regulator of myocardial contractility. Proc. Natl. Acad. Sci. USA **98:** 13889–13894.
65. REPPEL, M., P. SASSE, D. MALAN, *et al.* 2007. Functional expression of the Na^{+}/Ca^{2+} exchanger in the embryonic mouse heart. J. Mol. Cell. Cardiol. **42:** 121–132.

Phosphorylation of Na$^+$/Ca^{2+} Exchanger in TAB-Induced Cardiac Hypertrophy

YUKI KATANOSAKA,[a,b] BONGJU KIM,[c] SHIGEO WAKABAYASHI,[a] SATOSHI MATSUOKA,[c] AND MUNEKAZU SHIGEKAWA[a,d]

[a]*Department of Molecular Physiology, National Cardiovascular Center Research Institute, Suita, Osaka 565-8565, Japan*

[b]*Department of Cardiovascular Physiology, Graduate School of Medicine, Dentistry and Pharmaceutical Sciences, Okayama University, Okayama 700-8558, Japan*

[c]*Department of Physiology and Biophysics, Graduate School of Medicine, Kyoto University, Sakyo-ku, Kyoto 606-8501, Japan*

[d]*Department of Human Life Sciences, Senri-Kinran University, Suita, Osaka 565-0873, Japan*

ABSTRACT: Both protein kinase Cα-dependent Na$^+$/Ca^{2+} exchanger1 (NCX1) phosphorylation and calcineurin activity are required for the depression of NCX activity observed in chronically phenylephrine (PE)-treated hypertrophic neonatal rat cardiomyocytes. In this study, we explored the possibility that the same changes occur *in vivo* hypertrophy. In the hypertrophic hearts of thoracic aortic-banded (TAB) mice, NCX1 phosphorylation increased significantly compared with control hearts. Furthermore, the TAB-induced cardiac hypertrophy was much less prominent in transgenic mice overexpressing an NCX1 mutant having defective phosphorylation sites. These data suggest that the phosphorylation status of NCX1 may play an important role in the pathogenesis of load-induced cardiac hypertrophy.

KEYWORDS: Na$^+$/Ca^{2+} exchanger; phosphorylation; protein kinase Cα; cardiac hypertrophy; thoracic aortic banding

INTRODUCTION

In hypertrophic and failing hearts from human patients and animal models, sarcolemmal Na$^+$/Ca^{2+} exchanger1 (NCX1) expression has often been shown to be elevated.[1,2] However, whether such elevated NCX expression invariably leads to enhanced function in disease conditions is not clear, although

Address for correspondence: Yuki Katanosaka, 2-5-1 Shikata-cho, Okayama 700-8558, Japan. Voice: 81-86-235-7117; fax: 81-86-235-7430.
ytanigu@md.okayama-u.ac.jp

Ann. N.Y. Acad. Sci. 1099: 373–376 (2007). © 2007 New York Academy of Sciences.
doi: 10.1196/annals.1387.024

enhanced NCX expression and function has been reported in some previous studies. In our recent report,[3] we observed an elevated level of NCX1 expression but a marked depression of NCX activity in hypertrophic neonatal rat cardiomyocytes subjected to chronic phenylephrine (PE) treatment or chronic adenoviral infection of activated calcineurin Aβ. The depressed NCX activity in these myocytes is partially recovered by the addition of the calcineurin inhibitor FK506 or the protein kinase C (PKC) inhibitor Calphostin C. Since the effects of these two inhibitors are additive, calcineurin and PKC activities contribute independently to the observed reduction of NCX activity. We have subsequently studied the phosphorylation status of NCX1 protein and its relation to NCX activity in PE-treated or activated CnA-infected cardiomyocytes. We found that PKCα is involved in NCX1 phosphorylation, while calcineurin did not directly influence the latter.[4,5] We found that both PKCα-dependent NCX1 phosphorylation and calcineurin activity are required for the depression of NCX activity observed in PE-treated or activated CnA-infected hypertrophic cardiomyocytes.

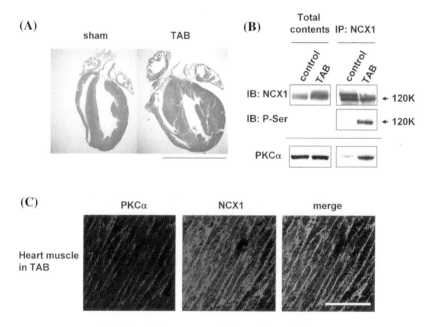

FIGURE 1. Phosphorylation status of NCX1 in the heart of TAB mouse. (**A**) Transverse section of hearts subjected to a 5-week TAB. Scale bar, 0.5 cm. (**B**) Lysates of hearts from control and TAB mice (*Total contents*) and materials immunoprecipitated from heart lysates with anti-NCX1 (*IP: NCX1*) were subjected to immunoblot analysis with anti-NCX1, anti-phosphor-Ser (*P:P-Ser*), or anti-PKCα. (**C**) Immunostaining of ventricular muscle of TAB mouse with anti-NCX1 or anti-PKCα. Bar, 20 μm.

To explore the possibility that the changes of NCX function described above, that is, phosphorylation and inhibition of NCX1, also occur in *in vivo* cardiac hypertrophy, we used the thoracic aortic-banded (TAB) mouse model. In mice subjected to 4-week TAB, the heart weight increased twofold compared with non-TAB controls. Immunoblot analysis revealed that NCX1 expression increased about twofold in TAB heart, whereas PKCα did not change its expression before and after TAB (FIG. 1 A, B). On the other hand, NCX1 phosphorylation, as estimated by immunoblot analysis with antiphosphoserine antibody, was significantly increased in TAB hearts compared with non-TAB controls, whereas NCX current density was reduced by about 40% in hypertrophic cardiomyocytes from TAB mice. The latter finding is consistent with the previous reports by others[6,7] that NCX currents are reduced in hypertrophic cardiomyocytes isolated from mice subjected to TAB or the transgenic mice overexpressing Gαq. Furthermore, we found that PKCα was co-localized with NCX1 in the peripheral sarcolemma of TAB cardiomyocytes and co-precipitated with anti-NCX1 from lysates of TAB hearts. Thus, association of NCX1 with PKCα is significantly increased in the hearts of TAB mice.

We then examined the effect of the transgenic overexpression of an NCX1 mutant having defective phosphorylation sites (NCX1-S249A/S250A/S357A, see Ref. 8) on the development of cardiac hypertrophy in mice subjected to TAB. After 4 weeks of TAB, we detected very low levels of NCX1 phosphorylation and association of NCX1 with PKCα in the hearts of these transgenic mice. In contrast, increased levels of NCX1 phosphorylation and association of NCX1 with PKCα were observed in control TAB mice as described above. Finally, we found that the TAB-induced cardiac hypertrophy was much less prominent in the NCX1 phosphorylation site mutant transgenic mice than in control TAB mice. It is known that calcineurin and PKC signaling is fully activated during the progression of pressure overload-induced cardiac hypertrophy in aortic banded animals.[9] All these results suggest that depressed NCX activity may contribute to the etiology of *in vivo* cardiac hypertrophy induced by pressure overload.

REFERENCES

1. STUDER, R., H. REINECKE, J. BIGER, *et al.* 1994. Gene expression of the cardiac Na$^+$-Ca^{2+} exchanger in end-stage human heart failure. Circ. Res. **75:** 443–453.
2. SHIGEKAWA, M. & T. IWAMOTO. 2001. Cardiac Na$^+$/Ca^{2+} exchange: molecular and pharmacological aspect. Circ. Res. **88:** 864–876.
3. KATANOSAKA, Y., Y. IWATA, Y. KOBAYASHI, *et al.* 2005. Calcineurin inhibits Na$^+$/Ca^{2+} exchange in phenylephrine-treated hypertrophic cardiomyocytes. J. Biol. Chem. **280:** 5764–5772.
4. SHIGEKAWA, M., Y. KATANOSAKA & S. WAKABAYASHI. 2007. Regulation of cardiac Na$^+$/Ca^{2+} exchanger by calcineurin and protein kinase C. Ann. N. Y. Acad. Sci. In press.

5. KATANOSAKA, Y., Y. IWATA, S. WAKABAYASHI, *et al.* 2006. Cardiac Na^+/Ca^{2+} exchanger is highly phosphorylated and inhibited by $PKC\alpha$ dependent on calcineurin activity in PE-treated hypertrophic cardiomyocytes. Biophys. J. January, 511a

6. ZHENGYI, W., N. BRIDGID, K. WILLIAM, *et al.* 2001. Na^+/Ca^{2+} exchanger remodeling in pressure overload cardiac hypertrophy. J. Biol. Chem. **276:** 17706–17711.

7. MITARAI, S., T.D. REED & A. YATANI. 2000. Changes in ionic currents and beta-adrenergic receptor signaling in hypertrophied myocytes overexpressing $G\alpha q$. Am. J. Physiol. Heart. Circ. Physiol. **279:** H139–H148.

8. IWAMOTO, T., Y. PAN, T.Y. NAKAMURA, *et al.* 1998. Protein kinase C-dependent regulation of Na^+/Ca^{2+} exchanger isoforms NCX1 and NCX3 does not require their direct phosphorylation. Biochemistry **37:** 17230–17238.

9. MOLKENTIN, J.D. & G.W. DORN, II. 2001 Annu. Rev. Physiol. **63:** 391–426.

Role of Ca²⁺ Transporters and Channels in the Cardiac Cell Volume Regulation

A. TAKEUCHI,[a,b] S. TATSUMI,[a] N. SARAI,[b,c] K. TERASHIMA,[b,d]
S. MATSUOKA,[a,b] AND A. NOMA[a,b]

[a]*Department of Physiology and Biophysics, Graduate School of Medicine, Kyoto University, Sakyo-ku Yoshida-Konoe, Kyoto, 606-8501, Japan*

[b]*Cell/Biodynamics Simulation Project Kyoto University, Shimogyo-ku Chudoji Awata, 600-8815, Japan*

[c]*Department of Nano-Medicine Merger Education Unit, Graduate School of Medicine, Kyoto University, Sakyo-ku Yoshida-Konoe, Kyoto, 606-8501, Japan*

[d]*Pharmacokinetics Research Laboratories, Dainippon Sumitomo Pharma Co., Ltd., 1-98, Kasugadenaka 3-chome, Konohana-ku, Osaka, 554-0022, Japan*

ABSTRACT: Na^+/K^+ pump is one of key mechanisms to maintain cell volume. When it is inhibited, cells are at risk of swelling. However, in guinea pig ventricular myocytes, the cell area as an index of cell volume was almost constant during 90 min Na^+/K^+ pump blockade with 40 μM ouabain despite the marked membrane depolarization. In this study, involvements of Ca^{2+} transporters and channels in the cardiac cell volume regulation were proposed by conducting the computer simulation in parallel with the experimental validation.

KEYWORDS: comprehensive cardiac cell model; cell volume regulation; Na^+/K^+ pump; PMCA; Na^+/Ca^{2+} exchanger

INTRODUCTION

The Na^+/K^+ pump is known as a constitutive protein in mammalian cells and has an essential role in regulating the cell volume.[1] The basic mechanisms for cell volume regulation have been explained theoretically as follows using simple mathematical models of cell volume regulation.[2,3] The [K^+] gradient, created by the Na^+/K^+ pump across the plasma membrane, settles the negative V_m. This negative V_m expels Cl^- out of the cell through Cl^- channels, compensating for the continuous Cl^- influx via Cl^--coupled transporters. Thereby

Address for correspondence: Ayako Takeuchi, Department of Physiology and Biophysics, Graduate School of Medicine, Kyoto University, Sakyo-ku Yoshida-Konoe, Kyoto, 606-8501, Japan. Voice: +81-75-753-4357; fax: +81-75-753-4349.

takeuchi@biosim.med.kyoto-u.ac.jp

Ann. N.Y. Acad. Sci. 1099: 377–382 (2007). © 2007 New York Academy of Sciences.
doi: 10.1196/annals.1387.020

the pump maintains cellular osmolarity to keep the cell volume intact. Accordingly, the time course of cell swelling caused by blocking the Na^+/K^+ pump largely depends on the membrane Cl^- permeability. It also depends on the membrane Na^+ permeability indirectly through affecting the redistribution of K^+ across the membrane and V_m.[3]

Interestingly, it is well known that the cardiac cell volume hardly changes during the Na^+/K^+ pump blockade.[4,5] Taken the above theoretical mechanisms into consideration, the cell volume maintenance in the cardiac myocytes may be partially explained by the low-membrane Na^+ and Cl^- permeabilities in these cells. However, role of Ca^{2+}, which should increase during the Na^+/K^+ pump blockade, in the cell volume regulation has not been well clarified. Recently, we implemented the Cl^- and water fluxes into an excitation–contraction coupling model of the guinea pig ventricular myocyte, Kyoto model.[3,6] In this study, we analyzed the mechanisms of cell volume regulation by conducting the computer simulation in parallel with the experimental validation. We propose novel roles of Ca^{2+} transporters and channels in the cardiac cell volume regulation.

MATERIALS AND METHODS

Single ventricular myocytes were obtained by treating guinea pig hearts with collagenase. To avoid spontaneous contractions caused by blocking Na^+/K^+ pump, a nominally Ca^{2+}-free Tyrode solution was used. The Ca^{2+}-free Tyrode solution contained (in mM): NaCl 140, NaH_2PO_4 0.33, KCl 5.4, $MgCl_2$ 0.45, glucose 5.5, and HEPES 5 (pH 7.4). Cell area was calculated by processing cell image captured by a CCD camera (Hamamatsu Photonics, Hamamatsu, Japan). V_m and $[Na^+]_i$ were monitored by the dual wavelength ratio imaging of di-8-ANEPPS and SBFI, respectively.

Computer simulation was carried out using the Kyoto model[3] with slight modifications. Ca^{2+} flux through PMCA was set to 9% of the total Ca^{2+} extrusion, based on a report by Mackiewicz and Lewartowski.[7]

RESULTS AND DISCUSSION

Maintenance of Cell Volume during the Membrane Depolarization Induced by Na^+/K^+ Pump Block

The membrane depolarization takes the pivotal role in mediating blockade of the Na^+/K^+ pump to cell swelling.[2,3] In experiments, the treatment with 40 μM ouabain caused membrane depolarization ($\Delta V_m = 51.7 \pm 4.3$ mV from resting V_m of -91.7 ± 6.4 mV). In contrast to V_m, no obvious cell swelling was observed (the cell area at 90 min was $96.9 \pm 0.7\%$ of the original area), which was in agreement with the previous reports.[4,5] The Kyoto model well simulates

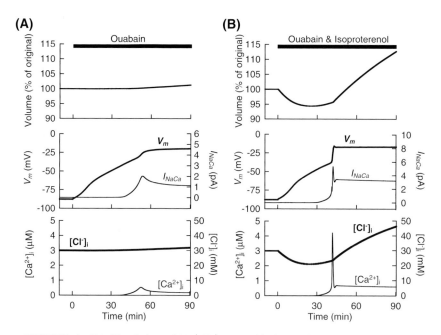

FIGURE 1. (**A**) Simulation of Na^+/K^+ pump block. At time 0, 40 μM ouabain is applied to the Kyoto model, and then time courses of cell volume (*top*), V_m and I_{NaCa} (*middle*), and $[Ca^{2+}]_i$ and $[Cl^-]_i$ (*bottom*) are plotted. (**B**) Simulation of Na^+/K^+ pump block with increased membrane Cl^- conductance. At time 0, 40 μM ouabain as well as 1 μM isoproterenol is applied to the Kyoto model. The time courses of cell volume (*top*), V_m and I_{NaCa} (*middle*), and $[Ca^{2+}]_i$ and $[Cl^-]_i$ (*bottom*) are plotted.

the time course of membrane depolarization and almost constant cell volume during the ouabain exposure (FIG. 1 A).

Role of Plasma Membrane Ca^{2+}-ATPase PMCA and Na^+/Ca^{2+} Exchanger in the Cell Volume Regulation

PMCA and Na^+/Ca^{2+} exchanger are the Ca^{2+} extrusion mechanisms in the cardiac myocyte. From the model analysis, it is proposed that these transporters are involved in maintaining the cell volume indirectly through regulating the Ca^{2+} homeostasis. In the simulation, $[Ca^{2+}]_i$ accumulation caused by the Na^+/K^+ pump block is attenuated to a low level (0.18 μM) by the active Ca^{2+} extrusion via PMCA (the bottom panel of FIG. 1 A). Low $[Ca^{2+}]_i$, in combination with the increased $[Na^+]_i$ and membrane depolarization, activates the reverse mode of the Na^+/Ca^{2+} exchange current (I_{NaCa}) (the middle panel of FIG. 1 A). Thereby Na^+ is extruded via reverse Na^+/Ca^{2+} exchange, and the cell volume change is relatively small. If PMCA was abolished, Na^+/K^+ pump block causes a remarkable $[Ca^{2+}]_i$ accumulation (19.50 μM), thus the period of reverse Na^+/Ca^{2+} exchange become shortened. Thereby the Na^+ accumulation occurs to the greater extent and so does the extent of cell swelling (not shown).

The above working hypothesis was experimentally verified. Application of a Na^+/Ca^{2+} exchange blocker, 1 μM SEA0400,[8] 60 min after the ouabain treatment resulted in the further increase of $[Na^+]_i$; namely, the increase in SBFI fluorescence ratio (F340/F380, $n = 3$). This result indicated that the mode of Na^+/Ca^{2+} exchange was reversed during the Na^+/K^+ pump block. Moreover, cells started to swell after the application of this drug ($n = 13$), confirming the theoretical prediction by the model that the reverse mode of Na^+/Ca^{2+} exchange in conjunction with PMCA plays a role in suppressing cell swelling during the Na^+/K^+ pump block. Although it has been considered that PMCA plays a minor role in regulating $[Ca^{2+}]_i$ in the cardiac myocyte,[7,9,10] the role of PMCA may become prominent as a compensatory mechanism for Na^+/K^+ pump dysfunction. Thus, it facilitates the extrusion of Na^+ out of the cells via the reverse mode of Na^+/Ca^{2+} exchange.

The Positive Feedback Cycle between L-Type Ca^{2+} Channel, I_{CaL}, and Ca^{2+}-Activated Background Cation Current, $I_{l(Ca)}$, in Determining the Cell Swelling

The cell swelling induced by the Na^+/K^+ pump block depends on the magnitude of membrane Cl^- conductance. Therefore, activation of CFTR Cl^- channel by the β-adrenergic stimulation should induce a cell swelling. The Kyoto model predicts that the CFTR Cl^- channel activation induces a cell swelling, but after a delay of ~40 min (top panel of FIG. 1 B). As clearly seen in the top and bottom panels of FIGURE 1 B, the cell volume and $[Cl^-]_i$ show similar time courses. The initial decrease in the cell volume is due to the enhanced Cl^- efflux through CFTR Cl^- channel activated from 0 to 1.3 nS. The Cl^- efflux, however, is truncated by the continuous membrane depolarization caused by the Na^+/K^+ pump block, and thereby the initial volume decrease is saturated. The minimum cell volume is 94% of the original volume (the top panel of FIG. 1 B). During the course of membrane depolarization, the window current of L-type Ca^{2+} channel, which is also magnified by threefold according to the β-adrenergic stimulation, is gradually activated, resulting in a rapid accumulation of $[Ca^{2+}]_i$ (the bottom panel of FIG. 1 B). This increase of $[Ca^{2+}]_i$ activates $I_{l(Ca)}$ and triggers a sudden jump of V_m to a level more positive than reversal potential for Cl^-, E_{Cl} (the middle panel of FIG. 1 B). Thereby, the rapid and marked cell swelling is induced through the increase in the Cl^- influx. Na^+/Ca^{2+} exchange operates in the reverse mode throughout the period of Na^+/K^+ pump block.

The above predictions were tested experimentally by the simultaneous application of 40 μM ouabain and 1 μM isoproterenol to the myocytes. The predicted biphasic volume change with a sudden V_m jump was actually observed; the cell area first gradually decreased to ~94% of the original cell area and then increased to ~105% 30 min after starting the swelling. The sudden V_m jump occurred at ~59 min, which is almost at the same time with this rapid swelling start, ~52 min. Moreover, the involvement of I_{CaL} in initiating the rapid swelling was confirmed by applying a specific blocker of L-type Ca^{2+} channel, nifedipine, simultaneously with ouabain and isoproterenol to the myocytes. The swelling phase was significantly depressed in the presence of 5 μM nifedipine leaving the first gradual shrinkage phase intact (data not shown). These results strongly supported the mechanisms predicted by the model simulation; that is, $[Ca^{2+}]_i$ accumulation via the window current of L-type Ca^{2+} channel caused the rapid swelling through the $I_{l(Ca)}$-triggered membrane depolarization.

The $I_{l(Ca)}$ has been frequently discussed in relation to the arrhythmic membrane excitation,[11] such as the delayed membrane depolarization, but rarely analyzed in the cell volume regulation. The present experimental and simulation study using the Kyoto model disclosed the positive feedback cycle leading to the final cell swelling during the Na^+/K^+ pump block; gradual membrane depolarization due to redistribution of K^+ across the membrane, opening of the window I_{CaL}, an increase in $[Ca^{2+}]_i$, activation of $I_{l(Ca)}$, accelerated depolarization, reversion from Cl^- efflux to influx through Cl^- channels, and final net influx of water.

CONCLUSION

The Kyoto model includes most of the ion channels and the major ion transporters on the cardiac cell membrane, mechanisms for the Ca^{2+} dynamics performed by SR, the intracellular Ca^{2+} buffers, and Cl^- and water fluxes through plasma membrane.[3] Thereby, we successfully derive a conclusion that Ca^{2+} dynamics has quite an important role in the cardiac cell volume regulation indirectly through affecting the Na^+ homeostasis. The feedback cycle of formulating a working hypothesis through simulation and validating the hypothesis by conducting new experiments, as exemplified in this study, does facilitate understanding complicated physiological and pathophysiological functions, which are performed by the interactions among numerous molecular mechanisms.

ACKNOWLEDGMENTS

This study was supported by the Leading Project for Biosimulation and a Grant-in-Aid for Scientific Research from the Ministry of Education, Culture, Sports, Science and Technology of Japan.

REFERENCES

1. BALSHAW, D.M. *et al.* 2001. Sodium pump function. *In* Cell Physiology Sourcebook: a Molecular Approach, 3rd ed. N. Sperelakis, Ed.: 261–269. Academic Press. London.
2. ARMSTRONG, C.M. 2003. The Na/K pump, Cl ion, and osmotic stabilization of cells. Proc. Natl. Acad. Sci. USA **100:** 6257–6262.
3. TERASHIMA, K. *et al.* 2006. Modelling Cl⁻ homeostasis and volume regulation of the cardiac cell. Phil. Trans. R. Soc. Lond. A. **364:** 1245–1265.
4. DREWNOWSKA, K. & C.M. BAUMGARTEN. 1991. Regulation of cellular volume in rabbit ventricular myocytes: bumetanide, chlorothiazide, and ouabain. Am. J. Physiol. Cell Physiol. **260:** C122–C131.
5. PINE, M.B. *et al.* 1980. Sodium permeability and myocardial resistance to cell swelling during metabolic blockade. Am. J. Physiol. Heart Circ. Physiol. **239:** H31–H39.
6. TAKEUCHI, A. *et al.* 2006. Ionic mechanisms of cardiac cell swelling induced by blocking Na^+/K^+ pump as revealed by experiments and simulation. J. Gen. Physiol. **128:** 495–507.
7. MACKIEWICZ, U. & B. LEWARTOWSKI. 2006. Temperature dependent contribution of Ca^{2+} transporters to relaxation in cardiac myocytes: important role of sarcolemmal Ca^{2+}-ATPase. J. Physiol. Pharmacol. **57:** 3–15.
8. MATSUDA, T. *et al.* 2001. SEA0400, a novel and selective inhibitor of the Na^+-Ca^{2+} exchanger, attenuates reperfusion injury in the in vitro and in vivo cerebral ischemic models. J. Pharmacol. Exp. Ther. **298:** 249–256.
9. BERS, D.M. *et al.* 1996. Na-Ca exchange and Ca fluxes during contraction and relaxation in mammalian ventricular muscle. Ann. N. Y. Acad. Sci. **779:** 430–442.
10. CHOI, H.S. & D.A. EISNER. 1999. The role of sarcolemmal Ca^{2+}-ATPase in the regulation of resting calcium concentration in rat ventricular myocytes. J. Physiol. **515:** 109–118.
11. CARMELIET, E. 1999. Cardiac ionic currents and acute ischemia: from channels to arrhythmias. Physiol. Rev. **79:** 917–1017.

NCX and NCKX Operation in Ischemic Neurons

LECH KIEDROWSKI

*Departments of Psychiatry and Pharmacology, The Psychiatric Institute,
University of Illinois at Chicago, Chicago, Illinois 60612, USA*

ABSTRACT: Within the first 2 min of global brain ischemia, extracellular [K$^+$] ([K$^+$]$_o$) increases above 60 mM and [Na$^+$]$_o$ drops to about 50 mM, indicating a massive K$^+$ efflux and Na$^+$ influx, a phenomenon known as anoxic depolarization (AD). Similar ionic shifts take place during repetitive peri-infarct depolarizations (PID) in the area penumbra in focal brain ischemia. The size of ischemic infarct is determined by the duration of AD and PID. However, the mechanism of cytosolic [Ca^{2+}] ([Ca^{2+}]$_c$) elevation during AD or PID is poorly understood. Our data show that the exposure of cultured rat hippocampal CA1 neurons to AD-like conditions promptly elevates [Ca^{2+}]$_c$ to about 30 μM. These high [Ca^{2+}]$_c$ elevations depend on external Ca^{2+} and can be prevented by removing Na$^+$ or by simultaneously inhibiting NMDA and AMPA/kainate receptors. These data indicate that [Ca^{2+}]$_c$ elevations during AD result from Na$^+$ influx via either NMDA or AMPA/kainate channels. The mechanism of the Na-dependent [Ca^{2+}]$_c$ elevations may involve a reversal of plasmalemmal Na$^+$/Ca^{2+} (NCX) and/or Na$^+$/Ca^{2+} + K$^+$ (NCKX) exchangers. KB-R7943, an NCX inhibitor, suppresses a fraction of the Na-dependent Ca^{2+} influx during AD. Therefore, Ca^{2+} influx via NCX and a KB-R7943-resistant pathway (possibly NCKX) is involved. Inhibition of the Na-dependent Ca^{2+} influx is likely to decrease ischemic brain damage. No drugs are known that are able to inhibit the KB-R7943-resistant component of Na-dependent Ca^{2+} influx during AD. The present data encourage development of such agents as potential therapeutic means to limit ischemic brain damage after stroke or heart attack.

KEYWORDS: ischemia; NCX; NCKX; calcium overload; stroke; brain damage; neuroprotection

ANOXIC DEPOLARIZATION (AD) IN ISCHEMIC BRAIN

In the hippocampus, the partial oxygen pressure (PO$_2$) drops to zero within a few seconds of global brain ischemia.[1,2] The lack of oxygen arrests mitochondrial function[3] and depolarizes the mitochondria.[4] As a result, the glutamate

Address for correspondence: Lech Kiedrowski, Ph.D., The Psychiatric Institute, 1601 W. Taylor St., Room 334W, Chicago, IL 60612. Voice: 312-413-4559; fax: 312-413-4544.
Lkiedr@psych.uic.edu

Ann. N.Y. Acad. Sci. 1099: 383–395 (2007). © 2007 New York Academy of Sciences.
doi: 10.1196/annals.1387.035

released during ischemia[5] acts on neurons with mitochondria that are unable to buffer Ca^{2+} influx as they would under normoxic conditions.[6] In the presence of oxygen, inhibiting mitochondrial Ca^{2+} buffering through mitochondrial depolarization protects cultured neurons from glutamate excitotoxicity,[7] involving activation of NMDA and AMPA/kainate receptors by the released glutamate.[8–10] However, the principal factor of ischemia is the lack of oxygen. The mechanisms of glutamate excitotoxicity and cytosolic Ca^{2+} accumulation in anoxic neurons remain unclear.

During *in vivo* ischemia, rapid oxygen depletion is followed by AD, during which extracellular K^+ ($[K^+]_o$) rapidly increases above 60 mM, $[Na^+]_o$ drops to about 50 mM, and $[Ca^{2+}]_o$ decreases about 10-fold.[11,12] AD is initiated by Na^+ influx via voltage-gated Na^+ channels and AMPA/kainate channels.[13–17] The role of NMDA receptors in the initiation of AD is unclear; one study indicated that blocking NMDA channels accelerates the onset of AD,[18] but other studies showed that blocking NMDA receptors has no effect on AD initiation[19,20] or delays it.[15,21] AD can be effectively blocked by application of a single drug, dextromethorphan, a sigma receptor ligand.[22] However, dextromethorphan also inhibits a broad range of ionic channels, including voltage-gated Na^+ channels, voltage-gated Ca^{2+} channels, and NMDA receptors.[23]

During AD, extensive extracellular acidification[24] inhibits NMDA receptors much more than AMPA/kainate receptors[25] and activates acid-sensing ion channels (ASIC) that have been linked to ischemic neuronal death.[26,27] AD takes place during the first 2 min of global brain ischemia, that is, a time when ischemia is not yet toxic.[28] The extent of ischemic brain damage is determined by the duration of AD; the longer AD lasts, the greater the damage.

The ionic shifts of AD are expected to compromise the operation of a number of phenomena that depend on the Na^+ concentration gradient across the plasma membrane. For example, during brain ischemia, plasmalemmal glutamate transporters release glutamate to the extracellular space,[29–32] which implies that the released glutamate acts on neurons with profoundly disturbed Na^+ homeostasis. The plasmalemmal Na^+/Ca^{2+} (NCX) and $Na^+/Ca^{2+} + K^+$ (NCKX) exchangers are also affected by ionic shifts of AD. These exchangers normally extrude cytosolic Ca^{2+}.[33] However, when cytosolic $[Na^+]$ increases and plasma membrane potential depolarizes, the equilibrium potential of the exchangers may become more negative than the plasma membrane potential, causing reversal of the exchange.

Experiments performed using Ca-sensitive microelectrodes in ischemic hippocampal CA1 neurons *in vivo* showed that within 5 min of ischemia, $[Ca^{2+}]_c$ increases to about 30 μM.[2,34] Although the exact mechanism of this $[Ca^{2+}]_c$ elevation remains unclear, our recent data (see below) indicate that the $[Ca^{2+}]_c$ elevations during AD critically depend on external Na^+ and Ca^{2+}, suggesting that NCX, NCKX, and/or other Na-dependent mechanisms of Ca^{2+} influx are involved.

SIMULATION OF AD *IN VITRO*

Recently, we tested the mechanism of $[Ca^{2+}]_c$ elevation in cultured hippocampal CA1 neurons exposed to AD-like conditions.[35] The AD-simulating medium was glucose free and contained K^+, Na^+, Ca^{2+}, and H^+ concentrations adjusted to mimic the extracellular milieu during AD *in vivo*: $[K^+]_o$ was increased to 65 mM, $[Na^+]_o$ reduced to 50 mM, $[Ca^{2+}]_o$ decreased to 0.13 mM, and pH was reduced to 6.6.[36] Moreover, the AD-simulating medium was supplemented with 100 μM glutamate to mimic the ischemic elevation in extracellular glutamate.[5] To ensure optimal conditions for the activation of NMDA receptors, 10 μM glycine, an NMDA receptor co-agonist,[37,38] was also included. To decrease PO_2 below 2 mmHg, the AD-simulating medium was bubbled for 30 min with 95% Ar/5% CO_2. Application of this medium to cultured CA1 neurons failed to induce a major $[Ca^{2+}]_c$ increase detectable by fura-2FF unless the O_2 was completely removed by adding 2 mM $Na_2S_2O_4$ (FIG. 1 A). $Na_2S_2O_4$ rapidly removes oxygen from the medium,[39] thus inhibiting the respiratory chain in mitochondria. The $Na_2S_2O_4$-induced $[Ca^{2+}]_c$ elevation (FIG. 2 B, C) can be interpreted to indicate that oxygen-deprived mitochondria are no longer able to buffer Ca^{2+} and/or produce ATP to support ATP-dependent Ca^{2+} pumps. Indeed, the effect of $Na_2S_2O_4$ on $[Ca^{2+}]_c$ has been mimicked by other means of inhibiting the respiratory chain in the mitochondria: antimycin A1, which inhibits electron transfer from cytochrome b to cytochrome c1[40] (FIG. 1 B), or NaCN, which inhibits cytochrome oxidase, also promptly increased $[Ca^{2+}]_c$ to about 30 μM. These data can be interpreted to

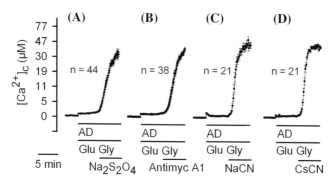

FIGURE 1. Simulation of AD in cultured hippocampal CA1 neurons. (**A**) Application of AD-simulating medium (AD) supplemented with 100 μM glutamate (Glu) and 10 μM glycine (Gly) and depleted of O_2 by 30 min of bubbling with 95% Ar/ 5% CO_2 failed to elevate $[Ca^{2+}]_c$ unless O_2 was completely removed by adding 2 mM $Na_2S_2O_4$. The effect of $Na_2S_2O_4$ on $[Ca^{2+}]_c$ was mimicked by inhibiting the respiratory chain with 1 μM antimycin A1 (**B**), 5 mM NaCN (**C**), and 5 mM CsCN (**D**). The data are averages ± SEM from n neurons monitored in representative experiments that were repeated with similar results at least three times. (From Kiedrowski,[35] with permission.)

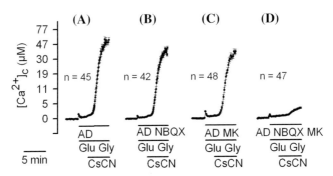

FIGURE 2. Either NMDA or AMPA/kainate receptor activation suffices to elicit high $[Ca^{2+}]_c$ elevations during AD simulation. **(A)** Control experiment; cultured CA1 neurons were exposed to AD simulating medium supplemented with 100 μM glutamate and 10 μM glycine. When indicated, 5 mM CsCN was added. A 10 μM NBQX **(B)** or 10 μM MK-801 (MK) **(C)** failed to prevent CsCN-induced $[Ca^{2+}]_c$ elevations. However, simultaneous inhibition of NMDA and AMPA/kainate receptors using 10 μM MK-801 plus 10 μM NBQX significantly suppressed them **(D)**. In all experiments, a vehicle (0.1% DMSO) was present during AD simulation. The data are averages ± SEM from n neurons monitored in representative experiments that were repeated with similar results at least three times. (From Kiedrowski,[35] with permission.)

indicate that, in cultured neurons, 30 min of bubbling with 95% Ar/5% CO_2 does not decrease PO_2 sufficiently to compromise mitochondrial function and consequently, Ca^{2+} homeostasis. Thus, to study the effects of a rapid PO_2 drop to zero during brain ischemia in cultured neurons, it is necessary to supply the AD-simulating medium with $Na_2S_2O_4$ or to chemically block the respiratory chain. Under such conditions, $[Ca^{2+}]_c$ promptly increases to about 30 μM (FIG. 1 A–D), which resembles the ischemic $[Ca^{2+}]_c$ elevation observed in CA1 neurons *in vivo*.[2,34]

In depolarized and glucose-deprived neurons, Na^+ plays a significant role in NMDA-induced $[Ca^{2+}]_c$ elevations and excitotoxicity.[41,42] To address the role of Na^+ in $[Ca^{2+}]_c$ elevations during AD, we induced chemical anoxia using CsCN, thus avoiding adding extra Na^+ or K^+ to the medium; a CsCN application elevates $[Ca^{2+}]_c$ during AD in the same manner as a NaCN application (compare FIG. 1 C versus D).

The CsCN-induced $[Ca^{2+}]_c$ elevations during AD (FIG. 2 A) could not be prevented by supplying the AD-simulating medium with 10 μM NBQX to inhibit AMPA/kainate receptors (FIG. 2 B), with 10 μM MK-801 to block NMDA channels (FIG. 2 C), or with 10 μM Gd^{3+} to block transient receptor potential (TRP) channels[43] and voltage-gated Ca^{2+} channels.[44] Interestingly, a simultaneous application of MK-801 and NBQX significantly suppressed these $[Ca^{2+}]_c$ elevations (FIG. 2 D).[35] The data show that operation of either AMPA/kainate or NMDA channels during AD suffice for these $[Ca^{2+}]_c$ elevations to occur.

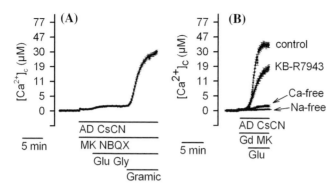

FIGURE 3. High $[Ca^{2+}]_c$ elevations during AD simulation in CA1 neurons are Na-dependent. **(A)** A combined application of 10 μM MK-801 (MK) plus 10 μM NBQX inhibited $[Ca^{2+}]_c$ elevations during AD simulation but this inhibition could be overridden by adding 5 μM gramicidin (Gramic) to induce Na^+ influx. **(B)** Effects an NCX inhibitor, KB-R7943, and Na^+ or Ca^{2+} removal on glutamate-induced $[Ca^{2+}]_c$ elevations during AD simulation. Although the AD-simulating medium was supplemented with 10 μM $GdCl_3$ and 10 μM MK-801 to inhibit voltage-gated Ca^{2+} channels, TRP channels, and NMDA channels, application of 100 μM glutamate (without glycine) promptly elevated $[Ca^{2+}]_c$. This $[Ca^{2+}]_c$ elevation could be prevented by substituting Na^+ with Li^+ (Na free) or removing external Ca^{2+} (Ca free). Inhibiting NCX with 10 μM KB-R7943 partly suppressed these $[Ca^{2+}]_c$ elevations. The data are averages ± SEM from over 20 neurons monitored in representative experiments that were repeated with similar results at least three times. (From Kiedrowski,[35] with permission.)

A common feature of AMPA/kainate channels and NMDA channels is Na^+ permeability.[45] Therefore, we tested the hypothesis that the $[Ca^{2+}]_c$ increase during AD simulation results from Na^+ influx. To this end, NMDA and AMPA/kainate channels were blocked during AD simulation (using MK-801 plus NBQX) and Na^+ influx was artificially induced by applying gramicidin, an antibiotic, which forms channels that permeate Na^+ and K^+, but not Ca^{2+}.[46] Gramicidin promptly increased $[Ca^{2+}]_c$ (FIG. 3 A), supporting the above hypothesis.

Consistent with the idea that during AD, the activation of AMPA/kainate receptors by glutamate is sufficient to elevate $[Ca^{2+}]_c$ to 30 μM, inhibition of NMDA channels, voltage-gated Ca channels, and Gd-sensitive TRP channels (with 10 μM MK-801 plus 10 μM Gd^{3+}) failed to prevent these $[Ca^{2+}]_c$ elevations and the presence of glycine in the medium (to facilitate NMDA receptor activation) was not necessary (FIG. 3 B). Na^+ influx played a critical role in $[Ca^{2+}]_c$ elevations because substituting Na^+ with Li^+ (FIG. 3 B) or Cs^{+35} completely prevented these elevations.

The AD-induced $[Ca^{2+}]_c$ elevations, which reached 30 μM, could be prevented by removing external Ca^{2+}, however, a small $[Ca^{2+}]_c$ elevation that could not be precisely measured using fura-2FF was observed under Ca-free conditions (FIG. 3 B). These data suggest that the large 30 μM $[Ca^{2+}]_c$ increase is mediated by Ca^{2+} influx and that Ca^{2+} release from internal stores plays a minor role.

Replacement of Na^+ with Li^+ or Cs^+ prevents Ca^{2+} influx via reversed NCX and NCKX because neither Li^+ nor Cs^+ substitutes for Na^+ in NCX or NCKX operations.[47] To test the extent to which NCX might be involved, KB-R7943, which inhibits NCX but not NCKX reversal in CA1 neurons,[47] was applied. KB-R7943 significantly reduced the rate of the Na-dependent $[Ca^{2+}]_c$ elevations during AD (FIG. 3 B).[35] The inhibitory effects of Li^+ and KB-R7943 (FIG. 3 B) on the glutamate-induced $[Ca^{2+}]_c$ elevations in the presence of MK-801 cannot be explained in terms of the inhibition of AMPA/kainate channels because neither Li^{+48} nor KB-R7943[49] blocks these channels.

MECHANISMS OF THE NA$^+$-DEPENDENT $[Ca^{2+}]_c$ ELEVATION DURING AD

The above data show that NMDA and AMPA/kainate channels must be simultaneously blocked to prevent the high $[Ca^{2+}]_c$ elevations induced by glutamate during AD simulation (FIG. 3 A), and further, that these $[Ca^{2+}]_c$ elevations critically depend on Na^+ influx (FIG. 3 B). The data imply a crucial role for Na^+ influx in high $[Ca^{2+}]_c$ elevations during AD and are consistent with the idea that depolarization of the plasma membrane combined with Na^+ influx leads to reversal of NCX and/or NCKX.[42,47,50] The data show that a KB-R7943-sensitive pathway, that is, NCX, plays a significant role in AD-induced $[Ca^{2+}]_c$ elevations, although a KB-R7943-resistant pathway contributed to these $[Ca^{2+}]_c$ elevations (FIG. 3 B). Further studies are necessary to elucidate the identity of the KB-R7943-resistant pathway. Nevertheless, considering that an earlier study showed the significant contribution of NCKX to the Na-dependent Ca^{2+} influx in CA1 neurons, and that the NCKX-mediated Ca^{2+} influx cannot be blocked by KB-R7943,[47] it is possible that NCKX reversal plays a role in the KB-R7943-resistant $[Ca^{2+}]_c$ elevations during AD.

NCX and NCKX mediate Ca^{2+} influx as long as their equilibrium potentials, E_{NCX} and E_{NCKX}, respectively, are more negative than the plasma membrane potential (E_m). During AD, E_m stabilizes at about -15 mV.[51,52] By estimating $[Ca^{2+}]_c$, at which E_{NCX} and E_{NCKX} equal -15 mV, one can approximate how high the exchangers can elevate $[Ca^{2+}]_c$ during AD. Taking into account the ionic shifts of AD and using the Nernst equation, E_{NCX} and E_{NCKX} can be calculated as a function of $[Ca^{2+}]_c$ (FIG. 4). This theoretical analysis shows that during AD-like conditions, NCX and NCKX may elevate $[Ca^{2+}]_c$ to 55 μM and 26 μM, respectively, which agrees with the 30 μM $[Ca^{2+}]_c$ elevations in CA1 neurons *in vivo*[2,34] and *in vitro* (FIG. 1) and confirms that the reversed

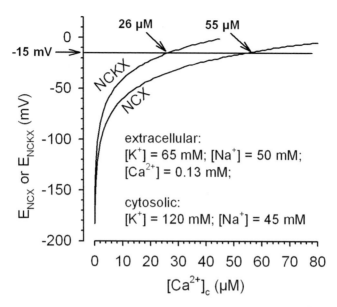

FIGURE 4. During AD, NCKX and NCX are expected to clamp $[Ca^{2+}]_c$ in the 26–55 µM range. NCKX and NCX mediate Ca^{2+} influx as long as the equilibrium potentials of NCKX (E_{NCKX}) and NCX (E_{NCX}) are lower than the plasma membrane potential (E_m). During AD, E_m stabilizes at about –15 mV.[51,52] E_{NCKX} and E_{NCX} in the figure were calculated according to the equations: $E_{NCKX} = 4E_{Na} - 2E_{Ca} - E_K$,[69] $E_{NCX} = 3 E_{Na} - 2E_{Ca}$,[33] where E_{Na}, E_{Ca} and E_K are equilibrium potentials for Na^+, Ca^{2+}, and K^+, respectively, calculated from the Nernst equation using the ionic concentrations indicated in the figure. These concentrations reflect the ionic shifts of AD.[16,36]

exchangers may mediate the $[Ca^{2+}]_c$ elevations during AD. One should consider, however, that the mechanism of the Na-dependent $[Ca^{2+}]_c$ elevations during AD may also include a direct Ca^{2+} influx via Na-gated cationic channels similar to those identified by Zhainazarov and Ache in lobster olfactory receptor neurons.[53] Whether such channels are expressed in hippocampal CA1 neurons is yet unclear.

Although Na^+ influx via AMPA/kainate or NMDA channels plays a dominant role in $[Ca^{2+}]_c$ elevation during AD simulation (FIG. 2), additional routes of Na^+ influx operate in ischemic neurons[54,55] and include voltage-dependent Na^+ channels,[56] Na^+/H^+ exchange,[57] and ASIC.[26,27] These additional pathways of Na^+ influx likely contribute to the ischemic $[Ca^{2+}]_c$ increase by activating the Na-dependent mechanisms of $[Ca^{2+}]_c$ elevation. Although homomeric ASIC1a permeate Na^+ and Ca^{2+},[58–60] a direct Ca^{2+} influx via these channels cannot explain the Na-dependent $[Ca^{2+}]_c$ elevations. First, the $[Ca^{2+}]_c$ increase during AD simulation does not coincide with the pH drop from 7.4 to 6.6 when the AD medium is applied. Second, substitution of external Na^+

with Li^+ does not block ASIC1a[58] but does prevent glutamate-induced $[Ca^{2+}]_c$ elevations during AD (FIG. 3).

NCX AND NCKX OPERATION BEFORE AND AFTER AD—A HYPOTHESIS

The following scenario likely takes place at the early stages of global brain ischemia. Prior to the ionic shifts of AD, glutamate is released at synapses and removed from the extracellular space by glial and neuronal glutamate transporters that rely on the Na^+ concentration gradient to transport glutamate into the cells.[61] At this stage of ischemia, the released glutamate activates synaptic NMDA and AMPA/kainate receptors, while NCX and NCKX still operate in the forward mode and extrude Ca^{2+} (FIG. 5, upper). At later stages of ischemia, following the ionic shifts of AD, however, the glutamate transporters reverse and spill glutamate beyond the space of the synaptic clefts. The spilled glutamate activates extrasynaptic NMDA and AMPA/kainate receptors, causing a persistent Na^+ influx that reverses NCX and NCKX. As a result, $[Ca^{2+}]_c$ is greatly elevated in a Na-dependent manner (FIG. 5, lower).

Although NCX and NCKX may reverse during AD, Na^+ and K^+ concentration gradients are quickly reestablished during reperfusion.[36] This is expected to cause NCKX and NCX to return to the forward mode in which they restore Ca^{2+} homeostasis and promote neuronal recovery from ischemia. For this reason, only inhibition of the reverse but not the forward mode of the exchangers may be expected to diminish ischemic brain damage. Theoretically, the reverse mode inhibition can be accomplished using drugs able to stabilize a desensitized conformation undertaken by NCX upon cytosolic Na^+ binding.[62] Via this mechanism, KB-R7943 appears to preferentially inhibit the reverse mode of NCX operation.[63,64]

CAN NCX AND/OR NCKX REVERSE DURING STROKE?

The AD-induced and Na-dependent $[Ca^{2+}]_c$ elevations described here may contribute to Ca^{2+} overload not only during global but also during focal brain ischemia. During the latter, in the area penumbra, repetitive peri-infarct depolarizations (PID) take place and are associated with similar ionic shifts as those observed during AD.[65] It is likely that a Na-dependent Ca^{2+} influx occurs during each PID. Since the duration of PID is directly proportional to the size of the infarct,[66-68] inhibiting Ca^{2+} influx via the Na-dependent pathways during PID should be neuroprotective. The therapeutic window for application of such inhibitors during stroke may be as long as PID continue to occur during focal brain ischemia, that is, several hours.[68] As inhibitors of certain Na-dependent pathways of Ca^{2+} influx, such as reversed NCKX do not yet exist, the present data emphasize the need for their development.

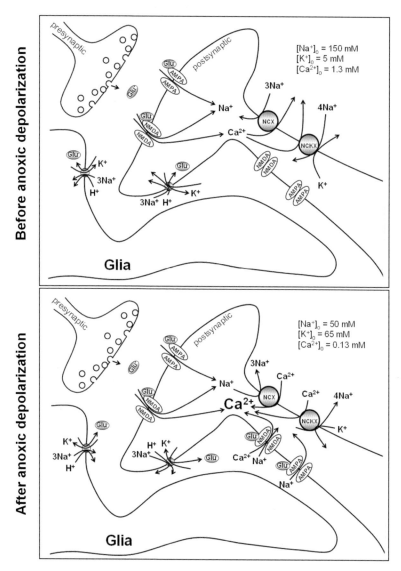

FIGURE 5. NCX and NCKX operation before and after AD. At the early stages of brain ischemia before AD, Na^+ and K^+ concentration gradients across the plasma membrane support the forward mode on NCX and NCKX, as well as glutamate uptake by plasmalemmal glutamate transporters. At that time, glutamate is released in a vesicular manner and activates only synaptic AMPA/kainate and NMDA receptors (upper panel). The drop in $[Na^+]_o$ and increase in $[K^+]_o$ during AD compromises the normal (forward) operation of the glutamate transporters, NCX and NCKX. The glutamate transporters reverse and spill glutamate into the extracellular space where it activates extrasynaptic NMDA and AMPA/kainate receptors. As a result, Na^+ influx is greatly increased causing NCX and NCKX reversal (lower panel).

ACKNOWLEDGMENTS

This work was supported by the National Institutes of Health Grant NS 37390.

REFERENCES

1. NAIR, P.K., D.G. BUERK & J.H. HALSEY, JR. 1987. Comparisons of oxygen metabolism and tissue PO2 in cortex and hippocampus of gerbil brain. Stroke **18:** 616–622.
2. SILVER, I.A. & M. ERECIŃSKA. 1990. Intracellular and extracellular changes of [Ca^{2+}] in hypoxia and ischemia in rat brain *in vivo*. J. Gen. Physiol. **95:** 837–866.
3. WILSON, D.F. *et al.* 1988. The oxygen dependence of mitochondrial oxidative phosphorylation measured by a new optical method for measuring oxygen concentration. J. Biol. Chem. **263:** 2712–2718.
4. PAQUET-DURAND, F. & G. BICKER. 2004. Hypoxic/ischaemic cell damage in cultured human NT-2 neurons. Brain Res. **1011:** 33–47.
5. BENVENISTE, H. *et al.* 1984. Elevation of the extracellular concentrations of glutamate and aspartate in rat hippocampus during transient cerebral ischemia monitored by intracerebral microdialysis. J. Neurochem. **43:** 1369–1374.
6. NICHOLLS, D.G. & S.L. BUDD. 2000. Mitochondria and neuronal survival. Physiol. Rev. **80:** 315–360.
7. STOUT, A.K. *et al.* 1998. Glutamate-induced neuronal death requires mitochondrial calcium uptake. Nat. Neurosci. **1:** 366–373.
8. SIMON, R.P. *et al.* 1984. Blockade of N-methyl-D-aspartate receptors may protect against ischemic damage in the brain. Science **226:** 850–852.
9. SHEARDOWN, M.J. *et al.* 1990. 2,3-Dihydroxy-6-nitro-7-sulfamoyl-benzo(F)-quinoxaline: a neuroprotectant for cerebral ischemia. Science **247:** 571–574.
10. LEE, J.M., G.J. ZIPFEL & D.W. CHOI. 1999. The changing landscape of ischaemic brain injury mechanisms. Nature **399:** A7–A14.
11. HANSEN, A.J. 1978. The extracellular potassium concentration in brain cortex following ischemia in hypo- and hyperglycemic rats. Acta Physiol. Scand. **102:** 324–329.
12. HANSEN, A.J. & T. ZEUTHEN. 1981. Extracellular ion concentrations during spreading depression and ischemia in the rat brain cortex. Acta Physiol. Scand. **113:** 437–445.
13. XIE, Y. *et al.* 1994. Effects of the sodium channel blocker tetrodotoxin (TTX) on cellular ion homeostasis in rat brain subjected to complete ischemia. Brain Res. **652:** 216–224.
14. XIE, Y. *et al.* 1995. Ion channel involvement in anoxic depolarization induced by cardiac arrest in rat brain. J. Cereb. Blood Flow Metab. **15:** 587–594.
15. TANAKA, E. *et al.* 1997. Mechanisms underlying the rapid depolarization produced by deprivation of oxygen and glucose in rat hippocampal CA1 neurons *in vitro*. J. Neurophysiol. **78:** 891–902.
16. MÜLLER, M. & G.G. SOMJEN. 2000. Na^+ and K^+ concentrations, extra- and intracellular voltages, and the effect of TTX in hypoxic rat hippocampal slices. J. Neurophysiol. **83:** 735–745.

17. MÜLLER, M. & G.G. SOMJEN. 2000. Na$^+$ dependence and the role of glutamate receptors and Na+ channels in ion fluxes during hypoxia of rat hippocampal slices. J. Neurophysiol. **84**: 1869–1880.
18. MARRANNES, R. *et al.* 1988. NMDA antagonists inhibit cortical spreading depression but accelerate the onset of neuronal depolarisation induced by asphyxia. *In* Mechanisms of Cerebral Hypoxia and Stroke. G. Somjen, Ed.: 303–304. Plenum Press. New York.
19. HERNANDEZ-CACERES, J. *et al.* 1987. Systemic ketamine blocks cortical spreading depression but does not delay the onset of terminal anoxic depolarization in rats. Brain Res. **437**: 360–364.
20. AITKEN, P.G., M. BALESTRINO & G.G. SOMJEN. 1988. NMDA antagonists: lack of protective effect against hypoxic damage in CA1 region of hippocampal slices. Neurosci. Lett. **89**: 187–192.
21. MARTIN, R.L. 1999. Block of rapid depolarization induced by *in vitro* energy depletion of rat dorsal vagal motoneurones. J. Physiol. **519**: 131–141.
22. ANDERSON, T.R. *et al.* 2005. Blocking the anoxic depolarization protects without functional compromise following simulated stroke in cortical brain slices. J. Neurophysiol. **93**: 963–979.
23. NETZER, R., P. PFLIMLIN & G. TRUBE. 1993. Dextromethorphan blocks N-methyl-D-aspartate-induced currents and voltage-operated inward currents in cultured cortical neurons. Eur. J. Pharmacol. **238**: 209–216.
24. SIEMKOWICZ, E. & A.J. HANSEN. 1981. Brain extracellular ion composition and EEG activity following 10 minutes ischemia in normo- and hyperglycemic rats. Stroke **12**: 236–240.
25. TRAYNELIS, S.F. & S.G. CULL-CANDY. 1991. Pharmacological properties and H$^+$ sensitivity of excitatory amino acid receptor channels in rat cerebellar granule neurones. J. Physiol. (Lond.) **433**: 727–763.
26. XIONG, Z.G. *et al.* 2004. Neuroprotection in ischemia: blocking calcium-permeable acid-sensing ion channels. Cell **118**: 687–698.
27. GAO, J. *et al.* 2005. Coupling between NMDA receptor and acid-sensing ion channel contributes to ischemic neuronal death. Neuron. **48**: 635–646.
28. SMITH, M.L., R.N. AUER & B.K. SIESJÖ. 1984. The density and distribution of ischemic brain injury in the rat following 2-10 min of forebrain ischemia. Acta Neuropathol. **64**: 319–332.
29. PHILLIS, J.W. *et al.* 1994. Characterization of glutamate, aspartate, and GABA release from ischemic rat cerebral cortex. Brain Res. Bull. **34**: 457–466.
30. PHILLIS, J.W., J. REN & M.H. O'REGAN. 2000. Transporter reversal as a mechanism of glutamate release from the ischemic rat cerebral cortex: studies with DL-threo-beta-benzyloxyaspartate. Brain Res. **868**: 105–112.
31. SZATKOWSKI, M. & D. ATTWELL. 1994. Triggering and execution of neuronal death in brain ischemia: two phases of glutamate release by different mechanisms. Trends Neurosci. **17**: 359–365.
32. ROSSI, D.J., T. OSHIMA & D. ATTWELL. 2000. Glutamate release in severe brain ischaemia is mainly by reversed uptake. Nature **403**: 316–321.
33. BLAUSTEIN, M.P. & W.J. LEDERER. 1999. Sodium/calcium exchange: its physiological implications. Physiol. Rev. **79**: 763–854.
34. ERECIŃSKA, M. & I.A. SILVER. 2001. Tissue oxygen tension and brain sensitivity to hypoxia. Respir. Physiol. **128**: 263–276.
35. KIEDROWSKI, L. 2007. Critical role of sodium in cytosolic [Ca^{2+}] elevations in cultured hippocampal CA1 neurons during anoxic depolarization. J. Neurochem. **100**: 915–923.

36. HANSEN, A.J. 1985. Effect of anoxia on ion distribution in the brain. Physiol. Rev. **65:** 101–148.
37. JOHNSON, J.W. & P. ASCHER. 1987. Glycine potentiates the NMDA response in cultured mouse brain neurons. Nature **325:** 529–531.
38. KLECKNER, N.W. & R. DINGLEDINE. 1988. Requirement for glycine in activation of NMDA-receptors expressed in *Xenopus oocytes*. Science **241:** 835–837.
39. LAMBETH, D.O. & G. PALMER. 1973. The kinetics and mechanism of reduction of electron transfer proteins and other compounds of biological interest by dithionite. J. Biol. Chem. **248:** 6095–6103.
40. SLATER, E.C. 1973. The mechanism of action of the respiratory inhibitor antimycin. Biochim. Biophys. Acta **301:** 129–154.
41. CZYŻ, A., G. BARANAUSKAS & L. KIEDROWSKI. 2002. Instrumental role of Na^+ in NMDA excitotoxicity in glucose-deprived and depolarized cerebellar granule cells. J. Neurochem. **81:** 379–389.
42. CZYŻ, A. & L. KIEDROWSKI. 2002. In depolarized and glucose-deprived neurons, Na^+ influx reverses plasmalemmal K^+-dependent and K^+-independent Na^+/Ca^{2+} exchangers and contributes to NMDA excitotoxicity. J. Neurochem. **83:** 1321–1328.
43. XIONG, Z.G., W.Y. LU & J.F. MACDONALD. 1997. Extracellular calcium sensed by a novel cation channel in hippocampal neurons. Proc. Natl. Acad. Sci. USA **94:** 7012–7017.
44. YU, S.P. *et al.* 1997. Mediation of neuronal apoptosis by enhancement of outward potassium current. Science **278:** 114–117.
45. MAYER, M.L. & G.L. WESTBROOK. 1987. The physiology of excitatory amino acids in the vertebrate central nervous system. Progr. Neurobiol. **28:** 197–276.
46. HLADKY, S.B. & D.A. HAYDON. 1972. Ion transfer across lipid membranes in the presence of gramicidin A. I. Studies of the unit conductance channel. Biochim. Biophys. Acta **274:** 294–312.
47. KIEDROWSKI, L. 2004. High activity of plasmalemmal K^+-dependent Na^+/Ca^{2+} exchangers in hippocampal CA1 neurons. Neuroreport **15:** 2113–2116.
48. KARKANIAS, N.B. & R.L. PAPKE. 1999. Subtype-specific effects of lithium on glutamate receptor function. J. Neurophysiol. **81:** 1506–1512.
49. HOYT, K.R. *et al.* 1998. Reverse Na^+/Ca^{2+} exchange contributes to glutamate-induced intracellular Ca^{2+} concentration increases in cultured rat forebrain neurons. Mol. Pharmacol. **53:** 742–749.
50. KIEDROWSKI, L. *et al.* 2004. Differential contribution of plasmalemmal Na^+/Ca^{2+} exchange isoforms to sodium-dependent calcium influx and NMDA excitotoxicity in depolarized neurons. J. Neurochem. **90:** 117–128.
51. CZEH, G., P.G. AITKEN & G.G. SOMJEN. 1993. Membrane currents in CA1 pyramidal cells during spreading depression (SD) and SD-like hypoxic depolarization. Brain Res. **632:** 195–208.
52. COLLEWIJN, H. & A.V. HARREVELD. 1966. Membrane potential of cerebral cortical cells during reading depression and asphyxia. Exp. Neurol. **15:** 425–436.
53. ZHAINAZAROV, A.B. & B.W. ACHE. 1997. Gating and conduction properties of a sodium-activated cation channel from lobster olfactory receptor neurons. J. Membr. Biol. **156:** 173–190.
54. FRIEDMAN, J.E. & G.G. HADDAD. 1994. Anoxia induces an increase in intracellular sodium in rat central neurons *in vitro*. Brain Res. **663:** 329–334.
55. SHELDON, C. *et al.* 2004. Sodium influx pathways during and after anoxia in rat hippocampal neurons. J. Neurosci. **24:** 11057–11069.

56. CARTER, A.J. *et al*. 2000. Potent blockade of sodium channels and protection of brain tissue from ischemia by BIII 890 CL. Proc. Natl. Acad. Sci. USA **97:** 4944–4949.

57. LUO, J. *et al*. 2005. Decreased neuronal death in Na^+/H^+ exchanger isoform 1-null mice after in vitro and *in vivo* ischemia. J. Neurosci. **25:** 11256–11268.

58. WALDMANN, R., *et al*. 1997. A proton-gated cation channel involved in acid-sensing. Nature **386:** 173–177.

59. CHU, X.P. *et al*. 2002. Proton-gated channels in PC12 cells. J. Neurophysiol. **87:** 2555–2561.

60. YERMOLAIEVA, O. *et al*. 2004. Extracellular acidosis increases neuronal cell calcium by activating acid-sensing ion channel 1a. Proc. Natl. Acad. Sci. USA **101:** 6752–6757.

61. GREWER, C. & T. RAUEN. 2005. Electrogenic glutamate transporters in the CNS: molecular mechanism, pre-steady-state kinetics, and their impact on synaptic signaling. J. Membr. Biol. **203:** 1–20.

62. HILGEMANN, D.W. *et al*. 1992. Steady-state and dynamic properties of cardiac sodium-calcium exchange. Sodium-dependent inactivation. J. Gen. Physiol. **100:** 905–932.

63. IWAMOTO, T., T. WATANO & M. SHIGEKAWA. 1996. A novel isothiourea derivative selectively inhibits the reverse mode of Na^+/Ca^{2+} exchange in cells expressing NCX1. J. Biol. Chem. **271:** 22391–22397.

64. WATANO, T. *et al*. 1996. A novel antagonist, No. 7943, of the Na^+/Ca^{2+} exchange current in guinea-pig cardiac ventricular cells. Br. J. Pharmacol. **119:** 555–563.

65. NEDERGAARD, M. & A.J. HANSEN. 1993. Characterization of cortical depolarizations evoked in focal cerebral ischemia. J. Cereb. Blood Flow Metab. **13:** 568–574.

66. MIES, G., T. IIJIMA & K.A. HOSSMANN. 1993. Correlation between peri-infarct DC shifts and ischaemic neuronal damage in rat. Neuroreport **4:** 709–711.

67. DIJKHUIZEN, R.M. *et al*. 1999. Correlation between tissue depolarizations and damage in focal ischemic rat brain. Brain Res. **840:** 194–205.

68. OHTA, K. *et al*. 2001. Calcium ion transients in peri-infarct depolarizations may deteriorate ion homeostasis and expand infarction in focal cerebral ischemia in cats. Stroke **32:** 535–543.

69. DONG, H. *et al*. 2001. Electrophysiological characterization and ionic stoichiometry of the rat brain K^+-dependent Na^+/Ca^{2+} exchanger, NCKX2. J. Biol. Chem. **276:** 25919–25928.

Na$^+$/Ca^{2+} Exchange and Ca^{2+} Homeostasis in Axon Terminals of Mammalian Central Neurons

SUK-HO LEE,[a] MYOUNG-HWAN KIM,[a] JU-YOUNG LEE,[a]
SANG HUN LEE,[a] DOYUN LEE,[a] KYEONG HAN PARK,[b]
AND WON-KYUNG HO[a]

[a]National Research Laboratory for Cell Physiology, Department of Physiology,
Seoul National University College of Medicine, Chongno-Ku,
Seoul 110-799, South Korea

[b]Department of Anatomy, Kangwon National University College of Medicine,
Chunchon 200-701, South Korea

ABSTRACT: We investigated Ca^{2+} clearance mechanisms (CCMs) at the
axon terminals of mammalian central neurons: neurohypophysial (NHP)
axon terminals and calyces of Held. Ca^{2+} transients were evoked by applying a short depolarization pulse via a patch pipette containing Ca^{2+}
indicator dye. Quantitative analysis of the Ca^{2+} decay phases revealed
that Na$^+$/Ca^{2+} exchange (Na/CaX) is a major CCM at both axon terminals. In contrast, no Na/CaX activity was found in the somata of NHP
axon terminals (supraoptic magnocellular neurons), indicating that the
distribution of Na$^+$/Ca^{2+} exchangers is polarized. Intracellular dialysis
of axon terminals with a K$^+$-free pipette solution attenuated the Na/CaX
activities by 90% in the NHP axon terminals and by 60% at the calyx
of Held, indicating that K$^+$-dependent Na$^+$/Ca^{2+} exchangers are involved. Studying the effects of specific inhibitors of smooth endoplasmic
reticulum Ca^{2+}-ATPase (SERCA) and plasma membrane Ca^{2+}-ATPase
(PMCA) on the Ca^{2+} decay rate revealed that PMCA contributed 23% of
total Ca^{2+} clearance, but that SERCA made no contribution at the calyx
of Held. The contribution of mitochondria was negligible for small Ca^{2+}
transients, but became apparent at peak Ca^{2+} levels higher than 2.5 μM.
When mitochondrial function was inhibited, the dependence of CCMs
on [Ca^{2+}]$_i$ at the calyx of Held showed saturation kinetics with $K_{1/2} =$
1.7 μM, suggesting that the Na/CaX activity is saturated at high [Ca^{2+}]$_i$.
The presynaptic Na$^+$/Ca^{2+} exchanger activity, which competes for cytosolic Ca^{2+} with mitochondria, may contribute to nonplastic synaptic
transmission at these axon terminals.

KEYWORDS: neurohypophysial axon terminals; calyx of Held; supraoptic
magnocellular neuron; NCX; NCKX; mitochondria

Address for correspondence: Suk-Ho Lee, M.D., Ph.D., Department of Physiology, Seoul National
University College of Medicine, Chongno-Ku, Yongon-Dong 28, Seoul 110-799, Korea. Voice: +82-
2-740-8222; fax: +82-2-763-9667.
leesukho@snu.ac.kr

Ann. N.Y. Acad. Sci. 1099: 396–412 (2007). © 2007 New York Academy of Sciences.
doi: 10.1196/annals.1387.011

INTRODUCTION

Ca^{2+} is a mediator of neurotransmitter release and plays a central role in activity-dependent changes in synaptic strength in axon terminals. Most axon terminals of mammalian central neurons are submicroscopic, and thus electrophysiological studies on axon terminals are limited. However, two exceptionally large axon terminals in mammalian central neurons do exist: neurohypophysial (NHP) axon terminals and calyces of Held in the medial nucleus of the trapezoidal body.[1,2] These axon terminals provide unique opportunities to study axon terminals using patch clamp techniques combined with microfluorometry.

Extrusion of cytosolic Ca^{2+} across the plasma membrane is mediated by plasma membrane Ca^{2+}-ATPase (PMCA) and Na^+/Ca^{2+} exchange (Na/CaX). Although Na/CaX activity, which is catalyzed by NCX and/or NCKX, is electrogenic, isolation of Na/CaX currents is not practical in most primary cells because many cells express other kinds of Ca^{2+}-activated currents. In particular, isolation of NCKX-mediated current, which involves K^+, is not possible in cells that are expressing Ca^{2+}-activated K^+ channels. Thus, electrophysiological studies of NCX and NCKX have been performed extensively in cardiac myocytes and in retinal photoreceptors, respectively, in which little Ca^{2+}-activated ionic current is expressed. To circumvent this obstacle in the study of Na/CaX in native cells, we measured Ca^{2+} transients instead of the exchanger current and estimated the Na/CaX activity from extracellular Na^+-dependent changes in the Ca^{2+} extrusion rate. Using this technique, we were able to identify the role of NCX and NCKX at the axon terminals of mammalian central neurons.

QUANTIFICATION OF Ca^{2+} CLEARANCE FROM A Ca^{2+} TRANSIENT

When the intracellular Ca^{2+} is in dynamic equilibrium with cytosolic Ca^{2+} buffer, the amount of total Ca^{2+} is the sum of free-form and bound-form Ca^{2+}:

$$[Ca^{2+}]_T = [Ca^{2+}] + [CaB]. \tag{1}$$

For convenience, let $y \equiv [Ca^{2+}]$, $y_B \equiv [CaB]$, and $y_T \equiv [Ca^{2+}]_T$. Because $y_B = y \cdot [B]_T/(K_D + y)$, differentiation of both sides of Eq. (1) with respect to $[Ca^{2+}]$ yields

$$dy_T/dy = 1 + K_D \cdot B_T/(y + K_D)^2, \tag{2}$$

where the second term is referred to as "Ca^{2+} binding ratio (κ)" of buffer B (denoted by κ_B).[3] Generally, an increment of y (free Ca^{2+}) in a compartment can be converted to an increment of y_T using Eq. (2) as follows:

$$\Delta y_T = \Delta y \cdot (1 + \kappa_B + \kappa_S), \tag{3}$$

where the subscripts B and S represent a Ca^{2+} indicator dye and endogenous Ca^{2+} buffer, respectively.

Suppose a compartment whose Ca^{2+} clearance rate is linearly dependent on cytosolic free $[Ca^{2+}]$ level. When Ca^{2+} is perturbed from an equilibrium state by a pulse-like influx of Ca^{2+} at $t = 0$, the time-dependent decay of total Ca^{2+} $[y_T(t)]$ can be expressed as:

$$dy_T(t)/dt = -\gamma \cdot y(t). \tag{4}$$

The γ in this equation is the definition of "clearance," and represents the sum of activities of Ca^{2+} clearance mechanisms (CCM) of the compartment. Combining Eqs. (3) and (4) yields

$$dy_T(t)/dt = dy(t)/dt \cdot (1 + \kappa_B + \kappa_S) = -\gamma \cdot y(t). \tag{5}$$

Above equation can be simplified as

$$dy(t)/dt = -\lambda \cdot y(t), \tag{6}$$

where

$$\lambda = \gamma/(1 + \kappa_B + \kappa_S). \tag{7}$$

Eq. (6) indicates that the decay of $y(t)$ follows a mono-exponential function with a rate constant λ.

The decay phase of most Ca^{2+} transients was best fitted with a bi-exponential function, $y(t) = A_0 + A_1 \cdot \exp(-t \cdot \tau_1^{-1}) + A_2 \cdot \exp(-t \cdot \tau_2^{-1})$, rather than mono-exponential function, indicating that γ is not constant but depends on $y(t)$. Because the decay rate at the peak (or $-dy(t)/dt \,|_{t=0}$) is $A_1 \cdot \tau_1^{-1} + A_2 \cdot \tau_2^{-1}$, we defined a decay rate constant at the peak as follows:

$$\lambda_{t=0} = \frac{-dy/dt|_{t=0}}{y(0)} = \frac{(A_1 \cdot \tau_1^{-1} + A_2 \cdot \tau_2^{-1})}{(A_1 + A_2)} \tag{8}$$

in analogy with Eq. (6). According to this equation, $\lambda_{t=0}$ can be calculated from the weighted average of the two rate constants of the bi-exponential fit to a Ca^{2+} decay phase.

Because $\lambda_{t=0}$ depends on the peak of $y(t)$, $\lambda_{t=0}$ values obtained from two different Ca^{2+} transients can be compared only if the two Ca^{2+} transients have similar peak amplitudes. To find out acceptable range of the difference in the peak amplitude, we obtained a set of Ca^{2+} transients with different amplitudes from the same cell (FIG. 1 A), and found that $\lambda_{t=0}$ values are fairly constant when $y(0) > 400$ nM (FIG. 1 B). The dependence of $\lambda_{t=0}$ on $\Delta[Ca^{2+}]_i$ was estimated in two different ways. First, the values for $\lambda_{t=0}$ were calculated by Eq. (8) from five Ca^{2+} transients with different peak amplitudes. Second, the $dy(t)/dt$ was calculated from the decay phase of one Ca^{2+} transient with the highest peak, and it was divided by $y(t)$ to obtain λ according to Eq. (6) (solid

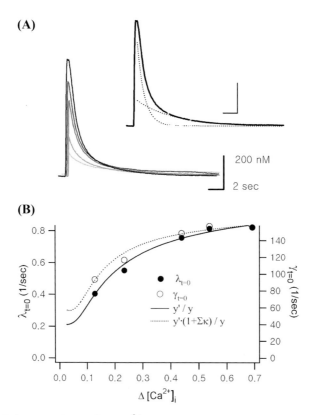

FIGURE 1. Dependence of the Ca^{2+} decay rate constant ($\lambda_{t=0}$) on $\Delta[Ca^{2+}]$. **(A)** A set of Ca^{2+} transients recorded in the same axon terminal. Each transient was evoked by a depolarization pulse of various durations ($10 \sim 200$ ms). *Inset*: The decay phase of the Ca^{2+} transient evoked by a 200 ms depolarization pulse was fitted with a bi-exponential function (*thin gray line*, inset). *scale bars*, 2 s, 200 nM. **(B)** The time derivative ($-d[Ca^{2+}]_i/dt$) of the decay phase of the highest Ca^{2+} transient in A was divided by $\Delta[Ca^{2+}](t)$ to obtain λ (according to Eq. (7)), and plotted as a function of $\Delta[Ca^{2+}]$ (*solid line*). Values for $\lambda_{t=0}$ were calculated according to Eq. (8) from bi-exponential fits to the five Ca^{2+} transients shown in A. These $\lambda_{t=0}$ values as a function of the peak $\Delta[Ca^{2+}]$ (*filled circles*) were superimposed. The λ values were converted to γ using $\kappa_S = 180$ (see Fig. 2 A) according to Eq. (6) or (9), and superimposed as a function of $\Delta[Ca^{2+}]$ (*open circles and dotted line, right ordinate*). Modified from Lee *et al.*[16] with permission.

line, Fig. 1 B). The $(-dy(t)/dt)/y(t)$ curve and the values of $\lambda_{t=0}$ as a function of $\Delta[Ca^{2+}]_i$ was superimposed on Figure 1 B, showing that similar λ values were obtained irrespective of the way of calculation. This result illustrates that the dependence of λ on $\Delta[Ca^{2+}]$ can be calculated from the decay phase of one big Ca^{2+} transient. The clearance at the peak, $\gamma_{t=0}$, can be calculated according to Eq. (7) as follows:

$$\gamma_{t=0} = \lambda_{t=0} \cdot (1 + \kappa_B + \kappa_S). \tag{9}$$

When two compartments have different κ, it is essential to use γ for comparing Ca^{2+} clearance. On the other hand, we used $\lambda_{t=0}$ for comparing two Ca^{2+} transient obtained from the same compartment, assuming that $\Sigma\kappa$ is the same.

MEASUREMENT OF ENDOGENOUS Ca^{2+} BINDING RATIOS IN NHP AXON TERMINALS AND IN SOMATA OF SUPRAOPTIC MAGNOCELLULAR NEURONS

Because the genuine Ca^{2+} clearance power of a cell is represented by γ rather than λ, estimates for κ of Ca^{2+} buffers are prerequisites for measurement of the Ca^{2+} clearance of a given compartment. To investigate the subcellular localization of CCMs, the κ_S was separately estimated in the supraoptic magnocellular neurons (SMNs) and in their pinched-off axon terminals dissociated from the NHPs.

The values for the κ of Ca^{2+} indicator dye (κ_B) can be calculated from the $[B]_T$ and K_D of the dye. The κ_S, however, should be estimated from the relationship between $\Delta y_{t=0}$ and κ_B or between λ and κ_B, as shown in Eqs. (3) and (7), respectively. After patch break-in, intracellular $[B]_T$ increases with a time constant that depends on the cell volume and the access resistance of the patch pipette.[4] A set of Ca^{2+} transients at various $[B]_T$ can be obtained by applying a short depolarization pulse during the increasing phase of $[B]_T$. Assuming that Δy_T and γ are constant among this set of Ca^{2+} transients, Eqs. (3) and (7) predict that initial amplitudes ($\Delta y_{t=0}$ or A) and decay rates (λ or τ^{-1}) of Ca^{2+} transients decrease as cytosolic $[B]_T$ increases. The κ_S can be estimated from the linear fit to the plot of $1/A$ or τ as a function of κ_B according to Eqs. (3) and (7), respectively. Mean values of κ_S were 187 ± 19 in axon terminals (FIG. 2 A) and 79 ± 2.6 in somata of SMNs (FIG. 2 B).

POLARIZED DISTRIBUTION OF Na^+/Ca^{2+} EXCHANGERS IN SMNs

We compared CCMs in somata and in axon terminals of SMNs using K^+-rich pipette solution containing 50 μM fura2. We examined the decay rate of Ca^{2+} transients evoked by a short depolarization pulse under normal and low $[Na^+]$ (20 mM, replaced by equimolar Li^+) conditions. The decay phases of Ca^{2+} transients were well fitted with a bi-exponential function both in the soma (*black line* in FIG. 3 B, *left*) and in the axon terminal (*black line* in FIG. 3 B, *right*). From the decay phases of these Ca^{2+} transients, values for $-d[Ca^{2+}]_i/dt$ were plotted as a function of $[Ca^{2+}]_i$ (*insets* of FIG. 3 B). The axon terminal showed faster $-d[Ca^{2+}]_i/dt$ than the soma in the control condition. The Ca^{2+} transients under the low external $[Na^+]$ and normal $[Na^+]$ conditions were

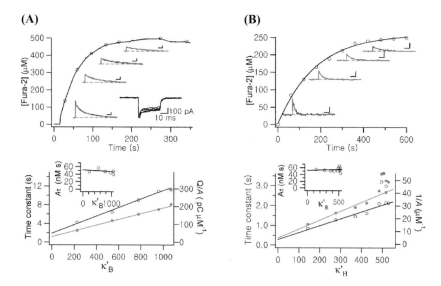

FIGURE 2. Estimation of endogenous Ca^{2+} binding ratios (κ_S) in the NHP axon terminal (A) and in the soma of a SMN (B). (A) *Upper,* Time course of fura-2 (500 μM) loading in the NHP axon terminal. Four exemplar Ca^{2+} transients (scale bars: 2 s, 20 nM) are shown. *Lower,* Plots of time constants (τ, *open circles, left ordinate*) and inverse of normalized amplitudes (Q/A, *gray filled circles, right ordinate*) versus incremental Ca^{2+} binding ratio of fura-2 (κ'_B). *Inset*: plot of Aτ versus κ'_B. κ_S were estimated from *x*-axis intercepts of regression lines. (B) *Upper,* Time course of fura-2 (250 μM) loading into the soma of a SMN (*abscissa*: elapsed time after break-in). Four exemplar Ca^{2+} transients are shown along the fura-2 loading curve (scale bars: 2 s, 20 nM). Time constants were estimated from mono-exponential fits to the decay phases of Ca^{2+} transients. *Lower,* Plots of τ (*open circles,* left ordinate) and inverse of initial amplitudes (1/A; gray-filled circles, *right ordinate*) versus κ'_B. Modified from Kim *et al.*[5] with permission.

superimposed (FIG. 3 B, *gray lines*). In somata of SMNs, lowering external [Na$^+$] had no effect on the decay rate of the Ca^{2+} transient (FIG. 3 B, *left*), indicating little contribution of Na/CaX in the soma. In contrast, the Ca^{2+} decay phase in NHP axon terminals was significantly slowed by the reduction of extracellular [Na$^+$] to 20 mM (FIG. 3 B, *right*).

To take into account the difference in κ_S in the two loci, $-d[Ca^{2+}]_i/dt$ was converted into $-d[Ca^{2+}]_T/dt$ according to Eq. (5). In FIGURE 3 C, plots of $-d[Ca^{2+}]_T/dt$ as a function of $[Ca^{2+}]_i$ obtained from the soma (*rectangles)* and the axon terminal (*circles*) under normal (*open symbols*) and low [Na$^+$] (*closed symbols*) conditions are superimposed. The $-d[Ca^{2+}]_T/dt$ plot of the axon terminal under the low Na$^+$ condition overlapped almost completely with that of the soma under the normal Na$^+$ condition, indicating that the Ca^{2+} clearance power of the axon terminal in the absence of Na/CaX was equivalent to that of the soma. Accordingly, under the normal $[Na^+]$ condition, the mean values for $\gamma_{t=0}$ in somata and axon terminals were 81.3 ± 5.0 s^{-1} ($n = 9$) and 250.3 ± 23.5 s^{-1} ($n = 7$), respectively, indicating that Ca^{2+} clearance was about threefold more efficient in the axon terminals (FIG. 3 D). While low extracellular $[Na^+]$ (20 mM) had almost no effect on $\gamma_{t=0}$ in somata, the same reduction of $[Na^+]_o$ decreased $\gamma_{t=0}$ in axon terminals by 65% (87.5 ± 5.1 s^{-1}, $n = 7$, $P < 0.01$; FIG. 3 D). These results suggest that the greater Ca^{2+} clearance power in axon terminals than in somata can be attributed to the polarized distribution of Na^+/Ca^{2+} exchangers, and that Ca^{2+} clearance in the somata is mediated by mechanisms other than Na^+/Ca^{2+} exchanger.[5]

DEPENDENCE OF Na/CaX ON INTRACELLULAR K$^+$ IN NHP AXON TERMINALS

The studies on synaptosomal fractions[6,7] and on presynaptic boutons[8,9] have suggested that the Na/CaX is one of the major CCMs in axon terminals together with PMCA. A previous study, however, reported that external Na$^+$ did not influence the Ca^{2+} decay rate at NHP axon terminals when they were intracellularly perfused with a K$^+$-free pipette solution.[10] Discoveries of new members of the NCKX family, which are widely distributed in the brain, prompted us to characterize the function of NCKX at NHP axon terminals.[11–13]

To test the possible involvement of NCKX in Ca^{2+} clearance from NHP axon terminals, internal K$^+$ was replaced with N-methyl-D-glucamine (NMG$^+$) or Cs$^+$ (FIG. 4 A). Under the internal NMG$^+$ condition, the decay rate of the Ca^{2+} transient was significantly slower than that obtained with a K$^+$-pipette, and it was not different from $\lambda_{t=0}$ obtained with a K$^+$-pipette under the low $[Na^+]_o$ condition (FIG. 4 B). Furthermore, the inhibitory effect of low $[Na^+]_o$ on the Ca^{2+} decay, which was clearly observed with the K$^+$-pipette, was negligible or markedly reduced under the NMG$^+$-pipette conditions (FIG. 4 A).

To determine the dependence of Na/CaX on cytosolic $[K^+]$, we quantified Na/CaX activity as the difference between $\lambda_{t=0,Na}$ and $\lambda_{t=0,Li}$ (denoted by λ_{Na-Li}) at various $[K^+]_i$, which were attained by intracellular dialysis using pipette solutions with K$^+$ replaced with various concentrations of NMG$^+$. Mean values of λ_{Na-Li} are plotted as a function of cytosolic $[K^+]$ in FIGURE 4 C. The best fit of the Michaelis–Menten equation to the plot was obtained at a K$_m$ value of 30.3 mM, indicating that NCKX in NHP axon terminals show a lower affinity for K$^+$ than NCKX of retinal rod cells.[14] The K$_m$ value was closer to the value reported in the study of heterologously expressed rat brain NCKX2.[15] Consistent with this result, our *in situ* hybridization study revealed that NCKX2 transcripts were clearly more abundant than NCKX3 transcripts in the supraoptic nucleus.[16]

In most terminals, the Na$^+$-dependent Ca^{2+} clearance was almost abolished under the condition of internal NMG$^+$. On average, λ_{Na-Li} estimated using the

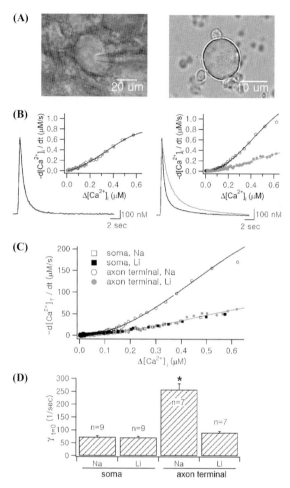

FIGURE 3. Effects of external Na^+ reduction (replaced with Li^+) on Ca^{2+} transients of somata and axon terminals. (**A**) Micrograph of the SMN soma in supraoptic nucleus (*left*) and that of axon terminals dissociated from NHP (*right*). (**B**) *Left*, Two Ca^{2+} transients from the soma were superimposed in control (*black line*) and in low $[Na^+]$ (24 mM) condition (*gray line*). Ca^{2+} transients were recorded using K^+-pipette solution containing 50 μM fura-2. *Right*, Two Ca^{2+} transients recorded in the axon terminal (*black line*: control, *gray line*: low Na^+ condition) with the same internal solution. *Insets*: time derivatives of decay phases of Ca^{2+} transients ($-d[Ca^{2+}]_i/dt$) as a function of $\Delta[Ca^{2+}]_i$, and polynomial fits (*open circles* and *black lines*: control condition, *filled circles* and *gray lines*: low Na^+ condition). (**C**) Plot of decay rates of total Ca^{2+} ($-d[Ca^+]_T/dt$) as a function of $\Delta[Ca^{2+}]$ obtained from Ca^{2+} transients in B (*rectangles*: soma, *circles*: axon terminal). Values for $-d[Ca^{2+}]_T/dt$ were calculated according to Eq. (5), assuming that $\kappa_S = 79$ for soma and $\kappa_S = 187$ for axon terminal, and that K_D of fura-2 is 201 nM. (**D**) Mean values for Ca^{2+} clearance at the peak ($\gamma_{t=0}$) calculated using Eq. (9) in normal (indicated by 'Na') and low $[Na^+]$ conditions (indicated by 'Li'). Asterisk indicates statistical significance ($P < 0.01$). Modified from Kim *et al.*[5] with permission.

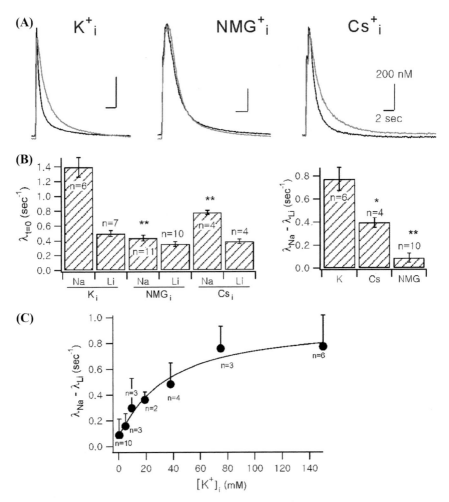

FIGURE 4. Intracellular K^+ dependence of Ca^{2+} decay rate in NHP axon terminals. (**A**) Representative Ca^{2+} transients recorded with K^+- (*left*), NMG^+- (*middle*) and Cs^+-pipettes (*right*). Ca^{2+} transients recorded at normal $[Na^+]_o$ (145 mM, *solid lines*) and at low $[Na^+]_o$ (20 mM, *gray lines*; Na^+ was replaced by Li^+) in each pipette condition were superimposed. (**B**) *Left*, Mean Ca^{2+} decay rate constants at the peak ($\lambda_{t=0}$) of Ca^{2+} transients under conditions indicated below the abscissa. Subscripts 'o' and 'i' represent external and internal major cations, respectively. *Right*, Mean values for the difference between $\lambda_{t=0}$ at normal (λ_{Na}) and at low $[Na^+]$ (λ_{Li}) in different pipette conditions (*abscissa*). Replacement of internal K^+ with Cs^+ or with NMG^+ reduced the difference ($\lambda_{Na} - \lambda_{Li}$) from 0.773 ± 0.1 ($n = 6$) to 0.396 ± 0.04 ($n = 4$, $P < 0.05$) and to 0.088 ± 0.04 ($n = 10$, $P < 0.01$), respectively. (**C**) Dependence of Na/CaX on intracellular $[K^+]$. Activity of Na/CaX was estimated from difference between λ_{Na} and λ_{Li}. Mean values for the difference were plotted as a function of $[K^+]_i$ (n = number of terminals studied; error bars, SEM), which was fitted with the Michaelis–Menten equation (*solid line*) : $0.07 \text{ s}^{-1} + (0.90 \text{ s}^{-1} \cdot [K^+]_i)/([K^+]_i + 30.3 \text{ mM})$. Modified from Lee *et al.*[16] with permission.

NMG$^+$-pipette ([K$^+$]$_i$ = 0.25 mM) was 11.3% of λ_{Na-Li} under the K$^+$- internal condition. When internal K$^+$ was replaced with Cs$^+$, however, λ_{Na-Li} was about half of the control value, and thus the Ca^{2+} decay rate at normal [Na$^+$]$_o$ (λ_{Na}) was between the values under the K$^+$-internal and NMG$^+$-internal conditions. To account for the discrepancy of λ_{Na} in Cs$^+$- and NMG$^+$-internal conditions, we hypothesized that NCKX is not completely selective for K$^+$, but that Cs$^+$ is able to substitute for the role of K$^+$ to some extent in the exchange process. To test this hypothesis, we compared the amplitude of outward NCKX2 currents (I$_{NCKX2}$) activated by bath applications of K$^+$ + Ca^{2+} with that activated by Cs$^+$ + Ca^{2+} in HEK293 cells, where NCKX2 was overexpressed, and found that the amplitude of I$_{NCKX2}$ evoked by Cs$^+$ was about 20% of that by K$^+$ (See Lee JY, Ho WK, Lee SH of this volume). Assuming that about 20% of NCKX is functional under the internal Cs$^+$ condition, the results in NHP axon terminals can be interpreted as showing that K$^+$-dependent Na/CaX comprises about 63% of the total Na/CaX activity (FIG. 4 B, right). We also tried to estimate Na/CaX activity with a Li$^+$-rich pipette because Li$^+$ was the poorest substitute for K$^+$ among the monovalent cations we tested. The intracellular dialysis of NHP axon terminals with Li$^+$-rich pipette solution, however, greatly reduced the Ca^{2+} current for unknown reasons, which made further experiments impossible.

In the SMN of an adult rat, it was difficult to clarify the dependence of Na/CaX activity in the soma on intracellular K$^+$ because the Na/CaX activity was too small in the soma.[5] Recently, we found that Na/CaX activity is larger in the SMN somata of young rats during the postnatal week (PW) 2–4 than in adult rats (PW10). The Na/CaX activity measured using the K$^+$-free Li$^+$-pipette solution was not different from those recorded using the K$^+$-pipette solution.[17] This result indicates that Na/CaX activity is mediated by NCX, rather than NCKX in the somata of young SMNs, supporting the idea that NCKX activity is localized to the axon terminals of SMNs.[5]

ROLE OF Na/CaX IN Ca^{2+} HOMEOSTASIS AT THE CALYX OF HELD

Most axon terminals of central neurons make synapses in conjunction with their postsynaptic target, and the CCMs at presynaptic terminals are of great importance for the mechanism of short-term synaptic plasticity of presynaptic origin.[18,19] To determine the role of Na/CaX in the Ca^{2+} clearance at presynaptic axon terminals, we investigated CCMs at the calyx of Held located in the auditory pathway of the rat brain stem by employing similar techniques and the quantification strategy that was used for NHP axon terminals.

Under the normal [Na$^+$]$_o$ condition, the mean value for $\lambda_{t=0}$ of the Ca^{2+} transients recorded with K$^+$-rich pipette solution was 12.01 $\pm$ 0.72 s^{-1} (FIG. 5 B, *solid lines*). Calculation of $\gamma_{t=0}$ with κ_S taken into account (κ_S = 40)[20] revealed that Ca^{2+} clearance at the calyx of Held was about twofold higher than that in NHP axon terminals. We evaluated the dependence of Na/CaX on intracellular K^+ by intracellular dialysis with Li^+ or TEA^+. Under the normal $[Na^+]_o$ condition, the decay rate constant ($\lambda_{t=0}$) of the Ca^{2+} transients recorded with Li^+- or TEA^+-pipettes was 58% of that under the control K^+-pipette condition (FIG. 5 B and C). Under the K^+-free internal condition, $\lambda_{t=0}$ was further reduced by lowering the external $[Na^+]$ (FIG. 5 Bb and Bc; *gray traces*), indicating that K^+-independent NCX also contributed to Ca^{2+} clearance at the calyx of Held. The mean values for $\lambda_{t=0}$ at the

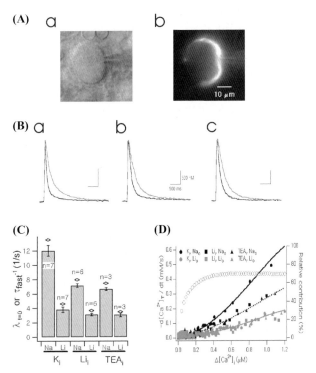

FIGURE 5. Contribution of Na/CaX to Ca^{2+} clearance at the calyx of Held. (**A**) DIC and fluorescence images of the calyx of Held that surrounds the postsynaptic principal neuron in medial nucleus of trapezoidal body. (**B**) Ca^{2+} transients evoked by a 50 ms depolarizing pulse were recorded with a K^+-pipette (**Ba**), with a Li^+-pipette (**Bb**), and with a TEA^+-pipette (**Bc**). Two Ca^{2+} transients recorded at $[Na^+]_o = 145$ mM (*black line*) and at $[Na^+]_o = 25$ mM (*gray line*; Na^+ was replaced by Li^+) are superimposed in each panel. (**C**) Mean values for $\lambda_{t=0}$ (bar graph) and for τ_{fast} (*diamonds*) of Ca^{2+} transients. (**D**) Total Ca^{2+} clearance rate ($-d[Ca^{2+}]_T/dt$) calculated from the time derivatives of the decay phases of six Ca^{2+} transients shown in B (*circles*, K^+- pipette; *squares*, Li^+-pipette, *triangles*, TEA^+-pipette). Relative contributions of Na/CaX to total Ca^{2+} clearance were plotted as a function of $\Delta[Ca^{2+}]_i$ (*open circles, right axis*). Modified from Kim *et al.*[21] with permission.

low $[Na^+]_o$ were statistically similar, irrespective of internal ionic conditions (FIG. 5 C).

The total Ca^{2+} clearance rate $(-d[Ca^{2+}]_T/dt)$ as a function of the $\Delta[Ca^{2+}]_i$ level was calculated from the time derivative of the decay phase of a Ca^{2+} transient under various conditions. We were able to estimate the relative contribution of Na/CaX to total Ca^{2+} clearance as a function of the $\Delta[Ca^{2+}]_i$ level from the difference in the $-d[Ca^{2+}]_T/dt$ curve before (FIG. 5 D, *black solid line*) and after (FIG. 5 D, *gray solid line*) $[Na^+]_o$ reduction under the K^+-pipette condition. The relative contribution of Na/CaX increased as $\Delta[Ca^{2+}]_i$ increased, reaching a maximum value (68%) at $\Delta[Ca^{2+}]_i$ >400 nM. These results indicate that Na/CaX is the primary CCM in the calyx of Held when the elevation of $\Delta[Ca^{2+}]_i$ is in the range of 1 to 2 μM, and that NCKXs and NCXs contribute to 42% and 26% of total Ca^{2+} clearance, respectively.

The relative contributions of other putative CCMs were assessed using specific inhibitors. Inhibition of PMCA decreased the Ca^{2+} decay rate by 23%, whereas inhibition of smooth endoplasmic reticulum Ca^{2+}-ATPase (SERCA) had no effect.[21]

INTERPLAY BETWEEN Na/CaX AND MITOCHONDRIA IN PRESYNAPTIC CA^{2+} CLEARANCE

We investigated the contribution of mitochondrial Ca^{2+} uptake to presynaptic CCM using carbonylcyanide m-chlorophenylhydrazone (CCCP), which inhibits mitochondrial Ca^{2+} uptake consequent to mitochondrial depolarization. To prevent CCCP from depolarizing the plasma membrane potential and elevating the resting Ca^{2+} level, the calyx was voltage-clamped at -70 mV and supplied with 5 mM ATP via patch pipette. Bath application of 2 μM CCCP had no effect on Ca^{2+} transients under the K^+-pipette condition, whereas CCCP significantly slowed the Ca^{2+} decay rate under the TEA^+-pipette condition (FIG. 6 A). We tested whether a mitochondrial contribution also occurs when Na/CaX is inhibited under the internal K^+ condition. After the Ca^{2+} decay had been slowed by $[Na^+]_o$ reduction (FIG. 6 B; *solid line*: normal $[Na^+]$, *gray line*: low $[Na^+]$), the addition of 2 μM CCCP to the low $[Na^+]_o$ bath solution led to a further slowing of the Ca^{2+} decay (FIG. 6 B; *gray dotted line*). When the normal external $[Na^+]$ bath solution was reintroduced in the presence of CCCP, the Ca^{2+} transient was completely restored to the level under the control condition (FIG. 6 B; *solid line*: normal $[Na^+]_o$, *dotted line*: normal $[Na^+]_o$ + 2 μM CCCP), indicating that mitochondria competes for cytosolic Ca^{2+} with Na/CaX and take part in the Ca^{2+} clearance only when Na/CaX is inhibited.

We next examined the mitochondrial contribution to Ca^{2+} clearance at higher $\Delta[Ca^{2+}]_i$ levels using low affinity Ca^{2+}-dye, fura2-FF. Whereas CCCP had no noticeable effect on Ca^{2+} transients whose peak $\Delta[Ca^{2+}]_i$ was lower than 2.5 μM (FIG. 6 C, *left traces*), CCCP slowed Ca^{2+} decay of the Ca^{2+}

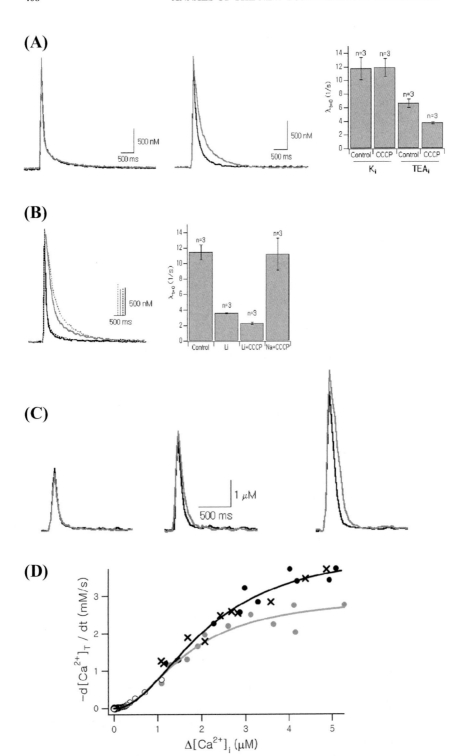

transient when its peak $\Delta[Ca^{2+}]_i$ was higher than 3 μM, and the effect of CCCP became more obvious as the peak $\Delta[Ca^{2+}]_i$ further increased (FIG. 6C, *middle* and *right*). We analyzed 11 pairs of Ca^{2+} transients and estimated the total Ca^{2+} clearance rate at the peak $[(-d[Ca^{2+}]_T/dt)_{t=0}]$. The values for $(-d[Ca^{2+}]_T/dt)_{t=0}$ before (*black circles*) and after (*gray circles*) CCCP treatment are plotted as a function of the peak $\Delta[Ca^{2+}]_i$ in FIGURE 6 D. In addition, the data of $-d[Ca^{2+}]_T/dt$ obtained with fura-4F in the lower $\Delta[Ca^{2+}]_i$ range (<1 μM) are superimposed (*open circles*). The composite graph of $-d[Ca^{2+}]_T/dt$ provides the dependence of $-d[Ca^{2+}]_T/dt$ on $\Delta[Ca^{2+}]_i$ over a wide range. The $-d[Ca^{2+}]_T/dt$ values in the presence of thapsigargin (2 μM, *crosses*) were not different from those under the control condition, indicating that SERCA did not contribute to Ca^{2+} clearance over the entire $\Delta[Ca^{2+}]_i$ range. Although no effect of CCCP was observed at $\Delta[Ca^{2+}]_i$ below 2 μM, CCCP downward shifted the $-d[Ca^{2+}]_T/dt$ curve in the higher range of $\Delta[Ca^{2+}]_i$ (FIG. 6 D, *gray circles*). As a result, in the presence of CCCP, the $-d[Ca^{2+}]_T/dt$ curve showed saturation in the high

---

FIGURE 6. Interaction between Na/CaX and mitochondrial Ca^{2+} uptake. (**A**) *Left,* Ca^{2+} transients recorded with a K^+-pipette solution in control condition (*black line*) and in the presence of 2 μM CCCP (*gray line*). *Middle,* Ca^{2+} transients recorded with a TEA^+-pipette (*black line: control condition, gray line: CCCP*). *Right,* mean values for $\lambda_{t=0}$ of Ca^{2+} transients under the conditions indicated below the abscissa. Mean values of $\lambda_{t=0}$ before and after the application of 2 μM CCCP were not different in internal K^+ condition, and those in TEA^+-pipette condition were significantly different (paired *t*-test, $n = 3$, $P < 0.05$). (**B**) *Left,* Effects of CCCP on Ca^{2+} clearance when Na/CaX was inhibited. Four Ca^{2+} transients recorded in the same cell sequentially using a K^+-pipette solution: (1) in normal $[Na^+]_o$ condition (*black line*), (2) after $[Na^+]_o$ reduction (replaced by 125 mM Li^+; *gray line*), (3) after the bath application of 2 μM CCCP (*gray-dotted line*) in the low $[Na^+]_o$ solution, (4) after the re-introduction of normal $[Na^+]_o$ in the presence of CCCP (*black dotted line*). Four Ca^{2+} transients were scaled to the same maximum value, and superimposed. *Right,* mean values for $\lambda_{t=0}$ in various conditions of similar experiments. (**C**) Effects of CCCP on Ca^{2+} clearance at various levels of $\Delta[Ca^{2+}]$. Each pair of Ca^{2+} transients were recorded with the K^+-pipette solution containing 50 μM fura2-FF in the control condition (*black line*) and in the presence of 2 μM CCCP (*gray line*). Three sets of Ca^{2+} transients were evoked by different levels of depolarization with a fixed 50 ms duration (*left,* −10 mV; *middle,* 0 mV; *right,* +10 mV). (**D**) Plot of $-d[Ca^{2+}]_T / dt$ values in control (*black circles*) and in CCCP condition (*gray circles*) as a function of $\Delta[Ca^{2+}]_i$. The values of $-d[Ca^{2+}]_T/dt$ were estimated from Ca^{2+} transients measured with fura-2FF in five different terminals (*closed circles* and *crosses*) and from those measured with fura-4F in four different terminals (*open circles*). The values in each set (set of control conditions or set of CCCP conditions) were pooled, and fitted using Hill's equation, $V_{max}/(1 + (K_{1/2}/[Ca^{2+}])^{HN})$ with the following parameters: $V_{max} = 4.30$ mM/s, $HN = 2$, $K_{1/2} = 2.14$ μM for the control condition (*black line*); $V_{max} = 2.96$ mM/s, $HN = 2$, $K_{1/2} = 1.7$ μM for the CCCP condition (*gray line*). The $-d[Ca^{2+}]_T/dt$ values determined in the presence of 2 μM thapsigargin were similar to those determined under the control condition (*crosses*). Modified from Kim *et al.*[21] with permission.

$\Delta[Ca^{2+}]_i$ range. Considering that Na/CaX is the major CCM in this high $\Delta[Ca^{2+}]_i$ range, the $-d[Ca^{2+}]_T/dt$ curve might represent the saturation of Na/CaX activity. The best fit of the Hill equation, $V_{max}/(1 + (K_{1/2} / [Ca^{2+}])^{HN})$, to the $-d[Ca^{2+}]_T/dt$ values in the presence of CCCP was obtained with the following parameters: $V_{max} = 2.96$ mM/s, $HN = 2$, and $K_{1/2} = 1.7$ μM (FIG. 6 C, *gray curve*). Interestingly, this $K_{1/2}$ value is similar to that estimated from the fit of the Michaelis–Menten equation to the dependence of NCKX currents recorded from retinal rod cells on $[Ca^{2+}]_i$ (1.6 μM).[22] In the absence of CCCP, the $-d[Ca^{2+}]_T/dt$ curve became more linear, indicating that the saturation of Na/CaX activity is partially compensated by the activation of mitochondrial Ca^{2+} uptake at $\Delta[Ca^{2+}]_i$ levels higher than 2.5 μM.

Mitochondria have been suggested to serve as a source for residual Ca^{2+}, which induces posttetanic synaptic enhancement.[18] The calyces of Held, which synapse onto the glycinergic principal cell of the medial nucleus of trapezoidal body, are thought to play the role of a sign-inverting relay in the neural circuit of sound localization.[23,24] Considering that faithful synaptic transmission is the main task for these synapses, Na/CaX might support a high-fidelity synaptic transmission by competing with mitochondria for presynaptic Ca^{2+}.

CONCLUSIONS

Na/CaX, mediated by NCX and NCKX, represents a major CCM both at NHP axon terminals and at the calyces of Held, the two giant axon terminals of mammalian central neurons. In SMN, the distribution of NCKX is polarized to the axon terminals. At the calyces of Held, Na/CaX competes with mitochondria in Ca^{2+} clearance. The rapid Ca^{2+} extrusion by Na/CaX may enable axon terminals to transmit information with high fidelity.

ACKNOWLEDGMENTS

This research was supported by a grant (M103KV010008-06K2201-00810) from Brain Research Center of the 21st Century Frontier Research Program and the grant for National Research Laboratory funded by the Ministry of Science and Technology, the Republic of Korea.

REFERENCES

1. CAZALIS, M., G. DAYANITHI & J.J. NORDMANN. 1987. Hormone release from isolated nerve endings of the rat neurohypophysis. J. Physiol. **390**: 55–70.

2. FORSYTHE, I.D. 1994. Direct patch recording from identified presynaptic terminals mediating glutamatergic EPSCs in the rat CNS, *in vitro*. J. Physiol. **479:** 381–387.
3. NEHER, E. 1995. The use of fura-2 for estimating Ca buffers and Ca fluxes. Neuropharmacology **34:** 1423–1442.
4. PUSCH, M. & E. NEHER. 1988. Rates of diffusional exchange between small cells and a measuring patch pipette. Pflugers Arch. **411:** 204–211.
5. KIM, M.H. *et al.* 2003. Distribution of K^+-dependent $Na^+/Ca2^+$ exchangers in the rat supraoptic magnocellular neuron is polarized to axon terminals. J. Neurosci. **23:** 11673–11680.
6. GILL, D.L., E.F. GROLLMAN & L.D. KOHN. 1981. Calcium transport mechanisms in membrane vesicles from guinea pig brain synaptosomes. J. Biol. Chem. **256:** 184–192.
7. FONTANA, G., R. ROGOWSKI & M. BLAUSTEIN. 1995. Kinetic properties of the sodium-calcium exchanger in rat brain synaptosomes. J. Physiol. **485:** 349–364.
8. REUTER, H. & H. PORZIG. 1995. Localization and functional significance of the $Na^+/Ca2^+$ exchanger in presynaptic boutons of hippocampal cells in culture. Neuron **15:** 1077–1084.
9. BLAUSTEIN, M.P. *et al.* 2002. Na/Ca exchanger and PMCA localization in neurons and astrocytes. Ann. N. Y. Acad. Sci. **976:** 356–366.
10. STUENKEL, E.L. 1994. Regulation of intracellular calcium and calcium buffering properties of rat isolated neurohypophysial nerve endings. J. Physiol. **481:** 251–271.
11. TSOI, M. *et al.* 1998. Molecular cloning of a novel potassium-dependent sodium-calcium exchanger from rat brain. J. Biol. Chem. **273:** 4155–4162.
12. KRAEV, A. *et al.* 2001. Molecular cloning of a third member of the potassium-dependent sodium-calcium exchanger gene family, NCKX3. J. Biol. Chem. **276:** 23161–23172.
13. LYTTON, J. *et al.* 2002. K^+-dependent $Na^+/Ca2^+$ exchangers in the brain. Ann. N. Y. Acad. Sci. **976:** 382–393.
14. PERRY, R.J. & P.A. MCNAUGHTON. 1993. The mechanism of ion transport by the Na^+-$Ca2^+$, K^+ exchange in rods isolated from the salamander retina. J. Physiol. **466:** 443–480.
15. DONG, H. *et al.* 2001. Electrophysiological characterization and ionic stoichiometry of the rat brain K(+)-dependent $Na^+/Ca2^+$ exchanger, NCKX2. J. Biol. Chem. **276:** 25919–25928.
16. LEE, S.H. *et al.* 2002. K^+-dependent $Na^+/Ca2^+$ exchanger is a major $Ca2^+$ clearance mechanism in axon terminals of rat neurohypophysis. J. Neurosci. **22:** 6891–6899.
17. LEE, S.H. *et al.* 2006. Postnatal developmental changes in $Ca2^+$ homeostasis in supraoptic magnocellular neurons. Cell Calcium Epub ahead of print (Sep. 2006).
18. TANG, Y. & R.S. ZUCKER. 1997. Mitochondrial involvement in post-tetanic potentiation of synaptic transmission. Neuron **18:** 483–491.
19. ZUCKER, R.S. & W.G. REGEHR. 2002. Short-term synaptic plasticity. Annu. Rev. Physiol. **64:** 355–405.
20. HELMCHEN, F., J.G.G. BORST & B. SARKMANN. 1997. Calcium dynamics associated with a single action potential in a CNS presynaptic terminal. Biophys. J. **72:** 1458–1471.
21. KIM, M.H. *et al.* 2005. Interplay between $Na^+/Ca2^+$ exchangers and mitochondria in $Ca2^+$ clearance at the calyx of Held. J. Neurosci. **25:** 6057–6065.

22. LAGNADO, L., L. CERVETTO & P.A. MCNAUGHTON. 1992. Calcium homeostasis in
     the outer segments of retinal rods from the tiger salamander. J. Physiol. **455**:
     111–142.
23. GROTHE, B. 2003. New roles for synaptic inhibition in sound localization. Nat.
     Rev. Neurosci. **4**: 540–550.
24. VON GERSDORFF, H. & J.G.G. BORST. 2001. Short-term plasticity at the calyx of
     Held. Nat. Rev. Neurosci. **3**: 53–64.

ncx1, *ncx2*, and *ncx3* Gene Product Expression and Function in Neuronal Anoxia and Brain Ischemia

L. ANNUNZIATO, G. PIGNATARO, F. BOSCIA, R. SIRABELLA, L. FORMISANO, M. SAGGESE, O. CUOMO, R. GALA, A. SECONDO, D. VIGGIANO, P. MOLINARO, V. VALSECCHI, A. TORTIGLIONE, A. ADORNETTO, A. SCORZIELLO, M. CATALDI, AND G. F. DI RENZO

Division of Pharmacology, Department of Neuroscience, School of Medicine, "Federico II" University of Naples, 80131 Naples, Italy

ABSTRACT: Over the last few years, although extensive studies have focused on the relevant function played by the sodium–calcium exchanger (NCX) during focal ischemia, a thorough understanding of its role still remains a controversial issue. We explored the consequences of the pharmacological inhibition of this antiporter with conventional pharmacological approach, with the synthetic inhibitory peptide, XIP, or with an antisense strategy on the extent of brain damage induced by the permanent occlusion of middle cerebral artery (pMCAO) in rats. Collectively, the results of these studies suggest that *ncx1* and *ncx3* genes could be play a major role to limit the severity of ischemic damage probably as they act to dampen $[Na^+]_i$ and $[Ca^{2+}]_i$ overload. This mechanism seems to be normally activated in the ischemic brain as we found a selective upregulation of NCX1 and NCX3 mRNA levels in regions of the brain surviving to an ischemic insult. Despite this transcript increase, NCX1, NCX2, and NCX3 proteins undergo an extensive proteolytic degradation in the ipsilateral cerebral hemisphere. All together these results suggest that a rescue program centered on an increase NCX function and expression could halt the progression of the ischemic damage. On the basis of this evidence we directed our attention to the understanding of the transductional and transcriptional pathways responsible for NCX upregulation. To this aim, we are studying whether the brain isoform of Akt, Akt1, which is a downstream effector of neurotrophic factors, such as NGF can, in addition to affecting the other prosurvival cascades, also exert its neuroprotective effect by modulating the expression and activity of *ncx1*, *ncx2*, and *ncx3* gene products.

KEYWORDS: NCX1; NCX2; NCX3; brain ischemia; anoxia

Address for correspondence: Lucio Annunziato, M.D., Division of Pharmacology, Department of Neuroscience, School of Medicine, "Federico II" University of Naples, via Sergio Pansini 5, 80131 Naples, Italy. Voice: +39-81-7463325; fax: +39-81-7463323.
lannunzi@unina.it

Ann. N.Y. Acad. Sci. 1099: 413–426 (2007). © 2007 New York Academy of Sciences.
doi: 10.1196/annals.1387.050

INTRODUCTION

A common feature of cerebral ischemia is the profound reduction of ATP production by mitochondria determined by oxygen and glucose deprivation. This leads to a failure of the Na^+/K^+ ATPase and consequently to a marked $[Na^+]$ overload that triggers a depolarization-induced opening of voltage-gated Ca^{2+} and Na^+ channels, which is followed by: (a) a rapid decrease in $[Na^+]_e$ and $[Ca^{2+}]_e$ and (b) a release of glutamate in the extracellular space. Glutamate, once released, further amplifies this process by opening Na^+ and Ca^{2+} permeable glutamate receptors and therefore causes a further lowering of $[Ca^{2+}]_e$ and $[Na^+]_e$.[1] This $[Ca^{2+}]_e$ decrease triggers the flux of more Na^+ and Ca^{2+} through TRPM7 channels.[2] The derangement of the ionic homeostasis that occurs in ischemic neurons as a consequence of the above described events is believed to represent a major causative factor in triggering anoxic neuronal cell death. In fact, the progressive accumulation of $[Na^+]_i$ and $[Ca^{2+}]_i$ can precipitate necrosis and/or apoptosis of selected vulnerable neurons. Therefore, any molecular factor that is able to counteract or dampen this ionic dysregulation is expected to exert a neurobeneficial effect in brain ischemia. In this context, a major role can be played by sodium–calcium exchanger (NCX) which, depending on $[Ca^{2+}]_i$ and $[Na^+]_i$ can operate either in the forward mode, coupling the uphill extrusion of Ca^{2+} to the influx of Na^+ ions, or in the reverse mode, mediating the extrusion of Na^+ and the influx of the Ca^{2+} ions.[3] Interestingly, the three genes of this antiporter NCX1, NCX2, and NCX3, are expressed in neurons, astrocytes, oligodendrocytes, and microglia.[4-6] In addition, a report from our laboratory has clearly indicated that both the transcripts and the proteins of the three NCX isoforms are selectively expressed in areas of the brain, which are critical for the development of the ischemic damage.[7,8] Over the last few years, although extensive studies have focused on the relevant function played by NCX during focal ischemia,[3] a thorough understanding of its role still remains a controversial issue.[9-14] However, data obtained in our laboratory using cellular anoxia or animal models mimicking human brain ischemia consistently suggest that an increase of NCX activity or expression may help neurons to survive after anoxic insults whereas a blockade of its function or expression leads to a worsening of brain damage thus pointing to NCX as a potential molecular target for therapeutical intervention in stroke.

CONSEQUENCES OF PHARMACOLOGICAL MODULATION OF NCX ACTIVITY WITH ANTISENSE DEOXYOLIGONUCLEOTIDE STRATEGY OR CONVENTIONAL DRUGS ON THE SEVERITY OF ISCHEMIC BRAIN DAMAGE

The pathophysiological considerations reported in the previous section provide a strong theoretical background to suggest that the fate of neurons after an ischemic insult could be also determined by how effective will be NCX in counteracting the $[Na^+]_i$ and $[Ca^{2+}]_i$ overload. To confirm this hypothesis we explored the consequences of the pharmacological inhibition of this antiporter on the extent of brain damage induced by the permanent occlusion of middle cerebral artery (pMCAO) in rats. This widely used animal model of stroke causes a large cerebral infarction involving basal ganglia and cortex, which closely reproduces the most commonly observed ischemic stroke in humans. As it occurs in human brain ischemia, two different regions can be identified in the brain of rats undergoing pMCAO: the *ischemic core* where the ischemic damage is so severe to determine a major loss of cells with a minority of surviving neurons, and the *penumbra region* where neurons, microglia, and glia cells are suffering but can be rescued by external pharmacological intervention.

A number of different strategies can be used to inhibit NCX activity.[3] First of all a large series of organic compounds that potently and quite selectively block NCX transport activity have been identified by *in vitro* screening over the years and are now available.[3,15] The diarylaminopropylamine derivative, Bepridil, is one of the first compounds identified as provided of an NCX inhibitory activity. When Bepridil was intracerebroventricularly (i.c.v.) injected 3, 6, and 22 h after pMCAO we found a marked increase of the ischemic lesion as compared with vehicle-injected pMCAO-bearing rats.[13] However, Bepridil has the serious limitation of not being very selective for NCX as it also blocks L-type[16] and T-type Ca^{2+} currents,[17] delayed-rectifier K^+ current, transient outward current,[18] as well as the K^+ current, activated by $[Na^+]_i$.[19] A group of much more selective NCX blockers was developed by Dr. Cragoe over a 20 years time span by modifying the structure of the precursor amiloride, a K^+ sparing diuretic. Whereas amiloride is per se provided of a significant NCX blocking activity, it lacks of specificity as it potently inhibits also the Na^+/H^+ exchanger. By introducing substituents on the terminal guanidino nitrogen atom a second class of compounds that potently inhibit NCX (K_i 1–10 μM) but are devoid of any significant effect on Na^+/H^+ exchanger has been developed. Among these compounds, [N-(4-chlorobenzyl)]2,4-dimethylbenzamyl (CB-DMB) appears to be the most specific inhibitor of NCX activity (K_i 7.3 μM), for it has no inhibitory properties against the Na^+/H^+ antiporter (K_i 500 μM) and the epithelial Na^+ channels (K_i 400 μM).[20] Taking advantage of this favorable pharmacodynamic profile, we used CB-DMB as a pharmacological tool to block NCX in the ischemic brain. What we found was that when CB-DMB, was continuously intracerebroventricularly infused with an osmotic minipump for 24 h after the beginning of the pMCAO, the volume of the ischemic region markedly increased.[13]

A much more selective and effective way to inhibit NCX activity became available when it was found that a specific 20 amino acid stretch (219–238), known as XIP, belonging to the so-called f-loop of the exchanger, exerts an autoinhibitory action thus limiting NCX activity.[21,22] Synthetic peptides with the same sequence were produced, indeed, and found to be potent NCX inhibitors *in vitro*.[23] Their usefulness as pharmacological tools to block NCX activity either in cells cultured *in vitro* or in living animals is strongly limited by small propensity of these molecules to cross the cell membrane. In order to increase XIP cell penetration the XIP molecule has been modified, such as it bears a glucose substituent attached to the Tyr-6 residue (FIG. 1 A) and therefore can be actively transported into the cell through the glucose transporters 1 and 3 and can block NCX activity (FIG. 1 C, D). We used this Tyr-6-glycosylated form of XIP as a tool to block NCX activity in rats undergoing pMCAO. In rat receiving the peptide by i.c.v. infusion with an Alzet osmotic minipump for 24 h

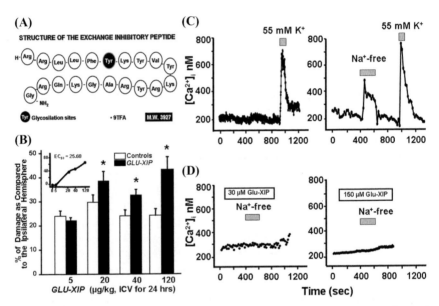

FIGURE 1. Panel **A** represents the structure of the 6-tyrosine glycosylated form of the exchange inhibitory peptide, XIP. Panel **B** shows effect of the specific NCX inhibitor GLU-XIP on infarct volume induced by pMCAO. Each column represents the mean ± SE of the percentages of damage as compared to the ipsilateral hemisphere. GLU-XIP (5, 20, 40, and 120 μg/kg) was continuously i.c.v. infused by an osmotic minipump at a delivery rate of 1 μL/h for 24 h. Control rats received vehicle. Each experimental group consisted of five animals. *$P < 0.05$ versus respective control group. Panel **C** shows a representative single cell trace of the effects exerted by extracellular Na$^+$ removal or K$^+$ addition on [Ca^{2+}]$_i$ in cortical neurons. Panel **D** shows a representative single cell trace of the effects exerted by GLU-XIP (30 and 100 μM) on NCX activation elicited by extracellular Na$^+$ removal. GLU-XIP was added after 120 s of baseline [Ca^{2+}]$_i$ monitoring and left in the chamber for the period as indicated by the bar. The traces are representative of 15–25 cells for each experimental group studied in at least three different experimental sessions. (Partially reproduced with modifications from Pignataro *et al.*[13] Neuropharmacology 2004 **46:** 439–48 with the permission of the Elsevier publisher.)

from the beginning of pMCAO, we observed a dramatic increase in infarct volume (FIG. 1 B).[13]

Both the results of the studies performed using organic NCX blockers and those of the experiments with glycosylated-XIP strongly indicate that when NCX activation is prevented in the ischemic brain, the severity of stroke markedly increases, thus suggesting that NCX limits the extension of ischemic brain damage. In order to further confirm this conclusion, we performed additional experiments using an antisense strategy to reduce NCX activity as a consequence of a downregulation of its protein expression. The great advantage of the antisense strategy over the above-mentioned conventional pharmacology approaches is that with the design of antisense molecule selectively directed against each of the three-exchanger transcripts we could identify differences in the pathophysiological role played by the three different isoforms. This molecular strategy added more specific information that could not be drawn from studies performed with conventional pharmacological compounds, which cannot discriminate among NCX gene products. As a matter of fact, our antisense studies provided the first evidence in support of the idea that the protective role against the ischemic damage could be not a general feature of all NCX isoforms. In fact, in rats receiving the continuous i.c.v. infusion, with an osmotic minipump, of AS-ODN directed against NCX1 or NCX3 there was a significant worsening of the ischemic damage. Conversely, NCX2 AS-ODN administration was ineffective (FIG. 2 A). These data strongly suggest that the functional integrity of NCX1 and NCX3 should be maintained or better potentiated in order to limit the extension of brain insult.

CHANGES IN THE EXPRESSION OF NCX1, NCX2, AND NCX3 TRANSCRIPTS AND PROTEINS IN BRAIN ISCHEMIA

The marked effect on the severity of experimental ischemic brain damage, observed when NCX was pharmacologically blocked, strongly suggests that NCX is activated after an ischemic insult and that this activation has the general meaning of a defense strategy against neuronal cell death. Therefore, it is tempting to assume that NCX could behave not differently from other "protective" proteins, such as VEGF, which are part of a more general "defense system" against ischemic cell death. As the expression of these protective proteins does increase in response to brain ischemia we were interested to study whether this assumption is also true for NCX. To this aim, we performed Western blot experiments with specific antibodies against each NCX isoform and quantitative *in situ radioactive hybridization* studies in brains from rats undergoing pMCAO. When we looked at protein expression by Western blot experiments we found that after pMCAO all three NCX proteins were downregulated in the ischemic core; NCX3 decreased in peri-infarctual area whereas NCX1 and NCX2 were unchanged (FIG. 2 B, C). However, by exploring with a more

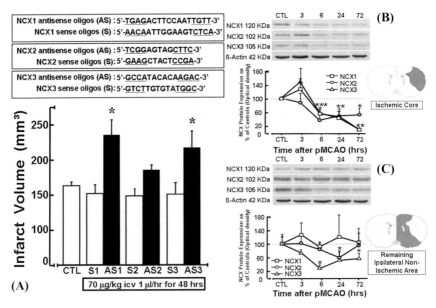

FIGURE 2. Panel **A** represents effect of AS1-, AS2-, and AS3-ODN on the volume of the ischemic area. Panels **B** and **C** show time course of NCX1, NCX2, and NCX3 protein expression after pMCAO in the ischemic core and the peri-infarct area. (Data are reproduced from Pignataro et al.[14] 2004 Stroke. **35:** 2566–70 with the permission of the Lippincott Williams & Wilkins publisher.)

morphologically defined technique, such as immunohistochemistry, it emerged that the few neurons surviving the ischemic insult in the core region overexpress NCX1 (FIG. 3). In order to understand whether these changes in protein expression were linked to changes in *ncx* gene transcription we performed quantitative studies by in situ radioactive hybridization[24] (FIGS. 4 and 5). Intriguingly, we found relevant differences in NCX mRNA levels in the *ischemic core* and *peri-infarct regions*. In particular, in the *ischemic core*, comprising divisions of the prefrontal, somatosensory, and insular cortices, all three NCX transcripts were downregulated. In the peri-infarct area, comprising part of the motor cortex and the lateral compartments of the caudate-putamen, NCX2 messenger ribonucleic acid (mRNA) was downregulated, whereas NCX3 mRNA was significantly upregulated. This NCX2 downregulation in the peri-ischemic region may be related to the transcriptional mechanisms controlling its expression. In effect, the repetitive spreading depression-like depolarization occurring during ischemia and originating in the ischemic core, propagates to the surrounding brain regions of the same hemisphere,[25] thus producing changes in the *ncx2* gene expression. This hypothesis is supported by data showing that in primary cerebellar granule cells (CGN) K^+-induced depolarization, followed by calcium influx, causes a selective downregulation of NCX2 mRNA in a

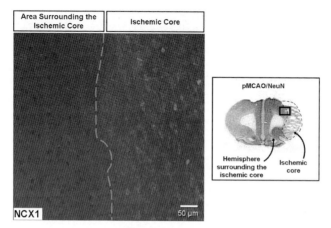

FIGURE 3. This picture shows NCX1 immunoreactivity at the border between the ischemic region (on the right) and the intact cortex (on the left). It is evident the increase of NCX1 immunoreactivity in the surviving neurons in the ischemic region.

calcineurin-dependent way.[26] In contrast, the expression of the other *ncx* gene, *ncx3*, seems to reveal an opposite regulatory mechanism, as demonstrated by its increase in the peri-infarct region. Consistently, in primary neuronal cultures, depolarization upregulates NCX3 mRNA in a calcineurin-independent manner.[26] We hypothesized that NCX3 may exert a neuroprotective response. Confirming this evidence, Bano *et al.*[27] showed that the silencing of NCX3 expression by RNA interference sensitizes CGN vulnerability to Ca^{2+} overload and excitotoxicity. However, since ischemia causes a relevant reduction of NCX proteins in the core and in the remaining ipsilateral nonischemic areas,[14] the possibility exists that all the changes observed in NCX family transcripts might be ascribed to regional differences in protein synthesis activity. As regards our hypothesis that *ncx* gene responses are not confined exclusively to core and peri-infarct areas, but can also be found in remote nonischemic brain regions, a large body of studies have, in fact, demonstrated that ischemic lesions consequent to MCAO induce a complex pattern of genomic responses in regions anatomically distant but functionally connected with the infarct areas.[28–30] Consistently, we found that NCX1 and NCX3 mRNA were upregulated in the deep layers of the infralimbic and prelimbic cortices. The upregulation of *ncx1* and *ncx3* gene expression in these two intact limbic cortical regions could be interpreted as a homeostatic response consequent to the reciprocal anatomic and functional connections existing between the insular cortex, damaged by pMCAO, and the infralimbic and prelimbic cortices, not involved in the insult.[31,32] Interestingly, neurons in these two limbic cortical regions project to brain stem nuclei, such as the nucleus of the solitary tract involved in the central cardiovascular control.[33,34] Therefore, the increase in NCX1 and NCX3 expression

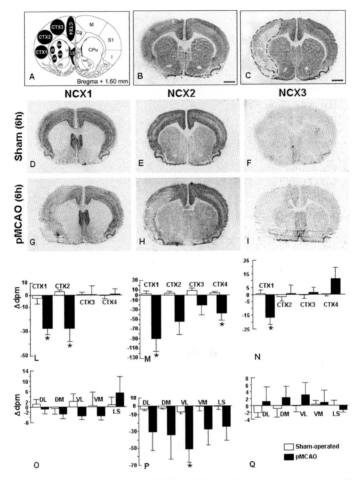

FIGURE 4. NCX1, NCX2, and NCX3 mRNA in the cerebral cortex, caudate putamen, and lateral septum of sham-operated and ischemic animals 6 h after sham surgery or
pMCAO. A schematic diagram of ROI at coronal bregma levels of +1.60 mm is shown
in panel **A**. Representative brain NeuN-immunohistochemistry-processed sections deriving from sham-operated and pMCAO rats are shown in panels **B** and **C**, respectively.
Panels **D–I** represent the *in situ* hybridization autoradiographic film images obtained
from sham-operated and ischemic brain sections with NCX1 (**D, G**), NCX2 (**E, H**),
and NCX3 (**F, I**) antisense probes. Messenger ribonucleic acid levels of NCX1, NCX2,
and NCX3, expressed as Δd.p.m. ± SEM and presented for each ROI analyzed, with
white columns for sham-operated rats and *black columns* for pMCAO rats, are shown in
panels **L** and **O**; **M** and **P**; **N** and **Q**, respectively. *$P < 0.05$ versus respective values in
sham-operated groups. AcbC = accumbens nucleus, core; AcbSh = accumbens nucleus,
shell; cc = corpus callosum; Cg = cingulate cortex; Cpu = caudate putamen; I = insular
cortex; LSI = lateral septal nucleus; LV = lateral ventricle; M = motor cortex; Pir = piriform cortex; S1 = primary somatosensory cortex; Shi = septohippocampal nucleus. Scale
bar in panels **B** and **C**: 2 mm. (From Boscia *et al.*[24] 2006. J Cereb Blood Flow Metab. **26:**
502–17 with the permission of Nature Publisher Group publisher.)

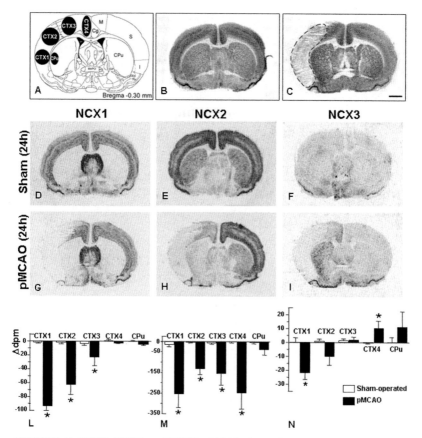

FIGURE 5. NCX1, NCX2, and NCX3 mRNA in the prefrontal cortex of sham-operated and ischemic animals 24 h after sham surgery or pMCAO. A schematic diagram of ROI at coronal bregma levels of +3.20 mm is shown in panel **A**. Representative brain NeuN-immunohistochemistry-processed sections deriving from sham-operated and pMCAO bearing rats are shown in panels **B** and **C**, respectively. Panels **D–I** represent the *in situ* hybridization autoradiographic film images obtained from sham-operated and ischemic brain sections with NCX1 (**D, G**), NCX2 (**E, H**), and NCX3 (**F, I**) antisense probes. Messenger ribonucleic acid levels of NCX1, NCX2, and NCX3, expressed as Δd.p.m. ± SEM and presented for each ROI analyzed, with *white columns* for sham-operated rats and *black columns* for pMCAO rats, are shown in panels **L**, **M**, and **N**, respectively. Δd.p.m. indicates the difference between the d.p.m. value of each ROI ipsilateral to the ischemic side and that of the contralateral corresponding side. *$P < 0.05$ versus respective values in sham-operated groups. aca = anterior commissure, anterior part; AID = agranular insular cortex, dorsal part; AIV = agranular insular cortex, ventral part; Cg1 = cingulate cortex, area 1; DP = dorsal peduncular cortex; DTT = dorsal tenia tecta; fmi = forceps minor of the corpus callosum; IL = infralimbic cortex; LO = lateral orbital cortex; M1 = primary motor cortex; M2 = secondary motor cortex; Pir = piriform cortex; PrL = prelimbic cortex; VO = ventral orbital cortex; VTT = ventral tenia tecta. Scale bar in panels **B** and **C**: 2 mm. (From Boscia *et al.*[24] 2006. J Cereb Blood Flow Metab. **26:** 502–17 with the permission of Nature Publisher Group publisher.)

may be related to a compensatory response activated in these circuits after stroke.[35–37] It is worth underlining that in humans, brain damage consequent to stroke is associated with sympathetic activity imbalance resulting in subsequent injury, arrhythmia, and sudden death.[38] Interestingly, NCX1 and NCX3 upregulation was also detected in the tenia tecti (TT), an area thought to be an anterior hippocampal rudiment and enriched in both serotonin[39] and orexin terminals and receptors.[40] Indeed, in the brain, the excitatory effect elicited by the stimulation of these two receptors is mediated by the activation of NCX exchanger currents.[41–45] In addition, in the same TT region, pMCAO induces NOS neuronal expression (nNOS),[46] whose gaseous endproduct NO activates NCX.[47] Like in the peri-infarct regions, in the four compartments of the caudate-putamen, an area that is not affected by the ischemic lesion, but is adjacent to the core, NCX2, displayed a downregulation. By contrast, in the same striatal subregions, the NCX3 isoform displayed an upregulation. The behavioral discrepancy between NCX2 and NCX3 suggests, once again, that when NCX2 expression decreases in response to hypoxic conditions, NCX3 may assume a replacing function. Consistently, biochemical studies have clearly demonstrated that whereas NCX2 activity is strictly dependent on ATP levels, which are lowered during the development of brain ischemia, NCX3 is the only *ncx* gene product that is independent of ATP.[48–50] On the whole, all these results indicate that NCX1, NCX2, and NCX3 mRNA expression is differentially regulated after pMCAO in the ischemic core, in the peri-infarct region as well as in anatomic and functional areas, located in the hemisphere ipsilateral to the lesion and related to the infarct regions. Furthermore, because over the past few years an increasing number of studies have endeavored to provide pharmacological agents able to selectively modulate NCX isoforms to gain a better understanding of their role and to develop more effective approaches for the treatment of ischemia, the data of this study could deepen our insights into NCX pathophysiological role.

MOLECULAR BASIS OF TRANSDUCTIONAL AND TRANSCRIPTIONAL ACTIVATION OF NCX AS A FUTURE PERSPECTIVE FOR MODULATING THE EXCHANGER EXPRESSION IN BRAIN ISCHEMIA: ROLE PLAYED BY THE AKT/PI3KINASE PATHWAYS

The general meaning of the above-reported considerations is that specific gene transcription cascades converging onto the promotors of *ncx1* and *ncx3* genes are activated during ischemia and cooperate to strongly induce *ncx* gene expression. We are currently working on NCX1 promotor in order to dissect the plethora of transcriptional factors that could play a role in regulating *ncx* gene transcription. Our preliminary data seem to suggest that additional signaling cascades besides those activated by ischemia could trigger NCX transcription and could be exploited to further increase NCX expression in the effort of reducing stroke-induced brain damage. In particular, our interest is currently focused on the transductional events following the activation of neurotrophin receptors since the activation of these receptors is known to elicit neuroprotection in ischemia.

NGF triggers the PI3K/Akt pathway and this transductional cascade could also be responsible for the increase in NCX expression occurring after NGF exposure. In accordance with this idea is the evidence that when we treated PC12 cells with NGF in the presence of the specific PI-3K inhibitor LY294002 (25 μM), this neurotrophin was much less effective in inducing NCX expression. To further investigate the role of the Akt/PI3K in *ncx* gene expression, we developed a stable PC12 subclone carrying either a dominant positive or a dominant negative Akt1 mutant gene under the control of a tet-off promotor system. Using this strategy we can either turn-on or switch-off Akt1 expression just removing or adding the tetracycline, doxycycline, in the medium and investigate changes in NCX protein expression. With this cellular model, we found that Akt1 activation can induce an increase in NCX1 and NCX3 expression (FIG. 6). This *ncx* gene product upregulation can in part contribute to Akt1 neuroprotective effects. This idea is supported by the evidence that NCX1 or

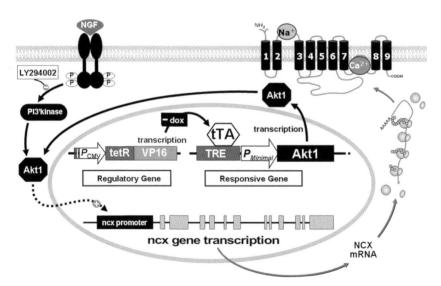

FIGURE 6. The picture schematically reports the experimental strategies used to investigate Akt1 effect on *ncx1* and *ncx3* gene expression. Tet-off strategy is detailed into the nucleus. NGF treatment and the use of drugs blocking the PI3-Akt pathway are also depicted on the left side of the figure. P_{CMV} = promotor of the cytomegalovirus gene; tetR = tetracycline repressor gene; VP16 = transcriptional activator domain tetR in an activator; TetR + VP16 = tetracyclin controlled trans activator (tTA); TRE = tetracycline responsive element; $P_{minimal}$ = minimal cytomegalovirus gene promotor.

NCX3 mRNA silencing partially decreases the ability of Akt1 to protect neurons from cell death induced by chemical hypoxia suggesting that part of the neuroprotective effect exerted by Akt1 can be ascribed to an Akt-induced synthesis of NCX1 or NCX3. Thus, the results of these studies represent a proof of concept of our idea that strategies increasing *ncx* gene expression and activity may protect the brain from ischemic damage and prompt us to search for additional transductional cascade candidates controlling NCX transcription and more suitable to be pharmacologically targeted *in vivo*. To this aim we are currently directing our efforts toward a more deep investigation of NCX1 promotor. Hopefully, these ongoing studies will pave the way for the development of new therapeutic strategies of stroke.

REFERENCES

1. CHOI, D.W., J.Y. KOH & S. PETERS. 1988. Pharmacology of glutamate neurotoxicity in cortical cell culture: attenuation by NMDA antagonists. J Neurosci. **8:** 185–196.
2. AARTS, M. *et al.* 2003. A key role for TRPM7 channels in anoxic neuronal death. Cell **115:** 863–877.
3. ANNUNZIATO, L., G. PIGNATARO & G.F. DI RENZO. 2004. Pharmacology of brain Na^+/Ca^{2+} exchanger: from molecular biology to therapeutic perspectives. Pharmacol. Rev. **56:** 633–654.
4. QUEDNAU, B.D., D.A. NICOLL & K.D. PHILIPSON. 1997. Tissue specificity and alternative splicing of the Na^+/Ca^{2+} exchanger isoforms NCX1, NCX2, and NCX3 in rat. Am. J. Physiol. **272:** C1250–C1261.
5. THURNEYSEN, T. *et al.* 2002. Sodium/calcium exchanger subtypes NCX1, NCX2 and NCX3 show cell-specific expression in rat hippocampus cultures. Brain Res. Mol. Brain Res. **107:** 145–156.
6. NAGANO, T. *et al.* 2004. Up-regulation of $Na(+)-Ca^{2+}$ exchange activity by interferon-gamma in cultured rat microglia. J. Neurochem. **90:** 784–791.
7. CANITANO, A. *et al.* 2002. Brain distribution of the Na^+/Ca^{2+} exchanger-encoding genes NCX1, NCX2, and NCX3 and their related proteins in the central nervous system. Ann. N. Y. Acad. Sci. **976:** 394–404.
8. PAPA, M. *et al.* 2003. Differential expression of the Na^+-Ca^{2+} exchanger transcripts and proteins in rat brain regions. J. Comp. Neurol. **461:** 31–48.
9. ANDREEVA, N. *et al.* 1991. Inhibition of Na^+/Ca^{2+} exchange enhances delayed neuronal death elicited by glutamate in cerebellar granule cell cultures. Brain Res. **548:** 322–325.
10. KIEDROWSKI, L. 1999. N-methyl-D-aspartate excitotoxicity: relationships among plasma membrane potential, $Na(+)/Ca(2+)$ exchange, mitochondrial $Ca(^{2+})$ overload, and cytoplasmic concentrations of $Ca(2+)$, $H(+)$, and $K(+)$. Mol. Pharmacol. **56:** 619–632.
11. AMOROSO, S. *et al.* 2000. Sodium nitroprusside prevents chemical hypoxia-induced cell death through iron ions stimulating the activity of the Na^+-Ca^{2+} exchanger in C6 glioma cells. J. Neurochem. **74:** 1505–1513.
12. MATSUDA, T. *et al.* 2001. SEA0400, a novel and selective inhibitor of the Na^+-Ca^{2+} exchanger, attenuates reperfusion injury in the *in vitro* and *in vivo* cerebral ischemic models. J. Pharmacol. Exp. Ther. **298:** 249–256.

13. PIGNATARO, G. *et al.* 2004. Evidence for a protective role played by the Na^+/Ca^{2+} exchanger in cerebral ischemia induced by middle cerebral artery occlusion in male rats. Neuropharmacology **46:** 439–448.

14. PIGNATARO, G. *et al.* 2004. Two sodium/calcium exchanger gene products, NCX1 and NCX3, play a major role in the development of permanent focal cerebral ischemia. Stroke **35:** 2566–2570.

15. IWAMOTO, T. & S. KITA. 2004. Development and application of Na^+/Ca^{2+} exchange inhibitors. Mol. Cell Biochem. **259:** 157–161.

16. YATANI, A., A.M. BROWN & A. SCHWARTZ. 1986. Bepridil block of cardiac calcium and sodium channels. J. Pharmacol. Exp. Ther. **237:** 9–17.

17. FISCHER, K.G. *et al.* 2002. Characterization of a $Na(+)$-$Ca(2+)$ exchanger in podocytes. Nephrol. Dial. Transplant. **17:** 1742–1750.

18. BERGER, F., U. BORCHARD & D. HAFNER. 1989. Effects of the calcium entry blocker bepridil on repolarizing and pacemaker currents in sheep cardiac Purkinje fibres. Naunyn Schmiedebergs Arch. Pharmacol. **339:** 638–646.

19. MORI, K. *et al.* 1998. Inhibitory effects of class I and IV antiarrhythmic drugs on the Na^+-activated K^+ channel current in guinea pig ventricular cells. Naunyn Schmiedebergs Arch. Pharmacol. **358:** 641–648.

20. SHARIKABAD, M.N., E.J. CRAGOE, JR. & O. BRORS. 1997. Inhibition by 5-N-(4-chlorobenzyl)-2′,4′-dimethylbenzamil of Na^+/Ca^{2+} exchange and L-type Ca^{2+} channels in isolated cardiomyocytes. Pharmacol. Toxicol. **80:** 57–61.

21. NICOLL, D.A., S. LONGONI & K.D. PHILIPSON. 1990. Molecular cloning and functional expression of the cardiac sarcolemmal $Na(+)$-Ca^{2+} exchanger. Science **250:** 562–565.

22. MATSUOKA, S. *et al.* 1997. Regulation of cardiac $Na(+)$-Ca^{2+} exchanger by the endogenous XIP region. J. Gen. Physiol. **109:** 273–286.

23. DIPOLO, R. & L. BEAUGE. 1994. Cardiac sarcolemmal Na/Ca-inhibiting peptides XIP and FMRF-amide also inhibit Na/Ca exchange in squid axons. Am. J. Physiol. **267:** C307–C311.

24. BOSCIA, F. *et al.* 2006. Permanent focal brain ischemia induces isoform-dependent changes in the pattern of Na^+/Ca^{2+} exchanger gene expression in the ischemic core, periinfarct area, and intact brain regions. J. Cereb. Blood Flow Metab. **26:** 502–517.

25. HOSSMANN, K.A. 1996. Periinfarct depolarizations. Cerebrovasc. Brain Metab. Rev. **8:** 195–208.

26. LI, L., D. GUERINI & E. CARAFOLI. 2000. Calcineurin controls the transcription of Na^+/Ca^{2+} exchanger isoforms in developing cerebellar neurons. J. Biol. Chem. **275:** 20903–20910.

27. BANO, D. *et al.* 2005. Cleavage of the plasma membrane Na^+/Ca^{2+} exchanger in excitotoxicity. Cell **120:** 275–285.

28. KIESSLING, M. & P. GASS. 1994. Stimulus-transcription coupling in focal cerebral ischemia. Brain Pathol. **4:** 77–83.

29. HATA, R. *et al.* 2000. Dynamics of regional brain metabolism and gene expression after middle cerebral artery occlusion in mice. J. Cereb. Blood Flow Metab. **20:** 306–315.

30. KURY, P., M. SCHROETER & S. JANDER. 2004. Transcriptional response to circumscribed cortical brain ischemia: spatiotemporal patterns in ischemic vs. remote non-ischemic cortex. Eur. J. Neurosci. **19:** 1708–1720.

31. SAPER, C.B. 1982. Convergence of autonomic and limbic connections in the insular cortex of the rat. J. Comp. Neurol. **210:** 163–173.

32. GABBOTT, P.L. *et al.* 2003. Areal and synaptic interconnectivity of prelimbic (area 32), infralimbic (area 25) and insular cortices in the rat. Brain Res. **993:** 59–71.
33. VERTES, R.P. 2004. Differential projections of the infralimbic and prelimbic cortex in the rat. Synapse **51:** 32–58.
34. VERBERNE, A.J. & N.C. OWENS. 1998. Cortical modulation of the cardiovascular system. Prog. Neurobiol. **54:** 149–168.
35. CHEUNG, R.T., V.C. HACHINSKI & D.F. CECHETTO. 1997. Cardiovascular response to stress after middle cerebral artery occlusion in rats. Brain Res. **747:** 181–188.
36. MCDOUGALL, S.J., R.E. WIDDOP & A.J. LAWRENCE. 2004. Medial prefrontal cortical integration of psychological stress in rats. Eur. J. Neurosci. **20:** 2430–2440.
37. RESSTEL, L.B., K.B. FERNANDES & F.M. CORREA. 2004. Medial prefrontal cortex modulation of the baroreflex parasympathetic component in the rat. Brain Res. **1015:** 136–144.
38. CHEUNG, R.T. & V. HACHINSKI. 2000. The insula and cerebrogenic sudden death. Arch. Neurol. **57:** 1685–1688.
39. SANTANA, N. *et al.* 2004. Expression of serotonin1A and serotonin2A receptors in pyramidal and GABAergic neurons of the rat prefrontal cortex. Cereb. Cortex **14:** 1100–1109.
40. TRIVEDI, P. *et al.* 1998. Distribution of orexin receptor mRNA in the rat brain. FEBS Lett. **438:** 71–75.
41. ERIKSSON, K.S. *et al.* 2001. Orexin/hypocretin excites the histaminergic neurons of the tuberomammillary nucleus. J. Neurosci. **21:** 9273–9279.
42. SERGEEVA, O.A. *et al.* 2003. Co-ordinated expression of 5-HT2C receptors with the NCX1 Na^+/Ca^{2+} exchanger in histaminergic neurones. J. Neurochem. **87:** 657–664.
43. WU, M. *et al.* 2002. Hypocretin increases impulse flow in the septohippocampal GABAergic pathway: implications for arousal via a mechanism of hippocampal disinhibition. J. Neurosci. **22:** 7754–7765.
44. WU, M. *et al.* 2004. Hypocretin/orexin innervation and excitation of identified septohippocampal cholinergic neurons. J. Neurosci. **24:** 3527–3536.
45. BURDAKOV, D., B. LISS & F.M. ASHCROFT. 2003. Orexin excites GABAergic neurons of the arcuate nucleus by activating the sodium–calcium exchanger. J. Neurosci. **23:** 4951–4957.
46. PENG, Z.C. *et al.* 1996. Induction of NADPH-diaphorase activity in the rat forebrain after middle cerebral artery occlusion. Exp. Neurol. **138:** 105–120.
47. ASANO, S. *et al.* 1995. Nitroprusside and cyclic GMP stimulate $Na(+)$-Ca^{2+} exchange activity in neuronal preparations and cultured rat astrocytes. J. Neurochem. **64:** 2437–2441.
48. LINCK, B. *et al.* 1998. Functional comparison of the three isoforms of the Na^+/Ca^{2+} exchanger (NCX1, NCX2, NCX3). Am. J. Physiol. **274:** C415–C423.
49. ANNUNZIATO, L. *et al.* 2007. Beyond excitotoxicity: TRPM7, ASICs and NCX in the scenario of glutamate-independent calcium toxicity. Stroke **38:** 661–664.
50. SECONDO, A. 2007. BHK cells transfected with NCX3 are more resistant to hypoxia followed by reoxygenation than those transfected with NCX1 and NCX2: possible relationship with mitochondrial membrane potential. Cell Calcium In press.

Sodium–Calcium Exchanger in Pulmonary Artery Smooth Muscle Cells

YUN-MIN ZHENG AND YONG-XIAO WANG

Center for Cardiovascular Sciences, Albany Medical College, Albany, New York 12208, USA

ABSTRACT: The expression and function of the Na^+/Ca^{2+} exchanger (NCX) in the regulation of intracellular Ca^{2+} homeostasis have been well studied in cardiac, skeletal, and systemic vascular myocytes, but not in pulmonary artery smooth muscle cells (SMCs). We have recently demonstrated that the NCX current is present in freshly isolated pulmonary artery SMCs using the patch-clamp technique. The current has a mean amplitude of 13 pA under near physiological resting conditions. The NCX may function in the forward mode to make a significant contribution to the decay of intracellular Ca^{2+} following Ca^{2+} release and/or depolarization. Hypoxic stimulation inhibits the NCX current, reduces the removal of intracellular Ca^{2+}, and enhances Ca^{2+} release from the sarcoplsamic reticulum. Using RT-PCR, subcloning and sequence analysis, we have shown that three NCX1 splice variants: NCX1.2 (containing exons B, C, and D), NCX1.3 (exons B and D), and NCX1.7 (exons B, D, and F) are expressed in pulmonary artery smooth muscle. Each of these splice variants expressed in HEK293 cells it likely to show a distinct activity in the removal of intracellular Ca^{2+}. Taken together, we provide clear evidence that NCX1 is functionally and molecularly expressed and plays a physiological role in pulmonary artery SMCs.

KEYWORDS: Na^+/Ca^{2+} exchanger; intracellular calcium; hypoxia; patch clamp; pulmonary artery smooth muscle cells

INTRODUCTION

The calcium ion (Ca^{2+}) serves as an important signal in the initiation and maintenance of numerous physiological responses in virtually all types of cells. The cell, therefore, precisely regulates the intracellular Ca^{2+} concentration ($[Ca^{2+}]_i$) by the concerted action of multiple Ca^{2+} entry and exit processes. An increase in $[Ca^{2+}]_i$ may occur owing to extracellular Ca^{2+} influx through plasmalemmal voltage-, store-, and receptor-operated Ca^{2+} channels. Another

Address for correspondence: Dr. Yong-Xiao Wang, Albany Medical College, Center for Cardiovascular Sciences (MC-8), 47 New Scotland Avenue, Albany, NY 12208. Voice: 518-262-9506; fax: 518-262-8101.

wangy@mail.amc.edu

Ann. N.Y. Acad. Sci. 1099: 427–435 (2007). © 2007 New York Academy of Sciences.
doi: 10.1196/annals.1387.017

important source for Ca^{2+} entry is Ca^{2+} release from intracellular stores, particularly from the sarcoplasmic reticulum via ryanodine receptors and inositol 1,4,5-triphosphate receptors. In turn, Ca^{2+} influx through voltage-dependent Ca^{2+} channels may induce Ca^{2+} release from the sarcoplasmic reticulum by activating ryanodine receptors, a process of Ca^{2+}-induced Ca^{2+} release. The Ca^{2+} removal is achieved by plasmalemmal and sarcoplasmic reticulum Ca^{2+}-ATPase and/or plasmalemmal Na^+/Ca^{2+} exchanger (NCX). The importance of NCX in the removal of cytosolic Ca^{2+} has been well established in several types of cells, particularly cardiac and skeletal muscle cells. In these striated muscle cells, plasmalemmal NCX can transport one Ca^{2+} out of the cell in exchange for three Na^+, which is called the *forward operation mode*. On the other hand, NCX can move Ca^{2+} into the cell to exit Na^+ under certain circumstances called the reverse operation mode. The NCX family consists of three isoforms, NCX1, NCX2, and NCX3. Each isoform is encoded by a distinct gene. NCX1 is widely distributed in the heart, blood vessel, kidney, brain, and other tissues, whereas NCX2 and NCX3 are predominantly expressed in the brain and skeletal muscle.[1–3]

Increasing evidence indicates that NCX1 is highly expressed and plays an important functional role in systemic vascular smooth muscle cells (SMCs). Several reports have shown that both NCX1 mRNAs and proteins are present in aortic SMCs.[4–7] Inhibition of NCX by replacing extracellular Na^+ with Li^+ or using antisense oligonucleotides-mediated gene knockdown significantly delays the removal of the elevated $[Ca^{2+}]_i$ following application of agonists and enhances agonist-induced increases in $[Ca^{2+}]_i$ in systemic vascular SMCs.[8–13] Similarly, pharmacological inhibition of NCX slows the relaxation of agonist-evoked contraction and increases agonist-evoked contraction in systemic vascular muscle tissues.[8,9,13–22] In addition, NCX may function in the reverse mode to cause or facilitate extracellular Ca^{2+} influx in systemic vascular SMCs and to enhance agonist-induced contraction in systemic vascular tissues.[23–27] Consistent with the established importance of NCX1 in systemic vascular SMCs, a number of studies have shown that the hypertension is associated with changes in the activity of NCX in various types of arterial beds.[28–34] Using the specific inhibitor of Ca^{2+} entry through NCX1 SEA0400, NCX1-deficient mice, and transgenic mice that specifically express NCX1.3 in smooth muscle, Iwamoto *et al*. have recently demonstrated that salt-sensitive hypertension is triggered by Ca^{2+} entry through NCX1 in arterial smooth muscle.[35]

Pulmonary circulation is structurally and functionally different from systemic circulation at the regional tissue, cellular, and molecular levels. Experimental studies have revealed that an increase in $[Ca^{2+}]_i$ is imperative for numerous physiological and pathological responses in pulmonary artery SMCs. However, little is known about the molecular expression and functional role of NCX in pulmonary artery SMCs. We have started to address this question by using patch-clamp, fluorescent imaging, and molecular biological approaches. In this review article, we highlight our recent work in the study of NCX in pulmonary artery SMCs. A part of the work has been published previously.[36]

NCX CURRENTS ARE PRESENT IN PULMONARY ARTERY SMCs

Using the protocol employed by Kimura *et al.* in heart cells,[37] we made direct measurements of the NCX current in freshly isolated pulmonary artery SMCs with the classic patch-clamp technique.[38] Single cells were obtained from rat resistance (external diameter < 300 μm) pulmonary arteries using a modified two-step enzymatic digestion method.[39] Myocytes were voltage-clamped at −60 mV, with $[Ca^{2+}]_i$ clamped at 1 μM to maximize the exchange current. Currents associated with voltage-dependent Ca^{2+} channels, K^+ channels, and Na^+–K^+ ATPase were blocked by nisoldipine, Ba^{2+}, and ouabain in the bath solution, and Cs^+ in the intracellular solution. A brief exposure from Na^+-free (Li^+ substitution) to 130 mM Na^+ solution induced an inward current, which was activated rapidly and decayed with a relatively slow kinetics. The mean current amplitude was about 25 pA. The NCX current was also detected under conditions, where $[Ca^{2+}]_i$ was maintained at physiological levels using the perforated patch-clamp method, although the average current amplitude was 13 pA. A smaller exchange current would be predicted from the shift of the current reversal potential due to the lower $[Ca^{2+}]_i$ (assuming $[Ca^{2+}]_i$ = 100 nM and Na_i = 5 mM, $E_{Ca/Na}$ = 0 mV).

To confirm that the inward current resulted from the electrogenic NCX, experiments were performed to determine the current reversal potential. Myocytes were dialyzed with intracellular solution containing 5 mM Na^+ and $[Ca^{2+}]_i$ clamped at 1 μM, and then exposed to Na^+-containing solution to activate the inward current. Ramp voltage protocols were imposed (−60 to 100 mV for 320 ms) before and during the activation of the Na^+-sensitive current. The subtracted ramp currents indicated that the Na^+-sensitive current was close to the calculated equilibrium potential of the NCX current (−60 mV), assuming the NCX stoichiometry of 3 Na^+/1 Ca^{2+}, as observed for the cardiac exchanger.[1–3] Similar results were observed from five other cells. Thus, a significant NCX current is present in freshly isolated pulmonary artery myocytes.

REMOVAL OF EXTRACELLULAR Na$^+$ SLOWS Ca^{2+} REMOVAL FROM THE CYTOSOL IN PULMONARY ARTERY MYOCYTES

To determine the role of the NCX in the removal of Ca^{2+} from the cytosol, we examined cytosolic Ca^{2+} decay rates in pulmonary artery SMCs in the absence of extracellular Na^+. Measurements of $[Ca^{2+}]_i$ were performed using fura-2 fluorescence.[38] The elevation of $[Ca^{2+}]_i$ was achieved by exposing my-ocytes to a caffeine/high K^+ (10/80 mM) solution for 2.5 s. The time constant for decay of $[Ca^{2+}]_i$ was markedly increased, and the peak $[Ca^{2+}]_i$ enhanced, when cells were exposed to the $[Ca^{2+}]_i$ elevation protocol in the absence of extracellular Na^+. The Ca^{2+} decay could be well fitted by single exponential equation. Removing the extracellular Na^+ significantly prolonged the time constant of Ca^{2+} decay. A second effect seen in these experiments is the marked augmentation of the initial peak $[Ca^{2+}]_i$ achieved by exposure to the caffeine/high K^+ solution. These data indicate that NCX may function in the forward mode to contribute to the cytosolic Ca^{2+} removal in freshly isolated pulmonary artery SMCs. Because a significant NCX current is mea-sured at basal levels of $[Ca^{2+}]_i$ (100 nM), the augmented amplitude of the peak increase in $[Ca^{2+}]_i$ by low Na^+ solutions is likely due to enhanced Ca^{2+} uptake into the sarcoplasmic reticulum associated with loss of Ca^{2+} transport across the plasmalemma.[1]

HYPOXIA SLOWS Ca^{2+} REMOVAL FROM THE CYTOSOL AND AUGMENTS PEAK $[Ca^{2+}]_i$ IN PULMONARY ARTERY SMCs

Our previous study has shown that the rate of Ca^{2+} decay following caffeine-induced Ca^{2+} release is markedly slower in pulmonary artery SMCs under severe hypoxia ($Po_2 < 40$ Torr) or chemical hypoxia than under nonhypoxic conditions.[40] This finding led us to hypothesize that hypoxia may inhibit one or more cellular Ca^{2+} removal processes. To test this hypothesis, we first examined the rate of Ca^{2+} removal before and during exposure of myocytes to mild hypoxia in cells isolated from pulmonary resistance artery myocytes. Mild hypoxic conditions (50–60 Torr) were chosen, because this level of hypoxia does not cause a significant elevation of $[Ca^{2+}]_i$. Application of caffeine/high K^+ solution to myocytes in a normal physiological solution produced a rapid $[Ca^{2+}]_i$ rise and typical decay with a time constant of approximately 6 s. In contrast, the $[Ca^{2+}]_i$ decay was much slower in the same myocytes after perfusion of with the hypoxic solution for 5 min. The mean increase in $[Ca^{2+}]_i$ was also greatly augmented following hypoxic stimulation.

To further explore the relationship between hypoxia and the removal of ex-tracellular Na^+, we sought to determine whether the hypoxic slowing of the rate of $[Ca^{2+}]_i$ removal from the cytosol and augmentation of the peak in-crease in $[Ca^{2+}]_i$ were affected by removing extracellular Na^+. Experiments were performed in which the $[Ca^{2+}]_i$ decay was examined under hypoxic con-ditions in the presence and absence of extracellular Na^+. $[Ca^{2+}]_i$ was elevated by applying caffeine/high K^+ solution first under hypoxia alone and then under conditions of hypoxia combined with Na^+-free bath solution. Removing extra-cellular Na^+ neither further slowed the rate of decay of $[Ca^{2+}]_i$ nor enhanced the peak increase in $[Ca^{2+}]_i$ following hypoxic stimulation. Thus, hypoxia and removal of extracellular Na^+ have similar and nonadditive effects on $[Ca^{2+}]_i$, suggesting that the mechanism underlying hypoxic effects on $[Ca^{2+}]_i$ may be associated with an inhibitory effect on the plasmalemmal NCX. Such a mechanism may explain previous findings that the slowed rate of relaxation of isolated pulmonary arteries by hypoxia no longer occurs in the absence of extracellular Na^+.[41]

HYPOXIA INHIBITS NCX CURRENTS IN PULMONARY ARTERY SMCs

Since hypoxia slowed the decay of $[Ca^{2+}]_i$ in a manner consistent with the NCX inhibition, we next sought to determine the effect of hypoxia on the NCX current. Myocytes were voltage-clamped at -60 mV using the perforated patch-clamp method, and the current was recorded before and during exposure to mild hypoxia. The NCX current could be repeatedly activated with little or no decline in its magnitude. In contrast, the current was markedly inhibited by mild hypoxia, compared to the current level upon initial exposure to Na^+ solution. These results suggest that hypoxia inhibits the plasmalemmal NCX in pulmonary artery myocytes, thereby resulting in a slowed rate of Ca^{2+} efflux from the cytosol, which would serve to sustain the elevation of $[Ca^{2+}]_i$ and produce a maintained vasoconstriction.

MULTIPLE NCX1 ISOFORMS ARE EXPRESSED IN PULMONARY ARTERY SMOOTH MUSCLE

Vascular muscle expresses the *NCX1* gene, which undergoes alternative splicing in its large intracellular loop region with six cassette exons: A, B, C, D, E, and F.[1–3,42] Thus, we also sought to confirm the expression of NCX1 mRNAs in pulmonary artery smooth muscle and determine the expression of specific splice variants using the reverse transcriptase polymerase chain reaction (RT-PCR). Total RNAs were isolated from endothelium-denuded rat pulmonary artery and heart muscle (as control) using the acid guanidinium thiocyanate-phenol-chloroform method. Single strand cDNA was synthesized using Superscript II reverse transcriptase. The resultant cDNA template was amplified with the sense and antisense oligonucleotide primers, designed to span the previously identified putative cytoplasmic region of variability in NCX1.[5] Two prominent RNAs were identified in pulmonary artery tissue, whereas one band was detected in heart muscle. These data are in agreement with previous studies that report expression of a single, larger RNA for NCX1 in the heart, but a second splice variant in other tissues.[5,7,43]

The expression of specific splice variants in pulmonary artery smooth muscle was further examined by using RT-PCR, subcloning, and sequence analysis. We performed PCR amplification on mouse pulmonary artery muscle cDNAs using specific primers designed on the basis of conserved amino acid sequences of putative transmembrane segments of the mammalian exchangers NCX1. PCR products were subcloned using the TA cloning kit with competent TOP10F' cells. PCR subclones were sequenced using general sequencing primers. We found that three splice variants containing exons B, C, and D (NCX1.2), exons B and D (NCX1.3), and exons B, D, and F (NCX1.7) were present in pulmonary artery smooth muscle. Their abundance (determined by the number of identified specific clones divided by the number of total analyzed clones) is as follows: NCX1.3, NCX1.2, and NCX1.7. Our results are consistent with previous reports that NCX1.3 is predominantly expressed in the aortic tissue.[5,7]

SPLICE VARIANTS OF PULMONARY ARTERY SMOOTH MUSCLE NCX1 IN HEK293 CELLS SHOW A DISTINCT ACTIVITY IN THE REMOVAL OF INTRACELLULAR Ca^{2+}

To test and compare the functional role of splice variants of pulmonary artery NCX1, stable transfection of NCX1.2, NCX1.3, and NCX1.7 in HEK293 cells were made using pCMV-FLAG vector. Transfected cells were loaded with fura-2/AM and then exposed to caffeine to cause an increase in $[Ca^{2+}]_i$. Our preliminary data indicate that the decay time of the elevated $[Ca^{2+}]_i$ was shorter in cells transfected with NCX1 splice variants than in control cells (transfected with vector alone); the mean time constant of the $[Ca^{2+}]_i$ decay was decreased by ∼50% in NCX1.7-transfected cells, and by ∼30% in NCX1.2 and NCX1.3. This distinct activity for different NCX1 splice variants in the removal of intracellular Ca^{2+} may play an important role in physiological and/or pathophysiological responses in pulmonary artery SMCs.

SUMMARY

We have demonstrated that NCX is functional under physiological resting conditions and plays an important role in the removal of cytosolic Ca^{2+} in freshly isolated rat pulmonary artery SMCs. This exchanger is inhibited during hypoxic stimuli, resulting in a slowing of Ca^{2+} removal from the cytosol and enhanced Ca^{2+} release, which may be an important mechanism in amplifying effects of hypoxia on Ca^{2+} influx or release.[44] Multiple splice variants of NCX1, NCX1.2, NCX1.3, and NCX1.7, are identified in pulmonary artery smooth muscle. Each of these splice variants is liable to exhibit a distinct activity in the extrusion of intracellular Ca^{2+}. In addition, a recent study has reported that NCX in cultured human pulmonary artery SMCs may mediate store depletion-induced Ca^{2+} entry and proliferation,[45] further suggesting that the functional importance of NCX in pulmonary artery SMCs.

ACKNOWLEDGMENTS

This work was supported by American Heart Association and National Institute of Health.

REFERENCES

1. BLAUSTEIN, M.P. & W.J. LEDERER. 1999. Sodium/calcium exchange: its physiological implications. Physiol. Rev. **79:** 763–854.
2. DIPOLO, R. & L. BEAUGE. 2006. Sodium/calcium exchanger: influence of metabolic regulation on ion carrier interactions. Physiol. Rev. **86:** 155–203.
3. IWAMOTO, T. 2006. Vascular Na^+/Ca^{2+} exchanger: implications for the pathogenesis and therapy of salt-dependent hypertension. Am. J. Physiol. Regul. Integr. Comp. Physiol. **290:** R536–R545.
4. JUHASZOVA, M., A. AMBESI, G.E. LINDENMAYER, et al. 1994. Na^+-Ca^{2+} exchanger in arteries: identification by immunoblotting and immunofluorescence microscopy. Am. J. Physiol. Cell Physiol. **266:** C234–C242.
5. NAKASAKI, Y., T. IWAMOTO, H. HANADA, et al. 1993. Cloning of the rat aortic smooth muscle Na^+/Ca^{2+} exchanger and tissue-specific expression of isoforms. J. Biochem. (Tokyo) **114:** 528–534.
6. SMITH, L. & J.B. SMITH. 1994. Regulation of sodium-calcium exchanger by glucocorticoids and growth factors in vascular smooth muscle. J. Biol. Chem. **269:** 27527–27531.
7. QUEDNAU, B.D., D.A. NICOLL & K.D. PHILIPSON. 1997. Tissue specificity and alternative splicing of the Na^+/Ca^{2+} exchanger isoforms NCX1, NCX2, and NCX3 in rat. Am. J. Physiol. Cell Physiol. **272:** C1250–C1261.
8. ASHIDA, T. & M.P. BLAUSTEIN. 1987. Regulation of cell calcium and contractility in mammalian arterial smooth muscle: the role of sodium-calcium exchange. J. Physiol. (Lond.) **392:** 617–635.
9. BOVA, S., W.F. GOLDMAN, X.J. YAUAN & M.P. BLAUSTEIN. 1990. Influence of Na^+ gradient on Ca^{2+} transients and contraction in vascular smooth muscle. Am. J. Physiol. Heart Circ. Physiol. **259:** H409–H423.
10. BORIN, M.L., R.M. TRIBE & M.P. BLAUSTEIN. 1994. Increased intracellular Na^+ augments mobilization of Ca^{2+} from SR in vascular smooth muscle cells. Am. J. Physiol. Cell Physiol. **266:** C311–C317.
11. SLODZINSKI, M.K., M. JUHASZOVA & M.P. BLAUSTEIN. 1995. Antisense inhibition of Na^+/Ca^{2+} exchange in primary cultured arterial myocytes. Am. J. Physiol. Cell Physiol. **269:** C1340–C1345.
12. SLODZINSKI, M.K. & M.P. BLAUSTEIN. 1998. Physiological effects of Na^+/Ca^{2+} exchanger knockdown by antisense oligodeoxynucleotides in arterial myocytes. Am. J. Physiol. Cell Physiol. **275:** C251–C259.
13. YAMANAKA, J., J. NISHIMURA, K. HIRANO & H. KANAIDE. 2003. An important role for the Na^+-Ca^{2+} exchanger in the decrease in cytosolic Ca^{2+} concentration induced by isoprenaline in the porcine coronary artery. J. Physiol. (Lond.) **549:** 553–562.
14. REUTER, H., M.P. BLAUSTEIN & G. HAEUSLER. 1973. Na-Ca exchange and tension development in arterial smooth muscle. Philos. Trans. R. Soc. Lond. B. Biol. Sci. **265:** 87–94.

15. OZAKI, H. & N. URAKAWA. 1979. Na-Ca exchange and tension development in guinea-pig aorta. Naunyn Schmiedebergs Arch. Pharmacol. **309:** 171–178.

16. PETERSEN, T.T. & M.J. MULVANY. 1984. Effect of sodium gradient on the rate of relaxation of rat mesenteric small arteries from potassium contractures. Blood Vessels **21:** 279–289.

17. SMITH, J.B. & L. SMITH. 1987. Extracellular Na^+ dependence of changes in free Ca^{2+}, $^{45}Ca^{2+}$ efflux, and total cell Ca^{2+} produced by angiotensin II in cultured arterial muscle cells. J. Biol. Chem. **262:** 17455–17460.

18. SMITH, J.B., E.J. CRAGOE, JR. & L. SMITH. 1987. Na^+/Ca^{2+} antiport in cultured arterial smooth muscle cells. Inhibition by magnesium and other divalent cations. J. Biol. Chem. **262:** 11988–11994.

19. BLAUSTEIN, M.P., A. AMBESI, R.J. BLOCH, et al. 1992. Regulation of vascular smooth muscle contractility: roles of the sarcoplasmic reticulum (SR) and the sodium/calcium exchanger. Jpn. J. Pharmacol. **58**(Suppl 2): 107P–114P.

20. ARNON, A., J.M. HAMLYN & M.P. BLAUSTEIN. 2000. Na^+ entry via store-operated channels modulates Ca^{2+} signaling in arterial myocytes. Am. J. Physiol. Cell Physiol. **278:** C163–C173.

21. SCHWEDA, F., B.K. KRAMER & A. KURTZ. 2001. Differential roles of the sodium-calcium exchanger in renin secretion and renal vascular resistance. Pflugers Arch. **442:** 693–699.

22. SCHWEDA, F., H. SEEBAUER, B.K. KRAMER & A. KURTZ. 2001. Functional role of sodium-calcium exchange in the regulation of renal vascular resistance. Am. J. Physiol. Renal Physiol. **280:** F155–F161.

23. KHOYI, M.A., R.A. BJUR & D.P. WESTFALL. 1991. Norepinephrine increases Na-Ca exchange in rabbit abdominal aorta. Am. J. Physiol. Cell Physiol. **261:** C685–C690.

24. IWAMOTO, T., K. HARADA, F. NAKAJIMA & T. SUKAMOTO. 1992. Effects of ouabain on muscle tension and intracellular Ca^{2+} level in guinea-pig aorta. Eur. J. Pharmacol. **224:** 71–76.

25. KHOYI, M.A., R.A. BJUR & D.P. WESTFALL. 1993. Time-dependent increase in Ca^{2+} influx in rabbit abdominal aorta: role of Na-Ca exchange. Am. J. Physiol. Cell Physiol. **265:** C1325–C1331.

26. ARNON, A., J.M. HAMLYN & M.P. BLAUSTEIN. 2000. Ouabain augments Ca^{2+} transients in arterial smooth muscle without raising cytosolic Na^+. Am. J. Physiol. Heart Circ. Physiol. **279:** H679–H691.

27. LEE, C.H., D. POBURKO, P. SAHOTA, et al. 2001. The mechanism of phenylephrine-mediated $[Ca^{2+}]_i$ oscillations underlying tonic contraction in the rabbit inferior vena cava. J. Physiol. (Lond.) **534:** 641–650.

28. ASHIDA, T., M. KURAMOCHI & T. OMAE. 1989. Increased sodium-calcium exchange in arterial smooth muscle of spontaneously hypertensive rats. Hypertension **13:** 890–895.

29. ASHIDA, T., Y. KAWANO, H. YOSHIMI, et al. 1992. Effects of dietary salt on sodium-calcium exchange & ATP-driven calcium pump in arterial smooth muscle of Dahl rats. J. Hypertens. **10:** 1335–1341.

30. NELSON, L.D., M.T. UNLAP, J.L. LEWIS & P.D. BELL. 1999. Renal arteriolar Na^+/Ca^{2+} exchange in salt-sensitive hypertension. Am. J. Physiol. Renal Physiol. **276:** F567–F573.

31. NELSON, L.D., N.A. MASHBURN & P.D. BELL. 1996. Altered sodium-calcium exchange in afferent arterioles of the spontaneously hypertensive rat. Kidney Int. **50:** 1889–1896.

32. WELLS, I.C. & A.J. BLOTCKY. 2001. Coexisting independent sodium-sensitive and sodium-insensitive mechanisms of genetic hypertension in spontaneously hypertensive rats (SHR). Can. J. Physiol. Pharmacol. **79:** 779–784.

33. HWANG, E.F., I. WILLIAMS, G. KOVACS, *et al.* 2003. Impaired ability of the Na^+/Ca^{2+} exchanger from the Dahl/Rapp salt-sensitive rat to regulate cytosolic calcium. Am. J. Physiol. Renal Physiol. **284:** F1023–F1031.

34. TANIGUCHI, S., K. FURUKAWA, S. SASAMURA, *et al.* 2004. Gene expression and functional activity of sodium/calcium exchanger enhanced in vascular smooth muscle cells of spontaneously hypertensive rats. J. Cardiovasc. Pharmacol. **43:** 629–637.

35. IWAMOTO, T., S. KITA, J. ZHANG, *et al.* 2004. Salt-sensitive hypertension is triggered by Ca^{2+} entry via Na^+/Ca^{2+} exchanger type-1 in vascular smooth muscle. Nat. Med. **10:** 1193–1199.

36. WANG, Y.X., P.K. DHULIPALA & M.I. KOTLIKOFF. 2000. Hypoxia inhibits the Na^+/Ca^{2+} exchanger in pulmonary artery smooth muscle cells. FASEB J. **14:** 1731–1740.

37. KIMURA, J., S. MIYAMAE & A. NOMA. 1987. Identification of sodium-calcium exchange current in single ventricular cells of guinea pig. J. Physiol. (Lond.) **384:** 199–222.

38. WANG, Y.X. & M.I. KOTLIKOFF. 1997. Inactivation of calcium-activated chloride channels in smooth muscle by calcium/calmodulin-dependent protein kinase. Proc. Natl. Acad. Sci. USA **94:** 14918–14923.

39. PEREZ, G.J., A.D. BONEV, J.B. PATLAK & M.T. NELSON. 1999. Functional coupling of ryanodine receptors to K_{Ca} channels in smooth muscle cells from rat cerebral arteries. J. Gen. Physiol. **113:** 229–238.

40. WANG, Q., Y.X. WANG, M. YU & M.I. KOTLIKOFF. 1997. Ca^{2+}-activated Cl^- currents are activated by metabolic inhibition in rat pulmonary artery smooth muscle cells. Am. J. Physiol. Cell Physiol. **273:** C520–C530.

41. SALVATERRA, C.G., L.J. RUBIN, J. SCHAEFFER & M.P. BLAUSTEIN. 1989. The influence of the transmembrane sodium gradient on the responses of pulmonary arteries to decreases in oxygen tension. Am. Rev. Respir. Dis. **139:** 933–939.

42. KOFUJI, P., W.J. LEDERER & D.H. SCHULZE. 1994. Mutually exclusive and cassette exons underlie alternatively spliced isoforms of the Na/Ca exchanger. J. Biol. Chem. **269:** 5145–5149.

43. LEE, S.L., A.S. YU & J. LYTTON. 1994. Tissue-specific expression of Na^+-Ca^{2+} exchanger isoforms. J. Biol. Chem. **269:** 14849–14852.

44. WEIR, E.K., H.L. REEVE, D.A. PETERSON, *et al.* 1998. Pulmonary vasoconstriction, oxygen sensing, and the role of ion channels: Thomas A. Neff lecture. Chest **114:** 17S–22S.

45. ZHANG, S., J.X. YUAN, K.E. BARRETT & H. DONG. 2005. Role of Na^+/Ca^{2+} exchange in regulating cytosolic Ca^{2+} in cultured human pulmonary artery smooth muscle cells. Am. J. Physiol. Cell Physiol. **288:** C245–C252.

Rapid Downregulation of NCX and PMCA in Hippocampal Neurons Following H$_2$O$_2$ Oxidative Stress

SERTAC N. KIP AND EMANUEL E. STREHLER

Department of Biochemistry and Molecular Biology, Mayo Clinic College of Medicine, Rochester, Minnesota 55905, USA

ABSTRACT: Na$^+$/Ca^{2+} exchangers (NCXs) and plasma membrane Ca^{2+} pumps (PMCAs) are crucial for intracellular Ca^{2+} homeostasis and Ca^{2+} signaling. Elevated [Ca^{2+}]$_i$ is a hallmark of neurodegenerative disease and stroke. Here we studied the short-term effect of oxidative stress on the plasma membrane Ca^{2+} extrusion systems in hippocampal neurons (HN) and found that after 2–3 h exposure to 300 µM H$_2$O$_2$, all NCXs and PMCAs were significantly downregulated at the RNA (NCX) and protein (PMCA) level. Rapid internalization and aggregation of the PMCA was also observed. Our data show that the plasma membrane calcium extrusion systems are sensitive early targets of neurotoxic oxidative stress.

KEYWORDS: calcium extrusion; hippocampal neurons; Na$^+$/Ca^{2+} exchanger; NCX; oxidative stress; PMCA

INTRODUCTION

Oxidative stress plays a crucial role in conditions such as aging, cardiac dysfunction, ischemic injuries, and neurodegenerative diseases including Alzheimer disease.[1,2] Neuronal death follows Ca^{2+} dyshomeostasis and the generation of oxidative stress.[3,4] However, despite studies linking neurotoxicity to Ca^{2+} overload, the relevance of altered Ca^{2+} efflux in the context of neurodegeneration is poorly understood. Because the maintenance and restoration of normal [Ca^{2+}]$_i$ is primarily dependent on the proper function of plasma membrane Ca^{2+} extrusion systems comprised of Ca^{2+} pumps (PMCA) and Na$^+$/Ca^{2+} exchangers (NCX/NCKX), these Ca^{2+} transporters are primary candidates as early targets of oxidative stress. We therefore determined the effect of exposure to H$_2$O$_2$ on the expression of NCX and PMCA in primary rat hippocampal neurons (HN).

Address for correspondence: Emanuel E. Strehler, Department of Biochemistry and Molecular Biology, Mayo Clinic College of Medicine, 200 First Street S.W., Rochester, MN 55905. Voice: 507-284-9372; fax: 507-284-2384.

strehler.emanuel@mayo.edu

Ann. N.Y. Acad. Sci. 1099: 436–439 (2007). © 2007 New York Academy of Sciences.
doi: 10.1196/annals.1387.005

MATERIALS AND METHODS

Primary embryonic rat (E18) HN and glial cultures were prepared as described.[5] The methods for reverse transciption polymerase chain reaction (RT-PCR), including the specific primers for amplification of NCX1, NCX2, and NCX3 transcripts, have also been reported.[5] Western blotting and immunofluorescence confocal microscopy were performed as described, using primary pan-PMCA antibody 5F10 at 1:2000 and 1:600, respectively.[5,6] To determine the effects of oxidative stress, cells were treated with 150–600 μM H_2O_2 added to the media for 1–3 h prior to analysis by RT-PCR, Western, or immunofluorescence microscopy.

RESULTS AND DISCUSSION

We analyzed the effect of the addition of H_2O_2 to the culture media of mature rat HN on the expression of NCX1-3 at the transcript level. Using RT-PCR, we found that a 3-h treatment with 300 μM H_2O_2 resulted in an almost complete depletion of the transcripts for NCX1 and NCX3, while NCX2 transcripts were much less affected (FIG. 1). NCX2 is highly expressed in the adult hippocampus, with transcript levels exceeding those of the other NCX isotypes by up to an order of magnitude.[7–9] The apparent protection of NCX2 mRNA from H_2O_2-induced oxidative damage may thus simply reflect

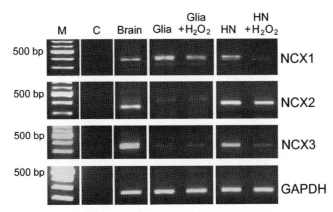

FIGURE 1. Effect of H_2O_2 on the expression of NCX transcripts in mature HN. RT-PCR was performed with NCX-specific primers using RNA from 21-day HN, and from confluent glia-enriched cultures and total rat brain as positive control. The cells were either left untreated, or incubated for 2–3 h in 300 μM H_2O_2 (Glia + H_2O_2, HN + H_2O_2) prior to RNA isolation. M, 100 bp marker ladder; C, negative control omitting template DNA. Glyceraldehyde-3-phosphate dehydrogenase (GAPDH) internal controls are shown in the *bottom panel*.

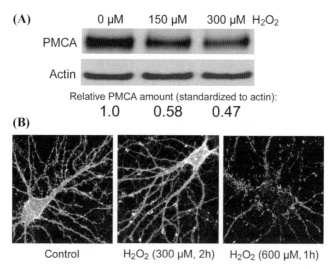

FIGURE 2. Effect of H_2O_2 on the expression and localization of PMCA in mature HN. (**A**) HN cultured for 14 days were treated for 2 h with 0, 150, and 300 μM H_2O_2 as indicated on top of each lane, and then harvested for SDS-polyacrylamide gel electrophoresis and Western blotting using antibody 5F10 against all PMCAs. The blot was also probed for β-actin as an internal protein loading control (*bottom panel*). The relative amount of PMCA standardized to the actin loading control is indicated beneath each lane. (**B**) Immunofluorescence localization of PMCAs in 14-day HN left untreated (control, *left panel*), or incubated for 2 h in the presence of 300 μM H_2O_2 (*middle panel*) or for 1 h in 600 μM H_2O_2 (*right panel*).

the elevated steady-state levels of NCX2 transcripts. However, the data also suggest that NCX2 mRNA levels (and thus perhaps protein function) may be at least partially resistant to moderate oxidative stress. FIGURE 1 also indicates that NCX transcripts in glia were much less affected by 300 μM H_2O_2 than in primary neurons. Differences in susceptibility to oxidative stress among the different cell types likely account for this effect.

We used Western blotting to determine the effect of H_2O_2 on the expression of the PMCAs in HN. FIGURE 2 A shows that there was a dose-dependent reduction in the relative PMCA level. After a 2-h treatment with 300 μM H_2O_2, PMCA levels decreased to about 50% of control. The oxidative stress also leads to a selective redistribution and gradual disappearance of PMCA immunoreactivity from the plasma membrane. PMCA was localized mainly in the soma membrane and in punctuate structures in dendrites in mature HN (FIG. 2 B, *left panel*). After a 2-h treatment with 300 μM H_2O_2, PMCA appeared aggregated in large clusters in what may be degenerating dendrites (FIG. 2 B, *middle panel*). At high levels (600 μM) of H_2O_2, cells showed reduced PMCA immunoreactivity already after 1 h, and the staining concentrated in large clusters.

CONCLUSION

H_2O_2-mediated oxidative stress leads to a rapid (within a few hours) decrease in the major plasma membrane Ca^{2+} extrusion systems. Functionally, this will result in reduced Ca^{2+} efflux capacity, and thus in a reduced capacity to counteract potentially harmful excitotoxic Ca^{2+} loads. Strategies to protect against loss of NCX and PMCA function may be useful to prevent the progression of the harmful biochemical cascades leading to neurodegeneration following oxidative stress and Ca^{2+} overload.

ACKNOWLEDGMENT

This work was supported by NIH grant RO1 GM28835 and the Mayo Foundation for Medical Education and Research.

REFERENCES

1. DHALLA, N.S., R.M. TEMSAH & T. NETTICADAN. 2000. Role of oxidative stress in cardiovascular diseases. J. Hypertens. **18:** 655–673.
2. COYLE, J.T. & P. PUTTFARCKEN. 1993. Oxidative stress, glutamate, and neurodegenerative disorders. Science **262:** 689–695.
3. CHOI, D.W. 1995. Calcium: still center-stage in hypoxic-ischemic neuronal death. Trends Neurosci. **18:** 58–60.
4. SATTLER, R. & M. TYMIANSKY. 2000. Molecular mechanisms of calcium-dependent excitotoxicity. J. Mol. Med. **78:** 3–13.
5. KIP, S.N. et al. 2006. Changes in the expression of plasma membrane calcium extrusion systems during the maturation of hippocampal neurons. Hippocampus **16:** 20–34.
6. DEMARCO, S.J. & E.E. STREHLER. 2001. Plasma membrane Ca^{2+}-ATPase isoforms 2b and 4b interact promiscuously and selectively with members of the membrane-associated guanylate kinase family of PDZ (PSD-95/Dlg/ZO-1) domain-containing proteins. J. Biol. Chem. **276:** 21594–21600.
7. YU, L. & R.A. COLVIN. 1997. Regional differences in expression of transcripts for Na^+/Ca^{2+} exchanger isoforms in rat brain. Mol. Brain Res. **50:** 285–292.
8. LI, X.-F. & J. LYTTON 2002. Differential expression of Na/Ca exchanger and Na/Ca+K exchanger transcripts in rat brain. Ann. N. Y. Acad. Sci. **976:** 64–66.
9. PAPA, M. et al. 2003. Differential expression of the Na^+-Ca^{2+} exchanger transcripts and proteins in rat brain regions. J. Comp. Neurol. **461:** 31–48.

Cleavage of the Plasma Membrane Ca^{2+}ATPase during Apoptosis

KATALIN PÁSZTY,[a] GÉZA ANTALFFY,[b] LUCA HEGEDÜS,[b]
RITA PADÁNYI,[b] ALAN R. PENHEITER,[c] ADELAIDA G. FILOTEO,[c]
JOHN T. PENNISTON,[d] AND ÁGNES ENYEDI[b]

[a]Membrane Research Group of the Hungarian Academy of Sciences,
Budapest H-1051, Hungary

[b]National Medical Center, Budapest H-1113, Hungary

[c]Department of Biochemistry and Molecular Biology, Mayo Foundation,
Rochester, Minnesota 55905, USA

[d]Neuroscience Center, Massachusetts General Hospital, Boston, Massachusetts
02114, USA

ABSTRACT: Maintenance of Ca^{2+} homeostasis is essential for normal
cellular function and survival. Recent evidences suggest that Ca^{2+} is
also an important player of apoptosis. We demonstrated that the plasma
membrane Ca^{2+} ATPase (PMCA) isoform 4b, a key element of cellular Ca^{2+} homeostasis, was cleaved by caspase-3 during the course of
apoptosis. This cleavage of PMCA removed the entire regulatory region
from the C terminus, leaving behind a 120-kDa catalytic fragment. Since
loss of PMCA activity could lead to intracellular Ca^{2+} overload and
consequently necrotic cell death, an important question is whether the
apoptotic fragment of PMCA retains full activity or it is inactivated. To
address this question, we constructed a C-terminally truncated mutant
that corresponded to the caspase-3 fragment of PMCA4b and showed
that it was fully and constitutively active. This mutant was targeted properly to the plasma membrane when it was expressed stably or transiently
in several different cell lines. We followed truncation of PMCA during
apoptosis induced by mitochondrial or receptor-mediated pathways and
found that a similar fragment of 120 kDa was formed and remained intact for several hours after treatment. We have also demonstrated that
the caspase-3 cleavage site is an important structural element of PMCA
and found that the accessibility of the caspase-3 site depended strongly
on the conformational state of the protein.

KEYWORDS: plasma membrane Ca^{2+} ATPase; calmodulin; apoptosis;
caspase-3; activation; localization; truncated mutant; structure function

Address for correspondence: Ágnes Enyedi, National Medical Center, Diószegi ut 64, Budapest,
Hungary H-1113. Voice/fax: 36-1-372-4353.
enyedi@biomembrane.hu

Ann. N.Y. Acad. Sci. 1099: 440–450 (2007). © 2007 New York Academy of Sciences.
doi: 10.1196/annals.1387.003

INTRODUCTION

Programmed cell death or apoptosis can be triggered by many different stimuli, including engagement of death receptors by cytokines (such as tumor necrosis factor [TNF] and Fas ligand),[1] detachment of endothelial cells from the extracellular matrix,[2,3] growth factor insufficiencies, toxins, oxidative stress,[4] and Ca^{2+} influx through the plasma membrane or release from the endoplasmic reticulum.[5] Most of these pathways end up with the activation of the major downstream effector caspase, caspase-3.[6] Calcium has been identified as a messenger that coordinates mitochondrial–endoplasmic reticulum interactions that drive apoptosis[7]; however, the precise role of Ca^{2+} in apoptosis remains controversial. Excessive Ca^{2+} overload can lead to apoptotic or rather necrotic cell death, whereas moderate elevation of Ca^{2+} can be either pro- or antiapoptotic depending on cell type and/or stimuli.[8-11]

The plasma membrane Ca^{2+} ATPases (PMCAs) are responsible for removing Ca^{2+} from the cell interior to the extracellular space, thus maintaining intracellular Ca^{2+} homeostasis.[12,13] Lack of proper function of PMCAs may lead to severe Ca^{2+} overload that may induce necrotic cell death as well as pathological conditions relevant to multiple sclerosis,[14] hypertension,[15,16] kidney disease, infertility,[17] and hearing loss.[18] The activity of PMCA is regulated in a complex way for which its carboxyl terminus is mostly responsible. This region has a high-affinity calmodulin-binding and auto-inhibitory sequence.[19] In addition, the C-terminal tail of the b splice forms of PMCAs interacts specifically with proteins that may modulate involvement of PMCAs in specific signaling microdomains.[20-26] Four genes encode mammalian PMCAs (PMCA1–4) and with alternative splicing that occurs at two separate sites over 20 different PMCA isoforms are generated.[27] PMCA1 and 4 are almost ubiquitously expressed, while PMCA2 and 3 are more specialized forms expressed in excitable tissues, such as skeletal muscle and brain.

Recently several experiments have suggested that PMCAs influence the course of apoptosis. Antisense inhibition of PMCA has been reported to induce apoptosis in vascular smooth muscle cells,[28] while another group showed that PMCA4 was required for TNF-induced apoptosis in L929 murine fibrosarcoma cells.[29] In granulosa cells, basic fibroblast growth factor prevents apoptosis most probably by a mechanism that increases Ca^{2+} efflux through promoting plasma membrane localization of the PMCA.[30] In addition, a novel functional interaction between PMCA and the tumor suppressor protein RASSF1 has been described that might link PMCA to apoptosis.[31]

Another recent finding is that PMCA4b is cleaved by caspase-3 during staurosporine (STS)-induced apoptosis.[32,33] In CHO cells Schwab *et al.*[33] showed that the expression of noncleavable mutants of PMCA4 delayed apoptosis and reduced the extent of secondary necrosis. When the cleavable wild-type PMCA4 was expressed instead, they suggested that cleavage by caspase-3 inhibited its activity and promoted internalization of PMCA4 stably expressed in CHO cells. In contrast, our group showed that in transiently transfected COS-7 cells this cleavage activated the pump.[32] Because the role of PMCA in apoptosis remains controversial, we decided to study further the cleavage of PMCA4b during the course of apoptosis induced by the intrinsic and extrinsic pathways. We also studied the localization and biochemical characteristics of the apoptotic fragment by making a truncated mutant of PMCA4.[34]

METHODS

Expression of Recombinant PMCAs

PMCA4b and PMCA4b-ct125 expressing MDCKII,[34] COS-7, and HEK293 cells were grown in Dulbecco's modified Eagle's medium supplemented with 10% fetal bovine serum, 100 units/mL penicillin, 100 μg/mL streptomycin, and 2 mM L-glutamine. All cells were kept at 37°C, 5% CO_2 in a humidified atmosphere. COS-7 and HEK293 cells were cultured to 80% confluency and transfected with pMM2 plasmid carrying the DNA for PMCA4b and PMCA4b-ct125 mutant[34] using FuGene 6 Transfection Reagent (Roche, Mannheim, Germany) according to the manufacturer's protocol.

Apoptosis Induction

COS-7 cells were grown on 6-well plates and transfected as above. Forty hours after transfection, apoptosis was induced by adding 1 μM STS in fresh tissue culture medium or by incubation on poly(2-hydroxyethyl methacrylate) (poly-HEMA)-treated plate (suspension-induced apoptosis or anoikis). Optionally, cells were pretreated with 50 μM W-7 calmodulin antagonist for an hour. MDCKII cell clones expressing the appropriate PMCA constructs were treated with either 2 μM STS or 10 ng/mL TNF-α and 1 μg/mL actinomycin D [ActD] (the addition of ActD facilitates cell death).[35] In all cases, cells were precipitated by 6% ice-cold trichloroacetic acid after different times of incubation, then resuspended in electrophoresis sample buffer containing 62.5 mM Tris–HCl, pH 6.8, 2% SDS, 10% glycerol, 5 mM EDTA, 125 mg/mL urea, and 100 mM dithiothreitol. Samples were electrophoresed on 7.5% polyacrylamide gel following Laemmli's procedure,[36] then subsequently electroblotted. Blots were immunostained by monoclonal anti-PMCA antibody 5F10 or polyclonal anti–poly(ADP-ribose) polymerase (PARP) antibody, as indicated.

Immunocytochemistry

COS-7 and HEK293 cells were plated on 8-well Nunc Lab-Tek Chambered Coverglass (Nalge Nunc International, Rochester, NY) previously coated with 0.03 mg/mL Vitrogen (Cohesion Technology, Palo Alto, CA). After 24 h (HEK293) and 48 h (COS-7) of transfection, cells were rinsed with Dulbecco's modified PBS (DPBS), fixed for 15 min at 37°C in 4% paraformaldehyde in DPBS. After three washes with DPBS, cells were permeabilized in prechilled methanol for 5 min at −20°C and blocked for 1 h at room temperature in blocking buffer (DPBS containing 2 mg/mL bovine serum albumin, 1% fish gelatin, 0.1% Triton-X 100, and 5% goat serum). Samples were then incubated for 1 h at room temperature with mouse monoclonal anti-PMCA antibody (5F10). After three washes in DPBS, cells were incubated for 1 h at room temperature with Alexa-488 conjugated goat anti-mouse IgG (H + L). After four further washes in DPBS, cells were incubated for 10 min with 1 μM 4′, 6-Diamidino-2-phenylindole dihydrochloride (DAPI). Samples were studied under an Olympus FV500-IX81 confocal laser-scanning microscope using an Olympus PLAPO 60 × (1.4) oil-immersion objective (Olympus Europa GmbH, Hamburg, Germany).

RESULTS AND DISCUSSION

PMCA4b Is Cleaved by Caspase-3 during Apoptosis

Previously we demonstrated that human PMCA4b is cleaved by caspase-3 during STS-induced apoptosis of COS-7 cells transiently expressing PMCA4b.[32] This cleavage produced a 120-kDa fragment of the pump. Alteration of residues of a caspase-3 consensus sequence (^{1077}DEID^{1080}) at the C terminus produced mutants that were resistant to caspase-3–mediated proteolysis. These experiments indicated that the cleavage occurred immediately after Asp^{1080} located five residues upstream of the calmodulin-binding motif. Thus, cleavage by caspase-3 removed the whole C-terminal calmodulin-binding inhibitory region but left the rest of the PMCA4b molecule intact. Independently, the formation of the 120-kDa fragment during STS-induced apoptosis of CHO cells stably expressing PMCA4b has been reported and the caspase-3 cleavage site has also been identified.[33] In contrast to the experiments on CHO cells, however, our functional assays using PMCA4b expressing COS-7 cells indicated that the fragment retained full activity and has an increased basal activity. Our data were in good correlation with previous findings demonstrating that removal of the regulatory region by site-directed mutagenesis[37] or by different proteases[38,39] activates the pump.

The 120-kDa Apoptotic Fragment of PMCA4b Is Constitutively Active and Targeted to the Plasma Membrane

To further establish that the 120-kDa apoptotic fragment retains full activity we constructed a truncated PMCA4 called PMCA4b-ct125 that lacks all residues (125 aa.) downstream of the caspase-3 cleavage site, and thus corresponds to the apoptotic fragment.[34] Then we generated MDCKII cell lines stably expressing wild-type and truncated PMCA4b. Both Ca^{2+} transport and EP formation activity measurements showed clearly that the truncated pump was fully and constitutively active.

Another important aspect of functionality of the truncated pump is its proper targeting. First, we studied the localization of the PMCA4b-ct125 mutant in fully polarized MDCKII cells by laser-scanning confocal microscopy. We found that both wild-type and truncated PMCA4b were localized to the basolateral membrane of MDCK cells stably expressing the PMCA constructs.[34] To study further the targeting properties of the apoptotic fragment, we transiently transfected HEK293 and COS-7 cells with wild-type and truncated pumps. FIGURE 1 shows clear plasma membrane localization of the apoptotic fragment in both cell lines confirming our previous findings that the carboxyl terminus is not required for proper targeting of PMCA4b. These data show that PMCA4b-ct125 is targeted to the plasma membrane regardless of cell type.

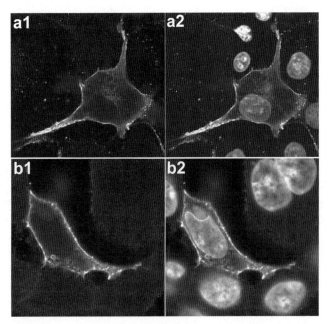

FIGURE 1. Cellular localization of PMCA4b-ct125 mutant. COS-7 cells (**a1**) and HEK293 cells (**b1**) were transfected with PMCA4b-ct125 mutant. Cells were fixed with 4% paraformaldehyde 48 h (COS-7 cells) and 24 h (HEK293 cells) after transfection and labeled with 5F10 anti-PMCA antibody and fluorescently labeled secondary antibody. (**a2 and b2**) Same cells were counterstained with DAPI to visualize the nuclei of the transfected and nontransfected cells.

Fragmentation of PMCA4b during Apoptosis Induced by Intrinsic or Extrinsic Pathways

Apoptosis can be initiated in different ways, including the mitochondrial (intrinsic) pathway[40] and the pathway through death receptors (extrinsic).[41] Both pathways finally lead to the activation of the executor protease, caspase-3. To test the degradation of PMCA4b in both kinds of apoptotic pathways we used (*a*) the general protein kinase inhibitor STS to induce apoptosis through a mitochondria-mediated pathway and (*b*) TNF-α (a multifunctional cytokine) in combination with ActD to kill cells through death receptors. ActD is generally used to promote apoptotic cell death by TNF-α.[35] FIGURE 2 A shows the treatment of MDCKII cells expressing PMCA4b with STS. We used the apoptotic degradation of a widely accepted caspase-3 substrate, poly(ADP-ribose) polymerase (PARP),[42] as an indicator of the course of apoptosis. Caspase-3 cleaves the 116-kDa PARP protein into 85- and 25-kDa fragments. As expected,

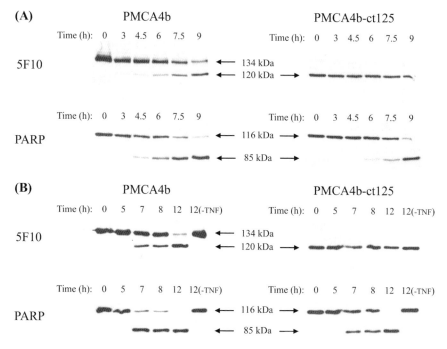

FIGURE 2. Cleavage of PMCA4b during STS and TNF-α-induced apoptosis of MDCKII cells. Apoptosis of PMCA4b or PMCA4b-ct125 expressing MDCKII cells was induced by 2 μM STS (**A**) or 10 ng/mL TNF-α along with 1 μg/mL ActD (**B**). Samples were taken at the indicated times. Immunoblots of 20 μg cell lysates were stained by either anti-PMCA antibody 5F10 or anti-PARP antibody that specifically recognizes both the full length and the 85-kDa fragment of the PARP protein, as indicated.

PARP degradation ensued along with the apoptotic death of the cells. Concurrently the 120-kDa apoptotic fragment of PMCA4b was formed. Similarly, when cell death was initiated by TNF-α/ActD through death receptors, the pump was cleaved and the same fragment was produced (FIG. 2 B). These experiments demonstrate that the specific cleavage of PMCA4b occurs when caspase-3 is activated irrespective of the upstream apoptotic pathways. It is also clear that the 120-kDa fragment remains intact for a relatively long time throughout the apoptotic process. This was confirmed when cells expressing the PMCA4b-ct125 mutant were treated with STS or TNF-α/ActD (FIG. 2 A and B, respectively). No further degradation of the mutant was detected even after a prolonged time of these treatments. The typical degradation of PARP protein indicates that the stability of PMCA4b-ct125 was not due to the lack of activation of caspase-3. However, there was a slight difference in the kinetics of PARP degradation between the wild-type and mutant PMCA expressing cells. The formation of the 85-kDa PARP fragment was slower in cells expressing the PMCA4b-ct125, which might be a result of the constitutively active calcium-pumping mechanism that in fact might decelerate the apoptotic process.

The Cleavage by Caspase-3 Depends on the Conformation of the Pump

We studied the effect of calmodulin binding on the fragmentation of PMCA4b by recombinant caspase-3. When we tested the digestion of PMCA4b expressing COS-7 cell membrane, the pump was fully converted to the 120-kDa fragment only in the presence of Ca^{2+}-calmodulin.[43] In the absence of calmodulin, the digestion was slow and incomplete even after prolonged incubation with the protease. In accordance with these findings here, we show that a known calmodulin antagonist, W-7, greatly reduced PMCA4b fragmentation during STS- (FIG. 3 A) or anoikis-induced apoptosis (FIG. 3 B) of COS-7 cells, while it has only a moderate effect on PARP fragmentation, a hallmark of caspase-3 activation. These results indicate that in the auto-inhibited state Asp^{1080} is shielded by intramolecular interactions between the C terminus and the catalytic core and suggest that at low intracellular Ca^{2+} (when calmodulin is not bound), PMCA4b is protected from caspase-3–like proteases. Binding of Ca^{2+}-calmodulin, on the other hand, will promote PMCA cleavage.

The Caspase-3 Cleavage Site of PMCA4b Is Also Involved in the Autoinhibition

The caspase-3 cleavage site (Asp^{1080}) of PMCA4b is located five residues upstream of the calmodulin-binding motif that is part of a highly acidic region connecting the regulatory C terminus with transmembrane domain 10. We

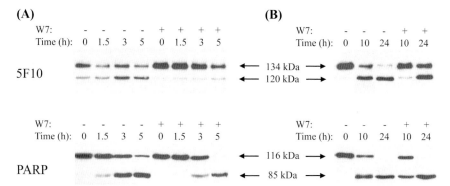

FIGURE 3. A calmodulin antagonist inhibits the cleavage of PMCA4b during apoptosis. PMCA4b expressing COS-7 cells were either treated with 1 μM STS (**A**) or incubated on poly-HEMA–coated plates (**B**) for the indicated times. Optionally, cells were preincubated with 50 μM W-7 calmodulin antagonist for 1 h prior to treatment. Samples were taken at the indicated times. Immunoblots of 2 μg cell lysates for PMCA and 25 μg for PARP protein were stained by anti-PMCA antibody 5F10 or anti-PARP antibody.

demonstrated that mutation of Asp^{1080} in this region not only protected the pump from caspase-3 cleavage but also greatly enhanced the basal activity so that the mutant was nearly fully active without calmodulin.[44] Our data suggested that Asp^{1080} is an essential residue in the connecting region that may assist in orienting the regulatory carboxyl terminus to the catalytic sites of PMCA to form a stable inhibited conformation. This is also consistent with the finding that Asp^{1080} is not exposed in the inhibited state as discussed earlier.

SUMMARY

In summary, our data show that: (*a*) a 120-kDa apoptotic fragment of PMCA4b is formed during apoptosis induced by either receptor mediated or mitochondrial pathways; (*b*) the fragment was relatively stable since no further degradation was detected during the apoptotic process; (*c*) the formation of the fragment was dependent on the binding of calmodulin; (*d*) the fragment was constitutively active; (*e*) it was targeted to the plasma membrane when expressed in three different cell lines; and (*f*) mutation of the caspase-3 cleavage site to an alanine results in a nearly fully active pump. Our data indicate that activation (i.e., loss of intramolecular inhibition) rather than localization is the important apoptotic change in PMCA properties.

The formation of a fully activated PMCA fragment at the executioner phase of the apoptotic program might be important in protecting apoptotic cells from excessive Ca^{2+} overload. Such an overload might have undesirable effects, such as impaired mitochondrial function and a loss of cellular ATP, leading to secondary necrosis.

Recently, internalization of PMCA4b and 2b and a corresponding decrease in Ca^{2+} clearance in hippocampal neurons treated with an excitotoxic concentration of glutamate has been reported.[45] In this scenario, however, calpain rather than caspase-3–mediated cleavage of PMCA seems responsible for the internalization. Internalization-mediated loss of function of PMCA might indeed represent a cross-talk between apoptosis and necrosis (as suggested by Schwab et al.[33]) in some specific cell death systems such as neuronal excitotoxicity. More detailed studies, however, focusing on the PMCA are needed to understand its fate in different cell death models.

ACKNOWLEDGMENTS

This work was supported in part by OTKA T049476, NKFP 060-1A-2004, and by National Institutes of Health GM28835 (to J.T.P.) grants as well as by OTKA Postdoctoral Fellowship Dφ48496 (to R.P.) and Bolyai Janos Research Fellowship BO/00585/05 (to K.P.).

REFERENCES

1. AGGARWAL, B.B. 2003. Signalling pathways of the TNF superfamily: a double-edged sword. Nat. Rev. Immunol. **3:** 745–756.
2. FRISCH, S.M. & R.A. SCREATON. 2001. Anoikis mechanisms. Curr. Opin. Cell. Biol. **13:** 555–562.
3. GROSSMANN, J. 2002. Molecular mechanisms of "detachment-induced apoptosis–Anoikis." Apoptosis **7:** 247–260.
4. HOIDAL, J.R. 2001. Reactive oxygen species and cell signaling. Am. J. Respir. Cell. Mol. Biol. **25:** 661–663.
5. WERTZ, I.E. & V.M. DIXIT. 2000. Characterization of calcium release-activated apoptosis of LNCaP prostate cancer cells. J. Biol. Chem. **275:** 11470–11477.
6. GRUTTER, M.G. 2000. Caspases: key players in programmed cell death. Curr. Opin. Struct. Biol. **10:** 649–655.
7. HAJNOCZKY, G., E. DAVIES & M. MADESH. 2003. Calcium signaling and apoptosis. Biochem. Biophys. Res. Commun. **304:** 445–454.
8. MATTSON, M.P. & S.L. CHAN. 2003. Calcium orchestrates apoptosis. Nat. Cell. Biol. **5:** 1041–1043.
9. ORRENIUS, S., B. ZHIVOTOVSKY & P. NICOTERA. 2003. Regulation of cell death: the calcium-apoptosis link. Nat. Rev. Mol. Cell Biol. **4:** 552–565.
10. DEMAUREX, N. & C. DISTELHORST. 2003. Cell biology. Apoptosis–the calcium connection. Science **300:** 65–67.
11. APATI, A. et al. 2003. Calcium induces cell survival and proliferation through the activation of the MAPK pathway in a human hormone-dependent leukemia cell line, TF-1. J. Biol. Chem. **278:** 9235–9243.

12. CARAFOLI, E., E. GARCIA-MARTIN & D. GUERINI. 1996. The plasma membrane calcium pump: recent developments and future perspectives. Experientia **52:** 1091–1100.

13. STREHLER, E.E. & M. TREIMAN. 2004. Calcium pumps of plasma membrane and cell interior. Curr. Mol. Med. **4:** 323–335.

14. KURNELLAS, M.P. *et al.* 2005. Plasma membrane calcium ATPase deficiency causes neuronal pathology in the spinal cord: a potential mechanism for neurodegeneration in multiple sclerosis and spinal cord injury. Faseb. J. **19:** 298–300.

15. KAMIJO, T. *et al.* 1996. Renal abnormality of calcium handling in spontaneously hypertensive rats. Kidney Int. Suppl. **55:** S166–S168.

16. LEHOTSKY, J. *et al.* 2002. The role of plasma membrane Ca^{2+} pumps (PMCAs) in pathologies of mammalian cells. Front Biosci. **7:** d53–d84.

17. SCHUH, K. *et al.* 2004. Plasma membrane Ca^{2+} ATPase 4 is required for sperm motility and male fertility. J. Biol. Chem. **279:** 28220–28226.

18. PRASAD, V. *et al.* 2004. Phenotypes of SERCA and PMCA knockout mice. Biochem. Biophys. Res. Commun. **322:** 1192–1203.

19. PENNISTON, J.T. & A. ENYEDI. 1998. Modulation of the plasma membrane Ca^{2+} pump. J. Membr. Biol. **165:** 101–109.

20. KIM, E. *et al.* 1998. Plasma membrane Ca^{2+} ATPase isoform 4b binds to membrane-associated guanylate kinase (MAGUK) proteins via their PDZ (PSD-95/Dlg/ZO-1) domains. J. Biol. Chem. **273:** 1591–1595.

21. DEMARCO, S.J., M.C. CHICKA & E.E. STREHLER. 2002. Plasma membrane Ca^{2+} ATPase isoform 2b interacts preferentially with Na^+/H^+ exchanger regulatory factor 2 in apical plasma membranes. J. Biol. Chem. **277:** 10506–10511.

22. DEMARCO, S.J. & E.E. STREHLER. 2001. Plasma membrane Ca^{2+}-ATPase isoforms 2b and 4b interact promiscuously and selectively with members of the membrane-associated guanylate kinase family of PDZ (PSD95/Dlg/ZO-1) domain-containing proteins. J. Biol. Chem. **276:** 21594–21600.

23. SCHUH, K. *et al.* 2001. The plasma membrane calmodulin-dependent calcium pump: a major regulator of nitric oxide synthase I. J. Cell Biol. **155:** 201–205.

24. SCHUH, K. *et al.* 2003. Interaction of the plasma membrane Ca^{2+} pump 4b/CI with the Ca^{2+}/calmodulin-dependent membrane-associated kinase CASK. J. Biol. Chem. **278:** 9778–9783.

25. GOELLNER, G.M., S.J. DEMARCO & E.E. STREHLER. 2003. Characterization of PISP, a novel single-PDZ protein that binds to all plasma membrane Ca^{2+}-ATPase b-splice variants. Ann. N. Y. Acad. Sci. **986:** 461–471.

26. SGAMBATO-FAURE, V. *et al.* 2006. The Homer-1 protein Ania-3 interacts with the plasma membrane calcium pump. Biochem. Biophys. Res. Commun. **343:** 630–637.

27. STREHLER, E.E. & D.A. ZACHARIAS. 2001. Role of alternative splicing in generating isoform diversity among plasma membrane calcium pumps. Physiol. Rev. **81:** 21–50.

28. SASAMURA, S. *et al.* 2002. Antisense-inhibition of plasma membrane Ca^{2+} pump induces apoptosis in vascular smooth muscle cells. Jpn. J. Pharmacol. **90:** 164–172.

29. ONO, K., X. WANG & J. HAN. 2001. Resistance to tumor necrosis factor-induced cell death mediated by PMCA4 deficiency. Mol. Cell Biol. **21:** 8276–8288.

30. PELUSO, J.J. 2003. Basic fibroblast growth factor (bFGF) regulation of the plasma membrane calcium ATPase (PMCA) as part of an anti-apoptotic mechanism of action. Biochem. Pharmacol. **66:** 1363–1369.

31. ARMESILLA, A.L. *et al.* 2004. Novel functional interaction between the plasma membrane Ca^{2+} pump 4b and the proapoptotic tumor suppressor Ras-associated factor 1 (RASSF1). J. Biol. Chem. **279:** 31318–31328.
32. PASZTY, K. *et al.* 2002. Plasma membrane Ca^{2+} ATPase isoform 4b is cleaved and activated by caspase-3 during the early phase of apoptosis. J. Biol. Chem. **277:** 6822–6829.
33. SCHWAB, B.L. *et al.* 2002. Cleavage of plasma membrane calcium pumps by caspases: a link between apoptosis and necrosis. Cell Death Differ. **9:** 818–831.
34. PASZTY, K. *et al.* 2005. The caspase-3 cleavage product of the plasma membrane Ca^{2+}-ATPase 4b is activated and appropriately targeted. Biochem. J. **391:** 687–692.
35. JONES, B.E. *et al.* 2000. Hepatocytes sensitized to tumor necrosis factor-alpha cytotoxicity undergo apoptosis through caspase-dependent and caspase-independent pathways. J. Biol. Chem. **275:** 705–712.
36. LAEMMLI, U.K. 1970. Cleavage of structural proteins during the assembly of the head of bacteriophage T4. Nature **227:** 680–685.
37. ENYEDI, A. *et al.* 1993. A highly active 120-kDa truncated mutant of the plasma membrane Ca^{2+} pump. J. Biol. Chem. **268:** 10621–10626.
38. SARKADI, B. *et al.* 1986. Molecular characterization of the in situ red cell membrane calcium pump by limited proteolysis. J. Biol. Chem. **261:** 9552–9557.
39. JAMES, P. *et al.* 1989. Modulation of erythrocyte Ca^{2+}-ATPase by selective calpain cleavage of the calmodulin-binding domain. J. Biol. Chem. **264:** 8289–8296.
40. GREEN, D.R. & G. KROEMER. 2004. The pathophysiology of mitochondrial cell death. Science **305:** 626–629.
41. THORBURN, A. 2004. Death receptor-induced cell killing. Cell Signal **16:** 139–144.
42. BOULARES, A.H. *et al.* 1999. Role of poly (ADP-ribose) polymerase (PARP) cleavage in apoptosis. Caspase 3-resistant PARP mutant increases rates of apoptosis in transfected cells. J. Biol. Chem. **274:** 22932–22940.
43. PADANYI, R. *et al.* 2003. Intramolecular interactions of the regulatory region with the catalytic core in the plasma membrane calcium pump. J. Biol. Chem. **278:** 35798–35804.
44. PASZTY, K. *et al.* 2002. Asp1080 upstream of the calmodulin-binding domain is critical for autoinhibition of hPMCA4b. J. Biol. Chem. **277:** 36146–36151.
45. POTTORF, W.J., 2nd. *et al.* 2006. Glutamate-induced protease-mediated loss of plasma membrane Ca^{2+} pump activity in rat hippocampal neurons. J. Neurochem. **98:** 1646–1656.

The Plasma Membrane Na$^+$/Ca^{2+} Exchanger Is Cleaved by Distinct Protease Families in Neuronal Cell Death

D. BANO, E. MUNARRIZ, H. L. CHEN, E. ZIVIANI, G. LIPPI, K. W. YOUNG, AND P. NICOTERA

MRC Toxicology Unit, University of Leicester, Hodgkin Building, Lancaster Road LE1 9HN, Leicester, United Kingdom

ABSTRACT: Neurodegenerative conditions commonly involve loss of neuronal connectivity, synaptic dysfunction with excessive pruning, and ionic imbalances. These often serve as a prelude to cell death either through the activation of apoptotic or necrotic death routines or excess autophagy. In many instances, a local or generalized Ca^{2+} deregulation is involved in signaling or executing cell death. We have recently shown that in brain ischemia, and during excitotoxicity triggered by excess glutamate, the irreversible Ca^{2+} deregulation leading to necrosis is due to calpain-mediated modulation of the plasma membrane Na$^+$/Ca^{2+} exchanger (NCX). Here we show that the NCX can also be cleaved by caspases in neurons undergoing apoptosis, which suggests that cleavage of the main Ca^{2+} extrusion pathway is a lethal event in multiple forms of cell death.

KEYWORDS: brain ischemia; excitotoxicity; NCX; calcium; calpains; caspase; cell death

INTRODUCTION

Neuronal degenerative diseases are complex disorders characterized by loss of connectivity and varying degrees of neuronal cell death.[1] The latter may be the result of apoptosis, necrosis, or excess autophagy depending on the type, and intensity, of insult and the local metabolic conditions. Brain ischemia is a leading cause of death in the Western world. During brain infarction, neuronal loss can occur by multiple mechanisms including oxidative and nitrosative stress, cytokine- and chemokine-mediated cell killing, metabolic cell death, and death caused by an uncontrolled Ca^{2+} overload.[2,3] The latter is brought about via excessive gating of ion channels by excitatory neurotransmitters

Address for correspondence: Dr. Daniele Bano, MRC Toxicology Unit, University of Leicester, Hodgkin Building, Lancaster Road LE1 9HN, Leicester, UK. Voice: +44-0116-2525571; fax: 0044 116 252 5616.

db81@le.ac.uk

Ann. N.Y. Acad. Sci. 1099: 451–455 (2007). © 2007 New York Academy of Sciences.
doi: 10.1196/annals.1387.006

accumulated during the ischemic period in a process known as excitotoxicity.[4] The unregulated opening of Ca^{2+}-permeable plasma membrane channels can trigger both caspase-dependent or -independent cell death, depending on the extent and the duration of the insult.[5] The excess entry of Ca^{2+} is not restricted to the ionotropic glutamate receptors, but involves voltage-operated channels and cation permeable channels, such as TRPM7 and ASICs.[6,7] The excitotoxic process in mammalian neurons in part resembles a cell death paradigm in *Caenorhabditis elegans* (*C. elegans*), where gating of the degenerin channels mediates Ca^{2+} overload and necrotic cell death following mechanic injury.[8] In the *C. elegans* death paradigm, Ca^{2+}-dependent proteases known as calpains mediate cell demise.[9] An analogous process is present in animal models as an overexpression of the calpain inhibitory protein, calpastatin ameliorates neuronal damage triggered by the excitotoxic injection of kainic acid in the hippocampus.[10]

We have previously demonstrated that one of the two major plasma membrane Ca^{2+} extruding systems, the plasma membrane Ca^{2+} ATPase, is cleaved[11] by caspases in apoptosis. In cells undergoing apoptosis, caspases can cleave the Ca^{2+} ATPase pumps at the C-terminal and inhibit Ca^{2+} extrusion. This enhances the cytosolic Ca^{2+} concentrations, which in turn amplifies the death signal and promotes membrane lysis. More recently, we have also shown in a collaborative work that calpains mediate the degradation of the neuronal Na^+/Ca^{2+} exchanger (NCX) during excitotoxic injury.[12] This triggers the irreversible Ca^{2+} elevation that ultimately causes rapid neuronal necrosis. Here, we report that cleavage of NCX can also be operated by caspases during apoptosis, linking therefore the inhibition of Ca^{2+} efflux with distinct death paradigms.

RESULTS AND DISCUSSION

To test whether NCX was modified during caspase-dependent cell death, primary dissociated cerebellar granule neurons (CGN) were exposed to colchicine or staurosporine (STS), or deprived of K^+ by switching the medium from 25 mM KCl to 5 mM KCl. As a control for caspase-independent cell death, we used our established glutamate model.[12] As shown in FIGURE 1 A, all stimuli caused NCX3 degradation. Notably, three fragments were generated during apoptosis: two at about 60 kDa, which were generated by calpains and similar to those found in excitotoxic settings and an additional fragment at 66 kDa, which was blocked by the polycaspases inhibitor Z-VAD-fmk. Neither cathepsins nor proteasome inhibitors prevented the formation of these degradation products (data not shown). Furthermore, in a human neuroblastoma cell line treated with STS to induce the activation of caspase-3 by the intrinsic apoptotic pathway, the 66-kDa fragment was the predominant fragmentation product. Again, formation of this fragment was blocked by Z-VAD-fmk, by Ac-DEVD-cho, and the

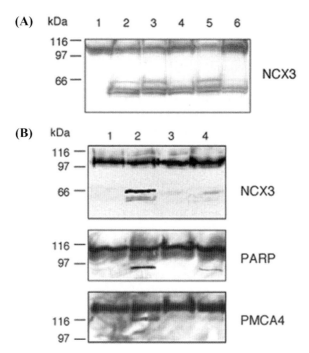

FIGURE 1. (**A**) CGN were incubated for 18 h in serum-free medium containing 5 mM KCl (2), 1 μM STS in the presence (4) or absence (3) of Z-VAD-fmk, 1μM Colchicine (5), or 60 μM glutamate for 3.5 h (6). Lane 1 = control. In caspase-3-dependent processes, Western blot with an antibody specific for NCX3 (kindly provided by Prof. K.D. Philipson) showed that the transporter was processed in three fragments, two at about 60 kDa not blocked by Z-VAD-fmk and one at 66 kDa, which was sensitive to the caspase inhibitor. (**B**) Human neuroblastoma SH-SY5Y cells were induced to apoptosis by 1 μM STS for 5 h. In order to block caspase activity, cells were pretreated for 30 min with 100 μM Z-VAD-fmk, 25 μM M791 (kindly provided by Dr. D.W. Nicholson) or alternatively 100 μM Ac-DEVD-cho. Cells were harvested and plasma membrane enriched samples loaded on 7.5% SDS-PAGE. Proteins were transferred onto nitrocellulose membrane and developed with antibody against NCX3. Neurons undergoing apoptosis by STS (2) showed a NCX3 fragment at 66 kDa, which was blocked by Z-VAD-fmk (3) and M791 (4). The same filter was also probed with antibody against PARP and PMCA4 (kindly provided by Prof. E. Carafoli) as a positive control for caspase-3 activation.

more selective inhibitor M791 (FIG. 1 B). NCX3 was cleaved by caspases to a similar extent as two known caspase substrates, poly-ADP-ribose polymerase (PARP) and the plasma membrane Ca^{2+} ATPase isoform 4 (PMCA). In line with our previous observations it appears that degradation of the Ca^{2+} extrusion pathways is a process required during central nervous system (CNS) injury regardless of the predominant death routine.

Cleavage of NCX by calpains is essential to achieve the lethal Ca^{2+} overload in excitotoxic settings. It is tempting to speculate that in neuronal apoptosis, removal of the major Ca^{2+} extrusion pathway by caspases can sensitize dying neurons to normal Ca^{2+} signals and help in promoting neuronal disassembly. This is supported by our observation that NCX3 downregulation by siRNA transforms nonlethal glutamate Ca^{2+} signals into an irreversible Ca^{2+} overload.[12] Interestingly, Wang and colleagues have recently provided genetic evidence that in sensory neurons loss of NCX activity is responsible for Ca^{2+} overload and retinal degeneration triggered by defects in TRP channels.[13] Studies on NCX3 knockout animals have also shown a significant skeletal muscle fiber necrosis due to the impairment in Ca^{2+} clearance and consequent calcium overload.[14] Similarly, it has been recently reported that prostaglandin E2 EP1 receptors contribute to neuronal degeneration by impairing the activity of the NCX.[15]

In conclusion, our findings illustrate the critical role of NCX in excitable cells and provide evidence that Ca^{2+} deregulation in pathological condition is enhanced by the cleavage of the plasma membrane extrusion systems. Pharmacological approaches able to prevent and reduce the modulation of the NCX and PMCA by proteases can be a powerful tool to delay neuronal loss during CNS injury.

REFERENCES

1. LIPTON, P. 1999. Ischemic cell death in brain neurons. Physiol. Rev. **79:** 1431–1568.
2. ORRENIUS, S., B. ZHIVOTOVSKY & P. NICOTERA. 2003. Regulation of cell death: the calcium-apoptosis link. Nat. Rev. Mol. Cell Biol. **4:** 552–565.
3. BERLIOCCHI, L., D. BANO & P. NICOTERA. 2005. Ca2+ signals and death programmes in neurons. Philos. Trans. R. Soc. Lond. B. Biol. Sci. **360:** 2255–2258.
4. CHOI, D.W. & S.M. ROTHMAN. 1990. The role of glutamate neurotoxicity in hypoxic-ischemic neuronal death. Annu. Rev. Neurosci. **13:** 171–182.
5. ANKARCRONA, M. *et al.* 1995. Glutamate-induced neuronal death: a succession of necrosis or apoptosis depending on mitochondrial function. Neuron **14:** 961–973.
6. AARTS, M. *et al.* 2003. A key role for TRPM7 channels in anoxic neuronal death. Cell **115:** 863–877.
7. XIONG, Z.G. *et al.* 2004. Neuroprotection in ischemia: blocking calcium-permeable acid-sensing ion channels. Cell **118:** 687–698.
8. BIANCHI, L. *et al.* 2004. The neurotoxic MEC-4(d) DEG/ENaC sodium channel conducts calcium: implications for necrosis initiation. Nature Neurosci. **7:** 1337–1344.
9. SYNTICHAKI, P. *et al.* 2002. Specific aspartyl and calpain proteases are required for neurodegeneration in C. Elegans. Nature **419:** 939–944.
10. HIGUCHI, M. *et al.* 2005. Distinct mechanistic roles of calpain and caspase activation in neurodegeneration as revealed in mice overexpressing their specific inhibitors. J. Biol. Chem. **280:** 15229–15237.
11. SCHWAB, B.L. *et al.* 2002. Cleavage of plasma membrane calcium pumps by caspases: a link between apoptosis and necrosis. Cell Death Differ. **9:** 818–831.
12. BANO, D. *et al.* 2005. Cleavage of the plasma membrane Na+/Ca2+ exchanger in excitotoxicity. Cell **120:** 275–285.

13. WANG, T. *et al.* 2005. Light activation, adaptation, and cell survival functions of the Na+/Ca2+ exchanger CalX. Neuron **45:** 367–378.
14. SOKOLOW, S. *et al.* 2004. Impaired neuromuscular transmission and skeletal muscle fiber necrosis in mice lacking Na/Ca exchanger 3. J. Clin. Invest. **113:** 265–273.
15. KAWANO, T. *et al.* 2006. Prostaglandin E(2) EP1 receptors: downstream effectors of COX-2 neurotoxicity. Nat. Med. **12:** 225–229.

Role of Na/Ca Exchange and the Plasma Membrane Ca²⁺–ATPase in β Cell Function and Death

ANDRÉ HERCHUELZ, ADAMA KAMAGATE, HELENA XIMENES, AND FRANÇOISE VAN EYLEN

Laboratory of Pharmacology, Brussels University School of Medicine, B-1070, Brussels, Belgium

ABSTRACT: Recent progresses concerning the Na/Ca exchanger (NCX) and the plasma membrane Ca²⁺–ATPase (PMCA) in the pancreatic β cell are reviewed. The rat β cell expresses two splice variants of NCX1 and six splice variants of the 4 PMCA isoforms. At the protein level, the most abundant forms are PMCA2 and PMCA3, providing the first evidence for the presence of these two isoforms in a non-neuronal tissue. Overexpression of NCX1 in an insulinoma cell line altered the initial rise in cytosolic-free Ca²⁺ concentration ($[Ca^{2+}]_i$) induced by membrane depolarization and the return of the $[Ca^{2+}]_i$ to the baseline value on membrane repolarization, indicating that NCX contributes to both Ca²⁺ inflow and outflow in the β cell. In contrast, overexpression of the PMCA markedly reduced the global rise in Ca²⁺ induced by membrane depolarization, indicating that the PMCA has a capacity higher than expected to extrude Ca²⁺. Glucose, the main physiological stimulus of insulin release from the β cell, has opposite effect on NCX and PMCA transcription, expression and activity, inducing an increase in the case of NCX and a decrease in the case of the PMCA. This indicates that when exposed to glucose, the β cell switches from a low-efficiency Ca²⁺ extruding mechanism, the PMCA, to a high-capacity system, the NCX, in order to better face the increase in Ca²⁺ inflow induced by the sugar. To our knowledge, this is the first demonstration of a reciprocal change in PMCA and NCX1 expression and activity in response to a given stimulus in any tissue.

KEYWORDS: sodium–calcium exchange; Na/Ca exchange; plasma membrane; Ca²⁺–ATPase; calcium; Ca²⁺; PMCA

Address for correspondence: André Herchuelz, Laboratoire de Pharmacodynamie et de Thérapeutique, Université Libre de Bruxelles, Faculté de Médecine, Route de Lennik, 808-Bâtiment GE, B-1070 Bruxelles, Belgium. Voice: 32-2-555-62-01; fax: 32-2-555-63-70.
herchu@ulb.ac.be

Ann. N.Y. Acad. Sci. 1099: 456–467 (2007). © 2007 New York Academy of Sciences.
doi: 10.1196/annals.1387.048

INTRODUCTION

Calcium (Ca^{2+}) plays a key role in the process of glucose-induced insulin release from the pancreatic β cell. When the β cell is exposed to the sugar, a complex series of events is initiated that culminate in a rise in cytosolic-free Ca^{2+} concentration ($[Ca^{2+}]_i$), which triggers the exocytosis of insulin.

Like most other cells, the β cell is equipped with a double system responsible for Ca^{2+} extrusion: a plasma membrane Ca^{2+}–ATPase (PMCA) and a Na/Ca exchange transport system.[1,2] The Na/Ca exchanger (NCX) is an electrogenic transporter located at the plasma membrane that couples the exchange of 3 Na^+ for 1 Ca^{2+}. In the cardiac myocyte, Na/Ca exchange is the major mechanism of Ca^{2+} extrusion, restoring basal Ca^{2+} levels between heartbeats. The exchanger may also reverse during the heart cycle and hence allow Ca^{2+} entry during the systole.[3,4] In the rat pancreatic β cell, Na/Ca exchange displays a quite high-capacity and participates in the control of $[Ca^{2+}]_i$ and of insulin release.[5] The PMCA belongs to the P-type family of transport ATPases that form a phosphorylated intermediate during the reaction cycle.[6] Ca^{2+} is not solely important in cell signaling but may also trigger apoptosis or necrosis. In this article we summarize some recent contributions on the role of Na/Ca exchange and the PMCA in β cell function and death.

NCX and PMCA Isoforms Expressed in the β Cell

In a previous review of the subject presented at the Fourth International Conference on Na/Ca exchange in Banff (Canada) 5 years ago, we showed that the β cell expressed two splice variants of the isoform NCX1, namely NCX1.3 et NCX1.7. PCR amplification did not yield any DNA fragment for NCX2, and NCX3 was not looked for.[7,8]

In further work, we studied the expression of the PMCA in three insulin-secreting preparations (a pure β cell preparation, RINm5F cells, and pancreatic islet cells), using reverse-transcribed PCR, RNase protection assay, and Western blotting.[9] The four main isoforms, PMCA1, PMCA2, PMCA3, and PMCA4 were expressed in the three preparations. Six alternative splice mRNA variants, characterized at splice sites A and C were detected in the three preparations (rPMCA1xb, 2yb, 2wb, 3za, 3zc, 4xb), plus one additional variant in pancreatic islet cells (PMCA4za). At the mRNA and protein level, five variants predominated (1xb, 2wb, 3za, 3zc, 4xb), while one additional isoform (4za), predominated at the protein level only. This provided the first evidence for the presence of PMCA2 and PMCA3 isoforms at the protein level in non-neuronal tissue. Hence, the pancreatic β cell is equipped with multiple PMCA isoforms with possible differential regulation, providing a full range of PMCA for $[Ca^{2+}]_i$ regulation.[9]

Contribution of Na/Ca Exchange and PMCA to Ca^{2+} Inflow and Outflow

To measure the contribution of the Na/Ca exchange to Ca^{2+} inflow and outflow, we used two complementary approaches: the antisense oligonucleotide (AS-oligos) strategy to knockout the NCX in pancreatic β cells[10] and the overexpression of the exchanger (splice variant NCX1.7) in an insulin-secreting cell line (BRIN-BD11 cells).[11]

The two methods led to complementary results. Thus, the treatment with AS-oligos induced two major effects on the K^+-induced $[Ca^{2+}]_i$ increase: first, it reduced both the rate of $[Ca^{2+}]_i$ rise and the maximal amplitude of the initial peak by about 30% and 50%, respectively; second, it reduced the rate of $[Ca^{2+}]_i$ decrease on membrane repolarization by about 70%. The latter data indicate that in rat β cells, Na/Ca exchange accounts for about 70% of the Ca^{2+} outflow and that during the upstroke of the action potential, the NCX may reverse and contribute to Ca^{2+} entry (about 25% of the initial peak).[10]

To further evaluate the role played by Na/Ca exchange in Ca^{2+} homeostasis, we examined the effect of NCX1.7 exchange overexpression on the changes in $[Ca^{2+}]_i$ induced by a similar rise in extracellular K^+ concentration. Overexpression could be assessed at the mRNA and protein level, with appropriate targeting to the plasma membrane assessed by immunofluorescence and the increase in Na/Ca exchange activity.[11] FIGURE 1 illustrates the effect of KCl (50 mmol/L) on $[Ca^{2+}]_i$. In control cells, K^+ induced a biphasic increase in $[Ca^{2+}]_i$ consisting in an initial peak followed by a plateau phase. The increase

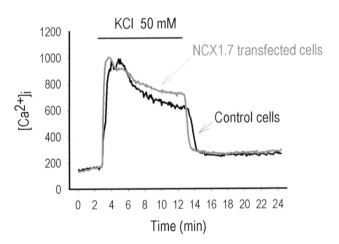

FIGURE 1. Effect of KCl on $[Ca^{2+}]_i$ in control and NCX1.7-overexpressing BRIN-BD11 cells. The *bar* above curves indicates the period of KCl exposure. The curves shown are the mean of 94 and 106 traces, respectively. (Van Eylen *et al.*[11]; Copyright © 2002 American Diabetes Association, from Diabetes, Vol. 51; 366–375. Reprinted with permission from The American Diabetes Association.)

in $[Ca^{2+}]_i$ was rapidly reversible upon removal of K^+ from the solution. In NCX1.7-overexpressing cells, several differences could be observed. First, K^+ induced a more rapid increase in $[Ca^{2+}]_i$. Thus, the maximum $[Ca^{2+}]_i$ was reached after 144 ± 5 and 112 ± 5 s in control and overexpressing cells, respectively ($P < 0.0001$). Similarly, whereas the rate of $[Ca^{2+}]_i$ increase averaged 10.1 ± 0.2 nmol/L/s ($n = 94$) in control cells, it averaged 17.8 ± 0.3 nmol/L/s in NCX1.7-overexpressing cells ($n = 106$, $P < 0.0001$). Second, the $[Ca^{2+}]_i$ observed at the steady state (before K^+ removal) was higher in transfected cells (725 ± 17 nmol/L) than in control cells (610 ± 15 nmol/L, $P < 0.0001$). Last, the decrease in $[Ca^{2+}]_i$ seen on membrane repolarization was more rapid in overexpressing cells (7.0 ± 0.2 nmol/L/s) than in control cells (5.2 ± 0.2 nmol/L/s, $P < 0.0001$) and occurred about 30 s earlier.[11] The latter data confirmed that Na/Ca exchange indeed contributes to both Ca^{2+} inflow and outflow in the β cell. Nevertheless, it is striking to see that with the exception of the parameters cited hereover, the global pattern of the $[Ca^{2+}]_i$ rise was not profoundly modified. This is in contrast with what we observed when we overexpress the PMCA (see below).

In order to evaluate the role played by the PMCA in Ca^{2+} outflow, the isoform PMCA 2wb was overexpressed in the same β cell line. Overexpression could be assessed at the mRNA and protein level, with appropriate targeting to the plasma membrane assessed by immunofluorescence and the increase in PMCA activity.[12] Three clones were isolated, two showing high levels of PMCA activity (clones 4 and 5) and one showing a lower level of overexpression (clone 2). Three clones transfected with the expression vector lacking the PMCA 2wb construct were generated.[12]

To evaluate the role played by PMCA on Ca^{2+} homeostasis, we examined the effect of PMCA2wb overexpression on changes in $[Ca^{2+}]_i$ induced by membrane depolarization. FIGURE 2 illustrates the effect of membrane depolarization induced by K^+ (50 mmol/L) on $[Ca^{2+}]_i$. First, basal $[Ca^{2+}]_i$ did not differ between control and overexpressing cells and averaged 124 ± 6 nmol/L ($n = 350$). Second, whether in control (FIG. 2 A) or overexpressing cells (clone 4, FIG. 2 B; clone 5, FIG. 2 C), K^+ induced a biphasic increase in $[Ca^{2+}]_i$ consisting in an initial peak followed by a plateau phase. The increase in $[Ca^{2+}]_i$ was rapidly reversible upon removal of K^+ from the solution. Cells transfected with the vector only, showed a K^+-induced increase in $[Ca^{2+}]_i$ similar to that of parental nontransfected cells ($P > 0.20$) (FIG. 2 A). Third, in PMCA2wb-overexpressing cells, the magnitude of the increase in $[Ca^{2+}]_i$ was markedly reduced. Thus, a two- and sixfold reduction of the rise in $[Ca^{2+}]_i$ was observed in clones 4 and 5, respectively, compared with vector-only transfected cells. Indeed, the increase in $[Ca^{2+}]_i$, as measured by the area under the curve during the time of K^+ stimulation, averaged 12.3 ± 0.5, 6.5 ± 0.3, and 2.1 ± 0.1 μmol/L/h in vector-only transfected cells, clone 4 cells, and clone 5 cells, respectively ($P < 0.0001$). In view of the inverse correlation between the level of protein expression and the increase in $[Ca^{2+}]_i$, the effect of K^+ on clone 2,

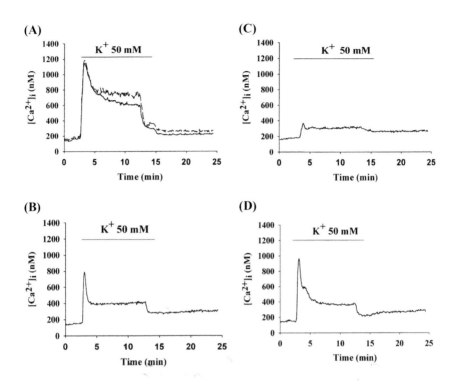

FIGURE 2. Effect of K^+ on $[Ca^{2+}]_i$ in control and PMCA2wb-overexpressing BRIN-BD11 cells. Panel (**A**) shows effect of K^+ (50 mM) on $[Ca^{2+}]_i$ in control cells nontransfected (continuous line) and transfected with the empty vector (*dotted line*). Panels B, C, and D show PMCA2wb-transfected cells clone 4 (**B**), clone 5 (**C**), and clone 2 (**D**). K^+ has been added after 2 min. The curves shown are the mean of more than 50 traces in each case. (Kamagate *et al.*[12]; Copyright © 2002 American Diabetes Association, from Diabetes, Vol. 51, 2002; 2773–2788. Reprinted with permission from The American Diabetes Association.)

which shows a lower overexpression, was also tested (FIG. 2 D). The result obtained was intermediate between control cells and clone 4 and 5 cells (8.7 ± 0.5 μmol /L/h), indicating the existence of an inverse correlation between the level of protein expression and the K^+-induced increase in $[Ca^{2+}]_i$. The reduction in the rise in $[Ca^{2+}]_i$ can be understood as the result of an increased outward Ca^{2+} transport in PMCA2wb-overexpressing compared with control cells. The results also indicate that there was no difference in PMCA activity between nontransfected and vector-only transfected cells.[12]

Taken as a whole, the latter data indicate that the PMCA contribute to Ca^{2+} outflow from the β cell. However, it appears difficult from the latter data to calculate the exact contribution of the PMCA to Ca^{2+} outflow. Indeed, in the clone showing the largest overexpression, the rise in $[Ca^{2+}]_i$ was almost completely abolished. This is striking because the PMCA is considered as a high-affinity, low-capacity system. Hence, the overexpression of such a low-capacity system was not expected to reduce the rise in $[Ca^{2+}]_i$ to such an extent. Here, it is important to notice that the PMCA overexpressed was PMCA2wb, namely a PMCA without any insert at splicing site B, which is located in the middle of the calmodulin-binding site. Splice variants displaying an insert in site B display a low affinity for Ca^{2+} and are active even in the absence of calmodulin.[13] In addition, as alluded to above, it is surprising and unexpected that the overexpression of a high-affinity–low-capacity system (PMCA) reduces by about 80% the global increase in $[Ca^{2+}]_i$ induced by K^+, while the overexpression of a low affinity–high capacity like the NCX only induces "cosmetic" changes in K^+-induced rise in $[Ca^{2+}]_i$ (see above).

With respect to PMCA, it is conceivable that its capacity to transport Ca^{2+} is greater and its affinity for the cation lower than usually estimated. With respect to Na/Ca exchange, the data maybe are not so surprising because the exchanger contributes to two opposite Ca^{2+} movements, Ca^{2+} inflow and outflow. Thus, when both movements (forward and reverse Na/Ca exchange) are altered at the same time, it is conceivable that a balanced effect could be obtained with only modest changes in the global rise in $[Ca^{2+}]_i$. This would then imply that the contribution of Na/Ca exchange to Ca^{2+} inflow is not so negligible, at least in the β cell, as indicated by the marked changes in the rise of $[Ca^{2+}]_i$ on membrane depolarization induced by the knockout or the overexpression of the exchanger (FIGS. 1 and 2).

Opposite Effects of Glucose on Plasma Membrane Ca^{2+}–ATPase and NCX Transcription, Expression, and Activity in Rat Pancreatic β Cells

Different lines of evidence suggest that glucose, the main physiological stimulus of insulin release, stimulates β cell Na/Ca exchange activity.[11,14] Previous work on the PMCA show, on the contrary, that glucose inhibits PMCA activity.[15] Thus, while two groups out of three found a direct inhibitory effect of glucose on enzyme activity (when added to the assay medium),[16–18] a fourth group found no direct effect, but showed that the activity of the ATPase was significantly inhibited when measured in islets previously incubated with glucose.[19] Although found by three groups, glucose-induced inhibition of PMCA activity, whether by a direct or indirect effect, was found somewhat surprising and/or unexpected. It was suggested that inhibition of PMCA could contribute to the increase in $[Ca^{2+}]_i$ that stimulates insulin release.[16,17,19] However, such inhibition is consistent with the canonical view on the respective roles of PMCA and NCX in Ca^{2+} homeostasis,[11] that in the case of the β cell, could be formulated as follows: when stimulated by glucose, the β cell is faced with a major increase in Ca^{2+} inflow, and therefore switches from a low-efficiency Ca^{2+} extruding mechanism, the PMCA, to a high-capacity system, the NCX.

In order to ascertain such a view, we measured the effect of glucose on PMCA and NCX1 isoforms transcription, expression, and activity in rat islet cells.[20]

Glucose (11.1 and 22.2 mM) induced a parallel decrease in PMCA transcription, expression, and activity. In contrast, the sugar induced a parallel increase in NCX transcription, expression, and activity. The effects of the sugar were mimicked by the metabolizable insulin secretagogue α-ketoisocaproate and persisted in the presence of the Ca^{2+} channel blocker nifedipine.[20]

Ca^{2+}–ATPase activity was measured by monitoring the release of ^{32}P from $[\gamma\text{-}^{32}P]$ ATP by islet homogenates in the presence of ouabain and thapsigargin to ensure complete inhibition of the Na^+/K^+–ATPase and sarco(endoplasmic) reticulum Ca^{2+}–ATPase activity, without affecting the PMCA. Na^+/Ca^{2+} exchange activity was measured as intracellular Na^+-dependent $^{45}Ca^{2+}$ uptake, in the presence of 10 μM of nifedipine to block the effect of glucose on voltage-sensitive Ca^{2+} channels. FIGURE 3 shows that Ca^{2+}–ATPase activity was decreased when the islet cells were cultured for 24 h in the presence of 11.1 mM and 22.2 mM glucose ($P < 0.025$), while Na^+/Ca^{2+} exchange activity was increased under the same experimental conditions ($P < 0.025$).[20]

Taken as a whole, the present data, confirm the view that in response to a stimulation by glucose, the β cell switches from a low-efficiency Ca^{2+} extruding mechanism, the PMCA, to a high-capacity system, the NCX, in order to better face the increase in Ca^{2+} inflow. To our knowledge, this is the first demonstration of a reciprocal change in PMCA and NCX1 expression and activity in response to a given stimulus in any cellular preparation or tissue. The upregulation of NCX1 by glucose may have further advantage because NCX1, in addition to be able to extrude Ca^{2+}, may also contribute to Ca^{2+} entry through reverse Na/Ca exchange and generate an inward current that may prolong the duration of the burst of spikes of electrical activity generated by glucose and hence enhance insulin release.[11] In addition, the upregulation of NCX1 could help to protect the β cell against the deleterious actions of elevated glucose levels. Indeed, the sugar, when used at a high concentration was observed to trigger apoptosis in β cells, a process that was Ca^{2+} dependent.[21]

Mechanism of Cytokine-Induced β Cell Death

Intracellular Ca^{2+} may also trigger apoptosis and regulate death-specific enzymes. In our previous review on the subject,[7] we showed that overexpression of the exchanger increased β cell death by apoptosis, a phenomenon resulting from the depletion of endoplasmic reticulum (ER) Ca^{2+} stores, that induced ER stress, with subsequent activation of caspase-12, an ER-specific caspase. The overexpression also reduced β cell growth.[22]

Evidences suggest that programmed cell death (apoptosis) represents the main mechanism of β cell death in animal models of type 1 diabetes mellitus

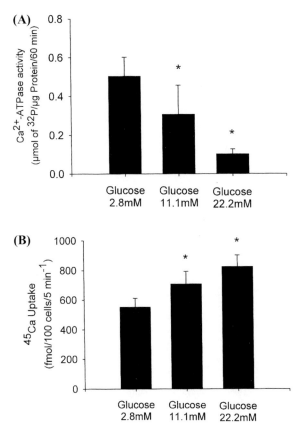

FIGURE 3. PMCA and NCX activity in islets cells after 24 h culture in the presence of various concentrations of glucose. (**A**) PMCA activity was measured as the release of ^{32}P from [γ-^{32}P] ATP (10 μg protein/sample). Means ± SEM refers to four determinations (*$P < 0.05$); (**B**) Na/Ca exchange was measured as intracellular Na$^+$-dependent ^{45}Ca^{2+} uptake. Means ± SEM refers to six determinations (*$P < 0.05$). (Reprinted with permission from Ximenes *et al.*[20])

(T1DM) and possibly also in human T1DM.[23] On the other hand, type 1 cytokines, such as interleukin-1β (IL-1β), tumor necrosis factor-α (TNF-α), and interferon-γ (IFN-γ), are early mediators of β cell death in T1DM.[23] Moreover, it has been observed by microarray analysis that IL-1β + IFN-γ decrease the expression the ER Ca^{2+} pump SERCA2b in the β cell.[24–26] Therefore, we wondered whether cytokines could not induce β cell death by depleting ER Ca^{2+} stores as induced by Na/Ca overexpression. Within the frame of a collaborative work with Décio Eizirik we could indeed show that

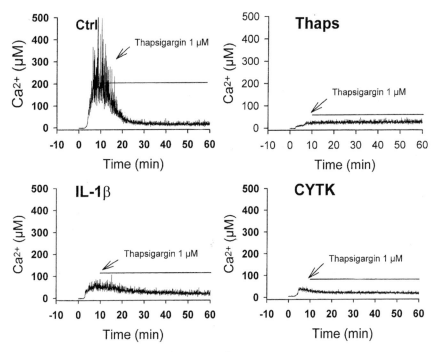

FIGURE 4. Effect of cytokines on ER Ca^{2+} concentration in primary β cells. FACS-purified rat β cells were exposed for 24 h to IL-1β (50 U/mL) or IL-1β + IFN-γ (0.036 μg/mL, CYTK, *lower panels*), or left untreated (control, *upper* and *left panel*), and then exposed acutely to thapsigargin (*bars* above the curves). Prolonged (24 h) preexposure to thapsigargin (1 μM, *upper* and *right panel*) was used as a positive control for depletion of ER Ca^{2+}. The figures are representative of three similar experiments.

cytokines induced ER Ca^{2+} depletion, with activation of diverse components of the ER stress response.[27]

The low-affinity Ca^{2+} indicator furaptra was used to monitor free Ca^{2+} concentration in the ER of individual β cells after controlled permeabilization of the plasma membrane.[12,22] After recording the fluorescence obtained by excitation at 340 and 380 nM, cells were permeabilized in "intracellular medium" containing digitonin 4 μM and 200 nM Ca^{2+}. Upon the sudden drop in fluorescence caused by the loss of cytoplasmic furaptra, the detergent was removed while continuing the measurement of the fluorescence at both wavelengths. The loss of cytoplasmic furaptra was associated with an inversion of the 340/380 nM fluorescence excitation ratio indicating that the remaining indicator was exposed to the higher concentrations of free Ca^{2+} prevailing in intracellular stores.[12,22] This corresponds to the initial increase in ER Ca^{2+} concentration ($[Ca^{2+}]_{ER}$) illustrated in FIGURE 4 (upper and left panel). This initial rise in $[Ca^{2+}]_{ER}$ was reduced by $81 \pm 7\%$ and $93 \pm 6\%$ in IL-1β and IL1β + IFN-γ-treated cells (FIG. 4, lower panels; $P \leq 0.05$, $n = 3$). Thus, while the $[Ca^{2+}]_{ER}$ averaged 200 μM in control cells, it averaged less than 100 μM, and less than 50 μM in islet cells exposed to IL-1β alone or to the mixture of cytokines, respectively. We have previously observed that thapsigargin released 50% of the Ca^{2+} pool sensed by furaptra under the present condition.[12] Therefore, thapsigargin was used to estimate Ca^{2+} stores in the ER. In control cells, the acute exposure to thapsigargin induced a 90% decrease in $[Ca^{2+}]_{ER}$ (FIG. 4, upper and left panel). This effect of thapsigargin was almost completely abolished in IL-1β and IL-1β + IFN-γ-treated cells, respectively (FIG. 4, lower panels), compared to control cells ($P \leq 0.05$ and $P \leq 0.01$, respectively; $n = 3$). In cells cultured for 24 h in the presence of the SERCA inhibitor, thapsigargin failed to reduce the $[Ca^{2+}]_{ER}$ (FIG. 4, upper and right panel). These data indicate that cytokine induced a severe depletion of ER Ca^{2+} stores, with activation of ER stress, like overexpression of the NCX.[27] This suggests that a reduction in Na/Ca exchange activity may lead to an increase in ER Ca^{2+} stores and perhaps confer resistance to the β cell against cytokines. Such hypothesis is presently tested in our laboratory.

ACKNOWLEDGMENTS

The authors thank D.L. Eizirik and A. K. Cardozo of the Laboratory of Experimental medicine for their collaboration in the work on apoptosis. We thank Christiane De Bruyne, Anne Van Praet, and Anissa Iabkriman for technical help. This work was supported by The Belgian Fund for Scientific Research (FRSM 3.4545.96, LN 9.4514.93, LN 9.4510.95, and 3.456.000), of which F.V.E. was a postdoctoral researcher; by the Concerted Action IREN in the Biomed 2 programme; and the ALFA programme IRELAN of the European Union of which H.X. was a grant holder coming from the Institute of Biomedical Science, University of São Paulo, São Paulo, Brazil.

REFERENCES

1. CARAFOLI, E. 1988. Membrane transport of calcium: an overview. Method Enzymol. **157:** 3–11.
2. BLAUSTEIN, M.P., W.F. GOLDMAN, G. FONTANA, *et al.* 1991. Physiological roles of the sodium-calcium exchanger in nerve and muscle. Ann. N Y Acad. Sci. **639:** 254–274.
3. EGGER, M. & E. NIGGLI. 1999. Regulatory function of Na-Ca exchange in the heart: milestones and outlook. J. Membr. Biol. **168:** 107–130.
4. BLAUSTEIN, M.P. & W.J. LEDERER. 1999. Sodium/calcium exchange: its physiological implications. Physiol. Rev. **79:** 763–854.

5. HERCHUELZ, A. & P.O. PLASMAN. 1991. Na/Ca exchange in the pancreatic β-cell. Ann. N Y Acad. Sci. **639:** 642–656.
6. CARAFOLI, E. 1994. Biogenesis: plasma membrane calcium ATPase: 15 years of work on the purified enzyme. FASEB J. **8:** 993–1002.
7. HERCHUELZ, A., O. DIAZ-HORTA & F. VAN EYLEN. 1991. Na/Ca exchange in function, growth and demise of pancreatic β-cells. Ann. N Y Acad. Sci. **976:** 315–324.
8. VAN EYLEN, F., M. SVOBODA & A. HERCHUELZ. 1997. Identification, expression pattern and potential activity of Na/Ca exchanger isoforms in rat pancreatic β-cells. Cell Calcium **21:** 185–193.
9. KAMAGATE, A., A. HERCHUELZ, A. BOLLEN & F. VAN EYLEN. 2000. Expression of multiple plasma membrane Ca^{2+}-ATPases in rat pancreatic islet cells. Cell Calcium **27:** 231–246.
10. VAN EYLEN, F., C. LEBEAU, J. ALBUQUERQUE-SILVA & A. HERCHUELZ. 1998. Contribution of Na/Ca exchange to Ca^{2+} outflow and entry in the rat pancreatic β-cell. Diabetes **47:** 1873–1880.
11. VAN EYLEN, F., O. DIAZ HORTA , A. BAREZ, et al. 2002. Overexpression of the Na/Ca exchanger shapes stimulus-induced cytosolic Ca^{2+} oscillations in insulin-producing BRIN-BD11 cells. Diabetes **51:** 366–375.
12. KAMAGATE, A., A. HERCHUELZ & F. VAN EYLEN. 2002. Plasma membrane Ca^{2+}-ATPase overexpression reduces Ca^{2+} oscillations and increases insulin release induced by glucose in insulin-secreting BRIN-BD11 cells. Diabetes **51:** 2773–2788.
13. ENYEDI, A., A.K. VERMA, R. HEIM, et al.1994. The Ca^{2+} affinity of the plasma membrane Ca^{2+} pump is controlled by alternative splicing. J. Biol. Chem. **269:** 41–43.
14. PLASMAN, P.O., P. LEBRUN & A. HERCHUELZ. 1990. Characterization of the process of sodium-calcium exchange in pancreatic islet cells. Am. J. Physiol. **259:** E844–E850.
15. GAGLIARDINO, J.J. & J.P.F.C. ROSSI. 1994. Ca^{2+}-ATPase in pancreatic islets: its possible role in the regulation of insulin secretion. Diabetes Metab. Rev. **10:** 1–17.
16. LEVIN, S.R., B.G. KASSON & J.F. DRIESSEN. 1978. Adenosine triphosphatases of rat pancreatic islets: comparison with those of rat kidney. J. Clin. Invest. **62:** 692–701.
17. HOENIG, M., R.J. LEE & D.C. FERGUSON. 1990. Glucose inhibits the high-affinity $(Ca^{2+} Mg^{2+})$-ATPase in the plasma membrane of a glucose-responsive insulinoma. Biochem. Biophys. Acta **1022:** 333–338.
18. KOTAGAL, N., J.R. COLCA, D. BUSCETTO & M.L. MACDANIEL. 1985. Effect of insulin secretagogues and potential modulators of secretion on a plasma $(Ca^{2+}$-$Mg^{2+})$-ATPase activity in islets of Langerhans. Arch. Biochem. Biophys. **238:** 161–169.
19. GRONDA, C.M., J.P. ROSSI & J.J. GALGIARDINO. 1988. Effect of different insulin secretagogues and blocking agents on islet cell Ca^{2+}-ATPase activity. Biochem. Biophys. Acta. **943:** 183–189.
20. XIMENES, H.M., A. KAMAGATE, F. VAN EYLEN, et al. 2003. Opposite effects of glucose on plasma membrane Ca^{2+}-ATPase and Na/Ca exchanger transcription, expression, and activity in rat pancreatic β-cells. J. Biol. Chem. **278:** 22956–22963.

21. EFANOVA, J.B., S.V. ZAITSEV, B. ZHIVOTOVSKY, *et al.* 1998. Glucose and tolbutamide induce apoptosis in pancreatic beta-cells. A process dependent on intracellular Ca^{2+} concentration. J. Biol. Chem. **273:** 33501–33507.

22. DIAZ-HORTA, O., A. KAMAGATE, A. HERCHUELZ & F. VAN EYLEN. 2002. Na/Ca exchanger overexpression induces endoplasmic reticulum-related apoptosis and caspase-12 activation in insulin-releasing BRIN-BD11 cells. Diabetes **51:** 1815–1824.

23. EIZIRIK, D.L. & T. MANDRUP-POULSEN. 2001. A choice of death: the signal-transduction of immune-mediated β-cell apoptosis. Diabetologia **44:** 2115–2133.

24. CARDOZO, A.K., H. HEIMBERG, Y. HEREMANS, *et al.* 2001. A comprehensive analysis of cytokine-induced and nuclear factor-KB-dependent genes in primary rat pancreatic β-cells. J. Biol. Chem. **276:** 48879–48886.

25. KUTLU, B., A.K. CARDOZO, M.I. DARVILLE, *et al.* 2003. Discovery of gene networks regulating cytokine-induced dysfunction and apoptosis in insulin-producing INS-1 cells. Diabetes **52:** 2701–2719.

26. CARDOZO, A.K., M. KRUHOFFER, R. LEEMAN, *et al.* 2001. Identification of novel cytokine-induced genes in pancreatic β-cells by high-density oligonucleotide arrays. Diabetes **50:** 909–920.

27. CARDOZO, A.K., F. ORTIS, J. STORLING, *et al.* 2005. Cytokines downregulate the sarcoendoplasmic reticulum pump Ca^{2+} ATPase 2b and deplete endoplasmic reticulum Ca^{2+}, leading to induction of endoplasmic reticulum stress in pancreatic β-cells. Diabetes **54:** 452–461.

Overexpression of Na/Ca Exchanger Reduces Viability and Proliferation of Gliosarcoma Cells

EVRARD NGUIDJOE AND ANDRÉ HERCHUELZ

Laboratory of Pharmacology, Brussels University School of Medicine, B-1070, Brussels, Belgium

In a recent work, we observed that overexpression of the Na/Ca exchanger1 (NCX1), increased apoptosis and reduced cellular proliferation in an insulin-secreting cell line.[1]

The aim of this work was to examine whether overexpression of the exchanger would produce similar effects in a poorly differentiated tumoral cell line and to further explore the effects of such overexpression.

The Na/Ca exchanger (isoform 1.7) was transfected in gliosarcoma cells and three clones showing different levels of expression were isolated. There was an inverse correlation ($P < 0.001$) between the level of expression and cellular viability as measured using the MTT (3-(4,5-dimethylthiazol-2yl)-2,5-diphenyl tetrazolium bromide) test, a reduction of 40% being observed in the clone showing the highest level of expression. Similarly, a reduction in proliferation rate was observed with a maximum of 50% in the latter clone. Cell death was due to apoptosis, and attended by a depletion of endoplasmic reticulum (ER) Ca^{2+} stores and reduction in ER Ca^{2+} concentration, suggesting that cellular death was due to ER stress, as induced by the increased extrusion of Ca^{2+} from the cell. Overexpression of the exchanger reduced Bcl-2 expression at the protein level but only in the clone showing the highest overexpression of the exchanger.

It is concluded that overexpression of the Na/Ca exchanger may represent a new approach in cancer gene therapy, and that the role of Ca^{2+} in the process of apoptosis is underestimated and deserves further investigation.

REFERENCE

1. DIAZ-HORTA O, A. KAMAGATE, A. HERCHUELZ & F. VAN EYLEN. 2002. Na/Ca exchanger overexpression induces endoplasmic reticulum-related apoptosis and caspase-12 activation in insulin-releasing BRIN-BD11 cells. Diabetes **51**: 1815–1824.

Address for correspondence: André Herchuelz, Laboratoire de Pharmacodynamie et de Thérapeutique, Université Libre de Bruxelles, Faculté de Médecine, Route de Lennik, 808-Bâtiment GE, b-1070 Bruxelles, Belgium. Voice: +32-2-555-62-01; fax 32-2-555-63-70.
herchu@ulb.ac.be

Ann. N.Y. Acad. Sci. 1099: 468 (2007). © 2007 New York Academy of Sciences.
doi: 10.1196/annals.1387.053

Redox Modulation of the Apoptogenic Activity of Thapsigargin

CLAUDIA CERELLA,[a] SIMONA COPPOLA,[a] MARIA D'ALESSIO,[a] MILENA DE NICOLA,[a] ANDREA MAGRINI,[b] ANTONIO BERGAMASCHI,[c] AND LINA GHIBELLI[a]

[a]Dipartimento di Biologia, Universita' di Roma Tor Vergata, 00133 Rome, Italy

[b]Cattedra di Medicina del Lavoro, Universita' di Roma Tor Vergata, 00133 Rome, Italy

[c]Cattedra di Medicina del Lavoro, Universita' Cattolica Sacro Cuore, 00168 Rome, Italy

ABSTRACT: Thapsigargin (THG), a selective inhibitor of endoplasmic reticulum (ER) Ca^{2+}-ATPases, causes the rapid emptying of ER Ca^{2+}; in some cell types, this is accompanied by apoptosis, whereas other cells maintain viability. In order to understand the molecular determinants of such a different behavior, we explored the role of oxygen versus nitrogen radicals, by analyzing the apoptogenic ability of THG in the presence of inhibitors of glutathione or nitric oxide (NO) synthesis, respectively. We observed that oxygen radicals play a sensitizing role whereas nitrogen radicals prevent THG-dependent apoptosis, showing that the apoptogenic effect of THG is redox sensitive.

KEYWORDS: thapsigargin; apoptosis; calcium; glutathione; nitric oxide

INTRODUCTION

Thapsigargin (THG) is a sesquiterpene alkaloid that poisons sarcoplasmic and endoplasmic reticulum (ER) Ca^{2+} ATPases (SERCA), thus determining a transient increase in cytosolic Ca^{2+} and a sustained depletion of the ER Ca^{2+} pool.[1] THG is a popular inducer of apoptosis,[2] due to either the transient cytosolic Ca^{2+} increase,[2] or the sustained ER Ca^{2+} pool depletion.[2] Since the two events occur simultaneously, it is quite difficult to assess the specific contribution of one or the other to apoptosis. In spite of an effective ER Ca^{2+} emptying, some cells are refractory to THG-induced apoptosis[2]; the molecular determinants of such difference remain unknown. Since SERCA are redox-regulated

Address for correspondence: Lina Ghibelli, Dipartimento di Biologia, Universita' di Roma Tor Vergata, via Ricerca Scientifica, 1, 00133 Roma (I), Italy. Voice: +39-06-7259-4323; fax: +39-06-2023500. ghibelli@uniroma2.it

Ann. N.Y. Acad. Sci. 1099: 469–472 (2007). © 2007 New York Academy of Sciences. doi: 10.1196/annals.1387.028

proteins, we addressed the question on whether a redox disequilibrium might modulate cells' sensitivity to the apoptogenic effect of THG. We found that the depletion of glutathione (GSH), the main intracellular antioxidant, and the consequent increase in oxygen radicals, on the one side, and a decrease in nitrogen radicals (i.e., by inhibiting nitric oxide [NO] production), on the other, sensitize U937 cells to THG.

MATERIALS AND METHODS

Cell Culture and Treatments

U937 cells (human tumoral promonocytes) and fresh explanted rat thymocytes were cultured as described.[3] *GSH depletion*: U937 cells were incubated for 24 h with the 1mM of the specific inhibitor of GSH synthesis buthionine sulfoximine (BSO).[4] *NO synthase inhibition*: U937 cells were pretreated 1 h before other treatments with 600 μM L-NAME. *Induction of apoptosis*: U937 cells and rat thymocytes were treated with 10 nM THG. *Apoptosis detection*: Apoptosis was evaluated by analyzing nuclear morphology in stained cell nuclei with Hoechst 33342, as described.[3,4]

RESULTS AND DISCUSSION

We analyzed the apoptogenic ability of 10 nM THG on two different cell systems, promonocytic U937 and fresh explanted rat thymocytes. THG, at this dose, is able to inhibit only one class of SERCA.[1] In our systems, THG determines a transient lowering of ER Ca^{2+} that is restored in 5 min, as revealed by image analysis of cells loaded with the specific ER Ca^{2+} indicator chlortetracycline (CTC; data not shown). FIGURE 1 A shows that THG induces apoptosis in rat thymocytes, a well-known example of sensitive cells, whereas it is unable to induce apoptosis in U937.

In order to understand the determinants of the insensitivity of U937 cells to the apoptogenic effect of THG, we investigated whether it might depend on the intracellular redox status. To this purpose, we used two complementary treatments.

First, we depleted U937 of GSH, by inhibiting its synthesis with BSO. At 24 h of BSO, GSH is undetectable[4] and oxygen radicals are increased by about 25% with respect to control, as revealed by dichlorofluorescein diacetate staining (not shown; see Ref. 4). In these conditions, THG turns into a strong apoptogenic agent on U937 (FIG. 1 B). Notably, the extent of Ca^{2+} released by THG is significantly increased in BSO-treated cells, whereas the time of clearance of the extra cytosolic Ca^{2+} derived from THG-induced ER emptying is not changed (not shown). These results indicate that an oxidant environment sensitizes U937 to THG-induced apoptosis.

(A)

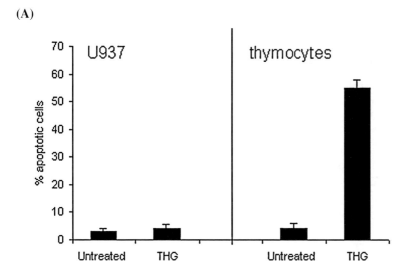

(B)

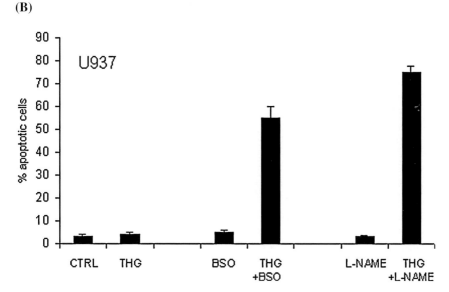

FIGURE 1. Redox modulation sensitizes U937 cells to apoptosis. (**A**) Differential sensitivity to apoptosis: 10 nM THG was added to U937 (left) or rat thymocytes (right). Apoptosis was measured after 6 h of treatment. Values are the average of three independent experiments ± SD. (**B**) U937 cells were pretreated with 1mM BSO for 24 h or with 600 μM L-NAME for 1 h; then cells were treated with 10 nM THG and apoptosis evaluated after 7 h or 3 h of THG treatment, respectively. Both BSO and L-NAME sensitizes U937 cells to THG-dependent apoptosis. The values are the average of three independent experiments ± SD.

As a second approach, we inhibited NO synthases with L-NAME. Also in this case, THG turned into a potent apoptogenic agent on U937 (FIG. 1 B).

NO plays a very different role with respect to oxygen radicals. Indeed, redox-sensitive proteins often have an exposed cysteine residue that can be oxidized; these oxidations are usually regulatory events. NO readily reacts with cysteines leading to their nitrosylation, a reversible event that also plays a major role in the regulation of protein activity. A nitrosylated cysteine cannot be oxidized; thus nitrosylations and oxidations are mutually exclusive events and an environment that favors one or the other can determine a different outcome of the activity of the target protein. SERCA are known to be sensitive to NO, which have a transiently inhibitory effect.[5] Possibly, a nitrosylation might protect SERCA from an irreversible THG poisoning.

In conclusion, GSH and NO seem to be able to contrast THG-induced apoptosis, since their depletion sensitized the otherwise insensitive U937 cells. Interestingly, these cells are surprisingly rich in both molecules,[4] thus possibly giving a rationale for their survival to THG.

REFERENCES

1. THASTRUP, O. *et al*. 1990. Thapsigargin, tumor promoter, discharges intracellular Ca^{2+} stores by specific inhibition of the endoplasmic reticulum Ca^{2+}-ATPase. Proc. Natl. Acad. Sci. USA **82:** 2466–2470.
2. HUAFENG, W. *et al*. 1998. Bcl-2 protects against apoptosis in neuronal cell line caused by thapsigargin-induced depletion of intracellular calcium stores. J. Neurochem. **70:** 2305–2314.
3. FANELLI, C. *et al*. 1999. Magnetic fields increase cell survival by inhibiting apoptosis via modulation of Ca^{2+} influx. FASEB J. **13:** 95–102.
4. D'ALESSIO, *et al*. 2004. Oxidative Bax dimerization promotes its translocation to mitochondria independently of apoptosis. FASEB J. **19:** 1504–1506.
5. RENNA, M. *et al*. 2006. Nitric oxide-induced endoplasmic reticulum stress activates the expression of cargo receptor proteins and alters the glycoprotein transport to the Golgi. Int. J. Biochem.Cell Biol. **38:** 2040–2048.

The Role of Na^+/Ca^{2+} Exchanger in Endothelin-1-Aggravated Hypoxia/Reoxygenation-Induced Injury in Renal Epithelial Cells

SATOMI KITA,[a] AYAKO FURUTA,[a,b] YUKIO TAKANO,[b] AND TAKAHIRO IWAMOTO[a]

[a]Department of Pharmacology, School of Medicine, Fukuoka University, Fukuoka 814-0180, Japan

[b]Department of Physiology and Pharmacology, Faculty of Pharmaceutical Sciences, Fukuoka University, Fukuoka 814-0180, Japan

ABSTRACT: We analyzed the role of the Na^+/Ca^{2+} exchanger (NCX) in endothelin-1-aggravated hypoxia/reoxygenation-induced injury in renal epithelial LLC-PK$_1$ cells. KB-R7943, a selective NCX inhibitor, suppressed hypoxia/reoxygenation-induced cell damage, whereas overexpression of NCX1 into cells enhanced it. Endothelin-1 significantly aggravated hypoxia/reoxygenation-induced injury in parental and NCX1-overexpressing LLC-PK$_1$ cells. Such aggravation by endothelin-1 was not observed in cells overexpressing a deregulated NCX1 mutant, which displays no protein kinase C-dependent activation. These results suggest that Ca^{2+} overload via NCX plays a critical role in hypoxia/reoxygenation-induced renal tubular injury, and that endothelin-1 aggravates the cell damage through the activation of NCX.

KEYWORDS: sodium–calcium exchange; endothelin-1; KB-R7943; hypoxia/reoxygenation; LLC-PK$_1$ cell

INTRODUCTION

Endothelin-1 (ET-1), a potent vasoconstrictor peptide, is considered to contribute to the pathogenesis of cardiovascular diseases and ischemia.[1] Levels of preproendothelin-1 and ET-1 are known to be elevated in the post-ischemic kidney.[2,3] Previous studies have revealed that endothelin type A (ET$_A$) receptor antagonists or ET$_A$/endothelin type B (ET$_B$) receptor antagonists improve ischemia/reperfusion-induced renal injury in animal models.[4,5] However, the role of ET-1 at the tubular level remains poorly understood.

Address for correspondence: Dr. Takahiro Iwamoto, Department of Pharmacology, School of Medicine, Fukuoka University, 7-45-1 Nanakuma Jonan-ku, Fukuoka 814-0180, Japan. Voice: +81-92-801-1011; ext.: 3261; fax: +81-92-865-4384.

tiwamoto@cis.fukuoka-u.ac.jp

Ann. N.Y. Acad. Sci. 1099: 473–477 (2007). © 2007 New York Academy of Sciences.
doi: 10.1196/annals.1387.041

The Na^+/Ca^{2+} exchanger (NCX) is an electrogenic transporter that exchanges three Na^+ for one Ca^{2+}. The physiological importance of NCX has been studied extensively in excitable tissues, such as cardiac muscle, vascular smooth muscle, and nerve fiber.[6] The exchanger plays the primary role in Ca^{2+} extrusion during cellular Ca^{2+} signaling. In the kidney, the NCX1 isoform is abundantly expressed and the NCX2 is slightly present. NCX1 is thought to play an important role in the active reabsorption of calcium in the distal nephron.[6] The transport activity of the exchanger is preferentially observed in the distal tubular cells, but the activity is also detectable in proximal tubular cells.[7] NCX1 is known to be regulated by several cytosolic factors[6]; it is activated by cytosolic Ca^{2+} (I_2 regulation) but is inhibited by cytosolic Na^+ (I_1 regulation). NCX1 is also upregulated by protein kinase C (PKC)-dependent pathway.[8,9] Recently, we reported the protective effects of NCX inhibitors on ischemia/reperfusion-induced renal injury, suggesting that Ca^{2+} overload via the reverse mode of NCX plays an important role in the pathogenesis of renal ischemic diseases.[7,10]

In this study, we demonstrated that ET-1 aggravates hypoxia/reoxygenation-induced cell injury in tubular epithelial cells through the activation of NCX.[11,12]

MATERIALS AND METHODS

LLC-PK$_1$ cells and their NCX transfectants were maintained in Dulbecco's modified Eagle's medium (DMEM) supplemented with 4% fetal calf serum (FCS), 50 U/mL penicillin, and 50 µg/mL streptomycin in a humidified incubator gassed with 5% CO_2-95% air at 37°C.

LLC-PK$_1$ cells and their transfectants were grown in 96-well microplates at 2×10^5 cells/well. After 2 days, the medium was changed to HEPES-buffered DMEM without glucose and FCS. The cells were exposed to hypoxic conditions in an Anaero Pack Pouch (Mitsubishi Gas Chemical, Tokyo, Japan), in which the oxygen concentration was less than 1% within 1 h, as described previously.[11] After 6 h of hypoxia, the cells were put in a humidified incubator gassed with 5%CO_2-95% air for 1 h in HEPES-buffered DMEM to which glucose was added at the beginning of reoxygenation. After the treatment of hypoxia/reoxygenation, lactate dehydrogenase (LDH) activity in the medium was measured using an LDH–Cytotoxic Test Kit (Wako Pure Chemicals, Osaka, Japan). LDH release expressed as a percentage of total cellular LDH activity.

RESULTS AND DISCUSSION

We examined the protective effect of KB-R7943 on hypoxia/reoxygenation-induced cell injury in LLC-PK$_1$ cells. In parental LLC-PK$_1$ cells, 6-h hypoxia followed by 1 h of reoxygenation produced a significant LDH release from damaged cells (FIG. 1). This cell damage was confirmed by observing morphological changes, such as bleb formation, suggesting that cell damage may be caused by Ca^{2+} overloading. When LLC-PK$_1$ cells were treated with 10 μM KB-R7943 during both hypoxia and reoxygenation, the hypoxia/reoxygenation-induced LDH release was suppressed by about 75%. Even when cells were treated with KB-R7943 (3 or 10 μM) during either hypoxia or reoxygenation, LDH release was also significantly reduced.

To evaluate the pathological importance of NCX1, we overexpressed the wild-type NCX1 and the deregulated mutant XIP-4YW,[11,12] which displays neither I$_1$ regulation nor PKC-dependent activation, in LLC-PK$_1$ cells. The protein expression of wild-type NCX1 or XIP-4YW mutant in transfectants was about 250–300% higher than that of endogenous NCX1. These transfectants were more susceptible to hypoxia/reoxygenation-induced injury than parental LLC-PK$_1$ cells. The hypoxia/reoxygenation-induced LDH release was two- to threefold enhanced in LLC-PK$_1$ cells overexpressing wild-type NCX1 or XIP-4YW mutant (FIG. 2).

We further examined the effect of ET-1 on hypoxia/reoxygenation-induced injury in LLC-PK$_1$ cells. As shown in FIGURE 2, ET-1 (0.01 or 1 μM) significantly enhanced hypoxia/reoxygenation-induced LDH release by ~40% in parental LLC-PK$_1$ cells and NCX1-overexpressing cells, but not in XIP-4YW-overexpressing cells. The effect of ET-1 in NCX1-overexpressing cells

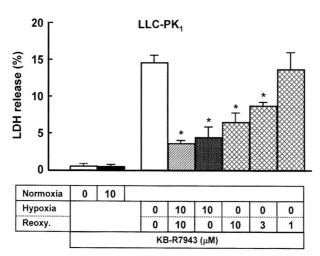

FIGURE 1. Effect of KB-R7943 on LDH release from LLC-PK$_1$ cells exposed to 7 h of normoxia or 6 h of hypoxia followed by 1 h of reoxygenation. KB-R7943 (1–10 μM) was added to the medium during normoxia and during hypoxia and/or reoxygenation (Reoxy.). Data are mean ± SE of six independent experiments. *$P < 0.01$ compared with hypoxia and reoxygenation without KB-R7943. (Data reproduced from Iwamoto *et al.*[11] with permission.)

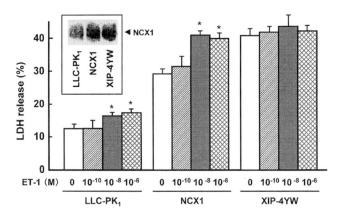

FIGURE 2. Effect of ET-1 on hypoxia/reoxygenation-induced LDH release in parental LLC-PK$_1$ cells and NCX1- and XIP-4YW-overexpressing cells. Cells were treated with ET-1 (10^{-10}, 10^{-8}, or 10^{-6} M) during both hypoxia and reoxygenation. Data are mean ± SE of five or six independent experiments. *$P < 0.05$ compared with hypoxia and reoxygenation without ET-1. Inset: Microsomes (30 μg) from cells were subjected to immunoblot analysis with anti-NCX1 antibody. (Data reproduced from Iwamoto *et al.*[11] with permission.)

was partially blocked by BQ788, a selective ET$_B$-receptor antagonist, but not by BQ123, a selective ET$_A$-receptor antagonist.[11]

In this study, we demonstrated that ET-1 significantly aggravated hypoxia/reoxygenation-induced cell injury (Ca^{2+} overload) in parental or NCX1-overexpressing LLC-PK$_1$ cells, but not in cells overexpressing XIP-4YW, which displays neither I$_1$ regulation nor PKC-dependent activation. We recently observed that the activity of NCX1 is upregulated by ET-1 in LLC-PK$_1$ cells.[12] Therefore, ET-1 seems to aggravate hypoxia/reoxygenation-induced renal injury via the PKC-dependent activation of NCX1.

In conclusion, these results support the view that ET-1 overproduction in the post-ischemic kidney accelerates Ca^{2+} overload via NCX, and that it then exaggerates the pathogenesis of hypoxic tubular damage.

ACKNOWLEDGMENTS

This work was supported by grants from the Takeda Science Foundation and the Inoue Foundation for Science.

REFERENCES

1. YANAGISAWA, M., H. KURIHARA, S. KIMURA, *et al.* 1988. A novel potent vasoconstrictor peptide produced by vascular endothelial cells. Nature **332:** 411–415.

2. KOHAN, D.E. 1997. Endothelins in the normal and diseased kidney. Am. J. Kidney Dis. **29:** 2–26.

3. WILHELM, S.M., M.S. SIMONSON, A.V. ROBINSON, *et al.* 1999. Endothelin up-regulation and localization following renal ischemia and reperfusion. Kidney Int. **55:** 1011–1018.

4. MINO, N., M. KOBAYASHI, A. NAKAJIMA, *et al.* 1992. Protective effect of a selective endothelin receptor antagonist, BQ123, in ischemic acute renal failure in rats. Eur. J. Pharmacol. **221:** 77–83.

5. KUSUMOTO, K., K. KUBO, H. KANDORI, *et al.* 1994. Effects of a new endothelin antagonist, TAK-044, on post-ischemic acute renal failure in rats. Life Sci. **55:** 301–310.

6. BLAUSTEIN, M.P. & W.J. LEDERER. 1999. Sodium/calcium exchange: its physiological implications. Physiol. Rev. **79:** 763–854.

7. YAMASHITA, J., S. KITA, T. IWAMOTO, *et al.* 2003. Attenuation of ischemia/reperfusion-induced renal injury in mice deficient in Na^+/Ca^{2+} exchanger. J. Pharmacol. Exp. Ther. **304:** 284–293.

8. IWAMOTO, T., Y. PAN, S. WAKABAYASHI, *et al.* 1996. Phosphorylation-dependent regulation of cardiac Na^+/Ca^{2+} exchanger via protein kinase C. J. Biol. Chem. **271:** 13609–13615.

9. IWAMOTO, T., Y. PAN, T.Y. NAKAMURA, *et al.* 1998. Protein kinase C-dependent regulation of Na^+/Ca^{2+} exchanger isoforms NCX1 and NCX3 does not require their direct phosphorylation. Biochemistry **37:** 17230–17238.

10. OGATA, M., T. IWAMOTO, N. TAZAWA, *et al.* 2003. A novel and selective Na^+/Ca^{2+} exchange inhibitor, SEA0400, improves ischemia/reperfusion-induced renal injury. Eur. J. Pharmacol. **478:** 187–198.

11. IWAMOTO, T., S. KITA & T. KATSURAGI. 2004. Endothelin-1 aggravates hypoxia/reoxygenation-induced injury in renal epithelial cells through the activation of a Na^+/Ca^{2+} exchanger. J. Cardiovasc. Pharmacol. **44**(Suppl 1): S462–S466.

12. KITA, S., T. KATSURAGI & T. IWAMOTO. 2004. Endothelin-1 enhances the activity of Na^+/Ca^{2+} exchanger type 1 in renal epithelial cells. J. Cardiovasc. Pharmacol. **44**(Suppl 1): S239–S243.

Involvement of Na^+/Ca^{2+} Exchanger Type-1 in Ischemia-Induced Neovascularization in the Mouse Hindlimb

YUKIKO MATSUI,[a] SATOMI KITA,[b] TAKESHI KATSURAGI,[b] ISSEI KOMURO,[c] TAKAHIRO IWAMOTO,[b] AND HIROYUKI OHJIMI[a]

[a]Department of Plastic and Reconstructive Surgery, Fukuoka University Hospital, Fukuoka 814-0180, Japan

[b]Department of Pharmacology, School of Medicine, Fukuoka University, Fukuoka 814-0180, Japan

[c]Department of Cardiovascular Science and Medicine, Chiba University Graduate School of Medicine, Chiba 260-8670, Japan

ABSTRACT: The Na^+/Ca^{2+} exchanger (NCX) is considered to be involved in endothelial nitric oxide (NO) production and endothelium-dependent vasorelaxation, but little is known about the physiological and pathological roles of endothelial NCX in these processes. We examined the role of NCX1 in neovascularization in mice with hindlimb ischemia. Unilateral hindlimb ischemia was induced surgically in wild-type and heterozygous NCX1 knockout mice ($NCX1^{+/-}$) mice. We found that in $NCX1^{+/-}$ mice, blood flow recovery was significantly augmented compared with that in wild-type mice. N^G-nitro-L-arginine methyl ester treatment eliminated enhanced angiogenesis observed in $NCX1^{+/-}$ mice. These results suggest that NCX1 is involved in eNOS-dependent angiogenesis.

KEYWORDS: sodium–calcium exchange; NCX1 knockout mice; angiogenesis; endothelial NO synthase; Lipo prostaglandin E1

INTRODUCTION

In endothelial cells, the presence of Na^+/Ca^{2+} exchanger (NCX) has been demonstrated by immunoblotting and immunofluorescence.[1] Furthermore, cDNA coding for NCX1 has been detected in endothelial cells.[2] Recent reports suggest that endothelial NCX is involved in $[Ca^{2+}]_i$ regulation[3,4] and nitric oxide (NO) production.[5,6] It is considered that NO is an important regulator of

Address for correspondence: Dr. Takahiro Iwamoto, Department of Pharmacology, School of Medicine, Fukuoka University, 7-45-1 Nanakuma Jonan-ku, Fukuoka 814-0180, Japan. Voice: +81-92-801-1011; ext.: 3261; fax: +81-92-865-4384.
tiwamoto@cis.fukuoka-u.ac.jp

Ann. N.Y. Acad. Sci. 1099: 478–480 (2007). © 2007 New York Academy of Sciences.
doi: 10.1196/annals.1387.042

endothelial cell growth, angiogenesis, and endothelium-dependent vasorelaxation. However, little is known about the physiological and pathological roles of endothelial NCX in these processes. In this experiment, we examined the role of NCX1 in neovascularization in mice with hindlimb ischemia.

MATERIALS AND METHODS

Unilateral hindlimb ischemia was induced in wild-type and heterozygous NCX1 knockout mice (NCX1$^{+/-}$).[7] Animals were anesthetized by intraperitoneal injection of pentobarbital (50 mg/kg). The ligature was performed on the left femoral artery, 0.5 cm proximal to the bifurcation of the saphenous and popliteal arteries. In some mice, intravenous injection of Lipo prostaglandin E1 (Lipo-PGE1), a potent vasodilator and platelet aggregation inhibitor, were done before surgery and once a day after surgery for 7 days. To study possible involvement of endothelial NO synthase (eNOS), we treated mice with hindlimb ischemia with or without N^G-nitro-L-arginine methyl ester (L-NAME) for 28 days. Hindlimb blood perfusion was measured using a Laser Doppler Imaging (Monte System Co., Tokyo, Japan) at post surgery and days 7, 14, 21, and 28. We used all animals in accordance with the Guidelines for Animal Experiments in Fukuoka University.

RESULTS AND DISCUSSION

Unilateral hindlimb ischemia was induced surgically in wild-type and NCX1$^{+/-}$ mice. Systolic blood pressure and heart rate did not differ between wild-type and NCX1$^{+/-}$ mice. The blood flow in the ischemic and nonischemic legs was monitored weekly by Laser Doppler Imaging for 28 days. Immediately after ligature of left femoral artery and vein, a marked reduction of blood flow was observed in the left leg. In wild-type mice, blood flow of the ischemic leg recovered gradually. In NCX1$^{+/-}$ mice, blood flow recovery was significantly augmented compared with that in wild-type mice. Lipo-PGE1 treatment (3 and 10 μg/kg/day, i.v.) enhanced blood flow recovery at day 28 both in wild-type and NCX1$^{+/-}$ mice (with similar degree).

NO is a critical angiogenic mediator. Previous studies have shown that the overexpression of eNOS in the endothelium enhanced angiogenesis in response to hindlimb ischemia[8] and the administration of NO donor stimulated proliferation of cultured rat endothelial cells.[9] Intriguingly, it is considered that endothelial NCX is involved in endothelial NO production.[5,6] Therefore, to study possible involvement of eNOS, we treated mice with hindlimb ischemia with or without L-NAME for 28 days. L-NAME treatment eliminated enhanced angiogenesis observed in NCX1$^{+/-}$ mice as well as in Lipo-PGE1-treated mice. These results suggest that NCX1 is involved in eNOS-dependent angiogenesis (probably through a change in $[Ca^{2+}]_i$ regulation). Endothelial NCX1 thus seems to be a new therapeutic target for angiogenesis.

ACKNOWLEDGMENTS

This work was supported by Grant-in-Aid 18059031 for Scientific Research on Priority Areas from the Ministry of Education, Sports, Science and Technology of Japan, Grant-in-Aid 18590251 for Scientific Research for Japan Society for the Promotion of Science, and grants from the Takeda Science Foundation and the Ichiro Kanehara Foundation.

REFERENCES

1. JUHASZOVA, M., A. AMBESI, G.E. LINDENMAYER, et al. 1994. Na^+-Ca^{2+} exchanger in arteries: identification by immunoblotting and immunofluorescence microscopy. Am. J. Physiol. **266:** C234–C242.
2. QUEDNAU, B.D., D.A. NICOLL & K.D. PHILIPSON. 1997. Tissue specificity and alternative splicing of the Na^+/Ca^{2+} exchanger isoforms NCX1, NCX2, and NCX3 in rats. Am. J. Physiol. **272:** C1250–C1261.
3. TEUBL, M., K. GROSCHNER, S.D. KOHLWEIN, et al. 1999. Na^+/Ca^{2+} exchange facilitates Ca^{2+}-dependent activation of endothelial nitric-oxide synthase. J. Biol. Chem. **274:** 29529–29535.
4. SCHNEIDER, J.-C., D. EL KEBIR, C. CHÉREAU, et al. 2002. Involvement of Na^+/Ca^{2+} exchanger in endothelial NO production and endothelium-dependent relaxation. Am. J. Physiol. **283:** H837–H844.
5. PALTAUF-DOBURZYNSKA, J., M. FRIEDEN, M. SPITALER, et al. 2000. Histamine-induced Ca^{2+} oscillations in a human endothelial cell line depend on transmembrane ion flux, ryanodine receptors and endoplasmic reticulum Ca^{2+}-ATPase. J. Physiol. **524:** 701–713.
6. BERNA, N., T. ARNOULD, J. REMACLE, et al. 2002. Hypoxia-induced increase in intracellular calcium concentration in endothelial cells: role of the Na^+–glucose cotransporter. J. Cell. Biochem. **84:** 115–131.
7. WAKIMOTO, K., K. KOBAYASHI, M. KURO-O, et al. 2000. Targeted disruption of Na^+/Ca^{2+} exchanger gene leads to cardiomyocyte apoptosis and defects in heartbeat. J. Biol. Chem. **275:** 36991–36998.
8. NAMBA, T., H. KOIKE, K. MURAKAMI, et al. 2003. Angiogenesis induced by endothelial nitric oxide synthase gene through vascular endothelial growth factor expression in a rat hindlimb ischemia model. Circulation **108:** 2250–2257.
9. GUO, J.P., M.M. PANDAY, P.M. CONSIGNY, et al. 1995. Mechanisms of vascular preservation by a novel NO donor following rat carotid artery intimal injury. Am. J. Physiol. **269:** H1122–H1131.

The Na$^+$/Ca^{2+} Exchanger Isoform 3 (NCX3) but Not Isoform 2 (NCX2) and 1 (NCX1) Singly Transfected in BHK Cells Plays a Protective Role in a Model of *in Vitro* Hypoxia

AGNESE SECONDO, ILARIA ROSARIA STAIANO,
ANTONELLA SCORZIELLO, ROSSANA SIRABELLA,
FRANCESCA BOSCIA, ANNAGRAZIA ADORNETTO,
LORELLA MARIA TERESA CANZONIERO,
GIANFRANCO DI RENZO, AND LUCIO ANNUNZIATO

*Division of Pharmacology, Department of Neuroscience, School of Medicine,
"Federico II" University of Naples, 80131 Naples, Italy*

ABSTRACT: Chemical hypoxia produces depletion of ATP, intracellular Ca^{2+} overload, and cell death. The role of Na$^+$/Ca^{2+} exchanger (NCX), the major plasma membrane Ca^{2+} extruding system, has been explored in chemical hypoxia using BHK cells stably transfected with the three mammalian NCX isoforms: NCX1, NCX2, and NCX3. Here we report that the three isoforms show similar activity evaluated as [Ca^{2+}]$_i$ increase evoked by Na$^+$-free medium exposure in Fura-2-loaded single cells and NCX3 transfected cells are less vulnerable to chemical hypoxia compared to NCX1- and NCX2-transfected cells, suggesting that NCX3 could play a more relevant protective role during chemical hypoxia.

KEYWORDS: Na$^+$/Ca^{2+} exchanger; chemical hypoxia; neuroprotection; [Ca^{2+}]$_i$ homeostasis

The Na$^+$/Ca^{2+} exchanger (NCX)[1] is a 9-transmembrane protein that couples the efflux of Ca^{2+} ions to the influx of Na$^+$ into the cell or *vice versa* by operating in a bidirectional way.[2,3] Under physiological conditions, the primary role of the three NCX isoforms is to extrude Ca^{2+} via a forward mode of operation in response to a depolarization or a receptor stimulation coupled to

Address for correspondence: Lucio Annunziato, M.D., Division of Pharmacology, Department of Neuroscience, School of Medicine, "Federico II" University of Naples, Via Pansini 5, 80131 Naples, Italy. Voice: +39-81-7463318; fax: +39-81-7463323.
lannunzi@unina.it

Ann. N.Y. Acad. Sci. 1099: 481–485 (2007). © 2007 New York Academy of Sciences.
doi: 10.1196/annals.1387.052

an increase in intracellular Ca^{2+} concentrations ($[Ca^{2+}]_i$).[4] Furthermore, when intracellular concentrations of Na^+ increase—like in the presence of Na^+-K^+ ATPase blockade—a profound Ca^{2+} entry occurs through the activation of the *reverse mode*.

Three different gene products of this exchanger have been cloned[5–7]: NCX1, which is ubiquitously expressed in several tissues, NCX2, and NCX3, both widely expressed in the brain[8] and in the skeletal muscle.[9] In the central nervous system (CNS) both the transcripts and the proteins of each of the three subtypes show a distinct distribution pattern in different brain regions, suggesting that this heterogeneity might serve selective functional roles in distinct regions of the brain according to cellular demands.[8,10,11] In the last decade, several studies have reported that NCX is implicated in several pathological conditions of the CNS, such as anoxia,[12,13] chemical hypoxia,[12–14] aging,[15] Alzheimer's disease,[16] and glutamate-induced neurotoxicity.[17,18] More interestingly, Pignataro *et al.*[19] and Boscia *et al.*[20] demonstrated that the three proteins and transcripts of NCX1, NCX2, and NCX3 display a different expression in the ischemic core and in the peri-infarct brain regions after permanent middle cerebral artery occlusion (pMCAO). In addition, the knocking down of NCX1 and NCX3, but not of NCX2, by the oligodeoxynucleotide strategy exacerbates ischemic damage. By contrast, further evidence has suggested that the pharmacological inhibition of NCX reduces brain injury.[13] In light of this peculiar pattern of expression characterizing the three NCX isoforms and the contradictory role played by the exchanger, it was of interest to investigate the activity of these three gene products singly transfected in baby kidney hamster (BHK) cells that constitutively do not express any of the three isoforms. More specifically, the overall goal of this study was to determine the following hypotheses: (*a*) whether $[Ca^{2+}]_i$ handling in resting conditions depends on the type of NCX isoforms transfected in BHK cells; and (*b*) which of the three BHK cell clones, each singly and stably transfected with one NCX isoform, is more resistant to chemical hypoxia plus reoxygenation.

The results of this study, performed with the help of cloned cells stably expressing each isoform of the NCX, demonstrated that the NCX3 gene product, specifically expressed in the brain, may play a more relevant role in cell resistance to hypoxic conditions. Indeed, NCX3 transfected in BHK cells, unlike NCX1, NCX2, and wild-type cells are able to better survive chemical hypoxia injury possibly by virtue of their Ca^{2+}-buffering properties in conditions of ATP depletion. During resting conditions, the activities of the three isoforms, assessed by using single cell Fura-2 microfluorimetry technique, are comparable. In particular, NCX activity was evaluated in the reverse mode of operation elicited by single pulse of Na^+-deficient $NMDG^+$ medium (Na^+-free) in the absence or in the presence of the SERCA inhibitor thapsigargin (FIG. 1 A). The degree of the exchange activity was similar for all the three isoforms, as revealed by the quantification of $[Ca^{2+}]_i$ increase elicited by Na^+-free in BHK-NCX1, BHK-NCX2, and BHK-NCX3 (data not shown).

To evaluate cell resistance of BHK-Wt, BHK-NCX1, BHK-NCX2, and BHK-NCX3 transfected cells to hypoxia, these clones were exposed to 45 min of hypoxia followed by 15 h of reoxygenation. At this time, BHK-NCX3 clone was more resistant to this insult as compared to BHK-Wt, BHK-NCX1, and BHK-NCX2 clones and as revealed by the staining with propidium iodide and fluorescein diacetate (FIG. 1 B). Here we present preliminary evidence that BHK cells stably expressing NCX3 isoform are more resistant to chemical hypoxia plus reoxygenation-induced cell death.[21]

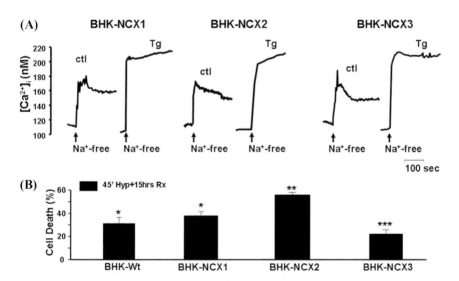

FIGURE 1. Effect of Na^+-free on $[Ca^{2+}]_i$ and effect of chemical hypoxia plus reoxygenation on cell death in BHK-Wt, BHK-NCX1, BHK-NCX2, and BHK-NCX3 cells. Panel **A** shows representative traces of the effect of Na^+-free solution perfused in the absence or in the presence of thapsigargin (Tg) (1 μM) on $[Ca^{2+}]_i$ in BHK-NCX1, BHK-NCX2, and BHK-NCX3 cells. $[Ca^{2+}]_i$ was evaluated with single cell Fura-2 (5 μM) video-imaging technique. Each trace is representative of circa 60–120 cells studied in at least three different experimental sessions. Panel **B** shows a bar graph of the effect of 45′ chemical hypoxia plus 15 h reoxygenation on cell death in BHK-Wt, BHK-NCX1, BHK-NCX2, and BHK-NCX3 cells double stained with 36 μM fluorescein diacetate and 7 μM propidium iodide. * $P < 0.05$ versus 45 min of chemical hypoxia; ** $P < 0.05$ versus 45 min of chemical hypoxia + 15 h of reoxygenation of BHK-Wt, BHK-NCX1, and BHK-NCX3; *** $P < 0.05$ versus 45 min of hypoxia + 15 h of reoxygenation of BHK-Wt, BHK-NCX1, and BHK-NCX2.

REFERENCES

1. PHILIPSON, K.D. & D.A. NICOLL. 2000. Sodium-calcium exchanger: a molecular perspective. Annu. Rev. Physiol. **62:** 111–113.

2. SANCHEZ-ARMASS, S. & M.P. BLAUSTEIN. 1987. Role of sodium-calcium exchanger in regulation of intracellular calcium in nerve terminals. Am. J. Physiol. **252:** C595–C603.

3. TAGLIALATELA, M. *et al.* 1990. Na^+-Ca^{2+} exchanger activity in central nerve endings. I. Ionic conditions that discriminate $^{45}Ca^{2+}$ uptake through the exchanger from that occurring through voltage-operated Ca^{2+} channels. Mol. Pharmacol. **38:** 385–392.

4. BLAUSTEIN, M.P. & W.J. LEDERER. 1999. Sodium/calcium exchange: its physiological implications. Physiol. Rev. **79:** 763–854.

5. NICOLL, D.A. *et al.* 1990. Molecular cloning and functional expression of the cardiac sarcolemmal Na^{2+}-Ca^{2+} exchanger. Science **250:** 562–565.

6. NICOLL, D.A. *et al.* 1996. Cloning of the third mammalian Na^{2+}-Ca^{2+} exchanger, NCX3. J. Biol. Chem. **271:** 24914–24921.

7. LI, Z. *et al.* 1994. Cloning of the NCX2 isoform of the plasma membrane Na^{2+}-Ca^{2+} exchanger. J. Biol. Chem. **269:** 17434–17439.

8. PAPA, M. *et al.* 2003. Differential expression of the Na^+-Ca^{2+} exchanger transcripts and proteins in rat brain regions. J. Comp. Neurol. **461:** 31–48.

9. LEE, S.L. *et al.* 1994. Tissue specific expression of Na^+-Ca^{2+} exchanger isoforms. J. Biol. Chem. **269:** 14849–14852.

10. YU, L. & R.A. COLVIN. 1997. Regional differences in expression of transcripts for Na^+/Ca^{2+} exchanger isoforms in rat brain. Brain Res. Mol. Brain Res. **50:** 285–292.

11. CANITANO, A. *et al.* 2002. Brain distribution of the Na^+-Ca^{2+} exchanger-encoding genes NCX1, NCX2 and NCX3 and their related proteins in the central nervous system. Ann. N. Y. Acad. Sci. **976:** 394–404.

12. AMOROSO, S. *et al.* 1997. Pharmacological evidence that the activation of the Na^+-Ca^{2+} exchanger protects C6 glioma cells during chemical hypoxia. Br. J. Pharmacol. **121:** 303–309.

13. MATSUDA, T. *et al.* 1996. Involvement of Na^+-Ca^{2+} exchanger in reperfusion-induced delayed cell death of cultured rat astrocytes. Eur. J. Neurosci. **8:** 951–958.

14. AMOROSO, S. *et al.* 2000. Sodium nitroprusside prevents chemical hypoxia-induced cell death through iron ions stimulating the activity of the Na^+-Ca^{2+} exchanger in C6 glioma cells. J. Neurochem. **74:** 1505–1513.

15. CANZONIERO, L.M. *et al.* 1992. The Na^+-Ca^{2+} exchanger activity in cerebrocortical nerve endings is reduced in old compared to young and mature rats when it operates as a Ca^{2+} influx or efflux pathway. Biochim. Biophys. Acta **1107:** 175–178.

16. WU, A. *et al.* 1997. Alzheimer's amyloid-beta peptide inhibits sodium/calcium exchange measured in rat and human brain plasma membrane vesicles. Neuroscience **80:** 675–684.

17. KIEDROWSKI, L. *et al.* 2004. Differential contribution of plasmalemmal Na^+/Ca^{2+} exchange isoforms to sodium-dependent calcium influx and NMDA excitotoxicity in depolarized neurons. J. Neurochem. **90:** 117–128.

18. BANO, D. *et al.* 2005. Cleavage of the plasma membrane Na^+-Ca^{2+} exchanger in excitotoxicity. Cell **120:** 275–285.

19. PIGNATARO, G. *et al.* 2004. Two Na^+-Ca^{2+} exchanger gene products, NCX1 and NCX3, play a major role in the development of permanent focal cerebral ischemia. Stroke **35:** 2566–2570.

20. BOSCIA, F. *et al.* 2005. Permanent focal brain ischemia induces isoform-dependent changes in the pattern of Na^+-Ca^{2+} exchanger gene expression in the ischemic core, periinfarct area, and intact brain regions. J. Cereb. Blood Flow Metab. **26:** 502–517.

21. SECONDO, A. *et al.* 2007. BHK cells transfected with NCX3 are more resistant to hypoxia followed by reoxygenation than those transfected with NCX1 and NCX2: possible relationship with mitochondrial membrane potential. Cell Calcium In press.

Involvement of the Potassium-Dependent Sodium/Calcium Exchanger Gene Product NCKX2 in the Brain Insult Induced by Permanent Focal Cerebral Ischemia

ORNELLA CUOMO,[a] GIUSEPPE PIGNATARO,[a] ROSARIA GALA,[a] FRANCESCA BOSCIA,[a] ANNA TORTIGLIONE,[a] PASQUALE MOLINARO,[a] GIANFRANCO DI RENZO,[a] JONATHAN LYTTON,[b] AND LUCIO ANNUNZIATO[a]

[a]Division of Pharmacology, Department of Neuroscience, School of Medicine, Federico II University of Naples, Via S. Pansini 5, 80131 Naples, Italy

[b]Department of Biochemistry and Molecular Biology, University of Calgary, Calgary, T2N 4N1 Alberta, Canada

ABSTRACT: Sodium/calcium exchangers are neuronal plasma membrane transporters, which by coupling Ca^{2+} and Na^+ fluxes, may play a relevant role in brain ischemia. The exchanger gene superfamily comprises two arms: the K^+-independent (NCX) and K^+-dependent (NCKX) exchangers. In the brain, three different NCX (NCX1, NCX2, NCX3) and three NCKX (NCKX2, NCKX3, NCKX4) family members have been described. Up to now, no sutides about the role played by NCKX proteins in cerebral ischemia have been published. The aim of the present study was to investigate the role of NCKX2 in an *in vivo* model of permanent middle cerebral artery occlusion (pMCAO). The role of this protein in the development of ischemic damage was assessed by knocking-down its expression with an antisense oligodeoxynucleotide (AS-ODN), intracerebroventricularly infused by an osmotic minipump for 48 h, starting from 24 h before pMCAO. The results showed that NCKX2 knocking-down by using antisense strategy increased the extent of the ischemic lesion. The results of this study suggest that NCKX2 could exert a neuroprotective effect during ischemic injury.

KEYWORDS: Na^+/Ca^{2+} exchanger; cerebral ischemia; NCKX2; antisense oligodeoxynucleotides

Address for correspondence: Lucio Annunziato, Division of Pharmacology, Department of Neuroscience, School of Medicine, Federico II University of Naples, Via S. Pansini 5, 80131 Naples, Italy. Voice: +39-81-7463325; fax: +39-81-7463323.
lannunzi@unina.it

Ann. N.Y. Acad. Sci. 1099: 486–489 (2007). © 2007 New York Academy of Sciences.
doi: 10.1196/annals.1387.051

INTRODUCTION

Recently, evidence has been provided on the involvement of some members of the sodium–calcium exchanger superfamily in the damage induced by brain ischemia. Detailed structural and functional studies have revealed an exchange gene superfamily comprising two arms: potassium-independent sodium–calcium exchangers (NCX)[1] and potassium-dependent sodium–calcium exchangers (NCKX) that can mediate Ca^{2+} and Na^+ fluxes across the plasma membrane.[2] Four genes of the NCKX family have been cloned, NCKX1 (SLC24A1),[3] NCKX2 (SLC24A2),[4] NCKX3 (SLC24A3),[5] and NCKX4 (SLC24A4).[6] All these proteins counter-transport in a bidirectional way 4 Na^+ versus 1 Ca^{2+} plus 1 K^+. Apart from NCKX1, which was initially characterized in retinal rod outer segments and then found in cells of hematopoietic origin, the other gene products were widely distributed throughout the brain.[4-6]

Previous studies from our laboratory showed that NCX gene products played a differential role in the ischemic rat brain and that NCX activation improved, whereas NCX inhibition impaired neuronal damage after permanent middle cerebral artery occlusion (pMCAO) in rats.[1,7,8]

Up to now, although studies performed in primary cultures of cerebellar granule cells showed that the inhibition of this potassium-dependent antiporter was required to significantly limit NMDA excitotoxicity in cerebellar granule cells,[9] no information about the role of NCKX in cerebral ischemia is available. The understanding of the specific role of NCKX2 isoform in the development of the ischemic damage could be useful to clarify its role in cerebral ischemia. To this aim, we knocked down NCKX2 protein expression with a specific antisense oligodeoxynucleotide (AS-ODN; FIG. 1). After this

+136   +155
NCKX2   AGCAGGAAAAAACTGAAGCTAATTCGAGTCATTGGCCTTGTCATGGGCCTGGTAGC

Antisense Oligo against NCKX2 5'-CCAATGACTCGAATTAGCTT-3'

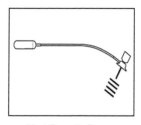

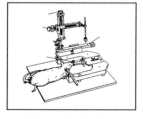

Alzet Osmotic Pump   Stereotaxic System

FIGURE 1. Sequence of the AS-ODN against NCKX2. NCKX2 AS-ODN (140 μg/kg) was continuously intracerebroventricularly (i.c.v.) infused by an osmotic minipump (1 μL/h) 24 h before and after pMCAO. Control rats received vehicle.

treatment, a remarkable enlargement of the infarct volume occurred. It can be hypothesized that the forward mode of NCKX2 is still operative in the penumbral region, as it occurs for NCX.[8] In this way, the exchanger will be extruding Ca^{2+} and K^+ and would promote the entry of Na^+. Therefore, the inhibition of NCKX2 at this stage would reduce the elimination of Ca^{2+} and K^+, thus enhancing Ca^{2+}- and K^+-mediated cell injury.[10,11] In agreement with this hypothesis, it has to be highlighted that the apparent affinity constant of K^+ for NCKX2 is much higher ($K_m = 29.3$ mM) than that for the other two isoforms ($K_m \sim 1$ mM), so its affinity to K^+ is lower. This peculiarity makes this isoform able to operate in the forward mode, even if the cell is strongly depolarized, as it might occur after an ischemic insult. NCKX2 protein expression inhibition, by blocking calcium extrusion, could cause a worsening of the ischemic damage.

By contrast, in the ischemic core region, in which a remarkable ATP loss occurs, there is a massive accumulation of intracellular Na^+, due to the failure of Na^+/K^+ ATPase. This intracellular Na^+ loading should promote the NCKX2 to operate in the reverse mode, leading probably to an accumulation of Ca^{2+} and K^+. Therefore, the NCKX2 inhibition in the core region would further worsen the necrotic lesion of the surviving glial and neuronal cells as intracellular Na^+ loading will increase.

In conclusion, these preliminary results suggest that the integrity of NCKX2 functional activity is a prerequisite for reducing brain insult induced by focal cerebral ischemia.

REFERENCES

1. ANNUNZIATO L., G. PIGNATARO & G.F. DI RENZO. 2004. Pharmacology of brain Na^+/Ca^{2+} exchanger: from molecular biology to therapeutic perspectives. Pharmacol. Rev. **56:** 633–654.
2. BLAUSTEIN M.P. & W.J. LEDERER. 1999. Sodium/calcium exchange: its physiological implications. Physiol. Rev. **79:** 763–840
3. REILANDER H., A. ACHILLES, U. FRIEDEL, et al. 1992. Primary structure and functional expression of the Na/Ca,K-exchanger from bovine rod photoreceptors. EMBO J. **11:** 1689–1695.
4. TSOI M., K.H. RHEE & D. BUNGARD. 1998. Chromosomal localization and genomic organization of the human retinal rod Na-Ca + K exchanger. Hum. Genet. **103:** 411-414.
5. KRAEV A., B.D. QUEDNAU, S. LEACH, et al. 2001. Molecular cloning of a third member of the potassium-dependent sodium-calcium exchanger gene family, NCKX3. J. Biol. Chem. **276:** 23161-23172.
6. LI X.F., A.S. KRAEV & J. LYTTON. 2002. Molecular cloning of a fourth member of the potassium-dependent sodium-calcium exchanger gene family, NCKX4. J. Biol. Chem. **277:** 48410–48417.
7. PIGNATARO G., A. TORTIGLIONE, A. SCORZIELLO, et al. 2004a. Evidence for a protective role played by the Na^+/Ca^{2+} exchanger in cerebral ischemia induced by middle cerebral artery occlusion in male rats. Neuropharmacology **46:** 439–448.

8. PIGNATARO G., R. GALA, O. CUOMO, *et al.* 2004b. Two sodium/calcium exchanger gene products, NCX1 and NCX3, play a major role in the development of permanent focal cerebral ischemia. Stroke **35:** 2566–2570.

9. KIEDROWSKI L., A. CZYZ, G. BARANAUSKAS, *et al.* 2004. Differential contribution of plasmalemmal Na/Ca exchange isoforms to sodium-dependent calcium influx and NMDA excitotoxicity in depolarized neurons. J. Neurochem. **90:** 117–128.

10. KRISTIAN T. & B.K. SIESJO. 1998. Calcium in ischemic cell death. Stroke **29:** 705–718.

11. YU S.P., C.H. YEH & S.L. SENSI. 1997. Mediation if neuronal apoptosis by enhancement of outward potassium current. Science **278:** 114–117.

Analysis of Calcium Changes in Endoplasmic Reticulum during Apoptosis by the Fluorescent Indicator Chlortetracycline

CLAUDIA CERELLA,[a] CRISTINA MEARELLI,[a] MILENA DE NICOLA,[a] MARIA D'ALESSIO,[a] ANDREA MAGRINI,[b] ANTONIO BERGAMASCHI,[c] AND LINA GHIBELLI[a]

[a]Dipartimento di Biologia, Universita' di Roma Tor Vergata, 00133 Rome, Italy

[b]Cattedra di Medicina del Lavoro, Universita' di Roma Tor Vergata, 00133 Rome, Italy

[c]Cattedra di Medicina del Lavoro, Universita' Cattolica "Sacro Cuore," 00168 Rome, Italy

ABSTRACT: Many studies suggest that endoplasmic reticulum (ER) Ca^{2+} pool rather than cytosolic Ca^{2+} may play a crucial role in triggering apoptosis. In this study, we performed an image analysis of cells loaded with the fluorescent dye chlortetracycline (CTC) to *in situ* analyze Ca^{2+} changes within the ER in apoptosing promonocytic U937 cells. The results, validated through the use of thapsigargin (THG) as ER Ca^{2+} depletor, confirm the findings that apoptotic cells have a Ca^{2+}-depleted ER, in contrast with treated but still viable cells.

KEYWORDS: apoptosis; endoplasmic reticulum; calcium; thapsigargin; chlortetracycline

INTRODUCTION

Endoplasmic reticulum (ER) is the main store of Ca^{2+} in the cell, involved in many intracellular signaling events including apoptosis. Many studies suggest that sensitivity to apoptosis correlates with the amount of Ca^{2+} releasable from ER, rather than cytosolic Ca^{2+}.[1] However, the actual role of ER Ca^{2+} in apoptosis is still uncertain. One problem is the availability of direct and reliable methods to directly analyze ER Ca^{2+} content changes in apoptosis. A second

Address for correspondence: Claudia Cerella, Dipartimento di Biologia, Universita' di Roma Tor Vergata, via Ricerca Scientifica, 1, 00133 Roma (I), Italy. Voice: +39-06-7259-4323; fax: +39-06-2023500.

cerella@uniroma2.it

Ann. N.Y. Acad. Sci. 1099: 490–493 (2007). © 2007 New York Academy of Sciences.
doi: 10.1196/annals.1387.021

problem is the unhomogeneity of the population of cells induced to apoptosis, due to the well-known asynchrony of the apoptotic process.

Chlortetracycline (CTC) is a cell-permeant fluorescent molecule, possessing a hydrophobic moiety that allows its targeting to intracellular membranes. This, combined with the ability to increase fluorescence upon Ca^{2+} binding ($K_d =$ ~400 nM), allows CTC to delineate intracellular Ca^{2+} containing vesicles, such as ER as a visual pattern upon fluorescence microscopy analysis. Thus, CTC seems an ideal molecule for a direct *in situ* analysis of ER Ca^{2+} changes.

In this study, we describe a non-invasive method that easily allows to monitor Ca^{2+} changes within ER, based on image analysis of cells loaded with CTC. We validated and applied this approach to apoptosing U937 cells, showing a low ER Ca^{2+} content in apoptotic cells. This *in situ* approach also allowed to show different Ca^{2+} states in apoptotic versus still viable cells.

MATERIALS AND METHODS

Cell Culture and Treatments

U937 cells (human tumoral promonocytes) were cultured as described.[2,3] Apoptosis was induced with 10 μg/mL puromycin (PMC) and evaluated by analyzing nuclear morphology in stained cell nuclei with 8 μg/mL 7-aminoactinomycin D.[2,3] *Analysis of CTC fluorescence*: (A) Fluorimetric analysis—Cells at the concentration of 2×10^6/mL were washed in phosphate saline buffer (PBS) and incubated at 37°C with 10 μM CTC for 15 min in the dark. Cells were washed twice and resuspended in modified Mg^{2+}-deprived Hank's balanced salt solution (HBSS) + 1 mg/mL glucose + 650 μM $CaCl_2$. CTC fluorescence was measured using a Perkin–Elmer (Waltham, MA) fluorimeter at 37°C, with lambda emission and lambda excitation tuned at 398 nm and 527 nm, respectively. (B) Fluorescence microscopy analysis—1×10^6 cells were fixed in 4% paraformaldehyde for 15 min, washed in PBS and incubated for 15 min at 37°C with 100 μM CTC and 8 μg/mL 7-aminoactinomycin D to visualize at the same time nucleus and ER. At the end of the double staining, cells were washed and resuspended in the same buffer at the concentration of 1×10^6/mL. The cells were loaded on slides and analyzed for fluorescence microscopy (filters D for CTC; M2 for 7-aminoactinomycin D). *Analysis of cytosolic Ca^{2+}*: cells were stained with Fluo3-AM and cytosolic Ca^{2+} measurements were performed as described.[3] *Modulation of ER Ca^{2+} content*: Thapsigargin (THG) was used at 10 nM or 1 μM, to differentially poison the different ER Ca^{2+} ATPases.[4]

RESULTS AND DISCUSSION

First of all, we checked whether CTC fluorescence was due to the ER Ca^{2+} content. To this purpose, we performed a fluorimetric analysis of CTC-loaded

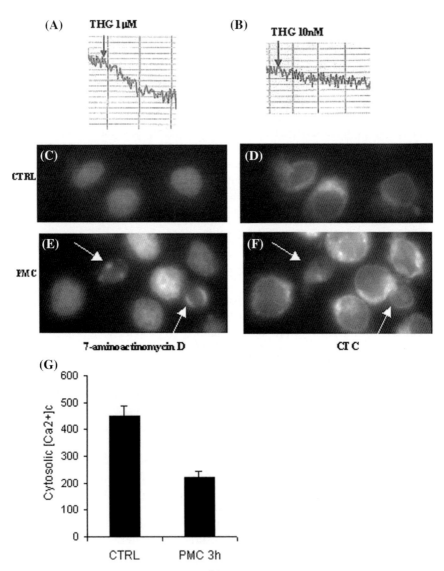

FIGURE 1. CTC reveals reticular Ca^{2+} content during apoptosis. (**A** and **B**) Fluorimetric analysis shows that CTC evidentiates ER Ca^{2+} content: 1 μM or 10 nM THG, as ER Ca^{2+} depletor, lower CTC fluorescence in a dose-dependent way. The height of every quadrant corresponds to fluorescence arbitrary units; the width corresponds to 1cm and the advance rate paper corresponds to 5 mm/min. It is shown in one of the three experiments. (**C, D, E, F**) Double-staining with 7-aminoactinomycin D, for morphological determination of apoptosis, and CTC in untreated and 3h PMC-treated U937. Upon PMC treatment, apoptotic cells present a very diminished CTC fluorescence with respect to untreated cells. (**G**) Quantification of cytosolic Ca^{2+} increase after THG-dependent due to ER Ca^{2+} pool depletion.[4]

cells in the presence or absence of THG, here used as a tool of ER Ca^{2+} emptying. THG significantly lowered CTC fluorescence in a dose-dependent way (FIG. 1 A, B), thus indicating that CTC actually reveals ER Ca^{2+}.

Next, we analyzed ER Ca^{2+} content. Cells induced to apoptosis by PMC are easily recognized by the fragmented nucleus,[2,3] here evidenced by 7-aminoactinomycin D (FIG. 1 C, E). Apoptotic cells show a weaker CTC staining with respect to either untreated cells, or treated cells that are still viable (i.e., not yet apoptotic; FIG. 1 D, F). The lower ER Ca^{2+} content was confirmed by the standard indirect method, based on the fluorimetric measurement of Ca^{2+} releasable by THG as the increase of cytosolic Ca^{2+}[3,4] (FIG. 1 G).

All these results indicate that apoptotic U937 cells have a depleted ER Ca^{2+} pool. In addition, the *in situ* CTC measurement, showing heterogeneity among treated cells, could reveal that this depletion is not a consequence of the damaging treatment, but is concomitant with apoptosis, as shown by the perfect overlapping of apoptotic nuclei and depleted ER. We propose that CTC could be an alternative approach to the well-known *in situ* analysis based on cell transfection with the recombinant Ca^{2+}-sensitive photoprotein aequorin.[1]

REFERENCES

1. PINTON, P. *et al.* 2001. The Ca^{2+} concentration of the endoplasmic reticulum is a key determinant of ceramide-induced apoptosis: significance for the molecular mechanism of Bcl-2 action. EMBO J. **20:** 2690–2701.
2. GHIBELLI, L. *et al.* 2006. NMR exposure sensitizes tumor cells to apoptosis. Apoptosis **16:** 19–43.
3. FANELLI, C. *et al.* 1999. Magnetic fields increase cell survival by inhibiting apoptosis via modulation of Ca^{2+} influx. FASEB J. **13:** 95–102.
4. THASTRUP, O. *et al.* 1990. Thapsigargin, tumor promoter, discharges intracellular Ca^{2+} stores by specific inhibition of the endoplasmic reticulum Ca^{2+}-ATPase. Proc. Natl. Acad. Sci. USA **82:** 2466–2470.

Sodium–Calcium Exchangers in the Nucleus

An Unexpected Locus and an Unusual Regulatory Mechanism

ROBERT W. LEDEEN AND GUSHENG WU

Department of Neurology and Neurosciences, New Jersey Medical School–UMDNJ, Newark, New Jersey 07103, USA

ABSTRACT: Whereas sodium–calcium exchangers (NCXs) have long been recognized as plasma membrane constituents that serve to maintain homeostatic concentrations of Ca^{2+} in the cytoplasm, they were recently shown to also occur in the nuclear envelope (NE) of neural and other cells where they function to regulate nuclear Ca^{2+}. A unique feature of NCXs in the NE is their high-affinity binding to GM1 ganglioside, this association being required for optimal exchanger activity. The NCX–GM1 complex occurs in the inner membrane of the NE and transfers Ca^{2+} from the nucleoplasm to the NE lumen. In neuronal cells, nuclear GM1 levels are low prior to differentiation but increase rapidly as axonal outgrowth progresses. Cells from genetically altered mice lacking GM1 have limited ability to regulate nuclear Ca^{2+}, and the mice themselves showed similar deficit as seen in their high susceptibility to kainite-induced seizures. These are attenuated by LIGA-20, a derivative of GM1 that enters the nuclear membrane and restores nuclear NCX activity to normal level.

KEYWORDS: sodium–calcium exchanger; nuclear calcium; gangliosides; GM1 ganglioside–NCX complex; nuclear membrane; cytoprotection by nuclear NCX–GM1 complex

INTRODUCTION

Sodium–calcium exchangers (NCXs) as well as gangliosides are generally perceived as plasma membrane components, and while that is indeed the predominant locus in most cells, both groups of molecules have also been detected in intracellular sites.[1–3] The list of intracellular loci was expanded when it was discovered that both gangliosides[4,5] and NCXs[6] occur in the nuclear envelope

Address for correspondence: Robert Ledeen, New Jersey Medical School–UMDNJ, Department of Neurology and Neurosciences MSB H506, 185 So. Orange Avenue, Newark, NJ 07103. Voice: 973-972-7989; fax: 973-972-5059.
ledeenro@umdnj.edu

Ann. N.Y. Acad. Sci. 1099: 494–506 (2007). © 2007 New York Academy of Sciences.
doi: 10.1196/annals.1387.057

(NE) of several cell types and form a molecular association. These somewhat unexpected findings raised the question of physiological function, to which at least a partial answer was provided in experiments demonstrating regulation of Ca^{2+} in various nuclear compartments.[6] Moreover, this modulatory function in the nucleus required high-affinity association of NCX and GM1, a feature not shared with these components in the plasma membrane. A summary of these and related findings are recounted in the following review. Additional reviews have appeared recently, which expand on this subject and emphasize different aspects of the phenomenon.[7–9]

DETECTION OF GM1 AND NCX IN THE INNER NUCLEAR MEMBRANE

Whereas the earliest studies of nuclear gangliosides employed nuclei from nonnervous tissues,[4,10,11] our investigations began with primary neurons from the central and peripheral nervous systems[5,12] and neuroblastoma cells capable of differentiation.[5,13,14] Using both chemical analysis applied to isolated nuclei and immunocytochemical staining of whole cells with cholera toxin B (CtxB) subunit linked to horseradish peroxidase (HRP), GM1 ganglioside was shown to occur in nuclei with major localization in the NE (FIG. 1). This applied to all the above neural cell types. Chemical analysis based on thin-layer chromatography (TLC) revealed a relatively simple ganglioside pattern, GD1a being the only other ganglioside that was well expressed in the NE of NG108-15 neuroblastoma cells. Prominent expression of GM1 in the NE occurred in such cells and primary neurons following differentiation, since relatively little GM1 was detected in the NE prior to process outgrowth (FIG. 1). Moreover, working with neuroblastoma cells in which axonal- and dendritic-like processes could be separately induced, developmental increase of GM1 in the NE was found to correspond to axon outgrowth.[13,14] Other cell types of the nervous system as well as nonneural cells were later found to contain nuclear GM1 (see below).

Concurrent immunoblot studies in our laboratory revealed coexistence of NCX with GM1 in the NE of neurons and neuroblastoma cells.[6] Using anti-NCX1 antibody (Ab), immunoblot analysis applied to isolated NE from NG108-15 neuroblastoma cells revealed a pattern of NCX isoforms generally similar to that obtained with plasma membrane from the same source (FIG. 2). This included prominent bands at 220, 160, and 120 kDa, and also multiple bands at ~70 kDa. The latter has been described as a possible proteolytic fragment(s),[15] but a subsequent report[16] described a new splicing pattern that yields a 70 kDa protein corresponding to the N-terminal portion of the mature protein (120 kDa). The band at ~220 kDa, similar to one of that size detected in chinese hamster ovary (CHO) cells transfected with canine cardiac exchanger,[17] could represent a dimer of the 120 kDa or trimer of the 70 kDa protein. A few differences were noted in the NCX1 immunoblot pattern obtained with NE from differentiated versus undifferentiated cells, such as a

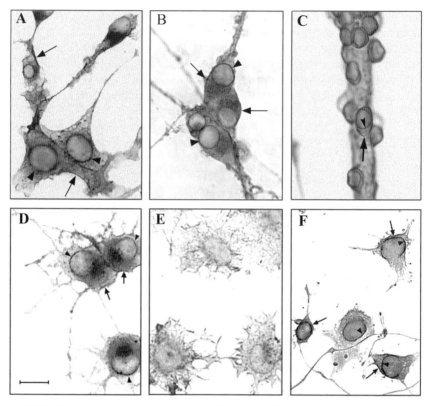

FIGURE 1. Cytochemical detection of GM1 and NCX in the NE of neural cell lines and primary neurons. Ctx B-HRP revealed GM1 in the NE of (**A**) differentiating Neuro2a cells, (**B**) cultured neurons from the superior cervical ganglion of embryonic rats, (**C**) cultured cerebellar granule neurons (CGN) from neonatal rat brain after 6 days *in vitro*, (**D**) differentiating NG108-15 cells, (**E**) undifferentiated NG108-15 cells. (**F**) Anti-NCX antibody linked to HRP was used to detect NCX in the NE of differentiating NG108-15 cells. In all figures, arrowheads indicate visible staining of NE and arrows staining of plasma membrane. Note the paucity of GM1 staining in the NE of undifferentiated NG108-15 cells (**E**) as compared to the distinctive presence of GM1 in the NE of differentiating neuronal cells (**A–D**). FIGURE 1 **A** and **C** are reproduced from FIGURE 3 of Wu *et al.* (Ref. 5) with permission of J. Neuroscience; FIGURE 1 **B** is reproduced from FIGURE 4 of Kozireski-Chuback *et al.* (Ref. 13) with permission of Elsevier; FIGURE 1 **D–F** are reproduced from FIGURE 1 of Xie *et al.* (Ref. 6) with permission of J. Neurochemistry.

more prominent band at 160 kDa in the former. The latter band was described as the mature NCX that shifted from 120 to 160 kDa gel position under nonreducing conditions; evidence suggested this involved a conformational change due to a disulfide bond.[18]

FIGURE 2 reveals another interesting feature obtained by running two parallel immunoblots following immunoprecipitation with anti-NCX1: one was stained

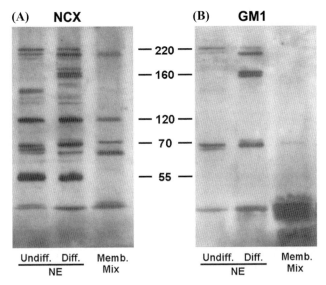

FIGURE 2. Immunoprecipitation and immunoblot analysis of the NCX associated with GM1 in the NE. The NE and mixture of nonnuclear cell membranes (Memb. Mix, including plasma membranes) were prepared from KCl-differentiated (Diff.) or undifferentiated (Undiff.) NG108-15 cells. Each was dispersed in 1% Triton X-100 and immunoprecipitated with monoclonal anti-NCX antibody plus protein L-agarose beads. The precipitated proteins were subjected to 7% nonreducing SDS-PAGE, transferred to PVDF membrane and blotted with polyclonal rabbit anti-NCX (**A**). A parallel SDS-PAGE gel was transferred to PVDF and blotted with Ctx B-HRP (**B**). NCX isoforms from the NE appearing at 220, 160, and 70 kDa remained associated with GM1 throughout SDS-PAGE and blotting. In contrast, NCX isoforms from the membrane mixture (containing plasma membrane) showed no retention of GM1 during SDS-PAGE, although the heavy staining at the migration front of Memb. Mix. (third lane of **B**) suggested GM1 might have been loosely associated with one or more of these isoforms. Reproduced from FIGURE 3 of Xie *et al.* (Ref. 6) with permission of J. Neurochemistry.

with anti-NCX1 and the other with CtxB-HRP. This revealed that not only did GM1 co-precipitate with NCX1, but also co-migrated with the latter during SDS-PAGE. This indicated high-affinity binding of NCX1 to GM1 in the NE, a property not shared with NCX1 in the plasma membrane; however, the presence of GM1 in the migration front of the latter (FIG. 2 B, third lane) suggests GM1 and NCX1 in the PM may associate with enough affinity for co-precipitation but not enough for co-migration during SDS-PAGE. We know of only one other example of such high-affinity binding of GM1 to a membrane protein, that being the Trk receptor in neural cells.[19] The parallel immunoblots of FIGURE 2 indicate GM1 is associated with the NCX1 isoforms at 70, 160, and 220 kDa but not the one at 120 kDa.

The NE is a double membrane structure, the outer membrane of which is continuous with the endoplasmic reticulum (ER) while the inner membrane is unique with an entirely different protein and lipid composition. The question thus arose as to which of the two membranes contains the NCX–GM1 complex. This was answered by employing a simple and effective procedure for separating the two membranes, based on selective stripping of the outer membrane with Na-citrate.[20] Successful separation was indicated by immunoblot analysis of Lamin B, which was abundantly present in the inner membrane fraction but only evident as a trace in the outer membrane.[8] Immunoblot analysis of the separated membranes following immunoprecipitation with anti-NCX showed the inner membrane as the locus for both NCX and GM1 (FIG. 3). Moreover, this experiment provided final proof that the NCX1–GM1 complex is truly intrinsic to the NE, and not part of another structure (e.g., Golgi apparatus, ER) that might have associated with the outer membrane of the NE. The results of TLC analysis of the separated membranes of the NE, involving no immunoprecipitation (FIG. 3 C), showed that gangliosides of the NE likely have other functions besides interaction with NCX; this was suggested in the observation that the outer membrane, which lacks NCX, has significant levels of both GM1 and GD1a. The same may apply to the inner membrane, since the ratio of GD1a to GM1 is much higher before (FIG. 3 C) compared to after (FIG. 3 D) immunoprecipitation. The role of GD1a is not known but has been proposed as that of reserve for GM1, undergoing conversion to the latter as needed by a sialidase detected in the NE.[21]

FUNCTIONAL ROLE OF NCX–GM1 COMPLEX IN THE NE

Location of NCX at the inner membrane of the NE suggested a role in transferring Ca^{2+} from nucleoplasm to the NE, the latter being a luminal Ca^{2+} storage site that is continuous with the ER lumen.[22] This was demonstrated by uptake experiments with isolated nuclei of NG108-15 cells incubated in the presence of buffered $^{45}Ca^{2+}$. This required first loading Na^+ into the NE lumen with appropriate ionophores, then exposing the nuclei to $^{45}Ca^{2+}$ in solutions with increasing ratio of choline to Na^+; after incubation for 15 s, the nuclei were rapidly washed and counted. As the choline to Na^+ ratio increased, more $^{45}Ca^{2+}$ was driven into the NE (FIG. 4). This experiment demonstrated the necessity for GM1 association in promoting optimal rate of transfer, as shown[6] in comparison to nuclei from undifferentiated NG108-15 cells, with little GM1, versus nuclei from differentiated cells whose GM1 in the NE had been upregulated (FIG. 1). The low level of exchange activity in nuclei from undifferentiated cells was significantly elevated by bath incubation of the nuclei in GM1-containing medium. Moreover, this was specific for GM1, no other ganglioside causing this kind of potentiation.[6] These *in vitro* experiments showed that Na^+/Ca^{2+} exchange in the nucleus is driven by a Na^+ gradient, as is the case for plasma membrane NCX. Such elevation of Na^+ in the NE is believed to occur naturally by a Na^+,K^+-ATPase in the inner membrane of the NE.[23]

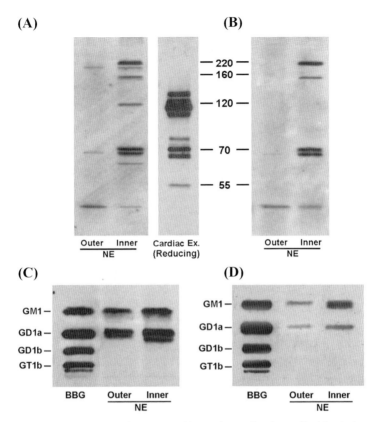

FIGURE 3. Comparison of NCX peptides and associated gangliosides in inner- and outer membranes of the NE from differentiated cells. Nuclei were isolated from KCl/db-cAMP-differentiated cells, and the outer membrane stripped away with cold Na-citrate; inner membrane was isolated from remaining nucleus by DNase/RNase treatment.[20] Exchanger proteins in both fractions were immunoprecipitated with C2C12 monoclonal antibody, separated on 7% nonreducing SDS-PAGE, transferred to PVDF membrane, followed by blot with polyclonal antibody (**A**) or Ctx B-HRP (**B**). Virtually all the exchanger protein(s) and associated GM1 are in the inner membrane. Each lane of *panels A and B* represents 100 μg of membrane protein subjected to immunoprecipitation. *Panel D* shows gangliosides extracted from equivalent amounts of immunoprecipitates with chloroform/methanol and subjected to HPTLC followed by Ctx B-HRP overlay, again showing predominance of exchanger-associated gangliosides in inner membrane of NE. *Panel C* shows gangliosides obtained by direct extraction of inner and outer membranes with chloroform-methanol (no immunoprecipitation) followed by HPTLC analysis with CtxB overlay. This revealed considerably more gangliosides than the immunoprecipitated samples (**D**), with predominance of GD1a (compared to predominance of GM1 associated with immunoprecipitated exchanger). Lanes 2 and 3 of *panels C and D* each represent 75 μg membrane protein directly extracted (**C**) or subjected to immunoprecipitation before extraction (**D**). These results suggest additional functions for GM1 (and GD1a) in these membranes besides association with NCX. BBG = bovine brain gangliosides. Reproduced from Xie *et al.* (Ref. 6) with permission of J. Neurochemistry.

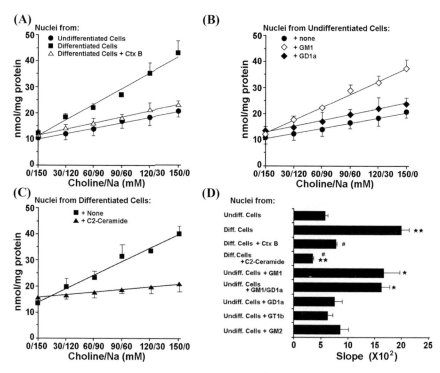

FIGURE 4. GM1 promotion of Na-mediated Ca^{2+} uptake in NE of isolated nuclei. Nuclei were isolated from undifferentiated or differentiated cells and studied for Na-driven $^{45}Ca^{2+}$ uptake into the NE (Ref. 6). Some experiments involved preincubation with: **(A)** Ctx B applied to nuclei from differentiated cells, **(B)** GM1 or GD1a (each at 10 µM) applied to nuclei from undifferentiated cells, or **(C)** C2-ceramide (inhibitor of NCX) applied to nuclei from differentiated cells. Exchange was carried out for 15 s in uptake buffer containing choline chloride/NaCl at the ratio shown (total = 150 mM). Data in *panels A–C* are average ± SEM from at least three independent experiments, each carried out in triplicate. *Panel D* indicates slopes (average ± SEM) determined by linear regression analysis; regression coefficients $R^2 \cong 0.88–0.98$. Statistical significance was determined by Student's two-tailed *t*-test: * $P < 0.05$, ** $P < 0.001$ compared to nuclei from undifferentiated cells; # $P < 0.001$ compared to nuclei from differentiated cells. All values were normalized to equivalent amounts of protein. GM1 was the only ganglioside of the four tested that potentiated Na-Ca exchange in nuclei from undifferentiated cells. Reproduced from Xie *et al.*, Ref. 6 with permission of J. Neurochemistry.

BASIS OF HIGH-AFFINITY ASSOCIATION OF NUCLEAR NCX WITH GM1

To understand the mechanism(s) involved in high-affinity binding of GM1 to NCX in the NE, previous work has revealed two related phenomena: (*a*) alternatively spliced isoforms of NCX, which favor targeting to the NE, and (*b*) topological features that allow charge–charge interaction of the associated molecules. The mammalian NCX forms a multigene family consisting of three subtypes, NCX1, NCX2, and NCX3, and various isoforms corresponding to splice variants have been identified for each.[24] The NCX1 subtype, which predominates in many neural cells, contains 6 exons (A–F) in the splicing region that code for a portion of the large intracellular peptide loop and give rise to the alternatively spliced isoforms of NCX1.[25] Exons A and B are believed to be expressed in a mutually exclusive manner in combination with one or more of the remaining (cassette) exons.[26] Three exon B-containing isoforms were shown to be the predominant transcripts in rat cortical astrocytes and C6 cells,[25] and that together with our finding that C6 cells contain only nuclear NCX[27] suggests that isoforms containing the B exon are preferentially expressed in the NE. Current studies in our laboratory support this hypothesis in that Jurkat cells, which lack NCX altogether, show NCX expression in the NE when transfected with plasmid containing the BCDEF isoform of NCX1 (work in preparation). The B exon was shown to contain four Arg residues,[16,28] the positive charge of which is proposed to interact with the negative charge of GM1 to form the high-affinity complex. This is in contrast to the A exon, hypothesized to be targeted to the plasma membrane, which contains only one Arg and might be less likely to associate tightly with GM1.

NCX in the plasma membrane, with nine transmembrane segments, mediates counter-transport of 3 Na^+ for 1 Ca^{2+} while promoting uphill extrusion of cytosolic Ca^{2+}. One topological requirement for this "forward" mode of exchange is that the large polypeptide loop between transmembrane segments 5 and 6 reside on the low Ca (cytosolic) side of the plasma membrane.[29] Assuming the same requirement for NCX of the NE, this peptide loop would extend into the nucleoplasm to facilitate transfer of nucleoplasmic Ca^{2+} across the inner nuclear membrane to the concentrated Ca^{2+} pool within the NE lumen. This would result in an important difference in GM1 orientation, the gangliotetraose oligosaccharide chain and NCX loop being on opposite sides of the plasma membrane but on the same side of the inner NE membrane (FIG. 5). Support for this orientation of GM1 in the inner nuclear membrane came from the observed availability of GM1 oligosaccharide to CtxB for cytochemical detection (FIG. 1) and blockage of Na^+/Ca^{2+} exchange (FIG. 4, Ref. 6). This topology would permit the negative charge of GM1 to interact with positive charge(s) in the alternatively spliced region of NCX; such charge–charge interaction was suggested in a recent study.[30]

CYTOPROTECTION BY NCX–GM1 COMPLEX OF THE NE

Calcium is known to have a critical role in apoptosis,[31] the nucleus being especially vulnerable due to the presence of various Ca^{2+}-dependent protease, nuclease, and lipase activities. Our studies suggest the NCX–GM1 complex

(A)   **(B)**

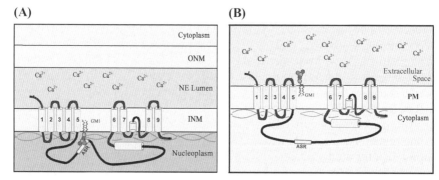

FIGURE 5. Proposed topology of GM1 and NCX in the NE (**A**) and plasma membrane (PM, **B**). In both cases, the large loop between transmembrane units 5 and 6 is located on the low Ca^{2+} side, that is, cytoplasm for PM and nucleoplasm for NE. This accords with demonstrated location of both GM1 and NCX in the inner nuclear membrane (INM) and occurrence of the large NCX loop in proximity to GM1 oligosaccharide chain. We have proposed (Ref. 30) that the high-affinity association of GM1 with NCX arises from the negative charge of N-acetylneuraminic acid in GM1 interacting with the alternative splice region (ASR) of the NCX loop, some of whose isoforms are enriched in positively charged amino acids (Ref. 26). Such association is not possible for the PM, since the NCX loop and GM1 oligosaccharide occur on opposite sides of the membrane. ONM = outer nuclear membrane. Reproduced from Ledeen and Wu, Ref. 9 with permission of Elsevier.

serves a cytoprotective role, for example, in shielding the nucleus against prolonged elevation of nucleoplasmic Ca^{2+}. Cultures of cerebellar granule neurons (CGN) from mice engineered to lack GM2/GD2 synthase, with resultant deficit of GM2, GD2, and all gangliotetraose gangliosides (including GM1), were found deficient in calcium regulatory capability that resulted in apoptotic death in the presence of high K^+.[32] That this was due to the absence of GM1 was suggested in the observation that CGN from these knockout mice could be rescued from elevated K^+ as well as excitotoxic levels of glutamate by GM1, and even more effectively by LIGA-20, a semisynthetic derivative of GM1.[33] This correlated with the known efficacy of LIGA-20, a membrane permeant analog of GM1, in restoring Ca^{2+} homeostasis in both normal CGN[34] and mutant (ganglioside-deficient) CGN.[33] The latter study revealed more facile entry of LIGA-20 into the cell interior and especially the NE than was the case for GM1. Involvement of the nucleus was also indicated by *in vivo* studies using the above ganglioside-deficient mice, which, when administered kainic acid, developed temporal lobe seizures of significantly greater severity/duration than was the case for normal mice.[35] The greater efficacy of LIGA-20 compared to GM1 in attenuating such seizes was correlated with the membrane-permeant properties of the former and its ability to enter brain cells, insert into the NE, and activate the subnormally active NCX of the NE. In that sense, it served as a functional replacement for the missing GM1 in the mutants.

CONCLUSIONS AND SPECULATIONS

The presence of NCX in the nuclear membrane was an unexpected finding, especially in view of the prevailing perception that Ca^{2+} diffuses freely through the numerous pore complexes in the NE. Independent regulation of nuclear Ca^{2+} was thus considered unlikely, although some studies pointed rather directly to this possibility.[36,37] It is now well accepted that both the inner and outer nuclear membranes contain mechanisms that regulate Ca^{2+} transport in the nucleus,[38] so the presence of a Ca^{2+} exchange mechanism may seem less surprising. Our studies also established a somewhat unique regulatory mechanism for this exchanger in revealing a requirement for firmly attached GM1, which in the case of neuronal cells occurs as a result of elevated expression of nuclear GM1 during the course of differentiation. In addition to neurons and glia, the NCX–GM1 complex occurs in some nonneural cells we have examined: NCTC, HeLa cells, and some lymphocytes. The Jurkat cell line was found to contain no NCX1 in either NE or plasma membrane.[30] In view of its antiapoptotic effect, it is worth considering the functional consequences to cells lacking this cytoprotective mechanism. This pertains, for example, to the subpopulation of lymphocytes we observed with this deficiency in the NE, analogous to Jurkat T cells.[30] Calcium signaling in T cells is recognized as highly complex, Ca^{2+} entry in such cells being long lasting and necessary for T cell function.[39,40] Attention is being directed to mechanisms by which immune effector cells disappear after eliminating foreign antigens and evidence was cited that such return of the immune system to rest is mainly due to programmed cell death of activated lymphocytes.[41] It may be speculated that cell death of this type could result from the absence of the nuclear NCX–GM1 complex, such cells being vulnerable to Ca^{2+}-induced apoptosis resulting from the prolonged Ca^{2+} elevation characteristic of activated T cells. Absence of the nuclear NCX–GM1 complex might also be a factor in maintaining unresponsiveness, or tolerance, to self-antigens. Further experimentation is needed to determine precisely which lymphocytes lack this nuclear mechanism and the manner in which this might relate to immune function.

ACKNOWLEDGMENT

This study was supported by National Institutes of Health Grant 2RO1 NS033912.

REFERENCES

1. CROMPTON, M., R. MOSER, H. LUDI & E. CARAFOLI. 1978. The interrelations between the transport of sodium and calcium in mitochondria of various mammalian tissues. Eur. J. Biochem. **82:** 25–31.

2. MALISAN, F. & R. TESTI. 2002. GD3 ganglioside and apoptosis. Biochim. Biophys. Acta **1585:** 179–187.
3. LEDEEN, R.W., S.M. PARSONS, M.F. DIEBLER. *et al.* 1988. Gangliosides composition of synaptic vesicles from *Torpedo* electric organ. J. Neurochem. **51:** 1465–1469.
4. MATYAS, G.R. & D.J. MORRE. 1987. Subcellular distribution and biosynthesis of rat liver gangliosides. Biochim. Biophys. Acta **921:** 599–614.
5. WU, G., Z.H. LU & R.W. LEDEEN. 1995. Induced and spontaneous neuritogenesis are associated with enhanced expression of ganglioside GM1 in the nuclear membrane. J. Neurosci. **15:** 3739–3746.
6. XIE, X., G. WU, Z.-H. LU & R.W. LEDEEN. 2002. Potentiation of a sodium-calcium exchanger in the nuclear envelope by nuclear GM1 ganglioside. J. Neurochem. **81:** 1185–1195.
7. LEDEEN, R.W. & G. WU. 2006. Gangliosides of the nuclear membrane: a crucial locus of cytoprotective modulation. J. Cell. Biochem. **97:** 893–903.
8. LEDEEN, R.W. & G. WU. 2006. GM1 ganglioside: another nuclear lipid that modulates nuclear calcium. GM1 potentiates the nuclear sodium-calcium exchanger. Can. J. Physiol. Pharmacol. **84:** 393–402.
9. LEDEEN, R.W. & G. WU. 2006. Sphingolipids of the nucleus and their role in nuclear signaling. Biochim. Biophys. Acta **1761:** 588–598.
10. KEENAN, T.W., D.J. MORRÉ & C.M. HUANG. 1972. Distribution of gangliosides among subcellular fractions from rat liver and bovine mammary gland. FEBS Lett. **24:** 204–208.
11. KATOH, N., T. KIRA & A. YUASA. 1993. Protein kinase C substrates and ganglioside inhibitors in bovine mammary nuclei. J. Dairy Sci. **76:** 3400–3409.
12. KOZIRESKI-CHUBACK, D., G. WU & R.W. LEDEEN. 1999. Developmental appearance of nuclear GM1 in neurons of the central and peripheral nervous systems. Dev. Brain Res. **115:** 201–208.
13. KOZIRESKI-CHUBACK, D., G. WU & R.W. LEDEEN. 1999. Upregulation of nuclear GM1 accompanies axon-like but not dendrite-like outgrowth in NG108-15 cells. J. Neurosci. Res. **55:** 107–118.
14. KOZIRESKI-CHUBACK, D., G. WU & R.W. LEDEEN. 1999. Axonogenesis of Neuro-2a cells correlates with GM1 upregulation in the nuclear and plasma membranes. J. Neurosci. Res. **57:** 541–550.
15. PHILIPSON, K.D., S. LONGONI & R. WARD. 1988. Purification of the cardiac Na^+-Ca^{2+} exchange protein. Biochim. Biophys. Acta **945:** 298–306.
16. VAN EYLEN, F., A. KAMAGATE & A. HERCHUELZ. 2001. A new Na/Ca exchanger splicing pattern identified *in situ* leads to a functionally active 70 kDa NH_2-terminal protein. Cell Calcium **30:** 191–198.
17. PIJUAN, V., Y. ZHUANG, L. SMITH, *et al.* 1993. Stable expression of the cardiac sodium-calcium exchanger in CHO cells. Am. J. Physiol. **264:** C1066–C1074.
18. SANTACRUZ-TOLOZA, L., M. OTTOLIA, D.A. NICOLL & K.D. PHILIPSON. 2000. Functional analysis of a disulfide bond in the cardiac Na^+-Ca^{2+} exchanger. J. Biol. Chem. **275:** 182–188.
19. MUTOH, T., A. TOKUDA, T. MIYADAI, *et al.* 1995. Ganglioside GM1 binds to the Trk protein and regulates receptor function. Proc. Natl. Acad. Sci. USA **92:** 5087–5091.
20. GILCHRIST, J.S.C. & G.N. PIERCE. 1993. Identification and purification of a calcium-binding protein in hepatic nuclear membranes. J. Biol. Chem. **268:** 4291–4299.

21. SAITO, M., L.L. FRONDA & R.K. YU. 1996. Sialidase activity in nuclear membranes of rat brain. J. Neurochem. **66:** 2205–2208.

22. PETERSEN, O.H., O.V. GERASIMENKO, J.V. GERASIMENKO, *et al.* 1998. The calcium store in the nuclear envelope. Cell Calcium **23:** 87–90.

23. GARNER, M.H. 2002. Na,K-ATPase in the nuclear envelope regulates Na^+: K^+ gradients in hepatocyte nuclei. J. Membr. Biol. **187:** 97–115.

24. THURNEYSEN, T., D.A. NICOLL, K.D. PHILIPSON & H. PORZIG. 2002. Sodium/calcium exchanger subtypes NCX1, NCX2, and NCX3 show cell-specific expression in rat hippocampus cultures. Mol. Brain Res. **107:** 145–156.

25. HE, S., A. RUKNUDIN, L.L. BAMBRICK, *et al.* 1998. Isoform-specific regulation of the Na^+/Ca^{2+} exchanger in rat astrocytes and neurons by PKA. J. Neurosci. **18:** 4833–4841.

26. KOFUJI, P., W.J. LEDERER & D.H. SCHULZE. Mutually exclusive and cassette exons underlie alternatively spliced isoforms of the Na^+/Ca^{2+} exchanger. J. Biol. Chem. **269:** 5145–5149.

27. XIE, X., G. WU & R.W. LEDEEN. 2004. C6 cells express a sodium-calcium exchanger/GM1 complex in the nuclear envelope but have no exchanger in the plasma membrane: comparison to astrocytes. J. Neurosci. Res. **76:** 363–375.

28. QUEDNAU, B.D., D.A. NICOLL & K.D. PHILIPSON. 1997. Tissue specificity and alternative splicing of the Na^+/Ca^{2+} exchanger isoforms NCX1, NCX2, and NCX3 in rat. Am. J. Physiol. **272:** C1250–C1261.

29. PHILIPSON, K.D. & D.A. NICOLL. 2000. Sodium-calcium exchange: a molecular perspective. Annu. Rev. Physiol. **62:** 111–133.

30. XIE, X., G. WU, Z.-H. LU, *et al.* 2004. Presence of sodium-calcium exchanger/GM1 complex in the nuclear envelope of non-neural cells: nature of exchanger-GM1 interaction. Neurochem. Res. **29:** 2135–2146.

31. MATTSON, M.P. & S.L. CHAN. 2003. Calcium orchestrates apoptosis. Nat. Cell Biol. **5:** 1041–1043.

32. WU, G., X. XIE, Z.-H. LU & R.W. LEDEEN. 2001. Cerebellar neurons lacking complex gangliosides degenerate in the presence of depolarizing levels of potassium. Proc. Natl. Acad. Sci. USA **98:** 307–312.

33. WU, G., Z.-H. LU, X. XIE & R.W. LEDEEN. 2004. Susceptibility of cerebellar granule neurons from GM2/GD2 synthase-null mice to apoptosis induced by glutamate excitotoxicity and elevated KCl: rescue by GM1 and LIGA20. Glycoconj. J. **21:** 305–313.

34. MANEV, H., M. FAVARON, S. VICINI, *et al.* 1990. Glutamate-induced neuronal death in primary cultures of cerebellar granule cells: protection by synthetic derivatives of endogenous sphingolipids. J. Pharmacol. Exp. Ther. **252:** 419–427.

35. WU, G., Z.-H. LU, J. WANG, *et al.* 2005. Enhanced susceptibility to kainite-induced seizures, neuronal apoptosis, and death in mice lacking gangliotetraose gangliosides. Protection with LIGA 20, a membrane-permeant analog of GM1. J. Neurosci. **25:** 11014–11022.

36. AL-MOHANNA, F.A., K.W.T. CADDY & S.R. BOLSOVER. 1994. The nucleus is insulated from large cytosolic calcium ion changes. Nature **367:** 745–750.

37. BADMINTON, M.N., J.M. KENDALL, C.M. REMBOLD & A.K. CAMPBELL. 1998. Current evidence suggests independent regulation of nuclear calcium. Cell Calcium **23:** 79–86.

38. GERASIMENKO, O. & J. GERASIMENKO. 2004. New aspects of nuclear calcium signaling. J. Cell Sci. **117:** 3087–3094.

39. WEISS, A., J. IMBODEN, D. SHOBACK & J. STOBO. 1984. Role of T3 surface molecules in human T-cell activation: T3-dependent activation results in an increase in cytoplasmic free calcium. Proc. Natl. Acad. Sci. USA **81:** 4169–4173.
40. LEWIS, R.S. 2001. Calcium signaling mechanisms in T lymphocytes. Annu. Rev. Immunol. **19:** 487–521.
41. PARIJS, L.V. & A.K. ABBAS. 1998. Homeostasis and self-tolerance in the immune system: turning lymphocytes off. Science **280:** 243–248.

Mitochondrial Ca^{2+} Flux through Na$^+$/Ca^{2+} Exchange

BONGJU KIM AND SATOSHI MATSUOKA

Department of Physiology and Biophysics, Graduate School of Medicine, Kyoto University, Kyoto 606-8501, Japan

ABSTRACT: To clarify the property of mitochondrial Na$^+$/Ca^{2+} exchange *in situ*, we measured mitochondrial Ca^{2+} using Rhod-2 in permeabilized rat ventricular myocytes. Cytoplasmic 300 nM Ca^{2+} (Ca$^{2+}_c$) augmented the Rhod-2 intensity by ∼ninefolds without cytoplasmic Na$^+$ (Na^+_c). Increasing Na^+_c attenuated the maximum level of Rhod-2 fluorescence, probably due to the activation of forward mode of mitochondrial Na$^+$/Ca^{2+} exchange. The Rhod-2 intensity decayed upon removing Ca$^{2+}_c$. The decay was dependent on Na^+_c (K$_{1/2}$ = ∼1 mM) and largely abolished by an inhibitor of mitochondrial Na$^+$/Ca^{2+} exchange, CGP-37157. It was suggested that Na$^+$ binding to the mitochondrial Na$^+$/Ca^{2+} exchange is saturated in the physiological concentration of Na^+_c.

KEYWORDS: mitochondrial Ca^{2+}; mitochondrial Na$^+$/Ca^{2+} exchanger; Ca^{2+} uniporter; ruthenium red; CGP-37157

INTRODUCTION

Ca^{2+} in cardiac mitochondria dynamically changes and modulates mitochondria function by activating Ca^{2+}-sensitive dehydrogenases.[1–3] Extramitochondrial Na$^+$-dependent Ca^{2+} release from mitochondria was first demonstrated by Carafoli *et al.*[4,5] Thereafter, the mitochondrial Na$^+$/Ca^{2+} exchange has been characterized mainly in the isolated mitochondria.[4–10] However, properties of the mitochondrial Na$^+$/Ca^{2+} exchange *in situ*, namely in the cardiomyocyte, have not well studied.

It has been reported that Na^+_c increases with increasing beating frequency.[11,12] Therefore, the change in Na^+_c conceivably affects mitochondrial Na$^+$/Ca^{2+} exchange activity and modulates mitochondrial Ca^{2+} concentration. In this study, we measured the mitochondrial Ca^{2+} in permeabilized rat ventricular myocytes under physiological concentrations of cytoplasmic Na$^+$

Address for correspondence: Satoshi Matsuoka, Department of Physiology and Biophysics, Graduate School of Medicine, Kyoto University, Yoshida-Konoe-cho, Sakyo-ku, Kyoto 606-8501, Japan. Voice: +81-75-753-4357; fax: +81-75-753-4349.

matsuoka@card.med.kyoto-u.ac.jp

Ann. N.Y. Acad. Sci. 1099: 507–511 (2007). © 2007 New York Academy of Sciences.
doi: 10.1196/annals.1387.027

(Na^+_c) and Ca^{2+} (Ca^{2+}_c), and studied the Na^+_c dependence of mitochondrial Na^+/Ca^{2+} exchange.

MATERIALS AND METHODS

Ventricular myocytes were isolated from Wistar rats by collagenase digestion. The myocytes were incubated for 1 h in a modified D-MEM solution containing 5 μM Rhod-2 AM (a cationic indicator for Ca^{2+}; Molecular Probes, Eugene, OR) at 4 °C.[3] Thereafter, the plasmalemma of myocyte was permeabilized with 0.1 mg/mL saponin for 1 min. The Rhod-2 fluorescence images were obtained with a confocal laser scanning microscope (FV500 Olympus, Tokyo, Japan; excitation at 531 nm and emission at > 560 nm). Bath solution contained (in mM) 68–118 KCl, 0–50 NaCl, 3 K_2ATP, 10 EGTA, 10 HEPES, 2 K-pyruvate, 1 K_2HPO_4, and 2 succinic acid (pH = 7.2/KOH). To yield 1.0 mM free Mg^{2+} and 300 nM free Ca^{2+}, 4.07–4.68 mM $MgCl_2$ and 6.71 mM $CaCl_2$ were added. The concentrations were calculated by Win-MAXC software (http://www.stanford.edu/~cpatton/maxc.html).[13] The permeabilized myocytes were first placed in a Na^+- and Ca^{2+}-free solution. To facilitate the Ca^{2+} influx into mitochondria, 300 nM Ca^{2+} was added to the bath solution. Then, Ca^{2+} efflux from mitochondria was activated by removal of Ca^{2+} from the bath solution. All experiments were performed at 36–37°C.

RESULTS AND DISCUSSION

Ca^{2+} Influx into Mitochondria

The application of 300 nM Ca^{2+} increased the Rhod-2 fluorescence. In the absence of Na^+_c, the Rhod-2 intensity increased up to 9.2 ± 2.8 folds ($n =$ 23), and it took about 10 min to reach a steady level (FIG. 1 A). The increase was abolished by an inhibitor of the Ca^{2+} uniporter, ruthenium red (7 μM).[14] This result is consistent with a view that the Ca^{2+} influx into mitochondria is mainly mediated by the Ca^{2+} uniporter. Pretreatment of the myocytes with 20 mM monensin and 50 mM Na^+_c did not enhance the Ca^{2+}_c-induced increase in the Rhod-2 fluorescence in the presence of ruthenium red. Therefore, Ca^{2+} influx into mitochondria through the reverse mode of Na^+/Ca^{2+} exchange is negligible, at least, under the present experimental condition.

When Na^+_c was increased, the maximum level of Rhod-2 fluorescence decreased (FIG. 1 B). The maximum level was 5.5 ± 1.6, 3.4 ± 0.7, 2.3 ± 0.6, and 1.4 ± 0.2 in the presence of 2, 6, 20, and 50 mM Na^+_c, respectively. The half inhibitory concentration (IC_{50}) of Na^+_c was 2 mM. The Na^+_c-dependent decrease is probably due to the forward mode of mitochondrial Na^+/Ca^{2+} exchange.

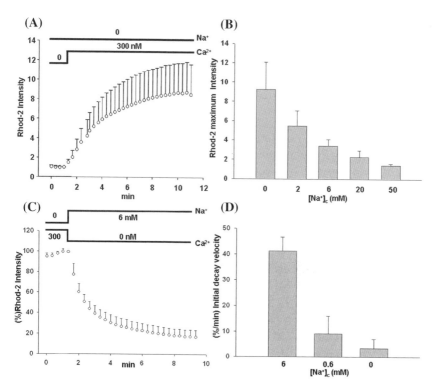

FIGURE 1. Ca^{2+} influx and efflux. (**A**) Ca^{2+} influx into mitochondria. The permeabilized myocyte was exposed to 300 nM Ca^{2+}_c. The Rhod-2 intensity was normalized to a base line before the Ca^{2+}_c addition ($n = 23$). (**B**) Na^+_c-dependence of Rhod-2 maximum intensity. (**C**) Mitochondrial Ca^{2+} efflux. The Rhod-2 intensity was normalized to a steady-state value before the removal of Ca^{2+}_c ($n = 8$). (**D**). Na^+_c-dependence of the initial velocity of Rhod-2 fluorescence decay ($n = 4$–8). Data represent mean ± SD.

Ca²⁺ Efflux from Mitochondria

To study the Ca^{2+} efflux from mitochondria, Ca^{2+}_c and Na^+_c were simultaneously changed to 0 and 6 mM, respectively, at the steady state after the 300 nM Ca^{2+} application (FIG. 1 C). The Rhod-2 fluorescence gradually decreased over several minutes. The fluorescence decay depended on Na^+_c ($K_{1/2}$ = ~1 mM) and most part of the decay was abolished by removing Na^+_c. The initial velocity of decay was 41.3 ± 5.4 ($n = 8$), 9.1 ± 6.7 ($n = 4$), and 3.4 ± 3.7 %/min ($n = 4$) in the presence of 6, 0.6, and 0 mM Na^+_c (FIG. 1 D). These data are consistent with a recent work using rat myocytes by Saotome et al.[15] Therefore, the Ca^{2+} efflux from mitochondria is largely mediated by the forward mode of mitochondrial Na^+/Ca^{2+} exchange. This view was further confirmed by applying an inhibitor of mitochondrial Na^+/Ca^{2+} exchange,

CGP-37157.[16] The fluorescence decay was abolished by 20 μM CGP-37157 (data not shown).

It was reported that the $K_{1/2}$ of $Na^+{}_c$ for the mitochondrial Ca^{2+} efflux was 2–12 mM in the isolated mitochondria[6–10] and reconstituted mitochondrial Na^+/Ca^{2+} exchange.[17] Our data suggested higher $Na^+{}_c$ affinity of mitochondrial Na^+/Ca^{2+} exchange. Despa et al.[11] demonstrated that resting $Na^+{}_c$ in the rat ventricular myocyte was $\sim$10 mM and that $Na^+{}_c$ increased linearly to 15 mM when stimulation frequency was increased to 2.5 Hz. Therefore, under the physiological concentration of $Na^+{}_c$, Na^+ binding to the mitochondrial Na^+/Ca^{2+} exchange is probably saturated, and the frequency-dependent change in $Na^+{}_c$ probably does not affect mitochondrial Na^+/Ca^{2+} exchange activity. The high $Na^+{}_c$ affinity of mitochondrial Na^+/Ca^{2+} exchange may contribute, at least in part, to the mitochondrial Ca^{2+} increase induced by high-beating frequency.[3]

Mg^{2+} was known to attenuate Ca^{2+} efflux from liver mitochondria ($IC_{50} = $ 1.0–1.5 mM).[9] However, changing free Mg^{2+} concentration (0.2–5 mM) did not significantly affect the Rhod-2 fluorescence decay induced by the $Ca^{2+}{}_c$ removal.

ACKNOWLEDGMENTS

This study was supported by the Leading Project for Biosimulation and a Grant-in-Aid for Scientific Research from the Ministry of Education, Culture, Sports, Science and Technology of Japan.

REFERENCES

1. McCormack, J.G. et al. 1990. Role of calcium ions in regulation of mammalian intramitochondrial metabolism. Physiol. Rev. **70:** 391–425.
2. Balaban, R.S. 2002. Cardiac energy metabolism homeostasis: role of cytosolic calcium. J. Mol. Cell Cardiol. **34:** 1259–1271.
3. Jo, H. et al. 2006. Calcium-mediated coupling between mitochondrial substrate dehydrogenation and cardiac workload in single guinea-pig ventricular myocytes. J. Mol. Cell Cardiol. **40:** 394–404.
4. Carafoli, E. et al. 1974. The release of calcium from heart mitochondria by sodium. J. Mol. Cell Cardiol. **6:** 361–371.
5. Crompton, M. et al. 1977. The calcium-induced and sodium-induced effluxes of calcium from heart mitochondria. Eur. J. Biochem. **79:** 549–558.
6. Crompton, M. et al. 1976. The sodium-induced efflux of calcium from heart mitochondria. Eur. J. Biochem. **69:** 453–462.
7. Crompton, M. et al. 1978. The interrelations between the transport of sodium and calcium in mitochondria of various mammalian tissues. Eur. J. Biochem. **82:** 25–31.

8. HAYAT, L.H. & M. CROMPTON. 1982. Evidence for the existence of regulatory sites for Ca^{2+} on the Na^+/Ca^{2+} carrier of cardiac mitochondria. Biochem. J. **202:** 509–518.
9. WINGROVE, D.E. & T.E. GUNTER. 1986. Kinetics of mitochondrial calcium transport. J. Biol. Chem. **261:** 15166–15171.
10. COX, D.A. & M.A. MATLIB. 1993. A role of the mitochondrial Na^+-Ca^{2+} exchanger in the regulation of oxidative phosphorylation in isolated heart mitochondria. J. Biol. Chem. **268:** 938–947.
11. DESPA, S. *et al.* 2002. Intracellular [Na^+] and Na^+ pump rate in rat and rabbit ventricular myocytes. J. Physiol. **539:** 133–143.
12. WANG, D.Y. *et al.* 1988. Role of aiNa in positive force-frequency staircase in guinea pig papillary muscle. Am. J. Physiol. **255:** C798–C807.
13. BERS, D.M. *et al.* 1994. A practical guide to the preparation of Ca^{2+} buffers. Methods Cell Biol. **40:** 3–29.
14. VASINGTON, F.D. *et al.* 1972. The effect of ruthenium red on Ca^{2+} transport and respiration in rat liver mitochondria. Biochim. Biophys. Acta **256:** 43–54.
15. SAOTOME, M. *et al.* 2005. Mitochondrial membrane potential modulates regulation of mitochondrial Ca^{2+} in rat ventricular myocytes. Am. J. Physiol. **288:** H1820–H1828.
16. COX, D.A. *et al.* 1993. Selectivity of inhibition of Na^+-Ca^{2+} exchange of heart mitochondria by benzothiazepine CGP-37157. J. Cardiovasc. Pharmacol. **21:** 595–599.
17. PAUCEK, P. & M. JABUREK. 2004. Kinetics and ion specificity of Na^+/Ca^{2+} exchange mediated by reconstituted beef heart mitochondrial Na^+/Ca^{2+} antiporter. Biochim. Biophys. Acta **1659:** 83–91.

Non-apoptogenic Ca²⁺-Related Extrusion of Mitochondria in Anoxia/Reoxygenation Stress

ANNALISA DORIO,[a] CLAUDIA CERELLA,[a] MILENA DE NICOLA,[a] MARIA D'ALESSIO,[a] GIAMPIERO GUALANDI,[b] AND LINA GHIBELLI[a]

[a]Dipartimento di Biologia, Università di Roma Tor Vergata, 00133 Rome, Italy

[b]DABAC, Università della Tuscia, 01011 Viterbo, Italy

ABSTRACT: Tumor cells often develop molecular strategies for survival to anoxia/reoxygenation stress as part of tumor progression. Here we describe that the B lymphoma Epstein–Barr-positive cells E2r survive reoxygenation in spite of a very high and long-lasting increase in cytosolic Ca²⁺ and the loss of about half of their mitochondria due to specific extrusion of the organelles from the cells. The extrusion typically occurs 3 days after reoxygenation, and a regular mitochondrial asset is regained after further 24 h.

KEYWORDS: calcium; apoptosis; anoxia; reoxygenation; mitochondria

INTRODUCTION

Hypoxic/anoxic stress plays a major role in two of the most important human pathologies, namely ischemia/reperfusion and tumors. On the one side, the rapid and strong blood flow stop occurring during accidental ischemia abruptly causes severe anoxia, leading to cell death by apoptosis or necrosis, to an extent that depends on the duration/strength of ischemia; this produces irreversible damage to poorly renewable tissues (brain or cardiac). The major cytotoxic phase occurs during reperfusion, when the reestablishment of tissue homeostasis produces extensive cell death by apoptosis, the extent of which still depends on the gravity of the ischemic phase. This is accompanied by (and perhaps due to) an uncontrolled increase in cytosolic-free Ca²⁺ ($[Ca^{2+}]c$), a potent inducer of apoptosis.

On the other side, during the growth of poorly vascularized tumor masses, a hypoxic environment partially impairs cell growth and viability, selecting cells for resistance to ischemic stress as part of tumor progression. Reoxygenation

Address for correspondence: Lina Ghibelli, Dipartimento di Biologia, Università di Roma "Tor Vergata," via Ricerca Scientifica 1, 00133 Roma, Italy. Voice: +39-06-7259-4323; fax: +39-06-2023500. ghibelli@uniroma2.it

Ann. N.Y. Acad. Sci. 1099: 512–515 (2007). © 2007 New York Academy of Sciences. doi: 10.1196/annals.1387.067

is gradually provided by neoangiogenesis; in culture this is a nontoxic process, suggesting that tumor cells have developed strategies for survival to reoxygenation.

Thus, return to homeostasis is more cytotoxic for nerve/cardiac cells after ischemia, than for tumor cells. Much effort is being posed in envisaging therapeutic protocols to get the reverse. Might we learn lessons from tumor cells strategies for survival?

MATERIALS AND METHODS

Cells: E2r are BL41 human B lymphoma cells converted by *in vitro* infection with Epstein–Barr virus.[1] Anoxia was provided by culturing cells in tissue culture flasks placed in an atmosbag (from Sigma, St. Louis, MO) filled with a controlled mixture of 5% CO_2; 95% N_2 (nominal anoxia). For reoxygenation, cells were placed in a regular CO_2 incubator. Mitochondria were labeled with mitotracker red, and visualized by a fluorescent microscope.[2] Cell viability and apoptosis were assessed by a fluorescent microscopy analysis of cells labeled with the cell-impermeant dye iodidium propide (viable cell are not stained) and with the cell-permeant dye Hoechst, which allow recognition of apoptotic cells, respectively.[3] Cytosolic Ca^{2+} concentration was assessed by loading cells with the specific dye Fluo3-AM; fluorescence was then quantified by flow cytometric analysis.[4] .

RESULTS AND DISCUSSION

During nominal anoxia, E2r cells suffer a slight cytotoxicity in terms of apoptosis and a drastic decrease in the proliferation rate (not shown). Upon reoxygenation, apoptosis is no longer detectable over the basal values found in control cultures, and the duplication rate is soon recovered, as shown in FIGURE 1 A. Thus, E2r fully and immediately recover in terms of viability and proliferation. To understand if this lack of cytotoxicity is due to the containment of $[Ca^{2+}]c$ or it occurs despite it, a time course of $[Ca^{2+}]c$ was performed at increasing times of reoxygenation. As shown in FIGURE 1 B, $[Ca^{2+}]c$ concentration very rapidly rises to values that are far beyond those normally compatible with cell survival (almost 2 mM); the extra Ca^{2+} is partially cleared at 24 h and 48 h, to return at normal levels at 72 h. Since promotion of $[Ca^{2+}]c$ increase is a major cytotoxic mechanism of reoxygenation, this indicates that survival to reoxygenation by E2r occurs in spite of high $[Ca^{2+}]c$, and implies mechanisms that act downstream to it. Interestingly, a surprising phenomenon accompanies the recovery to normal $[Ca^{2+}]c$, i.e., the extrusion of mitochondria. The image of a E2r cell in the process of extruding mitochondria is shown in FIGURE 1C

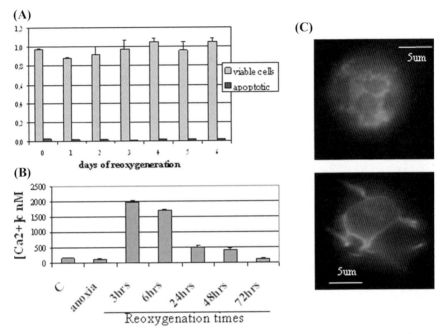

FIGURE 1. Cell proliferation and apoptosis in E2r cells during reoxygenation. Every day 500,000 cells were plated and counted after 24 h (**A**); average of four independent experiments ± SD. Cytosolic Ca^{2+} concentration ($[Ca^{2+}]c$) was measured at the indicated time points as described in "Materials and Methods"; data are the average of four independent experiments ± SD (**B**). Panel **C** shows the mitochondrial pattern of untreated (*top*) versus reoxygenated (72 h, *bottom*) E2r cells upon specific mitochondria labeling (see "Materials and Methods").

(*bottom*), compared with an untreated cell (*top*). The normal pattern of mitochondria unevenly distributed in the cytoplasm, changes to a pattern where mitochondria group into several (see picture) or unique (not shown) masses, to be released in the extracellular space. The amount of the released mitochondria is about half of the total cell mass; a normal amount is readily recovered after further 24 h. No loss of cell viability accompanies the phenomenon. To our knowledge, such a phenomenon was never described, and we are actively working in order to uncover the mechanisms through which this occurs, and the role it may play in the survival of tumor cells to anoxia/reoxygenation. Indeed, it will be important to determine whether cells survive in spite of mitochondria loss or due to it: since mitochondrial extrusion coincides with the recovery of normal $[Ca^{2+}]c$, it is tempting to speculate that mitochondria extrusion is a mean to dispose of the extra Ca^{2+} accumulated during reoxygenation.

REFERENCES

1. D'ALESSIO, M., M. DE NICOLA, S. COPPOLA, *et al.* 2005. Oxidative Bax dimerization promotes its translocation to mitochondria independently of apoptosis. FASEB J. **19:** 1504–1506.
2. DE NICOLA, M., G. GUALUANDI, A. ALFONSI, *et al.* 2006. Different fates of intracellular glutathione determine different modalities of apoptotic nuclear vesiculation. Biochem. Pharmacol. **72:** 1405–1416.
3. FANELLI, C., S. COPPOLA, R. BARONE, *et al.* 1999. Magnetic fields increase cell survival by inhibiting apoptosis via modulation of Ca^{2+} influx. FASEB J. **13:** 95–102.
4. CERELLA, C., M. D'Alessio, M. DE NICOLA, *et al.* 2003. Cytosolic and endoplasmic reticulum Ca^{2+} concentrations determine the extent and the morphological type of apoptosis, respectively. Ann. N. Y. Acad. Sci. **1010:** 74–77.

Na$^+$/Ca^{2+} Exchange as a Drug Target—Insights from Molecular Pharmacology and Genetic Engineering

TAKAHIRO IWAMOTO

Department of Pharmacology, School of Medicine, Fukuoka University, Fukuoka 814-0180, Japan

ABSTRACT: The Na$^+$/Ca^{2+} exchanger (NCX) is an ion transporter that exchanges Na$^+$ and Ca^{2+} in either Ca^{2+}-efflux or Ca^{2+}-influx mode, depending on membrane potential and transmembrane ion gradients. In myocytes, neurons, and renal tubular cells, NCX is thought to play an important role in the regulation of intracellular Ca^{2+} concentration. So far the benzyloxyphenyl derivatives (KB-R7943, SEA0400, SN-6, and YM-244769) have been developed as selective NCX inhibitors. These inhibitors possess different isoform selectivities, although they have similar properties, such as Ca^{2+}-influx mode selectivity and I$_1$ inactivation-dependence. Site-directed mutageneses have revealed that these inhibitors possess some molecular determinants (Phe-213, Val-227, Tyr-228, Gly-833, and Asn-839) for interaction with NCX1. These benzyloxyphenyl derivatives are expected to be useful tools to study the physiological roles of NCX. Interestingly, benzyloxyphenyl NCX inhibitors effectively prevent several ischemia–reperfusion injuries and salt-dependent hypertension in animal models. Furthermore, several experiments with genetically engineered mice provide compelling evidence that these diseases are triggered by pathological Ca^{2+} entry through NCX1. Thus, NCX inhibitors may have therapeutic potential as novel drugs for reperfusion injury and salt-dependent hypertension.

KEYWORDS: Na$^+$/Ca^{2+} exchange; benzyloxyphenyl NCX inhibitors; KB-R7943; SEA0400; SN-6; YM-244769; genetically engineered mice; reperfusion injury; salt-dependent hypertension

INTRODUCTION

The Na$^+$/Ca^{2+} exchanger (NCX) can transport Ca^{2+} either out of cells (the forward mode) or into cells (the reverse mode) in exchange for Na$^+$. NCX is

Address for correspondence: Dr. Takahiro Iwamoto, Department of Pharmacology, School of Medicine, Fukuoka University, 7-45-1 Nanakuma Jonan-ku, Fukuoka 814-0180, Japan. Voice: +81-92-801-1011; ext.: 3261; fax: +81-92-865-4384.

tiwamoto@cis.fukuoka-u.ac.jp

Ann. N.Y. Acad. Sci. 1099: 516–528 (2007). © 2007 New York Academy of Sciences.
doi: 10.1196/annals.1387.039

driven by membrane potential as well as Na^+ and Ca^{2+} concentration gradients.[1,2] In cardiomyocytes, NCX primarily pumps Ca^{2+} from inside to outside the cell during repolarization and diastole, which balances Ca^{2+} entry via L-type Ca^{2+} channels during cardiac excitation. NCX also mediates Ca^{2+} influx during the action potential upstroke, and helps maintain elevated $[Ca^{2+}]_{cyt}$ during the action potential plateau and systole. In arterial smooth muscle cells, NCX is mainly responsible for extruding Ca^{2+} to maintain Ca^{2+} homeostasis, but there is little information about vascular NCX compared to cardiac NCX. In the distal nephron, NCX is thought to play an important role in the active reabsorption of Ca^{2+}.

Mammalian NCX forms a multigene family comprising NCX1, NCX2, and NCX3. NCX1 is highly expressed in the heart, brain, and kidney, and at much lower levels in other tissues; whereas the expression of NCX2 and NCX3 is limited mainly to the brain and skeletal muscle.[1] These isoforms presumably have similar molecular topologies consisting of nine transmembrane segments and a large central cytoplasmic loop (FIG. 1).[2] The former part, particularly the α-repeat regions, may participate in ion transport; the latter part, possessing the exchanger inhibitory peptide (XIP) region and regulatory Ca^{2+}-binding

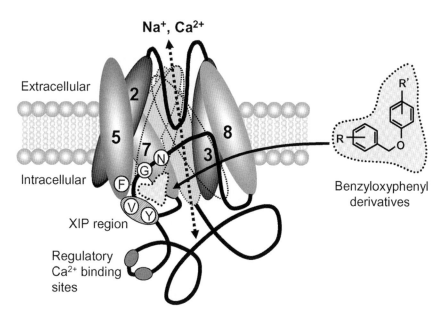

FIGURE 1. A nine-transmembrane model of NCX1 molecule and putative interaction domains for benzyloxyphenyl derivatives. Benzyloxyphenyl derivatives may interact with a specific receptor site, leading to blocking ion transport pore(s). Transmembrane helices, partially illustrated, are indicated by cylinders. The amino acid residues of NCX1 whose mutation alters the sensitivities to benzyloxyphenyl derivatives are indicated. XIP, exchanger inhibitory peptide; F, Phe-213; V, Val-227; Y, Tyr-228; G, Gly-833; N, Asn-839.

sites, is primarily involved in various regulatory properties.[1,2] NCX1 has been shown to be secondarily regulated by the transport substrates Na^+ and Ca^{2+}.[3] Intracellular Ca^{2+} at the submicromolar level activates NCX activity by promoting the recovery of the exchanger from the "I_2 inactivation state," whereas high $Na^+{}_i$ restrains the exchange by facilitating the entry of the exchanger into the "I_1 inactivation state" (Na^+-dependent inactivation).

Recently, the benzyloxyphenyl derivatives KB-R7943, SEA0400, SN-6, and YM-244769 have been developed as selective NCX inhibitors.[2] Potent and selective NCX inhibitors would be very useful for clarifying the physiological and pathophysiological roles of NCX. In this article, I will discuss the characteristics and therapeutic potential of NCX inhibitors.

DEVELOPMENT OF BENZYLOXYPHENYL
NCX INHIBITORS

Divalent and trivalent cations and organic compounds, such as $3',4'$-dichlorobenzamyl and bepridil can block NCX.[1] These inhibitors are, however, poorly specific to NCX, and are most effective when other ion transporters and ion channels are already blocked. In 1996, KB-R7943 was developed as a prototype selective NCX inhibitor.[4] This inhibitor was fairly specific to the NCX, because it exerted little influence on other ion transporters, such as the Na^+/H^+ exchanger, Na^+, K^+-ATPase, and Ca^{2+}-ATPases. It is now being widely used to study the physiological and pathological roles of NCX at the cellular and organ levels. However, KB-R7943 has been reported to possess some nonspecific actions against some ion channels and receptors.[2] In 2001, Matsuda et al. reported on SEA0400, a newly developed, more potent, and selective NCX inhibitor.[5] SEA0400 is more specific to the NCX, because it hardly inhibits other channels, receptors, and transporters. In 2002, SN-6, which is more specific than KB-R7943, was found from KB-R7943 derivatives.[6] Quite recently, YM-244769, a potent NCX inhibitor with lower cell toxicity, was introduced.[7] As shown in FIGURE 2, all these NCX inhibitors possess a benzyloxyphenyl structure, suggesting that this portion may be essential for their affinity to the NCX.

PROPERTIES OF BENZYLOXYPHENYL
NCX INHIBITORS

KB-R7943, SEA0400, SN-6, and YM-244769 inhibit dose-dependently the Ca^{2+} uptake via the reverse mode of NCX in fibroblasts expressing NCX isoforms (FIG. 3). These inhibitors have different isoform selectivities: KB-R7943 is more effective on NCX3 than on NCX1 and NCX2,[8] whereas SEA0400 predominantly blocks NCX1, only mildly blocks NCX2, and exerts almost

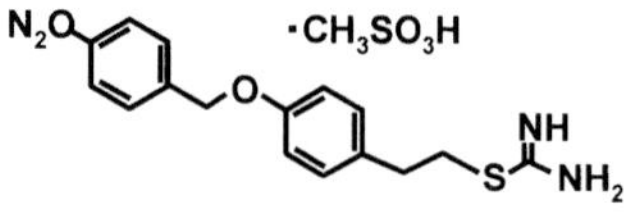

FIGURE 2. Chemical structures of benzyloxyphenyl NCX inhibitors.

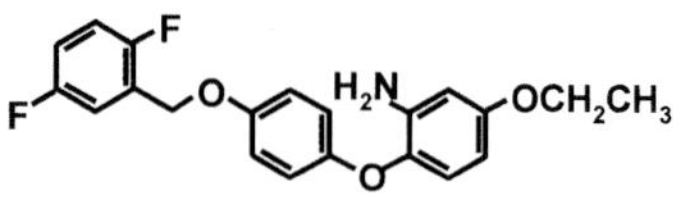

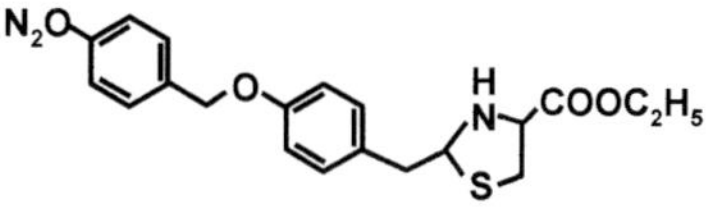

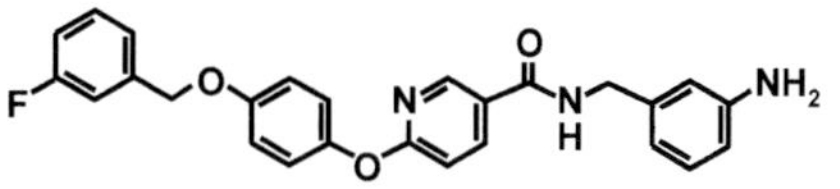

no influence upon NCX3.[9] SN-6 is more inhibitory to NCX1 than to NCX2 and NCX3.[6] On the other hand, YM-244769 is more effective on NCX3 than on NCX1 and NCX2.[10] Site-directed mutageneses reveal the important amino acids (Phe-213, Val-227, Tyr-228, Gly-833, and Asn-839 in NCX1) responsible for inhibition by benzyloxyphenyl derivatives.[6,9–11] These inhibitors probably interact with a specific receptor site, leading to blocking ion pore(s) formed within the membrane regions (FIG. 1).

Benzyloxyphenyl NCX inhibitors block the reverse mode of NCX1 much more effectively than the forward mode under unidirectional ionic conditions.[2] Recent mutational and electrophysiological analyses provide an explanation for the reverse mode selectivity of benzyloxyphenyl derivatives.[6,9,12] Intriguingly, the inhibitory potency of NCX inhibitors is directly coupled to the rate of I_1 inactivation (i.e., intracellular Na^+-dependent inactivation). Under unidirectional ionic conditions, the reverse mode is induced when $[Na^+]_{cyt}$ is high, whereas the forward mode is generated when $[Na^+]_{cyt}$ is reduced. NCX1 molecules thus tend to undergo I_1 inactivation in experimental conditions for the reverse mode, suggesting an apparent, but not substantial, reverse mode selectivity. These inhibitors likely stabilize the I_1 inactive state or accelerate the rate of I_1 inactivation. This suggests benzyloxyphenyl derivatives may be relatively dormant under normal conditions (low $[Na^+]_{cyt}$), but become effective under pathological conditions (high $[Na^+]_{cyt}$). This should be an ideal profile for therapeutic agents against intracellular Na^+-dependent cardiovascular diseases, such as ischemia–reperfusion injury and salt-sensitive hypertension.

THERAPEUTIC POTENTIAL OF BENZYLOXYPHENYL NCX INHIBITORS

Ischemia–Reperfusion Injury

Recent reports suggest benzyloxyphenyl NCX inhibitors efficiently prevent cardiac ischemia–reperfusion injury. KB-R7943 and SEA0400 strongly

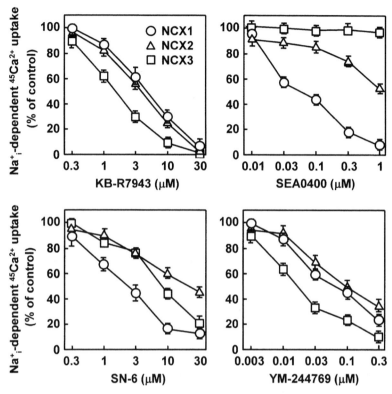

FIGURE 3. Dose-response curves for the effects of benzyloxyphenyl derivatives on Na^+_i-dependent $^{45}Ca^{2+}$ uptake in fibroblasts expressing NCX1, NCX2, and NCX3. The initial rates of $^{45}Ca^{2+}$ uptake into cells were measured in the presence or absence of indicated concentrations of NCX inhibitors. (Modified from Iwamoto et al.[6,8–10] with permission.)

suppress Ca^{2+} overload, hypercontracture, and cell damage induced by Ca^{2+} paradox and hypoxia(anoxia)/reoxygenation in cardiomyocytes.[4,13,14] Both inhibitors also significantly prevent cardiac ischemia–reperfusion injury (contractile dysfunction) in isolated perfused hearts treated with them either before or after ischemia.[14,15] Such effectiveness in postischemic treatment with NCX inhibitors supports the assumption that Ca^{2+} overload through NCX1 occurs primarily during reperfusion. The cardioprotective effect of SEA0400 is greater than that of KB-R7943.[14,16] SN-6 exhibits effective cardioprotection in the ischemia–reperfused heart by the dual actions of NCX1 inhibition and scavenging oxygen radicals.[17] KB-R7943 and SEA0400 have been shown to attenuate myocardial reperfusion injury in rat *in vivo* models.[14,15] "Myocardial stunning" is often observed in several clinical situations, such as after percutaneous transluminal coronary angioplasty, after thrombolysis, and after cardiopulmonary bypass. SEA0400 can significantly improve myocardial stunning in dog models.[18] Furthermore, KB-R7943, SEA0400, and SN-6 are also reported to be effective in ischemia–reperfusion (hypoxia/reoxygenation) injuries of the kidney[6,9,19,20] and brain.[5,21] Interestingly, YM-244769, which preferentially inhibits NCX3, more efficiently suppresses hypoxia/reoxygenation-induced neuronal cell damage, whereas SN-6, which preferentially inhibits NCX1, suppressed hypoxia/reoxygenation-induced renal cell damage to a greater extent.[10]

Recently, genetic manipulations in cells and mice provide compelling evidence that NCX1 inhibition protects against cardiac and renal reperfusion injuries. NCX1 inhibition by antisense oligonucleotides blocks Ca^{2+} overload during anoxia/reoxygenation in ventricular myocytes.[22] Heterozygous NCX1 knockout mice are useful as a low NCX1-expressing animal model.[23] Their hearts and kidneys are resistant to ischemia–reperfusion injuries.[19,24] Conversely, NCX1-overexpressing hearts from transgenic mice are hypersensitive to reperfusion injury.[25] Altogether, evidence suggests that inhibition of NCX1 by a reverse mode-specific inhibitor may have therapeutic potential for preventing cardiac and renal reperfusion injuries.

Salt-Dependent Hypertension

In arterial smooth muscle cells, NCX, like Ca^{2+}-ATPases, is thought to contribute to Ca^{2+} extrusion from the cytosol in the relaxation process.[1,2] Immunocytochemical staining of arterial smooth muscle cells indicated that the NCX protein is localized in the plasma membrane (PM) regions that are adjacent to junctional sarcoplasmic reticulum (SR).[1] This particular localization suggests that the exchanger may play a role in regulating the Ca^{2+} content of the SR stores, thereby modulating Ca^{2+} handling and vasoconstriction. Recent data obtained using antisense oligonucleotides indicate that NCX1 knockdown prolongs agonist responses by delaying the return of $[Ca^{2+}]_{cyt}$ to the resting level, and also inhibits ouabain-induced augmentation of agonist responses in $[Ca^{2+}]_{cyt}$ in cultured arterial smooth muscle cells.[1] Furthermore, reduced expression of NCX1 in aortas from NCX1 heterozygous mice decelerated Na^+-dependent relaxation and contraction.[23] These findings suggest that the NCX is involved in mediating both vasorelaxation and vasoconstriction depending on the condition of the blood vessel.

Recently, we examined the effects of SEA0400 on various hypertensive animal models. As shown in FIGURE 4, a single oral dose of SEA0400 (1–10 mg/kg) caused a dose-dependent decrease in arterial blood pressure in deoxycorticosterone acetate (DOCA)-salt hypertensive rats.[26] Furthermore, SEA0400 significantly decreased blood pressure in Dahl salt-sensitive rats and SHR when they were chronically loaded with high salt (FIG. 4). Notably, however, SEA0400 had no effect on blood pressure in normotensive Wistar Kyoto rats, SHR, stroke-prone SHR, salt-loaded or salt-unloaded Dahl salt-resistant

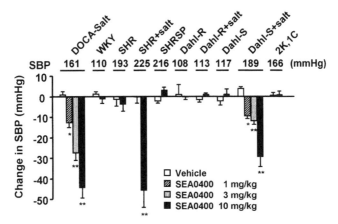

FIGURE 4. Antihypertensive effects of SEA0400 in various types of hypertensive rats. Spontaneously hypertensive rats (SHR), Dahl salt-resistant rats (Dahl-R), and Dahl salt-sensitive rats (Dahl-S) were fed a normal diet (0.3% NaCl) or high salt diet (8% NaCl; +salt) for 4–6 weeks. SEA0400 (1–10 mg/kg) or vehicle (5% gum Arabic) was orally administered. Peak change in systolic blood pressure (SBP) in mmHg for the respective rats is indicated. SBP was measured by tail cuff. Bars represent means ± SEM ($n = 4$–6). WKY, Wistar Kyoto rats; SHRSP, stroke-prone SHR; 2K,1C, two-kidney, one-clip renal hypertensive rats. *$P < 0.05$; **$P < 0.01$ compared with each vehicle group. (Modified from Iwamoto et al.[26] with permission.)

rats, salt-unloaded Dahl salt-sensitive rats, and two-kidney, one-clip renal hypertensive rats.[26] These findings suggest that the NCX inhibitor specifically suppresses salt-dependent hypertension.

To further determine the involvement of NCX1 gene in salt-dependent hypertension, we generated transgenic mice overexpressing NCX1.3, a major splicing isoform in arteries, under the smooth muscle α-actin promotor.[26] Intriguingly, these NCX1.3-transgenic mice were hypersensitive to salt, and the animals readily developed hypertension after high salt intake (FIG. 5 A).[26] Oral administration of SEA0400 reduced the elevated basal blood pressure of NCX1.3-transgenic mice and lowered the blood pressure of salt-loaded transgenic mice (FIG. 5 B). Very importantly, SEA0400 did not affect blood pressure in transgenic mice expressing a NCX1.3 mutant (G833C), which lacked the affinity to SEA0400, showing that SEA0400 specifically acts on the overexpressed NCX1.3 in arterial smooth muscle cells.

NCX1 heterozygous mice in which NCX1 function is reduced by ~50% (mild suppression) are also a very useful animal model for evaluating the physiological and pathological roles of NCX1.[23] Consistently, DOCA-salt treatment did not significantly alter the blood pressure of NCX1 heterozygous mice, whereas the same treatment produced a progressive elevation in blood pressure in wild-type mice.[26] On the other hand, hypertensive responses to chronic angiotensin II infusion were similar in NCX1 heterozygous mice and wild-type

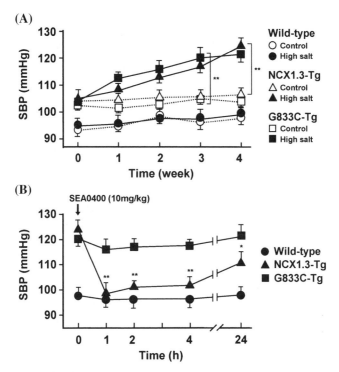

FIGURE 5. Salt-induced hypertension (**A**) and antihypertensive effects of SEA0400 (**B**) in transgenic mice (NCX1.3-Tg and G833C-Tg) and wild-type mice. Salt-loaded mice were given a high salt diet (8% NaCl) and tap water containing 1% NaCl for 4 weeks. Control mice were given a normal diet (0.3% NaCl) and tap water. SEA0400 (10 mg/kg) was orally administered into chronically salt-loaded mice. ** $P < 0.01$ versus control groups ($n = 5$ or 6). (Reproduced from Iwamoto et al.[26] with permission.)

mice. These results suggest that NCX1 heterozygous mice are preferentially resistant to salt-dependent hypertension. Taken together, our data provide compelling evidence that vascular NCX1 is a key mediator in the development of salt-dependent hypertension.

Endogenous cardiac glycosides contribute to the pathogenesis of salt-dependent hypertension, Cushing's syndrome, and primary aldosteronism in clinical patients and experimental animals.[27,28] We found that blood from DOCA-salt hypertensive rats contained humoral vasoconstrictors.[26] Importantly, arterial infusion of SEA0400 counteracted the vasoconstriction induced by humoral vasoconstrictors and exogenous ouabain in the femoral arteries of experimental animals.[26] These results indicate that excess dietary salt increases endogenous cardiac glycosides in plasma; the latter may contract peripheral blood vessels via vascular NCX1 and thereby result in hypertension.

In arterial smooth muscle cells, inhibition of Na^+, K^+-ATPase by cardiac glycosides should elevate local $[Na^+]$ just under the PM (FIG. 6). The restricted $[Na^+]$ accumulation facilitates Ca^{2+} entry through the vascular NCX1 isoform; this enhances arterial tone and causes hypertension. Notably, in arterial smooth muscle cells the NCX1 is co-localized with Na^+, K^+-ATPase α_2 and α_3 isoforms, which have high affinity for ouabain, in PM microdomains ("plasmerosomes") adjacent to the SR.[1] The concept of intracellular-linked Ca^{2+} and Na^+ transport at PM–SR junctions in arterial smooth muscle cells is also known as the "superficial buffer barrier" function.[29] Functional coupling between NCX and Na^+, K^+-ATPases has been reported in arterial smooth muscle cells and cardiomyocytes.[30,31]

Using knockin mice, Dostanic-Larson *et al.* recently demonstrated that the highly conserved cardiac glycoside-binding site of Na^+, K^+-ATPase (α_1 or α_2 isoform) plays an important role in the regulation of blood pressure, and that

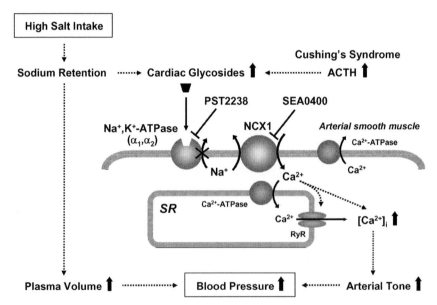

FIGURE 6. Proposed pathway responsible for salt-dependent hypertension. High salt intake (or Na^+ retention) and adrenocorticotropic hormone (ACTH)-dependent Cushing's syndrome cause the levels of endogenous cardiac glycosides that inhibit the Na^+, K^+-ATPase (α_1 and α_2 subtypes) to rise in the plasma (although Na^+ retention also increases plasma volume, resulting in elevated blood pressure). This results in the increase in sub-plasma membrane $[Na^+]$ of arterial smooth muscle. The restricted $[Na^+]$ accumulation elevates $[Ca^{2+}]_{cyt}$ by vascular NCX1-mediated Ca^{2+} entry. This enhances arterial tone and causes hypertension. SEA0400, an inhibitor for NCX1, as well as PST2238, a ouabain antagonist, block this Ca^{2+} entry and exhibit an antihypertensive effect in salt-dependent hypertension. SR, sarcoplasmic reticulum; RyR, ryanodine receptor (Modified from Iwamoto *et al.*[6] with permission).

it specifically mediates adrenocorticotropic hormone (ACTH)-induced hypertension, which is a model for hypertension related to the ACTH-dependent Cushing's syndrome.[32] They also clarified that the α_2 isoform of Na^+, K^+-ATPase can mediate ouabain-induced hypertension in mice.[33] These results provide a clear demonstration that an endogenous ligand for the Na^+, K^+-ATPase must be present in animals (FIG. 6). On the other hand, we found that nanomolar ouabain increases both $[Ca^{2+}]_{cyt}$ and myogenic tone in pressurized mouse small mesenteric arteries, and these effects are abolished by PST2238, a ouabain antagonist, and SEA0400, a specific NCX inhibitor.[26,34] Furthermore, the ouabain-induced $[Ca^{2+}]_{cyt}$ rise in arterial strips from NCX1.3-transgenic mice was greater than in those from wild-type mice, and SEA0400 then blocked these $[Ca^{2+}]_{cyt}$ rises in NCX1.3-transgenic mice and wild-type mice.[26] Taken together, these findings provide evidence that cardiac glycosides trigger Ca^{2+} entry through NCX1 in arterial smooth muscle cells by inhibiting the Na^+, K^+-ATPases (α_1 and α_2 isoforms) and elevating submembrane $[Na^+]$ (FIG. 6).

CONCLUSIONS

Recent studies using molecular pharmacological techniques and genetically engineered mice reveal functional and etiological implications of NCX1 in cardiac, arterial, neuronal, and renal tubular cells. NCX1 is multiregulated by intracellular Na^+ and Ca^{2+}, PIP_2, and protein kinases, and controls $[Ca^{2+}]_{cyt}$ and SR Ca^{2+} content mainly by Ca^{2+} extrusion from cells. Alteration of NCX1 activity (or expression) produces abnormalities in cellular Ca^{2+} regulation, resulting in several dysfunctions associated with ischemia–reperfusion injury and hypertension. During cardiac, renal, or neuronal reperfusion, NCX1 underlies the Ca^{2+} overload, resulting in cell injury. In addition, salt-dependent hypertension is triggered by Ca^{2+} entry via NCX1 in arterial smooth muscle. Benzyloxyphenyl NCX inhibitors may block these forms of abnormal Ca^{2+} transport by NCX1 and may then improve related morbid conditions. Several experiments suggest that NCX inhibitors have therapeutic potential for ischemia–reperfusion injury and salt-dependent hypertension. Treatment with NCX inhibitors will be an innovative therapeutic approach for several cardiovascular diseases. To reach this goal, however, further studies are necessary to develop new types of NCX inhibitors and precisely evaluate their efficacies.

ACKNOWLEDGMENTS

This work was supported by Grant-in-Aid 18059031 for Scientific Research on Priority Areas from the Ministry of Education, Sports, Science and

Technology of Japan, Grant-in-Aid 18590251 for Scientific Research for Japan Society for the Promotion of Science, and grants from the Takeda Science Foundation and the Ichiro Kanehara Foundation.

REFERENCES

1. BLAUSTEIN, M.P. & W.J. LEDERER. 1999. Sodium/calcium exchange: its physiological implications. Physiol. Rev. **79:** 763–854.
2. IWAMOTO, T. 2005. Sodium-calcium exchange inhibitors: therapeutic potential in cardiovascular diseases. Future Cardiol. **1:** 519–529.
3. HILGEMANN, D.W. 1996. The cardiac Na-Ca exchanger in giant membrane patches. Ann. N. Y. Acad. Sci. **779:** 136–158.
4. IWAMOTO, T., T. WATANO & M. SHIGEKAWA. 1996. A novel isothiourea derivative selectively inhibits the reverse mode of Na^+/Ca^{2+} exchange in cells expressing NCX1. J. Biol. Chem. **271:** 22391–22397.
5. MATSUDA, T., N. ARAKAWA, K. TAKUMA, et al. 2001. SEA0400, a novel and selective inhibitor of the $Na^+\text{-}Ca^{2+}$ exchanger, attenuates reperfusion injury in the *in vitro* and *in vivo* cerebral ischemic models. J. Pharmacol. Exp. Ther. **298:** 249–256.
6. IWAMOTO, T., Y. INOUE, K. ITO, et al. 2004. The exchanger inhibitory peptide region-dependent inhibition of Na^+/Ca^{2+} exchange by SN-6 [2-[4-(4-nitrobenzyloxy)benzyl]thiazolidine-4-carboxylic acid ethyl ester], a novel benzyloxyphenyl derivative. Mol. Pharmacol. **66:** 45–55.
7. KURAMOCHI, T., A. KAKEFUDA, H. YAMADA, et al. 2005. Synthesis and structure-activity relationships of benzyloxyphenyl derivatives as a novel class of NCX inhibitors: effects on heart failure. Bioorg. Med. Chem. **13:** 725–734.
8. IWAMOTO, T. & M. SHIGEKAWA. 1998. Differential inhibition of Na^+/Ca^{2+} exchanger isoforms by divalent cations and isothiourea derivative. Am. J. Physiol. **275:** C423–C430.
9. IWAMOTO, T., S. KITA, A. UEHARA, et al. 2004. Molecular determinant of Na^+/Ca^{2+} exchange (NCX1) inhibition by SEA0400. J. Biol. Chem. **279:** 7544–7553.
10. IWAMOTO, T. & S. KITA. 2006. YM-244769, a novel Na^+/Ca^{2+} exchange inhibitor that preferentially inhibits NCX3, efficiently protects against hypoxia/reoxygenation- induced SH-SY5Y neuronal cell damage. Mol. Pharmacol. **70:** 2075–2083.
11. IWAMOTO, T., S. KITA, A. UEHARA, et al. 2001. Structural domains influencing sensitivity to isothiourea derivative inhibitor KB-R7943 in cardiac Na^+/Ca^{2+} exchanger. Mol. Pharmacol. **59:** 524–531.
12. BOUCHARD, R., A. OMELCHENKO, H.D. LE, et al. 2004. Effects of SEA0400 on mutant NCX1.1 $Na^+\text{-}Ca^{2+}$ exchangers with altered ionic regulation. Mol. Pharmacol. **65:** 802–810.
13. LADILOV, Y., S. HAFFNER, C. BALSER-SCHAFER, et al. 1999. Cardioprotective effects of KB-R7943: a novel inhibitor of the reverse mode of Na^+/Ca^{2+} exchanger. Am. J. Physiol. **276:** H1868–H1876.
14. TAKAHASHI, K., T. TAKAHASHI, T. SUZUKI, et al. 2003. Protective effects of SEA0400, a novel and selective inhibitor of the Na^+/Ca^{2+} exchanger, on myocardial ischemia-reperfusion injuries. Eur. J. Pharmacol. **458:** 155–162.

15. NAKAMURA, A., K. HARADA, H. SUGIMOTO, et al. 1998. Effects of KB-R7943, a novel Na^+/Ca^{2+} exchange inhibitor, on myocardial ischemia/reperfusion injury. Folia. Pharmacol. Jpn. **111:** 105–115.

16. MAGEE, W.P., G. DESHMUKH, M.P. DENINNO, et al. 2003. Differing cardioprotective efficacy of the Na^+/Ca^{2+} exchanger inhibitors SEA0400 and KB-R7943. Am. J. Physiol. **284:** H903–H910.

17. HOTTA, Y., X. LU, M. YAJIMA, et al. 2002. Protective effect of SN-6, a selective Na^+-Ca^{2+} exchange inhibitor, on ischemia-reperfusion-injured hearts. Jpn. J. Pharmacol. **88:** 57 P.

18. TAKAHASHI, T., K. TAKAHASHI, M. ONISHI, et al. 2004. Effects of SEA0400, a novel inhibitor of the Na^+/Ca^{2+} exchanger, on myocardial stunning in anesthetized dogs. Eur. J. Pharmacol. **505:** 163–168.

19. YAMASHITA, J., S. KITA, T. IWAMOTO, et al. 2003. Attenuation of ischemia/reperfusion-induced renal injury in mice deficient in Na^+/Ca^{2+} exchanger. J. Pharmacol. Exp. Ther. **304:** 284–293.

20. OGATA, M., T. IWAMOTO, N. TAZAWA, et al. 2003. A novel and selective Na^+/Ca^{2+} exchange inhibitor, SEA0400, improves ischemia/reperfusion-induced renal injury. Eur. J. Pharmacol. **478:** 187–198.

21. SCHRÖDER, U.H., J. BREDER, C.F. SABELHAUS, et al. 1999. The novel Na^+/Ca^{2+} exchange inhibitor KB-R7943 protects CA1 neurons in rat hippocampal slices against hypoxic/hypoglycemic injury. Neuropharmacology **38:** 319–321.

22. EIGEL, B.N. & R.W. HADLEY. 2001. Antisense inhibition of Na^+/Ca^{2+} exchange during anoxia/reoxygenation in ventricular myocytes. Am. J. Physiol. **281:** H2184–H2190.

23. WAKIMOTO, K., K. KOBAYASHI, M. KURO-O, et al. 2000. Targeted disruption of Na^+/Ca^{2+} exchanger gene leads to cardiomyocyte apoptosis and defects in heartbeat. J. Biol. Chem. **275:** 36991–36998.

24. OHTSUKA, M., H. TAKANO, M. SUZUKI, et al. 2004. Role of Na^+-Ca^{2+} exchanger in myocardial ischemia/reperfusion injury: evaluation using a heterozygous Na^+-Ca^{2+} exchanger knockout mouse model. Biochem. Biophys. Res. Commun. **314:** 849–853.

25. CROSS, H.R., L. LU, C. STEENBERGEN, et al. 1998. Overexpression of the cardiac Na^+/Ca^{2+} exchanger increases susceptibility to ischemia/reperfusion injury in male, but not female, transgenic mice. Circ. Res. **83:** 1215–1223.

26. IWAMOTO, T., S. KITA, J. ZHANG, et al. 2004. Salt-sensitive hypertension is triggered by Ca^{2+} entry via Na^+/Ca^{2+} exchanger type-1 in vascular smooth muscle. Nat. Med. **10:** 1193–1199.

27. HAMLYN, J.M., B.P. HAMILTON & P. MANUNTA. 1996. Endogenous ouabain, sodium balance and blood pressure: a review and a hypothesis. J. Hypertens. **14:** 151–167.

28. GOTO, A. & K. YAMADA. 2000. Putative roles of ouabainlike compound in hypertension: revisited. Hypertens. Res. **23:** S7–S13.

29. POBURKO, D., K.H. KUO, J. DAI, et al. 2004. Organellar junctions promote targeted Ca^{2+} signaling in smooth muscle: why two membranes are better than one. Trends Pharmacol. Sci. **25:** 8–15.

30. FUJIOKA, Y., S. MATSUOKA, T. BAN, et al. 1998. Interaction of the Na^+-K^+ pump and Na^+-Ca^{2+} exchange via $[Na^+]_i$ in a restricted space of guinea-pig ventricular cells. J. Physiol. **509:** 457–470.

31. REUTER, H., S.A. HENDERSON, T. HAN, et al. 2002. The Na^+-Ca^{2+} exchanger is essential for the action of cardiac glycosides. Circ. Res. **90:** 305–308.

32. DOSTANIC-LARSON, I., J.W. VAN HUYSSE, J.N. LORENZ, *et al.* 2005. The highly conserved cardiac glycoside binding site of Na,K-ATPase plays a role in blood pressure regulation. Proc. Natl. Acad. Sci. USA **102:** 15845–15850.
33. DOSTANIC, I., R.J. PAUL, J.N. LORENZ, *et al.* 2005. The α_2-isoform of Na-K-ATPase mediates ouabain-induced hypertension in mice and increased vascular contractility *in vitro*. Am. J. Physiol. **288:** H477–H485.
34. ZHANG, J., M.Y. LEE, M. CAVALLI, *et al.* 2005. Sodium pump α_2 subunits control myogenic tone and blood pressure in mice. J. Physiol. **569:** 243–256.

Inhibitory Mechanism of SN-6, A Novel Benzyloxyphenyl Na$^+$/Ca^{2+} Exchange Inhibitor

SATOMI KITA AND TAKAHIRO IWAMOTO

Department of Pharmacology, School of Medicine, Fukuoka University, Fukuoka 814-0180, Japan

ABSTRACT: We investigated the pharmacological properties of SN-6, a new selective Na$^+$/Ca^{2+} exchanger (NCX) inhibitor. SN-6 preferentially inhibited the ^{45}Ca^{2+} uptake via NCX compared with the ^{45}Ca^{2+} efflux via NCX in NCX-transfected fibroblasts. SN-6 was three- to fivefold more inhibitory to the ^{45}Ca^{2+} uptake via NCX1 (IC$_{50}$ = 2.9 μM) than to that via NCX2 or NCX3. Our chimeric and site-directed mutageneses revealed that Phe-213, Val-227, Tyr-228, Gly-833, and Asn-839 in NCX1 are molecular determinants for interaction with SN-6. We also found that SN-6 potently protects against hypoxia/reoxygenation-induced cell damage in renal tubular cells.

KEYWORDS: sodium–calcium exchange; SN-6; mutagenesis; hypoxia/reoxygenation

INTRODUCTION

The Na$^+$/Ca^{2+} exchanger (NCX) is a bidirectional transporter that is controlled by membrane potential and transmembrane gradients of Na$^+$ and Ca^{2+}.[1,2] Mammalian NCX forms a multigene family comprising NCX1, NCX2, and NCX3. NCX1 is highly expressed in the heart, kidney, and brain and much lower levels in other tissues, whereas the expression of NCX2 and NCX3 is limited mainly to the brain and skeletal muscle. NCX1 has been shown to be secondarily regulated by the transport substrates Na$^+$ and Ca^{2+}.[3] Ca^{2+}$_i$ at the submicromolar level activates NCX activity by promoting the recovery of the exchanger from the "I$_2$ inactivation state," whereas high Na$^+$$_i$ restrains the exchange by facilitating the entry of the exchanger into the "I$_1$ inactivation state."

A selective NCX inhibitor will be very useful to study physiological and pathophysiological roles of NCX and to clarify the reaction mechanism of

Address for correspondence: Dr. Takahiro Iwamoto, Department of Pharmacology, School of Medicine, Fukuoka University, 7-45-1 Nanakuma Jonan-ku, Fukuoka 814-0180, Japan. Voice: +81-92-801-1011; ext.: 3261; fax: +81-92-865-4384.

tiwamoto@cis.fukuoka-u.ac.jp

Ann. N.Y. Acad. Sci. 1099: 529–533 (2007). © 2007 New York Academy of Sciences.
doi: 10.1196/annals.1387.040

this transporter. Moreover, such an inhibitor may have therapeutic potential as a new remedy for several ischemic diseases, arrhythmias, heart failure, and essential hypertension. KB-R7943, a benzyloxyphenyl derivative, was first introduced in 1996 as a selective NCX inhibitor.[4] In 2001, Matsuda *et al.* reported on SEA0400, a newly developed, potent, and selective inhibitor of NCX.[5] The inhibitory potency of SEA0400 is 80–100 times more powerful than that of KB-R7943. SEA0400 has an excellent specificity; KB-R7943 possesses nonspecific actions against ion channels, neuronal nicotinic acetyl-choline receptors, the N-methyl-D-aspartate receptor, and the norepinephrine transporter.[2]

We have recently developed SN-6, a new selective NCX inhibitor.[6] This drug was found by screening newly synthesized benzyloxyphenyl derivatives for inhibition of Na^+_i-dependent $^{45}Ca^{2+}$ uptake into NCX1-transfected fibroblasts. In this study, we investigated the inhibitory properties of SN-6 and searched for the structural domains responsible for its inhibition using site-directed mutagenesis. We further examined the protective effect of SN-6 on hypoxia/reoxygenation-induced injury using porcine tubular epithelial LLC-PK$_1$ cells.

MATERIALS AND METHODS

Chinese hamster lung fibroblasts (CCL39 cells) and their transfectants expressing wild-type, chimeric, and mutant exchangers[6] were maintained in Dulbecco's modified Eagle's medium (DMEM) supplemented with 7.5% fetal calf serum (FCS), 50 U/mL penicillin, and 50 μg/mL streptomycin in a humidified incubator gassed with 5% CO_2/95% air at 37°C. LLC-PK$_1$ cells and their NCX transfectants were grown in DMEM supplemented with 4% FCS, 50 U/mL penicillin, and 50 μg/mL streptomycin.

Na^+_i-dependent $^{45}Ca^{2+}$ uptake into cells expressing the wild-type or mutated exchangers were assayed as described in detail previously.[6] In brief, confluent cells in 24-well dishes were loaded with Na^+ by incubation at 37°C for 30 min in 0.5 mL of balanced salt solution (BSS) (10 mM HEPES/Tris, pH 7.4, 146 mM NaCl, 4 mM KCl, 2 mM $MgCl_2$, 0.1 mM $CaCl_2$, 10 mM glucose, and 0.1% bovine serum albumin) containing 1 mM ouabain and 10 μM monensin. $^{45}Ca^{2+}$ uptake was then initiated by switching the medium to Na^+-free BSS (replacing NaCl with equimolar choline chloride) or to normal BSS, both of which contained 0.1 mM $^{45}CaCl_2$ (370 kBq/mL) and 1 mM ouabain. After a 30-s or 1-min incubation,$^{45}Ca^{2+}$ uptake was terminated by washing cells four times with an ice-cold solution containing 10 mM HEPES/Tris, pH 7.4, 120 mM choline chloride, and 10 mM $LaCl_3$. Cells were then solubilized with 0.1 N NaOH, and aliquots were taken for determination of radioactivity and protein.

RESULTS AND DISCUSSION

We first compared the inhibitory effects of SN-6 on Na^+_i-dependent $^{45}Ca^{2+}$ uptake (i.e., the reverse mode) into CCL39 cells expressing NCX1, NCX2, and NCX3. SN-6 (up to 30 μM) inhibited dose-dependently the initial rate of $^{45}Ca^{2+}$ uptake into their transfectants with IC_{50} values of 2.9 ± 0.12, 16 ± 1.1, and 8.6 ± 0.27 μM ($n = 3$), respectively (FIG. 1), indicating that SN-6 has a selectivity to NCX1. On the other hand, SN-6 at 3–30 μM did not significantly affect Na^+_o-dependent $^{45}Ca^{2+}$ efflux (i.e., the forward mode), adenosine, adrenergic, glutamate, and bradykinin receptors, calcium, sodium, and potassium channels, Na^+/H^+ exchanger, Na^+, K^+-ATPase, and Ca^{2+}-ATPase.[6]

We searched for regions that may form the SN-6 receptor by NCX1/NCX3-chimeric analyses. We constructed two series of chimeras in which serial segments from NCX3 were transferred into NCX1 in exchange for the homologous segments and vice versa, and determined that amino acid regions 73 to 108 and 193 to 230 in NCX1 are mostly responsible for the differential drug response between NCX1 and NCX3. To identify the critical residues involved in drug sensitivity, unique residues in these critical regions of NCX1 were exchanged with the corresponding residues in NCX3. In addition, further mutageneses revealed that double substitutions of Val-227 and Tyr-228 in NCX1, which exist within the exchanger inhibitory peptide (XIP) region, mimicked

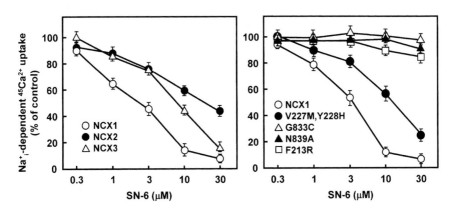

FIGURE 1. Dose-response curves for the effects of SN-6 on Na^+_i-dependent $^{45}Ca^{2+}$ uptake in cells expressing NCX1, NCX2, NCX3, or NCX1 mutants. The initial rates of $^{45}Ca^{2+}$ uptake into cells were measured in the presence or absence of indicated concentrations of SN-6. SN-6 was added 5 min before the start of uptake measurement. Data are presented as a percentage of the control values obtained in the absence of SN-6. Data are means ± SE of four independent experiments. (Modified from Iwamoto et al.[6] with permission.)

the different drug response (FIG. 1). In addition, F213R, G833C, and N839A mutations in NCX1 resulted in loss of drug sensitivity (FIG. 1). Therefore, we suppose that the XIP region (containing Val-227 and Tyr-228) and its neighboring regions, such as the fifth transmembrane (containing Phe-213) and the α-2 repeat (containing Gly-833 and Asn-839), may participate in the formation of the interaction domains with SN-6.

We next analyzed the effects of SN-6 on NCX1 mutants, which display the altered kinetics of Na^+-dependent inactivation (i.e., I_1 inactivation). XIP region mutants K229Q and F223E have been shown to exhibit completely eliminated and enhanced I_1 inactivation, respectively.[7] Interestingly, K229Q and F223E mutants exhibited a markedly reduced sensitivity and hypersensitivity, respectively, to inhibition by SN-6.[6] These data suggest that the inhibitory effect of SN-6 is related to the kinetics of I_1 inactivation. Similar results have also been observed in other benzyloxyphenyl derivatives, KB-R7943 and SEA0400.[8,9]

We examined the renoprotective effect of SN-6 in hypoxia/reoxygenation-induced injury in LLC-PK$_1$ cells. In parental LLC-PK$_1$ cells, 6-h hypoxia followed by 1 h of reoxygenation produced a significant lactate dehydrogenase (LDH) release from damaged cells. The hypoxia/reoxygenation-induced LDH release was 2.4-fold enhanced in LLC-PK$_1$ cells overexpressing wild-type NCX1. SN-6 (0.3–10 μM) dose-dependently protected against the hypoxia/reoxygenation-induced LDH release in parental LLC-PK$_1$ cells and NCX1 transfectants. In NCX1 transfectants, the IC$_{50}$ value of SN-6 for hypoxia/reoxygenation-induced LDH release was 0.63 ± 0.15 μM ($n = 3$). These results suggest that SN-6 predominantly works as a blocker of Ca^{2+} overload via NCX under hypoxic/ischemic conditions. Such a property of SN-6, which might be derived from its interaction with the endogenous XIP region, seems to be advantageous to developing it clinically as a new renal and cardiac protective anti-ischemic drug.

ACKNOWLEDGMENTS

This work was supported by grants from the Takeda Science Foundation and the Inoue Foundation for Science.

REFERENCES

1. BLAUSTEIN, M.P. & W.J. LEDERER. 1999. Sodium/calcium exchange: its physiological implications. Physiol. Rev. **79:** 763–854.
2. IWAMOTO, T. 2005. Sodium-calcium exchange inhibitors: therapeutic potential in cardiovascular diseases. Future Cardiol. **1:** 519–529.
3. HILGEMANN, D.W. 1996. The cardiac Na-Ca exchanger in giant membrane patches. Ann. N. Y. Acad. Sci. **779:** 136–158.

4. IWAMOTO, T., T. WATANO & M. SHIGEKAWA. 1996. A novel isothiourea derivative selectively inhibits the reverse mode of Na^+/Ca^{2+} exchange in cells expressing NCX1. J. Biol. Chem. **271:** 22391–22397.

5. MATSUDA, T., N. ARAKAWA, K. TAKUMA, et al. 2001. SEA0400, a novel and selective inhibitor of the Na^+-Ca^{2+} exchanger, attenuates reperfusion injury in the in vitro and in vivo cerebral ischemic models. J. Pharmacol. Exp. Ther. **298:** 249–256.

6. IWAMOTO, T., Y. INOUE, K. ITO, et al. 2004. The exchanger inhibitory peptide region-dependent inhibition of Na^+/Ca^{2+} exchange by SN-6 [2-[4-(4-nitrobenzyloxy)benzyl]thiazolidine-4-carboxylic acid ethyl ester], a novel benzyloxyphenyl derivative. Mol. Pharmacol. **66:** 45–55.

7. MATSUOKA, S., D.A. NICOLL, Z. HE, et al. 1997. Regulation of cardiac Na^+-Ca^{2+} exchanger by the endogenous XIP region. J. Gen. Physiol. **109:** 273–286.

8. IWAMOTO, T., S. KITA, A. UEHARA, et al. 2004. Molecular determinant of Na^+/Ca^{2+} exchange (NCX1) inhibition by SEA0400. J. Biol. Chem. **279:** 7544–7553.

9. BOUCHARD, R., A. OMELCHENKO, H.D. LE, et al. 2004. Effects of SEA0400 on mutant NCX1.1 Na^+-Ca^{2+} exchangers with altered ionic regulation. Mol. Pharmacol. **65:** 802–810.

Electrophysiological Effects of SN-6, a Novel Na$^+$/Ca^{2+} Exchange Inhibitor on Membrane Currents in Guinea Pig Ventricular Myocytes

CHUN-FENG NIU,[a] YASUHIDE WATANABE,[b] TAKAHIRO IWAMOTO,[c] KANNA YAMASHITA,[b] HIROSHI SATOH,[a] TUYOSHI URUSHIDA,[a] HIDEHARU HAYASHI,[a] AND JUNKO KIMURA[d]

[a] *Department of Internal Medicine III, Hamamatsu University School of Medicine, Hamamatsu 431-3192, Japan*

[b] *Division of Pharmacological Science, Department of Health Science, Hamamatsu University School of Medicine, Hamamatsu 431-3192, Japan*

[c] *Department of Pharmacology, School of Medicine, Fukuoka University, Fukuoka 814-0180, Japan*

[d] *Department of Pharmacology, School of Medicine, Fukushima Medical University, Fukushima 960-1295, Japan*

ABSTRACT: We examined the effect of SN-6 on the Na$^+$/Ca^{2+} exchanger (NCX) current (I_{NCX}) and other membrane currents in isolated guinea pig ventricular myocytes using the whole-cell voltage clamp technique. SN-6 suppressed the bidirectional I_{NCX} in a concentration-dependent manner. The IC$_{50}$ values of SN-6 were 2.3 µM and 1.9 µM for the outward and inward components of the bidirectional I_{NCX}, respectively. On the other hand, SN-6 suppressed the unidirectional outward I_{NCX} more potently than the inward I_{NCX}, with an IC$_{50}$ value of 0.6 µM. SN-6 at 10 µM inhibited the unidirectional inward I_{NCX} by only 22.4 $\pm$ 3.1%. SN-6 suppressed I_{NCX} more potentially when intracellular Na$^+$ concentration became higher. SN-6 inhibited I_{Na}, I_{Ca}, I_{Kr}, I_{Ks}, and I_{K1} by about 13%, 34%, 33%, 18%, and 13%, respectively. SN-6 shortened the action potential duration (APD) by about 34% and 25% at APD$_{50}$ and APD$_{90}$, respectively. These results indicate that SN-6 inhibits NCX in a similar manner to that of KB-R7943. SN-6 and KB-R7943 inhibit the unidirectional outward I_{NCX} more potently than the unidirectional inward I_{NCX}. Both drugs inhibit NCX in an intracellular Na$^+$ concentration-dependent manner. However, SN-6 affected other membrane currents less potently than KB-R7943.

Address for correspondence: Yasuhide Watanabe Ph.D., Division of Pharmacological Science, Department of Health Science, Hamamatsu University School of Medicine, 1-20-1 Hondayama, Hamamatsu, Shizuoka, 431-3192, Japan. Voice/fax: +81-53-435-2812.

w-yasu@hama-med.ac.jp

Ann. N.Y. Acad. Sci. 1099: 534–539 (2007). © 2007 New York Academy of Sciences.
doi: 10.1196/annals.1387.037

KEYWORDS: Na$^+$/Ca^{2+} exchange; SN-6; benzyloxyphenyl derivative; cardiac myocytes; whole-cell voltage clamp

INTRODUCTION

The Na$^+$/Ca^{2+} exchanger (NCX) is a bidirectional transporter that exchanges 3 Na$^+$ and 1 Ca^{2+} in either the Ca^{2+} exit mode or Ca^{2+} entry mode. NCX is driven by the electrochemical gradients of the substrate ions (Na$^+$ and Ca^{2+}) across the plasma membrane and membrane potentials.[1] Under physiological conditions of cardiac muscle, NCX mainly contributes to Ca^{2+} exit across the plasma membrane. In pathophysiological conditions, intracellular Ca^{2+} overload often occurs due to the Ca^{2+} entry mode of NCX and causes ischemia–reperfusion injury, arrhythmia, hypertrophy, and heart failure. Recently, KB-R7943 and SEA0400, benzyloxphenyl derivative NCX inhibitors have been reported to have cardioprotective effects against digitalis-induced arrhythmia and ischemia–reperfusion injury.[2–5]

Iwamoto *et al.* have reported that SN-6, a benzyloxphenyl derivative, is a new selective Na$^+$/Ca^{2+} exchange inhibitor.[6] SN-6 has a chemical structure with higher similarity to KB-R7943 than SEA0400. We examined the effect of SN-6 on the NCX current (I$_{NCX}$) and other membrane currents in isolated guinea pig ventricular myocytes using the whole-cell voltage clamp technique.

METHODS

Isolated single ventricular cells were digested from heart by Langendorff perfusion system with collagenase. Membrane currents were recorded by the whole-cell patch clamp method using pCLAMP8 software (Axon Instruments, Foster City, CA). The current–voltage (I–V) relationships for the bidirectional outward and inward I$_{NCX}$ were obtained by voltage ramp pulses between 30 mV and –150 mV, and unidirectional outward and inward I$_{NCX}$ between 30 mV and –50 mV, and between 30 mV and –100 mV, respectively, as described previously.[7,8] Other membrane currents: I$_{Na}$, I$_{Ca}$, I$_K$, and I$_{K1}$ were recorded as inward or outward currents by square pulses of depolarization or hyperpolarization from one holding potential to various membrane potentials. Current clamp experiments were also performed in the whole-cell recording mode. The "bidirectional" pipette solution contained 20 mM NaCl, 20 mM BAPTA, 10 mM CaCl$_2$ (free Ca^{2+} concentration 226 nM), 120 mM CsCl, 3 mM MgCl$_2$, 50 mM aspartic acid, 5 mM MgATP, and 10 mM HEPES (pH 7.2 with CsOH). The "bidirectional" extracellular solution contained 140 mM NaCl, 1 mM CaCl$_2$, 1 mM MgCl$_2$, 0.01 mM ryanodine, 0.02 mM ouabain, 0.01 mM nifedipine, and 5 mM HEPES-NaOH (pH 7.2). A unidirectional outward I$_{NCX}$ was induced by transiently changing [Ca^{2+}]$_o$ from 0 to

1 mM in the presence of 20 mM $[Na^+]_i$, 140 mM $[Na^+]_o$, and 96 nM free $[Ca^{2+}]_i$ (20 mM BAPTA and 6 mM $CaCl_2$). A unidirectional inward I_{NCX} was induced by changing $[Na^+]_o$, from 0 (substituted by 140 mM $[Li^+]_o$) to 140 mM in the presence of 226 nM free $[Ca^{2+}]_i$ (20 mM BAPTA and 10 mM $CaCl_2$) and 1 mM $[Ca^{2+}]_o$ without 20 mM $[Na^+]_i$. Other components of the pipette solution were essentially the same in both solutions as that for inducing the bidirectional I_{NCX}. The I_{Na} was measured at 28 mM Na^+ and 112 mM N-methyl-*d*-glucamine (NMG) in the external solution. The external solutions contained 28 mM NaCl, 112 mM NMG, 1 mM $CaCl_2$, 1 mM $MgCl_2$, 5 mM HEPES (pH 7.4), and 0.01 mM nifedipine. The I_{Ca} was measured at 140 mM NaCl external solution without nifedipine. The pH of the NMG solution was adjusted with HCl. The pipette solution contained 120 mM CsOH, 20 mM CsCl, 54 mM asparatic acid, 3 mM $MgCl_2$, 5 mM MgATP, 20 mM BAPTA, and 20 mM HEPES (pH 7.2). The I_K was measured using an external solution containing 145 mM choline-Cl, 1 mM $CaCl_2$, 1 mM $MgCl_2$, 5 mM HEPES, 0.33 mM KH_2PO_4, 5.5 mM glucose, 0.01 mM nifedipine, and 0.002 mM atropine. The I_{K1} was measured with the Tyrode external solution. The pipette solution contained 120 mM KOH, 20 mM KCl, 54 mM asparatic acid, 3 mM $MgCl_2$, 5 mM MgATP, 20 mM BAPTA, and 20 mM HEPES (pH 7.2).

RESULTS AND DISCUSSION

Bidirectional I_{NCX} was induced by 1 mM Ca^{2+} and 140 mM Na^+ in the external solution and 20 mM Na^+ and 226 nM free Ca^{2+} in the pipette solution. Under these ionic conditions, the reversal potential of the exchange current with a 3 Na^+:1 Ca^{2+} stoichiometry was calculated to be –68 mV. When the I_{NCX} became stable, the control external solution was switched to one containing SN-6. The I–V relationship recorded in the presence of SN-6 intersected with the control I–V curve at around –60 mV. The I_{NCX} magnitude was measured at +30 mV for the outward I_{NCX} component and at –150 mV for the inward I_{NCX} component, and the percentage of inhibition was calculated assuming that 5 mM Ni^{2+} completely inhibited each direction of I_{NCX}. SN-6 between 1 and 30 μM suppressed the bidirectional outward and inward I_{NCX} in a concentration-dependent manner. The IC_{50} values of SN-6 against the bidirectional outward and inward I_{NCX} were 2.3 μM and 1.9 μM, respectively, and the Hill coefficients were 1. Unidirectional outward I_{NCX} was induced by changing $[Ca^{2+}]_o$ from 0 mM to 1 mM in the presence of 20 mM Na^+ and 96 nM free Ca^{2+} in the pipette solution for 5–6 min. SN-6 between 0.1 and 10 μM suppressed unidirectional outward I_{NCX} in a concentration-dependent manner. The IC_{50} value for the unidirectional outward I_{NCX} was 0.6 μM with a Hill coefficient of 1. However, even 10 μM SN-6 inhibited the unidirectional inward I_{NCX} only by about 20%.

KB-R7943 and SEA0400 have been reported to inhibit both the directions of bidirectional I_{NCX} equally in a concentration-dependent manner. The IC_{50} values of KB-R7943 are approximately 0.3–1 μM and those of SEA0400 are 30–40 nM.[8,9] The inhibitory potencies on I_{NCX1} are SEA0400$\gg$SN-6=KB-R7943 in the heart.[8,9] The selectivity of NCX inhibitors for I_{NCX} over other membrane currents is SEA0400>SN-6>KB-R7943 in the heart.[8,10–12]

The three NCX isoforms demonstrate a different sensitivity to the NCX inhibitors. SEA0400 blocks NCX1 more potently than NCX2 and does not block NCX3.[13] KB-R7943 blocks NCX3 more potently than NCX1 and NCX2.[13] SN-6 inhibits NCX1 more potently than NCX2 or NCX3.[14] Putative interaction domains for benzyloxyphenyl derivatives include transmembrane segments 2, 3, 7, and 8 with the most influential amino acids including Phe-213, Val-221, Tyr-228, Gly-833, and Asn-839.[13]

Recently, it was reported that SEA0400 inhibited I_{NCX} in an intracellular Na^+ concentration-dependent manner in *Xenopus laevis* oocytes expressing cloned cardiac NCX.[11] SN-6 also inhibited bidirectional I_{NCX} in a $[Na^+]_i$ concentration-dependent manner. The IC_{50} values of SN-6 were 3.4 μM, 2.3 μM, and 1.1 μM at 10 mM, 20 mM, and 30 mM $[Na^+]_i$, respectively, with the Hill coefficients of 1. The Na^+-dependent block of I_{NCX} by SN-6 was statistically significant. Exchangers with a mutated XIP, which display undetectable I_1 inactivation, have shown a markedly reduced sensitivity to SEA0400.[15,16] It is probable that the inhibitory potency of NCX inhibitors is directly coupled to the rate of I_1 inactivation (i.e., intracellular Na^+-dependent inactivation). We examined whether SN-6 inhibited I_{NCX} from the extracellular side or from the cytoplasmic side by perfusing trypsin (2.5 μg/mL) through the pipette solution as described previously.[17–20] SN-6 inhibited I_{NCX} even in the presence of trypsin. Therefore SN-6 is a trypsin-insensitive NCX inhibitor.

In addition, we examined the effects of SN-6 at a concentration of 10 μM, which inhibited the bidirectional I_{NCX} by 80%, on other membrane currents and action potential duration (APD). SN-6 at 10 μM inhibited I_{Na}, I_{Ca}, I_{Kr}, I_{Ks}, and I_{K1} by about 13%, 34%, 33%, 18%, and 13%, respectively. SN-6 at 10 μM shortened the APD by about 34% and 25% at APD_{50} and APD_{90}, respectively.

In summary, we examined the effect of SN-6 on the I_{NCX} and other membrane currents in guinea pig single cardiac ventricular myocytes. We found that SN-6 inhibits I_{NCX} with a similar potency to that of KB-R7943. However, SN-6 affects other membrane currents less potently than KB-R7943. Both SN-6 and KB-R7943 inhibit the unidirectional outward I_{NCX} more potently than the unidirectional inward I_{NCX}. This effect of both drugs is intracellular Na^+ concentration-dependent. Lee *et al.* initially proposed the hypothesis that the direction-dependent difference in potencies of NCX inhibitors is ascribed to intracellular Na^+-dependent inhibition.[11] SN-6 may be a powerful tool for studying the physiological and pathological roles of NCX.

ACKNOWLEDGMENT

This work was supported by grants-in-aids from the Ministry of Education, Science, Sports and Culture of Japan (17590718) (Urushida T).

REFERENCES

1. BLAUSTEIN, M.P. & W.J. LEDERER. 1999. Sodium/calcium exchange: its physiological implications. Physiol. Rev. **79:** 763–854.
2. WATANO, T., Y. HARADA, K. HARADA, et al. 1999. Effect of Na^+/Ca^{2+} exchange inhibitor on ouabain-induced arrhythmias in guinea pigs. Br. J. Pharmacol. **127:** 1846–1850.
3. SEKI, S., M. TANIGUCHI, H. TAKEDA, et al. 2002. Inhibition by KB-R7943 of the reverse mode of the Na^+/Ca^{2+} exchanger reduces Ca^{2+} overload in ischemia-reperfused rat hearts. Circ. J. **66:** 390–396.
4. TAKAHASHI, K., T. TAKAHASHI, T. SUZUKI, et al. 2003. Protective effects of SEA0400, a novel and selective inhibitor of the $Na^+\text{-}Ca^{2+}$ exchanger, on myocardial ischemia-reperfusion injuries. Eur. J. Pharmacol. **458:** 155–162.
5. NAGASAWA, Y., B.-M. ZHU, J. CHEN, et al. 2005. Effects of SEA0400, a Na^+/Ca^{2+} exchange inhibitor, on ventricular arrhythmias in the in vivo dogs. Eur. J. Pharmacol. **506:** 249–255.
6. IWAMOTO, T., Y. INOUE, K. ITO, et al. 2004. The exchanger inhibitory peptide region-dependent inhibition of Na^+/Ca^{2+} exchange by SN-6 [2-[4-(4-nitrobenzyloxy)benzyl]thiazolidine-4-carboxylic acid ethyl ester], a novel benzyloxyphenyl derivative. Mol. Pharmacol. **66:** 45–55.
7. WATANO, T. & J. KIMURA. 1998. Calcium-dependent inhibition of the sodium-calcium exchange current by KB-R7943. Can. J. Cardiol. **14:** 259–262.
8. KIMURA, J., T. WATANO, M. KAWAHARA, et al. 1999. Direction-independent block of bi-directional Na^+/Ca^{2+} exchange current by KB-R7943 in guinea-pig cardiac myocytes. Br. J. Pharmacol. **128:** 969–974.
9. TANAKA, H., K. NISHIMURA, T. AIKAWA, et al. 2002. Effect of SEA0400, a novel inhibitor of sodium-calcium exchanger on myocardial ionic currents. Br. J. Pharmacol. **135:** 1096–1100.
10. WATANO, T., J. KIMURA, T. MORITA, et al. 1996. A novel antagonist, No. 7943, of Na-Ca exchange current in guinea-pig cardiac ventricular cells. Br. J. Pharmacol. **119:** 555–563.
11. LEE, C., N.S. VISEN, N.S. DHALLA, et al. 2004. Inhibitory profile of SEA0400 [2-[4-[(2,5-Difluorophenyl)methoxy]phenoxy]-5-ethoxyaniline] assessed on the cardiac $Na^+\text{-}Ca^{2+}$ exchanger, NCX1.1. J. Pharmacol. Exp. Ther. **311:** 748–757.
12. BIRINYI, P., K. ACSAI, T. BÁNYÁSZ, et al. 2005. Effects of SEA0400 and KB-R7943 on Na^+/Ca^{2+} exchange current in canine ventricular cardiomyocytes. Naunyn-Schmiedberg's Arch. Pharmacol. **372:** 63–70.
13. IWAMOTO, T. 2004. Forefront of Na^+/Ca^{2+} exchange studies: molecular pharmacology of Na^+/Ca^{2+} exchange inhibitors. J. Pharmacol. Sci. **96:** 27–32.
14. IWAMOTO, T. 2005. Sodium-calcium exchange inhibitors: therapeutic potential in cardiovascular diseases. Future Cardiol. **1:** 519–529.
15. IWAMOTO, T., S. KITA, A. UEHARA, et al. 2004. Molecular determinants of Na^+/Ca^{2+} exchange (NCX1) inhibition by SEA0400. J. Biol. Chem. **279:** 7544–7553.

16. BOUCHARD, R., A. OMELCHENKO, H.D. LE, *et al.* 2004. Effects of SEA0400 on mutant NCX1.1 Na^+-Ca^{2+} exchangers with altered ionic regulation. Mol. Pharmacol. **65:** 802–810.

17. WATANABE, Y. & J. KIMURA. 2000. Inhibitory effect of amiodarone on Na^+/Ca^{2+} exchange current guinea-pig cardiac myocytes. Br. J. Pharmacol. **131:** 80–84.

18. WATANABE, Y. & J. KIMURA. 2001. Blocking effect of bepridil on Na^+/Ca^{2+} exchange current in guinea pig cardiac ventricular myocytes. Jpn. J. Pharmacol. **85:** 370–375.

19. WATANABE, Y., T. IWAMOTO, I. MATSUOKA, *et al.* 2001. Inhibitory effect of 2,3-butanedione monoxime (BDM) on Na^+/Ca^{2+} exchange current in guinea-pig cardiac ventricular myocytes. Br. J. Pharmacol. **132:** 1317–1325.

20. WATANABE, Y., T. IWAMOTO, M. SHIGEKAWA, *et al.* 2002. Inhibitory effect of aprindine on Na^+/Ca^{2+} exchange current in guinea-pig cardiac ventricular myocytes. Br. J. Pharmacol. **136:** 361–366.

Directionality in Drug Action on Sodium–Calcium Exchange

D. NOBLE[a] AND M. P. BLAUSTEIN[b]

[a]Departments of Physiology, Anatomy, and Genetics, University of Oxford, Oxford, OX1 3PT, United Kingdom

[b]Department of Physiology, University of Maryland School of Medicine, Baltimore, Maryland 21201, USA

ABSTRACT: In pathological conditions, the exchanger may generate dele-terious calcium entry. A drug that inhibited calcium entry, while still allowing transport of calcium out of the cell would then seem attrac-tive. In fact, this is impossible for thermodynamic reasons. Inhibitors may appear to be more effective when the exchanger is operating in net calcium entry mode than in calcium exit mode. This is, however, always attributable to differences in conditions because there is strong internal sodium dependence of drug action on the exchanger. When the exchanger is operating near equilibrium, drug action is found to be equally effective in both directions.

KEYWORDS: drug action on sodium–calcium exchange; modes of sodium–calcium exchange; thermodynamics of sodium–calcium ex-change

In many of the papers given at this meeting and elsewhere reference is made to drugs acting on the sodium–calcium exchanger preferentially in one direction rather than the other. As an example, we may take the idea that the exchanger reverses direction during ischemia so that it is carrying calcium into the cell (net calcium entry/influx or "reverse" mode) rather than out of the cell (net calcium exit/efflux or "forward" mode).[a] A drug that then inhibits the exchanger in net calcium entry mode might limit such entry and so have a therapeutic effect.

There are some dangers in talking of directionality of drug action in this way. The purpose of this note is to clarify those dangers and to attempt to tighten up the language used. We can see at least three senses in which an agent might be said to have directionality in its action on the exchanger.

---

[a]Note: The terms "calcium exit" and "calcium entry" are preferred because they are descriptive and therefore more understandable to workers outside the exchanger field. The terms "efflux" and "influx" are correct, but are too easily confused or transposed in oral and written communications.

Address for correspondence: D. Noble, Departments of Physiology, Anatomy, and Genetics, University of Oxford, Parks Road, Oxford, OX1 3PT, UK. Voice: (+44)(0)1865 272533; fax: +44 1865 272554.

Denis.noble@physiol.ox.ac.uk

Ann. N.Y. Acad. Sci. 1099: 540–543 (2007). © 2007 New York Academy of Sciences.
doi: 10.1196/annals.1387.013

1. The site of action of a drug on a membrane-based transporter might be on one side of the membrane or the other. This may or may not have any connection with the direction in which the transporter is operating or with the question of whether the action is preferential as between the directions of transport. Of course, *transported* species, in this case sodium and calcium, have different actions on direction of movement depending on whether they are inside or outside when they are binding to the *transport* sites. But, when they are binding to other, *regulatory*, sites, their actions must depend on how the regulatory action works. So, except for the transported species, we should distinguish side of action from directionality in action.

2. The exchanger is "reversible" in that it can mediate net calcium movement in either direction across the membrane, depending upon the electrochemical driving force (the sodium and calcium concentrations on the two sides and the membrane potential). A drug may have a different degree or mechanism of action when the conditions in which the transporter is operating favor one direction compared to another. It is in this sense that people talk of drugs acting preferentially on net entry or net exit modes. Thus, a drug may produce more inhibition of the exchanger when it is operating in, say, net calcium entry mode compared to when it is operating in net calcium exit mode. Note, however, that even when the net calcium movement is inward there will be some exchangers operating in the calcium exit mode, and vice versa. How many are operating in the minority mode will depend on how far from equilibrium the exchanger may be. If the exchanger is inhibited, *both* of these modes will be inhibited. Otherwise it would be possible to break the laws of thermodynamics. Suppose for example that the exchanger is in net calcium entry mode with twice as much calcium moving inward as outward on the exchanger. If the calcium entry mode could be inhibited selectively by, say, 50% while the calcium exit mode is not affected, we would achieve a condition of no net flux even though the exchanger is not at equilibrium. That is impossible. Also, we cannot imagine the drug actually changing the net direction of movement. The exchanger would then move ions uphill against the net energy gradient, but that cannot occur without consumption of ATP.

3. This leads us to the third sense in which people may speak of directionality of action. Talk of action being preferential depending on whether the conditions favor the calcium exit or entry mode may lead people to imagine precisely what is forbidden, that is, that the drug can behave like a Maxwell's demon in acting more on one direction or another *in the same experimental conditions*. This misconception is encouraged by any diagrams indicating such preferential action. To see why this is forbidden, imagine that the exchanger is at equilibrium. If there really could be a drug that acted on, say, the calcium entry mode more than on the calcium

exit mode, then it could generate net movement of ions even though there is no energy gradient to generate such movement. We can be confident that, at equilibrium, net unidirectional transport does not occur.

The arguments here are *a priori* ones since they depend on respecting the laws of thermodynamics. Nevertheless, some people may prefer an actual experimental study to illustrate the point. Fortunately, a very good example is provided by the work of Kimura *et al.*[1] using the sodium–calcium exchange inhibitor KB-R7943. Previous work by the same authors had shown direction-dependent block in the sense that, when the conditions favored the calcium entry mode, the drug had a more powerful effect than when the conditions favored the calcium exit mode. In their 1999 paper, they set up conditions in which the equilibrium potential for the exchanger was in the middle of a repetitive voltage ramp protocol. Thus, the exchanger alternated between inward and outward net current (or flux) in exactly the same conditions except for the variation in membrane potential. The result was clear: there was no difference between the IC_{50} for the calcium exit and entry modes. Even this elegant experiment depended on the drug staying bound during the ramps. The authors therefore attributed their result to slow dissociation of the drug from its binding site. A drug that dissociated quickly and whose binding coefficient had been voltage- or current-dependent could have given a different result.

An explanation for the apparent directionality of drug action has been provided by Hryshko and colleagues.[2,3] They showed that inhibition of sodium–calcium exchange by another benzyloxyphenyl derivative, SEA0400, depends strongly on internal sodium: SEA0400 apparently stabilizes the internal sodium-dependent inactivated form of the exchanger. Affinity for the blocker was greatly reduced by mutating the internal sodium-dependent regulatory site to reduce sodium-dependent inactivation. The blockers therefore *appear* to exhibit "mode selectivity" because, especially in isolated membrane patches and membrane vesicles, exchanger-mediated net calcium entry is often measured with very high internal sodium, while low or zero internal sodium is used to measure exchanger-mediated net calcium exit. The high internal sodium condition, but *not* the low internal sodium condition, should favor binding of the blockers.

Finally, what terminology should we use? We favor statements like (for example): "Drug A appears to be (more) active when internal sodium is high so that net calcium entry is favoured at the (normal) resting membrane potential." The statement "Drug A preferentially inhibits the net calcium entry (/exit) mode" is ambiguous since it could lead to diagrams that imply breaking the laws of thermodynamics. Shorter reference to "reverse mode inhibition" (or net calcium entry mode inhibition) is even more likely to do that. When experiments show directionality in action we should always look for the differences in conditions, not for Maxwell's demons.

ACKNOWLEDGMENT

We thank Professor Junko Kimura for valuable comments and discussion.

REFERENCES

1. KIMURA, J., T. WATANO, M. KAWAHARA, E. SAKAI & J. YATABE. 1999. Direction-independent block of bi-directional Na^+/Ca^{2+} exchange current by KB-R7943 in guinea-pig cardiac myocytes. Br. J. Pharmacol. **128:** 969–974.
2. BOUCHARD, R., A. OMELCHENKO, H.D. LE, *et al.* 2004. Effects of SEA0400 on mutant NCX1.1 Na^+-Ca^{2+} exchangers with altered ionic regulation. Mol. Pharmacol. **65:** 802–810.
3. LEE, C., N.S. VISEN, N.S. DHALLA, *et al.* 2004. Inhibitory profile of SEA0400 [2-[4-[(2,5-difluorophenyl)methoxy]phenoxy]-5-ethoxyaniline] assessed on the cardiac Na^+-Ca^{2+} exchanger, NCX1.1. J. Pharmacol. Exp. Ther. **311:** 748–757.

Index of Contributors